Constraints and Adaptations to Global Change at the Land-Sea Interface: For a Shared Ecological and Energy Transition

From left to right: A part of the French and Japanese delegations at the "Abbaye aux Hommes" (Caen, France); An old small harbour on the Normandy Coast (Saint-Vaast-La-Hougue, France); Oyster beds on the Normandy Coast (Saint-Vaast-La-Hougue, France); A general view of the Mont Saint-Michel (France). Copyright Nicole Prouzet

Anne-Claire Bennis · Jean-Claude Dauvin ·
Eric Feunteun · Teruhisa Komatsu ·
Osamu Matsuda · Patrick Prouzet
Editors

Constraints and Adaptations to Global Change at the Land-Sea Interface: For a Shared Ecological and Energy Transition

Proceedings of the 19th French—Japanese Oceanography Symposium

Springer

Editors
Anne-Claire Bennis
Laboratoire Morphodynamique
Continentale et Côtière
University of Caen Normandie
Caen, France

Jean-Claude Dauvin
Laboratoire Morphodynamique
Continentale et Côtière
University of Caen Normandie
Caen, France

Eric Feunteun
Laboratoire de Biologie des Organismes et
Écosystèmes Aquatiques-BOREA, EPHE
PSL Research University
Dinard, France

Muséum National d'Histoire Naturelle
Dinard, France

Teruhisa Komatsu
Japanese-French Oceanographic Society
of Japan
Tokyo, Japan

Patrick Prouzet
SFJO France
Saint-Jean-de-Luz, France

Osamu Matsuda
International EMECS Center
Kobe, Japan

ISBN 978-3-031-90049-5 ISBN 978-3-031-90050-1 (eBook)
https://doi.org/10.1007/978-3-031-90050-1

This Springer imprint is published by the registered company Springer Nature Switzerland AG
The registered company address is: Gewerbestrasse 11, 6330 Cham, Switzerland

From left to right: Professor T. Komatsu (President SFJO Japan); Professor J.-Cl. Dauvin (Chairman Organization Committee Coast Caen 2023); Professeur D. Goutte (Vice-president Caen La Mer in charge of the Higher Education and Research); Professor O. Matsuda (Vice-president International EMECS Center); Dr I. Nakayama (President FRA Japan) and Dr P. Prouzet (President SFJO France). Copyright N. Prouzet

French and Japanese scientists visiting the harbour of Saint-Vaast-La-Hougue on the Normandy coast. At the forefront, a line of whelk pots. Copyright N. Prouzet

Organization Committee

Chairman: Jean-Claude Dauvin, Professor Emeritus, University Caen Normandie, France
Anne-Claire Bennis, Full Professor, University Caen Normandie
Marie-Pierre Bouet, IE Gestionnaire, CNRS
Mohamed Boutouil, Professor and Senior Researcher, ESITC
Clément Calvino, Post-Doctoral Researcher, University Caen Normandie
Mathilde Charbonnelle, Engineer, University Caen Normandie
Marc Delahaye, Director, CRPMEM Normandie
Éric Feunten, Full Professor, MNHN and Director of Research EPHE
Lucile Furgerot, Assistant-Professor, University Caen Normandy
Daniel Levacher, Professor, Emeritus University Caen Normandie
Christophe Mensens, GIS ECUME Coordinator, University Caen Normandie
Patrick Prouzet, Dr.-HDR, President, SFJO France
Teruhisa Komatsu, Professor, JFRCA President, SFJO, Japan
Manuel Savary, Director, CRC Normandie
Robin Vanpaemelen, Engineer, University Caen Normandie
Bernadette Tessier, Senior Researcher, CNRS

Scientific Committees

Symposium

Chairwoman: Full Professor, Anne-Claire Bennis, University Caen Normandie
Dr.-HDR Philippe Cury, France, IRD Sète
Dr. Masatsugu Takano, Japan, Fishery Research and Education Agency
Professor Simon Neil, UK, Bangor University
Professor Yusuke Uchiyama, Japan, Kobe University
Dr. Steven Degraer, Belgique, Royal Belgian Institute of Natural Sciences

Dr.-HDR Nathalie Niquil, France, CNRS-MNHN
Dr. Emmanuel Garnier, France, CNRS University de Franche-Comté
Professor Teruhisa Komatsu, Japan, President SFJO Japan
Professor Dano Roelvink, Pays-Bas, IHE Delft Institute for Water Education
Professor Waka Sato-Okoshi, Japan, Tohoku University
Dr. Alain Nadaï, France, CNRS

Forum

Chairman: Full Professor, Eric Feunteun MNHN, EPHE
Marc Delahaye, France, Director CRPMEM Normandie
Manuel Savary, France, Director CRC Normandie
Professor Yasuyuki Koike, Japan, scientific advisor SFJO
Dr. Hirotaka Nakahara, Japan, University Kanagawa
Professor Catherine Mariojouls, France, Agro Paris Tech.
Dr. Franck Lagarde, France, Ifremer Sète
Dr. Olivier Labussière, France, CNRS
Professor Emeritus Osamu Matsuda, Japan, International EMECS Center
Professor Kenji Okoshi, Japan, Toho University
Dr. Anne Souquière, France, Centre de Données et d'Expertise sur le Patrimoine Naturel
Dr. Lisa Garnier, France, RTE

Editorial Board

Foreword

The 19th Franco-Japanese Oceanography Symposium. Further Proof of the Vitality of the Links and the Continuity of Relations Between the French and Japanese Scientific Communities in the Field of Oceanography

This 19th Franco-Japanese Oceanography Conference is in line with previous ones and shows the vitality of the links established between the Japanese and French scientific communities. President Patrick Prouzet traces the history of these bonds of friendship established over time, the continuity of these relationships and the mutual aid movements, spontaneously appearing when certain specialized fields were in difficulty. These connections endure and have immense value. They are all the more precious because they are fragile, but they are essential in the face of the considerable problems that climate change will create in the world to come. Each country will approach these future difficulties with its experience and culture, with particular scientific and technical approaches. Their friendly confrontation and the exchanges that will result from it are always very rich, as all previous conferences have shown. They have often led to high quality joint work.

Thank you to President Komatsu Teruhisa, SFJO Japan and our many Japanese colleagues for coming to Normandy, both to discover this beautiful French province, and also to bring new Franco-Japanese cooperation to life.

Thank you to the major Japanese and French organizations for continuing to support these efforts, the future results of which will be fruitful, and often unexpected.

Thanks especially to President Patrick Prouzet, to his solid team: Prof. Jean-Claude Dauvin, Dr. Yves Henocque, Prof. Anne-Claire Bennis, Dr. Franck Lagarde and Prof. Eric Feunteun. With their colleagues and collaborators, they have carried out considerable work to make this 19th conference a great success and strengthen the links between the two scientific communities of the two countries.

The message I want to convey combines both science and friendship. Take advantage of the meetings during this conference to get to know your colleagues from the other country better personally. They will introduce you to another culture, other traditions, another way of thinking and approaching your own problems.

Success will be there when you have worked together, published together, created together. Do not hesitate to tackle original, even difficult subjects together: many discoveries are born in these unexplored avenues.

Make these connections continue later, experience will show you that they are priceless.

Prof. H. J. Ceccaldi
Président d'Honneur de la Société
franco-japonaise d'Océanographie
France, Lettres et Arts de Marseille
Académie des Sciences
Marseille, France
ceccaldi.hubert@orange.fr

Synthesis and Prospects

Adaptation to global change will require shared and cross-disciplinary knowledge. Limits of sectorial approach have been emphasised many times. Wind energy cannot be developed in isolation from other existing constraints.

The creation of Scientific Interest Groups is an interesting initiative, but the multiplication of these structures must not lead to a dispersion of human resources or duplication of research. Interconnection seems necessary in order to, at least, share knowledge. In this respect, feedback from Northern countries can help to resolve certain issues.

Some stakeholders remain particularly vulnerable to global change. This is the case for fishers and oyster farmers, but also for those involved in seaside tourism, with the proliferation of toxic species leading to the closure of bathing areas.

Academic research is involved and will have to step up its action with the appropriate resources in order to work on physical forcing, evolution of the coastline and proliferation of invasive and toxic species. The development of wind power cannot be disconnected from this context and will take place in an environment that is already largely exploited, where there is little space for flexibility if we do not want to further disrupt the sometimes very fragile equilibrium that characterises our aquatic ecosystems.

The summary of work developed as part of the Fisheries Research and Education Agency Japan (FRA Japan)/Ifremer MOU (Memorandum of Understanding) on the sustainability of fishing and aquaculture activities under the pressure of global change is an interesting initiative that structured session 2 of the Forum.

Alongside academic knowledge, there is traditional knowledge that is important to integrate, particularly at local and regional levels.

The importance of environmental education was emphasised. It is a component of the "Nature and Culture" project of the two SFJOs. In this context, the participation of teachers from French and Japanese high schools (Sète and Okayama high schools) is important and should be continued.

Session 4 of the meetings, co-organised with EMECS, was structured around the concept of "Sato-umi". The need for more bottom-up approaches was emphasised in order to better incorporate the opinions and actions of local communities and thus increase their degree of acceptance and adaptation to change. The eel approach as developed in France is very similar to a "Sato-umi" approach, but still comes up against far too many "top-down" decisions.

While the development of low-carbon energies is essential to reduce greenhouse gas emissions, the need to develop a less energy-consuming society has also been emphasised. In the fisheries sector, a technological leap must be made in order to develop fishing techniques that consume less fossil fuel and more carbon-free means of propulsion.

The President of the SFJO Japan announced that the 20th Franco-Japanese oceanography symposium would be held in Japan in November 2025 in Toba, prefecture of Mie. This is a place characterised not only by its Ama divers, but also as the place where pearl farming was invented by Mr. Mikimoto in 1893.

The Editorial Board

Opening Speeches

Introductory Speech—First Secretary Japanese Embassy in France

I congratulate the French-Japanese Society of Oceanography and the Japanese-French Oceanographic Society for the tremendous success of the 19th French-Japanese Symposium of Oceanography, which took place at Caen in October 2023.

Japan-France bilateral cooperation covers many fields, and relations between the two countries as "exceptional partners" are better than ever, from political and diplomatic to industrial spheres. Our science and technology cooperation has always occupied an important position as one of the major pillars of Japan-France cooperation.

In particular, the maritime field has been a driving force in our science and technology cooperation since the signing of the Japan-France Science and Technology Cooperation Agreement in 1974. So far, the two countries have held a series of dialogues on promoting joint projects through 27 Japanese-French Ocean Development Sub-Committees.

Today, we face diverse and complex social challenges such as climate change and biodiversity loss. Expectations and interest in promoting marine science and technology are increasing internationally due to the need to tackle these challenges beyond national boundaries. In this context, it is essential for innovation in the marine field to share the differences between Japanese and French approaches and mutually generate stimuli for research through interdisciplinary exchanges.

In this sense, it was highly significant to hold the Symposium, which brought together various participants from Japan and France. I am convinced that the outcome of the Symposium will be an impetus for further developing international research cooperation and, ultimately, lead to strengthening bilateral relations between Japan and France.

T. Ohno
Paris, France
e-mail: takahiro.ohno.s2w@cao.go.jp

Introductory Speech—Vice-President in Charge of International Affairs University Caen-Normandie

Mr. First secretary of the Japanese embassy in France, dear Mr. Ohno Takahiro,
Mr. President of the French-Japanese society of oceanography Japan, dear Professor Komatsu Teruhisa,
Mr. President of the French-Japanese society of oceanography France, dear Dr. Patrick Prouzet,
Mr. President of the Japan fisheries research and education agency, dear Dr. Nakayama Ichiro,
Dear Professor Matsuda Osamu, representative of International EMECS Center,
Mr. Director of the continental and coastal morphodynamics laboratory, dear Laurent Dezileau,
Madam Chairwoman of the Conference scientific committee, dear Professor Anne-Claire Bennis,
Mr. Chairman of the Conference scientific and technical committee, dear Professor Eric Feunteun,
Mr. Chairman of the Conference organizing committee, dear Professor Jean-Claude Dauvin,
Ladies and Gentlemen,
Dear colleagues,

On behalf of Prof. Lamri Adoui, President of the Université de Caen Normandie, it is a distinct privilege and an honour to welcome you to this international conference, Coast Caen—Between Land and Sea, and to the 19th Franco-Japanese symposium on oceanography, which is taking place here today in the Pierre Daure amphitheatre.

The University Caen-Normandie was founded in 1432, making it one of France's oldest universities. It is a higher education institution with an enrolment of 33,000 students across several campuses in Caen and the wider region. As a comprehensive University, it offers a wide range of degree programmes in five main fields. It welcomes students from around the world, with over three thousand three hundred and seventy (3370) international students having studied in Normandie over the last year, and over four hundred and eighty (480) students having participated in exchange programmes.

Furthermore, our institution is distinguished by its active engagement in research, which is structured into three pluridisciplinary axes.

(i) science and technology, (ii) humanities and social sciences, and (iii) biology, imaging, health, and environment. It has forty-one research units, making it a leader in innovation at the national and international levels. Many of its researchers are engaged in international research programs with foreign laboratories. Additionally, some of its postgraduate students benefit from international joint supervision of their Ph.D. thesis, which contributes to their recognition at an international level.

A University Involved in the Challenges of Ecological Transition

Furthermore, the University is engaged in addressing the challenges of ecological transition through its training and research programmes. In June 2023, the University was awarded the DD&RS (sustainable development and social responsibility) label. The label is awarded to higher education establishments that demonstrate a commitment to sustainable development. To be considered for the award, an institution must provide evidence that its approach to sustainable development is genuine, integrated and structured.

The concept of sustainable development represents a model of development that encompasses not only ecological considerations, but also the balance between environmental, economic and social factors. The Université de Caen Normandie is acutely aware of the pivotal role that sustainable development plays in the future careers of its students and is consistently refining its training programmes in this regard. The University currently offers 40 courses directly linked to the challenges of transitions, including seven University bachelor's degrees, six bachelor's degrees, seven professional bachelor's degrees, 19 master's courses, and one engineering degree. Additionally, one in five degrees incorporates the notion of sustainable development, and by the start of the 2025 academic year, all undergraduates will be trained in these issues.

Ultimately, a considerable proportion of the research projects conducted by our research laboratories and their associated partners fall within the overarching domain of sustainable development and social responsibility. All academic disciplines play a role in the pursuit of solutions to the challenges inherent in transitions.

A University Strong in Oceanography Research

The University is situated in the Normandie region, which boasts 600 km of coastline, and many of our research activities are focused on the fields of oceanography, marine biology and coastal environments. The Université de Caen Normandie is a leading institution in the field of oceanography. In the Shanghai thematic rankings, the University is ranked 151st in the world in this discipline.

A significant number of the University's labs and their researchers are engaged in research activities within this field. These include:

Continental and coastal morphodynamics—M2C, co-organizer of this symposium
Biology of aquatic organisms and ecosystems—BOREA
Identity and differentiation of space, environment and societies—IDEES
Human and animal ethology—ETHOS
Cherbourg Applied Sciences Laboratory—LUSAC

And of course, our Coastal Environment Research Centre—CREC

They are engaged in a number of significant national and international projects that serve to reinforce the dynamism of our University in the field of oceanography.

This international conference and symposium form part of this dynamic process, and we are pleased to offer our support to any collaborative work that may be identified during this conference.

We are grateful for your choice to visit our University. I would like to extend my gratitude and congratulations to all the organisers and partners of this event and wish you all fruitful exchanges. To our colleagues from other parts of France and abroad, we extend a warm welcome to our beautiful region.

Prof. C. Rochais
University Caen Normandie
Caen, France
e-mail: christophe.rochais@unicaen.fr

Introductory Speech—President of the French-Japanese Society of Oceanography France

The strong bond that exists between the two French-Japanese Oceanography Societies (SFJOs France and Japan) has been forged little by little over the course of a history spanning more than half a century.

Professor Sasaki Tadayoshi founded the SFJO Japan in 1960, when France and Japan were working together to explore the deep seabed off Japan with the FNRS-3 bathyscaphe.

In 1984, Prof. Hubert-Jean Ceccaldi and other pioneers of Franco-Japanese cooperation founded the French-Japanese Society of Oceanography France.

Since their creation, the objective of these two sibling societies has remained unchanged: to highlight, through a process of scientific outreach, the work carried out by the French and Japanese scientific communities in the fields of oceanography, fisheries and aquaculture, without overlooking the social and economic components, or the policies generated by the exploitation of marine and, more broadly, aquatic resources. In this context, we would like to mention the ongoing support of the French-Japanese Sasakawa Foundation, which works to develop friendly and cultural links between France and Japan.

In the course of this long and fruitful cooperation, two major events have taken place that have strengthened the ties of friendship between these two communities.

The first event dates from the late 1960s. It corresponded to severe epizootic crises that seriously affected the French oyster industry. The industry recovered thanks to the invaluable help of Japanese scientists and oyster farmers, who successfully imported and acclimatised Japanese oyster spat (*Crassostrea gigas*, now known as *Magallana gigas*) from the Sanriku coast in the Tohoku region.

The second one was in March 2011, when a tremendous tsunami hit the same coastline from which the spat that saved the French oyster industry originated. It destroyed everything in its path and devastated the Japanese Sanriku oyster industry. From the outset, French oyster farmers and scientists, notably through the SFJO, contributed to the revival of oyster farming in this region.

Since then, and in order to adapt to the current concerns of our societies: climate change, global change, sustainable development, adapting to change, the two societies have defined a project called: "Nature and Culture" to highlight the need for environmental governance (minimising the ecological footprint) and to incorporate a cultural component in the sense of traditional knowledge and know-how and environmental education that provides a link with future generations.

The general theme of this 19th Franco-Japanese Symposium on Oceanography, held at the University Caen-Normandie, should contribute to the implementation of a new paradigm, one that is no longer based on a sectoral approach in which each sector ignores the needs of the other, but on a more global approach, based on greater solidarity, in which the needs of everyone, and first and foremost of Nature, are taken into account by all stakeholders. It's a great challenge to take up if we want to adapt to global change, which is inevitably on the horizon, in a way that is well understood and shared, to ensure food, health and energy security for our future generations. Two renowned Japanese organisations have joined forces with the two SFJOs: the International EMECS Center and the Japan Fisheries Research and Education Agency. We would like to extend our warmest thanks to them and to all the partners and sponsors who helped make this event a reality.

Finally, I wouldn't want to end this speech without mentioning once again the strong involvement of our Japanese colleagues, which makes this event a highlight of scientific cooperation between France and Japan in the field of oceanography.

This is underlined by the sponsorship of the Japanese Embassy in France and the presence of Mr. Ohno Takahiro, First Secretary and Scientific Attaché of the Japanese Embassy in France.

Dr.-HDR P. Prouzet
Saint-Jean de Luz, France
e-mail: patrick.prouzet@orange.fr

Introductory Speech—President of French-Japanese Society of Oceanography Japan

On behalf of the 34 Japanese participants present and the Japanese-French Oceanographic Society of Japan, I would like to express my sincere gratitude to our French colleagues for their efforts in making this symposium possible. I would also like to thank the many sponsors, including the University of Caen Normandy, the CNRS, the Japanese Embassy in France, the French Embassy in Japan, the Normandy Region, the La Manche department, the Fondation franco-japonaise Sasakawa and all the companies that sponsored this event. In particular, I would like to thank Profs. Jean-Claude Dauvin, Anne-Claire Bennis, Eric Feunteun and Dr. Patrick Prouzet for their efforts in organizing this Symposium. We are honoured by the presence of Mr. Takahiro Ohno, Scientific Attaché of the Japanese Embassy in France today.

Coast Caen 2023 is also the 19th French-Japanese Oceanography Symposium. It takes two forms: the International Conference and the Forum which are hosted mainly by the two Japanese-French Oceanographic Societies of Japan and France. The Japan Fisheries Education and Research Organisation and Ifremer are holding a session at the Forum as part of the Japanese-French Scientific Cooperation Agreement. The concept of "Satoumi" was introduced at the 17th Japanese-French Oceanography Symposium held at the University of Bordeaux-II in 2017. This time, organisations practising satoumi activities as grassroots activities in local areas of Japan participate and discuss in a Forum session. This session is supported by the International EMECS (Environment Management of Enclosed Coastal Seas) Centre in Kobe, Japan.

On the 22nd of October, the Japanese delegation visited the Bay of Mont-Saint-Michel, guided by researchers from the M2C laboratory (continental and coastal morphodynamics), to gain a better understanding of coastal erosion, restauration efforts of salt marshes and the region's ecosystems.

On the 23rd of October, we visited the oyster beds, guided by the Normandy Regional Shellfish Farming Committee (Comité régional de la conchyliculture de Normandie), to gain a better understanding of the oyster farming situation in the region. The visits are used at the conference to complement the relevant presentations to understand the social, economic and environmental context and deepen the discussion.

By bringing together marine and fisheries researchers from two different cultures—France in the West and Japan in the Far East—to present, discuss and exchange the results of their research, discover each other's characteristics and apply them concretely in the field, we can not only develop and advance new research, but also contribute to achieving a sustainable marine environment and fishing industry for the future.

Prof. T. Komatsu
TUMSAT
Tokyo, Japan
e-mail: cymodocea@gmail.com

Introductory Speech—Vice-President of International EMECS Center, Japan

I have really appreciated the organizer for providing me of such a wonderful opportunity to attend Coast Caen 2023. I am at the International EMECS Center located in Kobe, Japan. EMECS stands for Environmental Management of Enclosed Coastal Seas, which is now crucial issue all over the world.

The International EMECS Center is established to construct an organized network linking governments, researchers, companies, private citizens and other entities to promote academic exchanges on an international level and conduct research, training and support activities, in order to conserve existing enclosed coastal sea environments and create new ones and help create a society capable of sustainable development in which human beings can coexist with the tremendous diversity of nature.

Aiming to establish an international and academic network to solve the environmental problems of enclosed coastal seas in an integrated way, the 1st International Conference on the Environmental Management of Enclosed Coastal Seas (EMECS1) was held in Kobe, Japan in 1990. Afterward, the EMECS Conference has been continuously held around the world every two or three years. Based on presentations and discussions in a wide range of fields held by participants such as researchers, administrative officials, enterprises, citizens, etc., the importance of the integrated management of enclosed coastal seas and their backlands, and "Shared Responsibility" has been transmitted to the world as "EMECS Conference Declaration" at each conference.

Among these International EMECS Conferences, it is a matter worthy of special mention that the 7th International EMECS Conference (EMECS7) was held here in Caen, Normandie, France in May 9–12, 2006, about 17 years ago. The main theme of the conference was "Sustainable Co-development of Enclosed Coastal Seas: Our Shared Responsibility" and it was attended by some 350 persons from 25 countries.

Theme of Technical Session 3, EMECS7 was "New Concept and Innovative Experiences in Coastal Management" in which the new concept, experiences and case studies of Satoumi were presented by the late Professor Tetsuo Yanagi, Osamu Matsuda (myself) and other Japanese scientists. These presentations were highly evaluated at the overall evaluation session as that Satoumi is a rational vision of coexistence designed to achieve symbiosis among human communities and coastal areas. As a result, EMECS7 held in Caen in 2006 became a good opportunity for international debut of Satoumi concept.

Therefore, I am very glad to be here to participate in the meeting held in Caen again. Since Satoumi related session will be also held in Coast Caen 2023, I am expecting that we will be able to recognize the progress of Satoumi related activities and discussion during the last 17 years.

It was also pointed out that progress in coastal management will become increasingly advanced and multifaceted, and that coastal management will require cooperation at the international, regional, national and local levels. Since subjects of Coast Caen 2023 to be discussed soon are very broad and diversified, I wish the conference will bear fruitful discussion and productive results.

Prof. O. Matsuda
Kobe, Japan
e-mail: matsuda036@go3.enjoy.ne.jp

Introductory Speech—President of the Organization Committee—Coast Caen 2023

After Bordeaux in 2017, Tokyo in 2020 but postponed by videoconference in 2021 to celebrate the 60th anniversary of the Franco-Japanese Oceanography Society Japan, due to the strong impact of Japan by COVID-19, it was decided by our two societies that the 19th Franco-Japanese Oceanography Symposium would be held in Caen and to entrust me with the organization. So, the 19th Franco-Japanese Symposium had been organised at the campus 1 of the Caen Normandy University which is in the centre of Caen near the Guillaume the conqueror Castel in October 2023.

The Normandy coast of the English Channel is a highly impacted ecosystem concerned by numerous human activities. It is subjected to heavy historic anthropogenic pressures and the emergence of new human activities: fishing, harbour extension, aggregate extraction, and deposition of dredged sediment, aquaculture and future offshore wind farms. By contrast, this area shows a high level of protection of its natural heritage, mainly owing to the Natural Reserve and Natura 2000 sites. All these human pressures in interactions influence the functioning and the dynamic of this complex ecosystem.

Four topics have been retained for the 19th Symposium corresponding to regional challenges of research, development, innovation and controversies between human activities in Normandy. All these topics are precisely described on the website of this international events: http://coastcaen2023.socfjp.com/.

Session 1—Exploitation of Living Resources: Necessity and Constraints;

Session 2—Development of Renewable energies and impact on socio-ecosystems;

Session 3—Coastline evolution and submersion of coastal territories.

A special topic was organized in the French-Japanese Forum on: The Satoumi concept and the management and Commons—an integrated approach, a link between land and sea and between Nature and Culture.

Before the beginning of this international event, two-day trips had been organised. The Sunday October 22, excursion has been organised for the visit of the famous abbey and Mount Saint-Michel Bay with a tidal range higher 12 m by Bernadette Tessier, Research Director at the CNRS. The second excursion was on the eastern coast of Cotentin at Saint-Vaast-La-Hougue to visit the harbour installation and the oyster parks, a lunch with the local products allowed us to test three Normandy oyster vintages, king scallops from the Bay of Seine and whelks and Venus clams from the Normand-Breton Gulf, and Normandy cider. Then, there were the visit of the SMEL laboratory and the shellfish Viking enterprise on the western coast of Cotentin. Thanks to the Regional Fisheries, Regional Shellfish Aquaculture Committees and the Manche Department for the day organization.

During the symposium, we have the opportunity to visit the Men Abbey (Abbaye–aux–Hommes), Caen City Hall on the 25th of October, this historic landmark of Caen that links us to William the Conqueror. In 2025, we will celebrate the 1000th anniversary of Caen, a regional anchorage to the sea with its port connected to the sea by its canal dug in the nineteenth century and completed in 1857. The Gala dinner had been held at the Peace Memorial of Caen on the 26th of October.

Finally, the last day on Friday the 27th of October, there was the visit of the maintenance centre of the "Parc Eolien en Mer du Calvados" at Ouistreham. Thanks to Caen La Mer and EDF Renewable for the visits and their hospitalities.

I would like to thank all the 31 sponsors who have given us their financial support, label or help. A special thanks to Marc Delahaye, director of the "Comité Régional des Pêches et des Elevages Marins de Normandie", for the submission of the application of European funding which made it possible to finance the interpreters who will accompany us during these four days.

Thanks to all the 24 members of the organizing committee, in particular to the staff and Direction of my UMR M2C Caen. A special thanks to Marie-Pierre Bouet for her precious, unfailing, efficient help, in a word, essential to the life of the laboratory. Thanks to Mrs. Chauvet for the management of the CNRS Azur Colloque tool.

Finally, a huge thank you to the University of Caen Normandy, its President Prof. Lamri Adoui, the logistics services, the printing press, the managers, the "Maison de la Recherche en Sciences Humaines" in which the two days of the international conference will take place. Without the help of UNICAEN and the provision of premises such as the prestigious Pierre Daure amphitheater in which we are staying during four days, the conference would not have been possible in this form.

We are 125 registrants, including a great delegation from Japan and a representation from the Japanese Embassy present to the open ceremony of the symposium. Thank you for your presence and for all your communications, thank to Prof. Teruhisa Komatsu, President of the Franco-Japanese Society Japan.

Thank to Ms. Reiko Vachot-Inukai, Ms. Miyu Mitake-Leroy and Ms. Anna Devaux-Tsuru for their interpreting work. The participants appreciated your professionalism. Thank you to the caterer the Durand Company of Caen for all your support throughout the conference from Saturday the 23rd to Friday the 27th of October, thanks to Mr. Duchemin and your team; all the guests appreciated the quality and the variety of the meals that were served. Thanks to MTCA for the installation of the sound system and your presence throughout the symposium.

I reiterate my thanks to all the participants who made this symposium a success. Thank you to the guest speakers in the plenary and thematic sessions and to all the poster and oral presentations for the diversity of the topics addressed and the quality of their presentations.

It seems to me that we have dealt more broadly with oceanographic issues than at previous conferences. In addition, the societal future of our coastal and coastal areas has taken a more important place in our research themes. The holistic vision, the only scale that allows us to try to understand, analyze and project ourselves on the future of our Earth System and our Global Ocean, is appropriate in the face of climate change and the well-being of our current and future human societies.

Prof. J.-C. Dauvin
Laboratoire Morphodynamique Continentale et Côtière
University of Caen Normandie
Caen, France
e-mail: Jean-claude.dauvin@unicaen.fr

Contents

Vulnerability and Resilience of Coastal Ecosystems to Global Change

Exploring Extreme Wave Propagation in Coastal Zones: A Combined Physical and Numerical Modeling Study

R. Matar, N. Abcha, N. Lecoq, E.-I. Turki, I. Abroug, and D. Dutykh

1 Introduction

Extreme events are a major concern for coastal planning and management due to their potential to cause widespread damage and disruption (Hansom et al. 2015). Therefore, studying and understanding these extreme events near the coast is crucial for coastal management, infrastructure planning, and disaster preparedness to mitigate the impact of such events and enhance the resilience of coastal regions (Dao et al. 2011, 2013; Seiffert and Ducrozet 2018). In the literature, extreme waves are described as large water waves with a crest to trough wave height, H, greater than twice the significant wave height, H_s, of the wave field (Dysthe et al. 2008; Zhang et al. 2019). The propagation of water waves in fluid mechanics is described by the Euler equations of ideal fluids, which are inviscid and irrotational (Whitham 1999; Katsaounis et al. 2020). Appropriate approximate mathematical models consisting of nonlinear and dispersive partial differential equations are often used for numerical simulations when studying the properties of water waves (Katsaounis et al. 2020). Dispersion and nonlinearity are the two fundamental parameters of the wave (Abroug et al. 2022). The Boussinesq equations are derived assuming non-hydrostatic pressure (Brocchini 2013), resulting in a system of dispersive equations capable of capturing

R. Matar (✉) · N. Abcha · I. Abroug
Normandy University, UNICAEN, UNIROUEN, CNRS, UMR 6143 M2C, Caen, France
e-mail: reine.matar@unicaen.fr

N. Lecoq · E.-I. Turki
Normandy University, UNIROUEN, UNICAEN, CNRS, UMR 6143 M2C, Rouen, France

D. Dutykh
Department of Mathematics, Khalifa University of Science and Technology, Abu Dhabi, United Arab Emirates

Causal Dynamics Pty Ltd., Perth, Australia

A.-C. Bennis et al. (eds.), *Constraints and Adaptations to Global Change at the Land-Sea Interface: For a Shared Ecological and Energy Transition*,
https://doi.org/10.1007/978-3-031-90050-1_1

more physical effects compared to the classical shallow water equations (Dutykh and Clamond 2016).

Experiments are performed using a controlled wave flume to monitor the evolution of wave characteristics from intermediate to shallow waters depths. In this context, the dispersive focusing technique for extreme wave generation (Whittaker et al. 2018) is used. As described in (Watanabe et al. 2020), this mechanism generates a compressed wave packet highly nonlinear near the focus and breakpoint (Baldock et al. 1996; Abroug et al. 2020a, b; Abroug et al. 2020a, b).

In the present study, laboratory experiments were performed on multiple wave trains and the experimental results were compared with numerical predictions. This comparison was made possible by using: (1) power spectral density analysis to examine the spatial evolution of spectral energy associated with waves propagating from an intermediate depth to the breaking and shoaling zones (2) wavelet transforms providing time–frequency localization, which gives insight into the changes in the contribution of different frequencies over time. To compare the experimental results with the numerical predictions, a simple hybrid Boussinesq-type shallow water model (Durán et al. 2018) has been applied. The numerical resolution of the problem relies on the finite volume method for spatial discretization with UNO2 reconstruction procedure and the 3rd order Runge–Kutta method for temporal discretization. Additional details can be found in (Durán et al. 2018; Dutykh et al. 2011; Khakimzyanov et al. 2020).

This paper is divided into several sections, Sect. 2 introduces the experimental and numerical setup. Section 3 discusses the results regarding the spectral energy and wavelet analysis. Lastly, Sect. 4 concludes with the main findings of the study.

2 Materials and Methods

2.1 Experimental Setup

The experiments were conducted in the wave flume of the Continental and Coastal Morphodynamics (M2C) laboratory in Caen. The dimensions of the flume are 22 m in length, 0.8 m in width, and 0.8 m in height, where the chosen water depth was $h_0 =$ 0.3 m. A slope gradient ($\beta = 1/25$) (Abroug et al. 2022), enabling wave transformation was positioned 9.5 m from the wave generator. A piston-type wave generator of Edinburgh Designs Ltd. was used to simulate various wave conditions (Fig. 1) with a wave focusing at 2 m from slope's toe. The temporal variation of the free surface elevation is measured with wave gauges having a sampling rate of 32 Hz. WG_R refers to the reference wave gauge located 4 m from the wave generator, as shown in Fig. 1. To cover the 10 m experimental area, an arrangement of five wave gauges strategically placed 0.2 m apart was replicated ten times (Abroug et al. 2020a, b; Xu et al. 2021).To account for the varying duration of extreme events, one, three, six, and nine wave trains were generated with time intervals of 45 s, 108 s, 208 s,

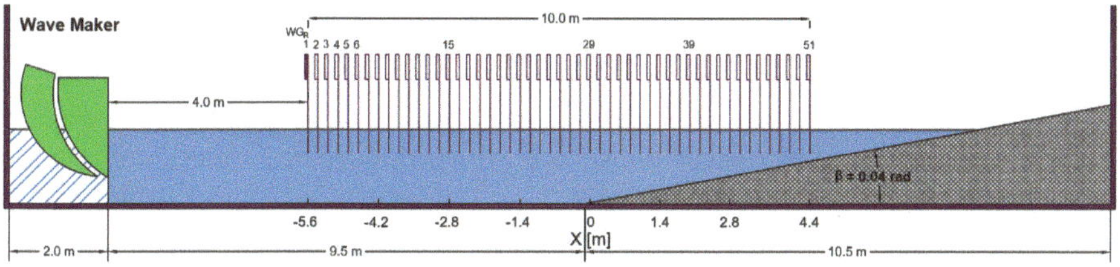

Fig. 1 Sketch of the wave flume, waves propagate from left to right

Table 1 Experimental parameters used

Spectrum	JONSWAP ($\gamma = 7$)
Peak frequency f_p	0.75 Hz
Non-linearity S_0	0.21
	0.27
	0.34
	0.41
	0.49
	0.58
	0.65
Number of trains	1–3–6–9
WG positions	51

and 302 s, respectively. The employed experimental parameters are summarized in Table 1.

In the current focused wave groups, the linear NewWave profile (Adcock and Taylor 2009; Tromans et al. 1991) is used as input. The free surface elevation is expressed using linear wave theory (LWT) as a sum of n sinusoidal wave components that are phased together at a single point in time and space as follows:

$$\eta(x, t) = \sum_{i=1}^{n} a_i \cos(k_i x - \omega_i + \varphi_i). \tag{1}$$

In this equation, x represents the distance to the mean wave-maker position, t represents time, a_i represents the wave amplitude of the ith component, k_i represents the wave number, ω_i represents the wave angular frequency, and φ_i represents the phase angle at the focusing point. The input parameters are the steepness, the wave spectrum type, the phase at focusing, the focus location, and the peak frequency (f_p). The wave trains generated are based on a JONSWAP spectrum (Hasselmann et al. 1973):

$$E_J(f) = \alpha \frac{g^2}{(2\pi)^4} f^{-5} \exp\left[-1.25\left(\frac{f}{f_p}\right)^{-4}\right] \gamma^{\delta}, \tag{2}$$

with $\alpha = \left(\frac{gF}{U_{10}^2}\right)^{-0.22}$; $\delta = \exp\left(\frac{-\left(\frac{f}{f_P}-1\right)^2}{2\sigma^2}\right)$ and $\sigma = 0.07$ if $f \leq f_p$ and 0.09 if $f > f_p$.

The peak enhancement factor γ determines the height of the peak and the narrowness of the spectrum, it ranges from 1 for a Pierson-Moskowitz spectrum to 7 for the narrowest spectrum. The JONSWAP ($\gamma = 7$) spectrum was used in this paper. Wave trains start to propagate at intermediate water depth ($k_p h_0 = 0.93$) (k_p is the wave number relative to the peak wave frequency f_p, and $h_0 = 0.3$ m is the constant water depth).The nonlinearity of the wave train is characterized by the initial local wave steepness (S_0), which is determined from the free surface elevation obtained by the reference wave gauge (WG_R) placed at $x = -5.6$ m. It is assumed that the wave train is well-formed at this point. The value of S_0 was calculated using Eq. (3), see (Abroug et al. 2020a, b, 2022; Tian et al. 2010). The focus of the wave train is mainly due to dispersion as shown in Eq. (4).

$$S_0 = k_{S0} \sum_{i=1}^{n} a_i, \tag{3}$$

$$(2\pi f_s)^2 = g k_s \tanh(k_s h_0). \tag{4}$$

Using the dispersion relation at $x = -5.6$ m, the spectrally weighted wave number k_{S0} is calculated. The surface elevation at the focal point is represented by $\sum a_i$ according to LWT. k_s represents the characteristic wavenumber, while f_s represents the characteristic frequency.

2.2 *Wavelet Analysis*

Wavelet analysis was used to analyse the free surface elevation of each wave train. The objective is to break down a signal into different frequency components over time. This allows us to determine the frequencies at which it oscillated and, even more importantly, the times at which these oscillations occurred (Tiscareno and Hedman 2018). The continuous wavelet transform $WT(a, \tau)$ of a time series of free surface elevation (t) is given by:

$$WT(a, \tau) = \int_{-\infty}^{+\infty} (t)\psi_{a,\tau}^*(t)dt, \tag{5}$$

$$\psi_{a,\tau}(t) = |a|^{-0.5}\psi\left(\frac{t-\tau}{a}\right), \tag{6}$$

where the asterisk (.*) indicates the complex conjugate and $\psi_{a,\tau}(t)$ [Eq. (6)] denotes the mother wavelet function which is expanded by a factor τ and is scaled by a factor

a (where frequency $f = \frac{1}{a}$). The Morlet wavelet function was chosen as the mother wavelet due to its ability to provide information about both phase and amplitude and it is defined as:

$$\psi(t) = \pi^{-1/4} e^{-\frac{t^2}{2}} e^{(iw_0 t)}, \tag{7}$$

where w_0 denotes the peak frequency of the wavelet. It is usually set to six [see (Grinsted et al. 2004; Dong et al. 2008) for more details]. It is convenient to express the scale 'a' as fractional powers of two [Eq. (8)] (Torrence and Compo 1998):

$$a_i = a_0 2^{i\delta}, \quad i = 0, 1, 2, \ldots, M, \tag{8}$$

$$M = \frac{1}{\delta} \log_2\left(\frac{N\Delta t}{a_0}\right), \tag{9}$$

where a_0 is the smallest resolvable scale and M [Eq. (9)] the largest. N is the number of points in the time series, and Δt is the time sampling. The a_0 parameter should be chosen equal to $2\Delta t$. The scale factor δ must be sufficiently small to achieve high resolution and adequate sampling in scale. Also, with the Morlet wavelet, a scale factor of $\delta = 0.5$ is the maximum value that provides adequate sampling (Dong et al. 2008). By modifying its scale and position, the function adjusts to different segments of the initial signal.

2.3 *Modified Peregrine Model*

The laboratory experiments are reproduced using a new Boussinesq type system with variable bottom topography (Durán et al. 2018), which is based on Peregrine's system (Peregrine 1972). This system is more suitable for practical applications, such as wave runup on non-uniform shores, as it employs higher order nonlinear terms to maintain a correct linear dispersion. The fluid is assumed to be incompressible, and the flow irrotational. To ensure invariance under vertical translations, the equations have been modified, for complete mathematical details, please refer to references (Durán et al. 2018; Dutykh et al. 2011). Based on these assumptions, the following system of equations is derived:

$$H_t + Q_x = 0, \tag{10}$$

$$Q_t + \left(\frac{Q^2}{H} + \frac{g}{2}H^2\right)_x - \left[\frac{H^2}{3}Q_{xxt} - \left(\frac{1}{3}H_x^2 - \frac{1}{6}HH_{xx}\right)Q_t + \frac{1}{3}HH_xQ_{xt}\right] = gHh_x, \tag{11}$$

where $H(x, t) = \eta(x, t) + h(x)$ is the total water depth, $h(x)$ is the depth below the mean sea level, $Q(x, t) = H(x, t)u(x, t)$ is the horizontal momentum, $u(x, t)$ describes the depth-averaged fluid velocity, while g is the constant acceleration due to gravity.

To define the initial conditions, the free surface experimental signals acquired at $x = 4$ m are used. To account for the effects of friction associated with the propagation of the wave, a dissipative source is introduced into the conservation equation [Eq. (11)]:

$$F(u, H) = -c_m g \frac{u|u|}{H^{\frac{1}{3}}}. \tag{12}$$

The calibration of the model was carried out by modifying the Manning coefficient c_m [Eq. (12)] until a good agreement was obtained between predictions and experimental results. For simplification purposes, the coefficient c_m was chosen to be a constant, $c_m = 1.2 \times 10^{-4}$ s m$^{-1/3}$ ($c_m = 2 \times 10^{-4}$ s m$^{-1/3}$ in Dutykh et al. 2011).

3 Results and Discussion

Numerical simulations were conducted to assess the accuracy of the mPeregrine model in terms of spatial variations in spectral energy. The model correctly predicts the spectrum computed on flat and sloping bottoms, upstream of the breaking point. However, deviations are amplified in the vicinity of the wave breaking. Figure 2 presents the experimental signals in black and the numerical signals in blue, along with their corresponding dimensionless frequency spectrum (PSD). Spectral estimation was done by the Welch method. The signals were divided into multiple segments with a 50% overlap. Each segment, approximately 16 s long, was subjected to the Hann window before undergoing a 2^{10}-points fast Fourier transform (FFT), which results in a high level of spectral resolution ($\Delta f = 0.03125$ Hz). Five specific positions of the flume are chosen in Fig. 2: $x = -4$ m (far from the slope), $x = 0$ m (toe of the slope), $x = 2$ m (wave breaks), $x = 3$ m, and $x = 4$ m (after wave breaking). It is important to note that while there is good agreement between the experimental and predicted free surface elevation time series, there is less agreement for the frequency spectrum. The reason for this is that the spectrum is related to the square of the wave amplitude, which amplifies the mismatch.

It is evident that in the downward direction (Fig. 2), the gap between the two signals becomes more significant. For example, the deviation of the numerical power spectral density (PSD) from the experimental PSD was almost negligible for $x \in [-5.6$ m, 0 m$]$. However, the mPeregrine model is unable to reproduce the same PSD at positions posterior to wave breaking ($x = 3$ m and $x = 4$ m), and the deviation is more pronounced. One explanation for this is that Boussinesq systems often fail to accurately predict near-breaking waves, which occur outside of the regime of long and small amplitude waves (Katsaounis et al. 2020). To address the varying

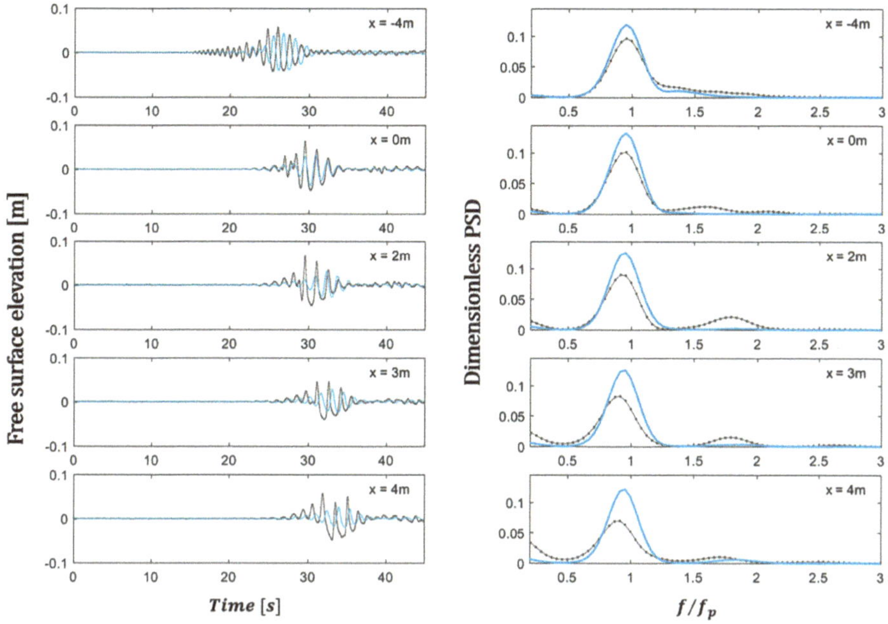

Fig. 2 On the left is shown the free surface elevation and, on the right, the dimensionless frequency spectrum obtained experimentally (black) and numerically (blue) at x = − 4 m (h = 0.3 m), x = 0 m (h = 0.3 m), x = 2 m (h = 0.22 m), x = 3 m (h = 0.18 m), and x = 4 m (h = 0.14 m) using a signal of one wave train and $S_0 = 0.41$

duration of extreme events, one, three, six, and nine wave trains were generated with time intervals of 45 s, 108 s, 208 s, and 302 s, respectively. This approach allows for the investigation of how the mPeregrine model responds to different durations of extreme events, ranging from shorter bursts to longer-lasting phenomena. Figure 3 corresponds to a signal of nine-wave trains, including both experimental and numerical signals, with their corresponding dimensionless frequency spectra.

Despite the variation in signal duration, the spectra in Fig. 3 exhibit similar trends to those observed in Fig. 2. This suggests that the mPeregrine model maintains consistency in predicting spectral energy variations across different durations of extreme events. Nonlinearity plays a crucial role in shaping the behavior of wave dynamics (Constantin 2018; Didenkulova and Pelinovsky 2012), especially in regions where wave amplitudes are high, such as near wave breaking. Figure 4 illustrates how divergences between experimental and numerical results increase as nonlinearity increases while keeping the position unchanged ($x = 3$ m). For instance, the first row of figures corresponds to a nonlinearity $S_0 = 0.21$ where the prediction aligns well with the experimental results. However, the last row corresponding to an $S_0 = 0.65$ exhibits a significant deviation between the experimentally and numerically predicted spectra.

Many computational algorithms encounter numerical issues when a large amplitude wave reaches the run-up area (Madsen et al. 1997). These problems arise

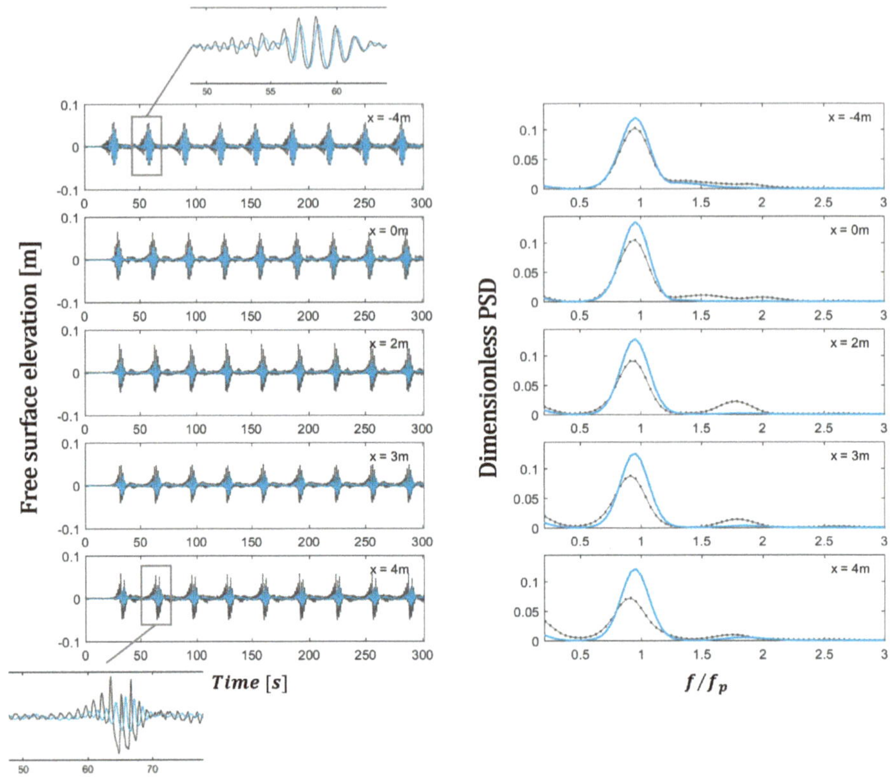

Fig. 3 On the left is shown the free surface elevation and, on the right, the dimensionless frequency spectrum obtained experimentally (black) and numerically (blue) at x = − 4 m (h = 0.3 m), x = 0 m (h = 0.3 m), x = 2 m (h = 0.22 m), x = 3 m (h = 0.18 m), and x = 4 m (h = 0.14 m) using a signal of nine wave trains and $S_0 = 0.41$

from uncontrolled numerical instabilities resulting from the discretization of dispersive terms (see Bellotti and Brocchini 2001; Durán et al. 2018). This observation underscores the importance of accurately capturing nonlinear effects, particularly in regions where wave amplitudes are amplified, such as in the run-up area after wave breaking. Furthermore, the wavelet transform analysis complemented the insights gained from the PSD analysis by identifying the individual contributions of different frequency components within each signal (Fu et al. 2021; Massel 2001). This method not only confirmed the results of the PSD analysis but also provided a more detailed view of the wave spectrum, providing valuable insights into the temporal and spatial variations of wave dynamics. The wavelet function is shifted throughout the signal, capturing the correlation between the wavelet and the signal at each specific point. This technique allows for the identification of frequency components within the signal at different temporal locations. In this context, five specific wave gauges were selected and two scenarios were investigated, a single wave train signal (Fig. 5) and a nine wave trains signal (Fig. 6).

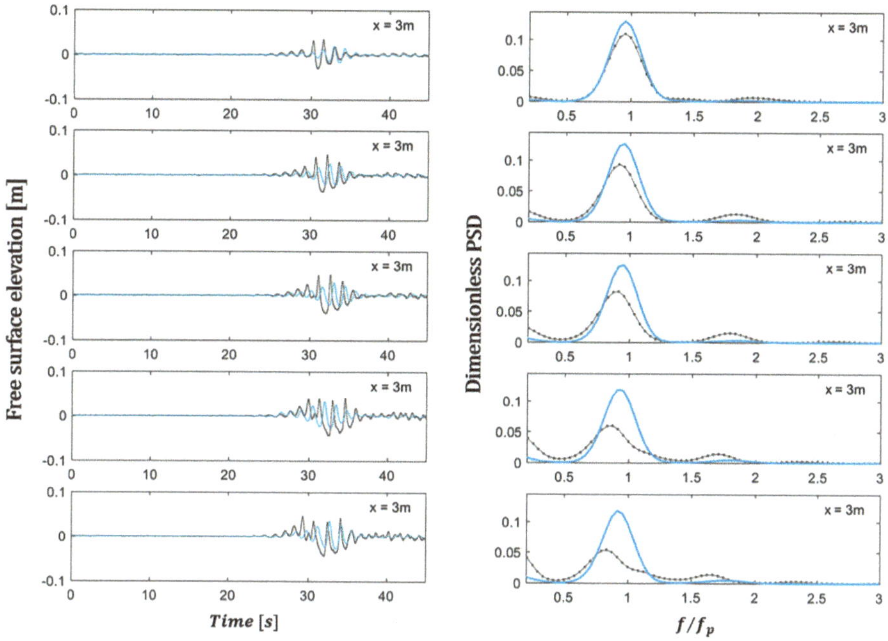

Fig. 4 On the left is shown the free surface elevation and, on the right, the dimensionless frequency spectrum obtained experimentally (black) and numerically (blue) at x = 3 m, with nonlinearity increasing from $S_0 = 0.21$ to $S_0 = 0.65$ downward

By applying the wavelet transform to the collected data at these different positions, frequency components of the wave signal were identified. This temporal perspective makes it easier to spot changes in the frequency content over time. Based on Fig. 5, the numerical simulations satisfactorily estimate the spectral changes for $0.5f_p < f < 1.5f_p$. However, the mismatch between predictions and measurements is significant elsewhere especially for $x > 2$ m (after wave breaking). These conclusions are consistent with the results obtained in (Abroug et al. 2020a, b). Similar conclusions regarding the performance of the numerical simulations can be drawn when analyzing a longer signal composed of nine wave trains, as depicted in Fig. 6. This indicates that the observed discrepancies persist even with a longer time interval, reinforcing the limitations of the model in accurately capturing wave dynamics outside of the range $0.5f_p < f < 1.5f_p$.

Figures 7 and 8 present the dimensionless spatial evolution of the spectrum, along with detailed spectra at three specific positions, presented both experimentally and numerically for $S_0 = 0.41$. Two cases were considered, a single wave train signal (Fig. 7) and a nine wave trains signal (Fig. 8). The positions are x = − 5.6 m, x = 2 m, and x = 4 m (where 0 is the toe of the slope), providing insights into the wave dynamics upstream and downstream of the wave breaking point. The Welch method was used for spectral estimation. All spectrograms indicate that the energy is focused at the peak frequency $f_p = 0.75$ Hz. Experimentally obtained Figs. 7a and 8a,

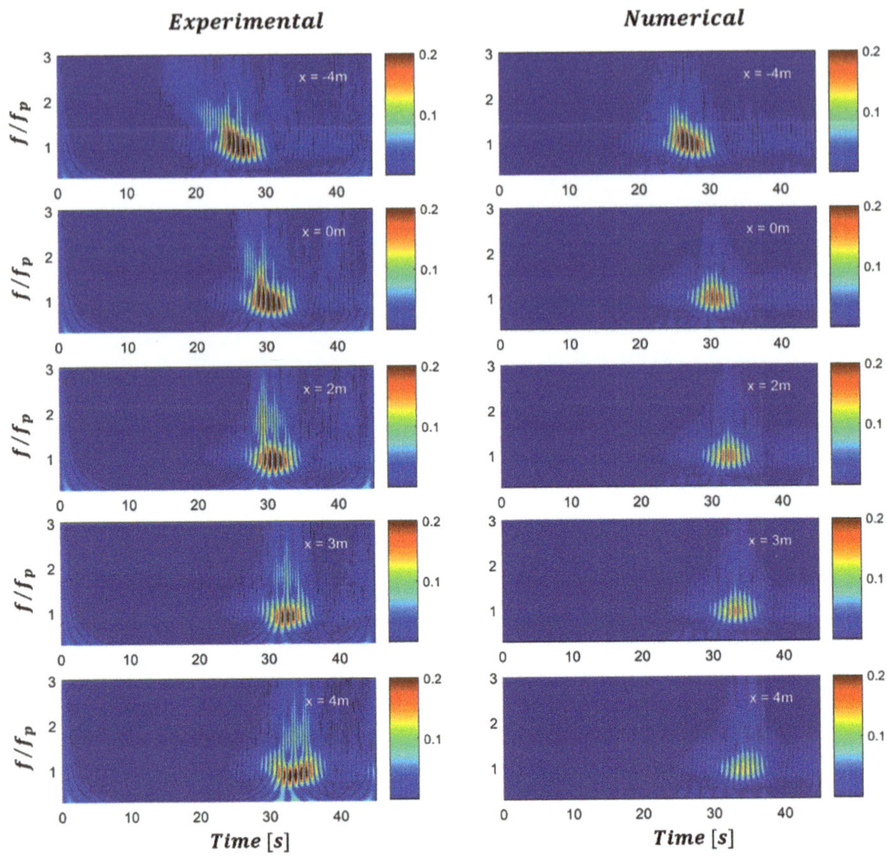

Fig. 5 Temporal evolution of the wavelet energy for five different positions along the flume: x = − 4 m (h = 0.3 m), x = 0 m (h = 0.3 m), x = 2 m (h = 0.22 m), x = 3 m (h = 0.18 m), and x = 4 m (h = 0.14 m) using a signal of one wave train and $S_0 = 0.41$

reveal a significant dissipation of energy post-wave breaking, particularly around x = 2 m. This dissipation is a characteristic behavior observed in natural wave processes (Khait et al. 2022). In contrast, Figs. 7c and 8c, obtained numerically, do not exhibit the same energy dissipation pattern post-wave breaking. The observed difference implies a limitation in the ability of the numerical model to precisely reproduce the dissipation process in the breaking zone.

Figures 7b and 8b illustrate that the wave spectrum widens as it approaches the shore due to the dissipation process, resulting in a decrease in total spectral energy. This evolution is influenced by boundary friction and nonlinear wave interactions, which play critical roles in shaping wave dynamics near the coast (Janssen 2004). Despite varying the position, Figs. 7d and 8d show no significant differences in the spectra obtained numerically. This inconsistency may indicate a simplification in the numerical model representation of spatial wave energy variations.

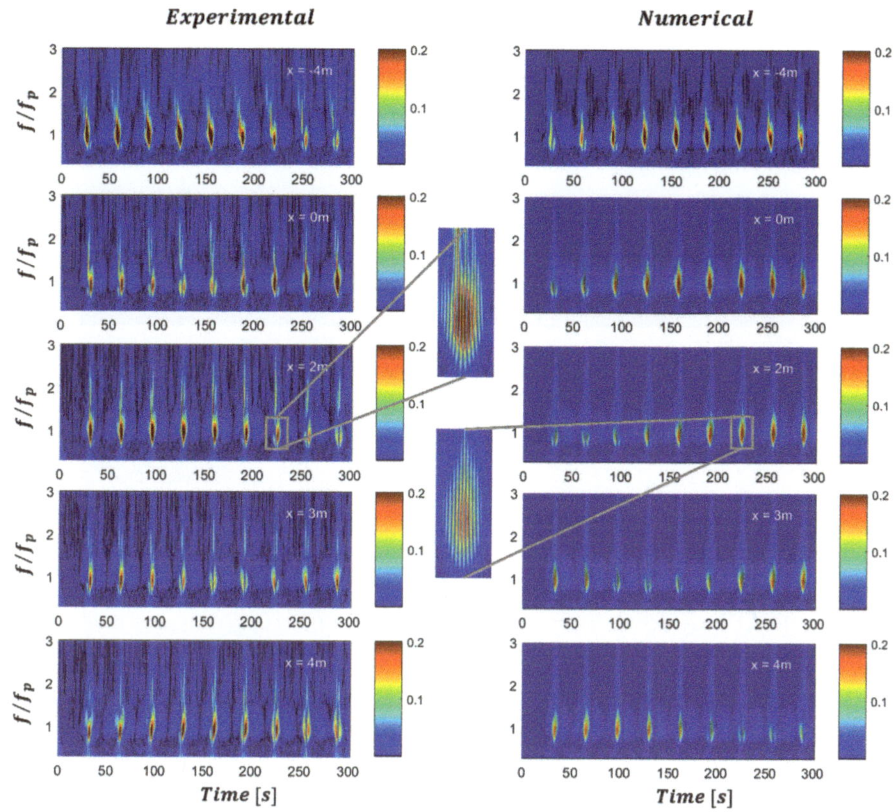

Fig. 6 Temporal evolution of the wavelet energy for five different positions along the flume: $x = -4$ m ($h = 0.3$ m), $x = 0$ m ($h = 0.3$ m), $x = 2$ m ($h = 0.22$ m), $x = 3$ m ($h = 0.18$ m), and $x = 4$ m ($h = 0.14$ m) using a signal of nine wave trains and $S_0 = 0.41$

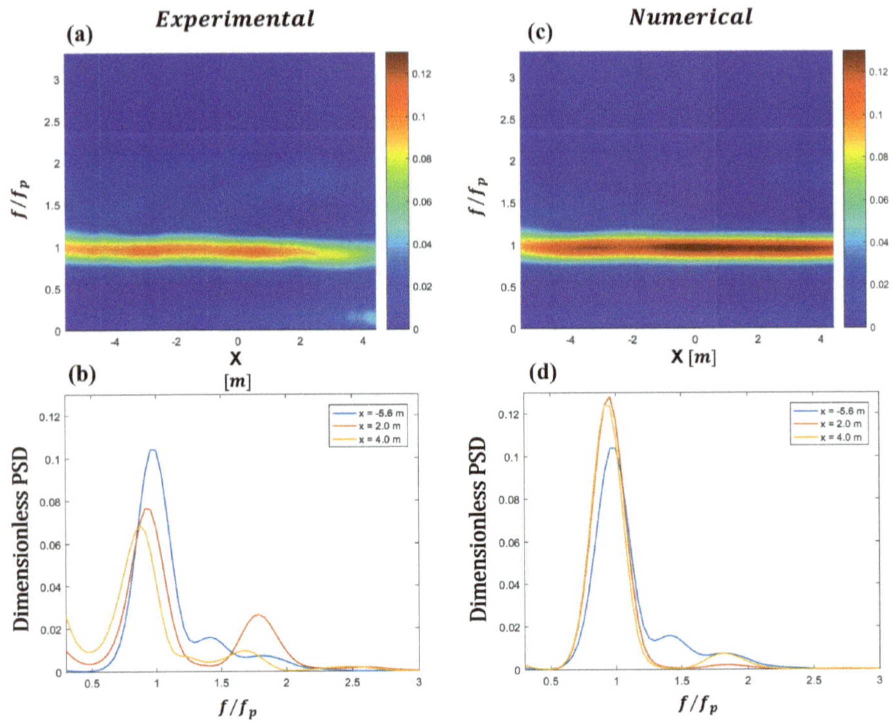

Fig. 7 Spatial evolution of the spectrum and Spectra of three positions using a signal of one wave train and $S_0 = 0.41$ for **a**, **b** experimental results; **c**, and **d** numerical results, respectively

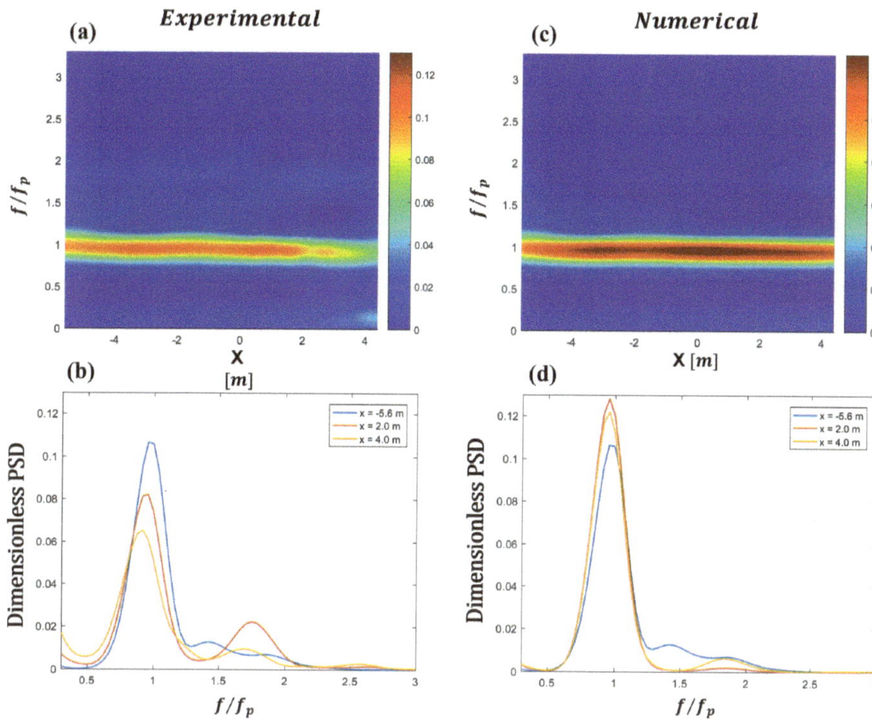

Fig. 8 Spatial evolution of the spectrum and Spectra of three positions using a signal of nine wave trains and $S_0 = 0.41$ for **a**, **b** experimental results; **c**, and **d** numerical results, respectively

4 Conclusion

The purpose of this paper was to evaluate the accuracy of the mPeregrine model in predicting spectral energy variations during the propagation of focused wave groups from intermediate to shallow waters depths. A series of physical experiments and numerical simulations are performed in this context, with a focus on areas upstream and downstream of the wave breaking.

The model demonstrated satisfactory performance in predicting spectral energy on flat and sloping bottoms upstream of wave breaking. This suggests that the model detects the underlying dynamics of wave propagation reasonably well under these conditions. Despite the overall agreement between experimental and numerical signals, discrepancies become amplified in the vicinity of wave breaking. This observation is crucial as it highlights a limitation of the model, particularly in accurately identifying the complex dynamics associated with wave breaking phenomena. Although the extreme event duration varies, the mPeregrine model remains consistent in predicting spectral energy variations, using a signal of one and nine wave trains. The impact of nonlinearity on the mPeregrine simulation, particularly evident after wave breaking, highlights the sensitivity of the model to nonlinear effects.

This is supported by the increasing divergences between simulated and experimental results as wave conditions become more nonlinear. This information is crucial for understanding the limitations of the simulation method and for choosing appropriate scenarios and conditions for the application of the model. Moreover, the wavelet transform technique allowed for a comprehensive frequency decomposition of the signals. This provided a detailed understanding of how individual frequency components contribute to the total wave energy over time. It was found that as the spectrum becomes narrower ($0.5f_p < f < 1.5f_p$), the alignment between predictions and actual measurements improves.

The analysis has provided valuable insights into the strengths and limitations of the mPeregrine model, particularly in the challenging context of wave breaking. It is important to choose appropriate scenarios and conditions for the application of the model. This understanding is essential for refining modeling approaches and improving predictions in coastal and ocean engineering applications.

Acknowledgements This publication is based upon work supported by the Khalifa University of Science and Technology under Award No. FSU-2023-014.

References

Abroug I, Abcha N, Jarno A, Marin F (2020a) Laboratory study of non-linear wave–wave interactions of extreme focused waves in the nearshore zone. Nat Hazards Earth Syst Sci 20(12):3279–3291. https://doi.org/10.5194/nhess-20-3279-2020

Abroug I, Abcha N, Dutykh D, Jarno A, Marin F (2020b) Experimental and numerical study of the propagation of focused wave groups in the nearshore zone. Phys Lett A 384(6):126144. https://doi.org/10.1016/j.physleta.2019.126144

Abroug I, Matar R, Abcha N (2022) Spatial evolution of skewness and kurtosis of unidirectional extreme waves propagating over a sloping beach. JMSE 10(10):1475. https://doi.org/10.3390/jmse10101475

Adcock TAA, Taylor PH (2009) Estimating ocean wave directional spreading from an Eulerian surface elevation time history. Proc R Soc A 465(2111):3361–3381. https://doi.org/10.1098/rspa.2009.0031

Baldock TE, Swan C, Taylor PH (1996) A laboratory study of nonlinear surface waves on water. Phil Trans R Soc Lond A 354(1707):649–676. https://doi.org/10.1098/rsta.1996.0022

Bellotti G, Brocchini M (2001) On the shoreline boundary conditions for Boussinesq-type models. Numer Methods Fluids 37(4):479–500. https://doi.org/10.1002/fld.194

Brocchini M (2013) A reasoned overview on Boussinesq-type models: the interplay between physics, mathematics and numerics. Proc R Soc A 469(2160):20130496. https://doi.org/10.1098/rspa.2013.0496

Constantin A (2017) Nonlinear water waves: introduction and overview. Phil Trans R Soc A 376(2111):20170310. https://doi.org/10.1098/rsta.2017.0310

Dao MH, Xu H, Chan ES, Tkalich P (2011) Numerical modelling of extreme waves by smoothed particle hydrodynamics. Nat Hazards Earth Syst Sci 11(2):419–429. https://doi.org/10.5194/nhess-11-419-2011

Dao MH, Xu H, Chan ES, Tkalich P (2013) Modelling of tsunami-like wave run-up, breaking and impact on a vertical wall by SPH method. Nat Hazards Earth Syst Sci 13(12):3457–3467. https://doi.org/10.5194/nhess-13-3457-2013

Didenkulova I, Pelinovsky E (2012) Nonlinear wave effects at the non-reflecting beach. Nonlinear Process Geophys 19(1):1–8. https://doi.org/10.5194/npg-19-1-2012

Dong G, Ma Y, Perlin M, Ma X, Yu B, Xu J (2008) Experimental study of wave–wave nonlinear interactions using the wavelet-based bicoherence. Coast Eng 55(9):741–752. https://doi.org/10.1016/j.coastaleng.2008.02.015

Durán A, Dutykh D, Mitsotakis D (2018) Peregrine's system revisited. In: Abcha N, Pelinovsky E, Mutabazi I (eds) Nonlinear waves and pattern dynamics. Springer International Publishing, Cham, pp 3–43. https://doi.org/10.1007/978-3-319-78193-8_1

Dutykh D, Clamond D (2016) Modified shallow water equations for significantly varying seabeds. Appl Math Model 40(23–24):9767–9787. https://doi.org/10.1016/j.apm.2016.06.033

Dutykh D, Katsaounis T, Mitsotakis D (2011) Finite volume schemes for dispersive wave propagation and runup. J Comput Phys 230(8):3035–3061. https://doi.org/10.1016/j.jcp.2011.01.003

Dysthe K, Krogstad HE, Müller P (2008) Oceanic rogue waves. Annu Rev Fluid Mech 40(1):287–310. https://doi.org/10.1146/annurev.fluid.40.111406.102203

Fu R, Ma Y, Dong G, Perlin M (2021) A wavelet-based wave group detector and predictor of extreme events over unidirectional sloping bathymetry. Ocean Eng 229:108936. https://doi.org/10.1016/j.oceaneng.2021.108936

Grinsted A, Moore JC, Jevrejeva S (2004) Application of the cross wavelet transform and wavelet coherence to geophysical time series. Nonlinear Process Geophys 11(5/6):561–566. https://doi.org/10.5194/npg-11-561-2004

Hansom JD, Switzer AD, Pile J (2015) Extreme waves. In: Coastal and marine hazards, risks, and disasters. Elsevier, pp 307–334. https://doi.org/10.1016/B978-0-12-396483-0.00011-X

Hasselmann K et al (1973) Measurements of wind-wave growth and swell decay during the joint north sea wave project (JONSWAP). Deut Hydrogr Z 8:1–95

Janssen P (ed) (2004) Nonlinear wave–wave interactions and wave dissipation. In: The interaction of ocean waves and wind. Cambridge University Press, Cambridge, pp 129–208. https://doi.org/10.1017/CBO9780511525018.005

Katsaounis T, Mitsotakis D, Sadaka G (2020) Boussinesq-Peregrine water wave models and their numerical approximation. J Comput Phys 417:109579. https://doi.org/10.1016/j.jcp.2020.109579

Khait A, Ma Z, Qian L, Bai W, Lin Z (2022) Energy dissipation and nonpotential effects in wave breaking. Int J Offshore Polar Eng 32(1):87–96. https://doi.org/10.17736/ijope.2022.ak43

Khakimzyanov G, Dutykh D, Fedotova Z, Gusev O (2020) Dispersive shallow water waves: theory, modeling, and numerical methods. In: Lecture notes in geosystems mathematics and computing. Springer International Publishing. https://doi.org/10.1007/978-3-030-46267-3

Madsen PA, Sørensen OR, Schäffer HA (1997) Surf zone dynamics simulated by a Boussinesq type model. Part I. Model description and cross-shore motion of regular waves. Coast Eng 32(4):255–287. https://doi.org/10.1016/S0378-3839(97)00028-8

Massel SR (2001) Wavelet analysis for processing of ocean surface wave records. Ocean Eng 28(8):957–987. https://doi.org/10.1016/S0029-8018(00)00044-5

Peregrine DH (1972) Equations for water waves and the approximation behind them. In: Waves on beaches and resulting sediment transport. Elsevier, pp 95–121. https://doi.org/10.1016/B978-0-12-493250-0.50007-2

Seiffert BR, Ducrozet G (2018) Simulation of breaking waves using the high-order spectral method with laboratory experiments: wave-breaking energy dissipation. Ocean Dynam 68(1):65–89. https://doi.org/10.1007/s10236-017-1119-3

Tian Z, Perlin M, Choi W (2010) Energy dissipation in two-dimensional unsteady plunging breakers and an eddy viscosity model. J Fluid Mech 655:217–257. https://doi.org/10.1017/S0022112010000832

Tiscareno MS, Hedman MM (2018) A review of Morlet wavelet analysis of radial profiles of Saturn's rings. Phil Trans R Soc A 376(2126):20180046. https://doi.org/10.1098/rsta.2018.0046

Torrence C, Compo GP (1998) A practical guide to wavelet analysis. Bull Amer Meteor Soc 79(1):61–78. https://doi.org/10.1175/1520-0477(1998)079%3c0061:APGTWA%3e2.0.CO;2
Tromans P, Anaturk AR, Hagemeijer P (1991) A new model for the kinematics of large ocean waves-application as a design wave. In: Proceedings of the ISOPE-91, vol 3
Watanabe Y, Tsuda Y, Saruwatari A (2020) Wave packet focusing in shallow water. Coast Eng J 62(2):336–349. https://doi.org/10.1080/21664250.2020.1756033
Whitham GB (1999) Linear and nonlinear waves, 1st edn. Wiley. https://doi.org/10.1002/9781118032954
Whittaker CN, Fitzgerald CJ, Raby AC, Taylor PH, Borthwick AGL (2018) Extreme coastal responses using focused wave groups: overtopping and horizontal forces exerted on an inclined seawall. Coast Eng 140:292–305. https://doi.org/10.1016/j.coastaleng.2018.08.004
Xu J, Liu S, Li J, Jia W (2021) Experimental study of wave height, crest, and trough distributions of directional irregular waves on a slope. Ocean Eng 242:110136. https://doi.org/10.1016/j.oceaneng.2021.110136
Zhang J, Benoit M, Kimmoun O, Chabchoub A, Hsu H-C (2019) Statistics of extreme waves in coastal waters: large scale experiments and advanced numerical simulations. Fluids 4(2):99. https://doi.org/10.3390/fluids4020099

Spatial and Temporal Monitoring of a Dune System Evolution Using Drone-Based Lasergrammetry: The Case Study of Anse Du Guesclin (Brittany)

D. James, A. Collin, N. Lamontagne, I. Jeuffard, O. Le Gal, É. Feunteun, G. Duthion, and J.-F. Lebas

1 Introduction

1.1 Coastal Global Changes' Context

Coastal risks, such as marine flooding and erosion, have become a major concern for stakeholders in the face of increasing frequency and intensity of climatic phenomena. As climate change progresses, coastal areas are subject to more frequent and severe events, including sea level rise, extreme storms, and changing precipitation patterns (IPCC 2023). Such factors contribute to increased vulnerability of coastal regions, affecting not only natural ecosystems but also human settlements, infrastructures and socio-economic activities (Mousavi et al. 2011; Pérez et al. 2002). The potential for extensive environmental and economic consequences underscores the urgency of addressing coastal risks.

D. James (✉) · A. Collin · N. Lamontagne · O. Le Gal · É. Feunteun
Centre de GéoÉcologie Littorale, École Pratique des Hautes Études, Université PSL, Dinard, France
e-mail: dorothee.james@ephe.psl.eu

I. Jeuffard
IGN, Saint-Mandé, France

É. Feunteun
UMR BOREA (MNHN, CNRS, SU, IRD, UCN, UA), Dinard, France

G. Duthion · J.-F. Lebas
Département d'Ille-et-Vilaine, Rennes, France

A.-C. Bennis et al. (eds.), *Constraints and Adaptations to Global Change at the Land-Sea Interface: For a Shared Ecological and Energy Transition*,
https://doi.org/10.1007/978-3-031-90050-1_2

1.2 Enhancing Costal Resilience

In responding to such challenges, stakeholders are increasingly focusing on comprehensive strategies to mitigate and adapt to coastal risks (Robert et al. 2023). Among these are the implementation of protective measures such as seawalls, dikes, and beach nourishment projects to combat erosion and submersion. Additionally, sustainable land-use planning and ecosystem-based approaches are emerging as effective means of enhancing coastal resilience. Collaboration among governments, communities, and scientific institutions is essential for developing and implementing robust policies and practices that address both immediate threats and long-term impacts of coastal risks (Molino et al. 2020).

1.3 Improved Coastal Monitoring

Unmanned aerial vehicles (UAVs) have become an essential tool for monitoring sediment evolution using passive photogrammetry at centimeter and fine temporal resolutions (James et al. 2020). Recently, drones have been enhanced by the integration of active LiDAR (Light Detection and Ranging) sensors, opening up a new era of precision and accuracy in the observation and monitoring of coastal geo-ecosystems using lasergrammetry (James and Collin 2023). This technological breakthrough represents a remarkable advance in remote sensing capabilities, enabling more detailed and nuanced analysis of topographic changes over time. By combining the agility and accessibility of UAVs with the high-resolution data provided by LiDAR sensors, researchers and environmental professionals can now delve deeper and understand the dynamics of landscapes and environmental processes.

The high spatial and temporal resolutions of these new tools enable us to understand landscape dynamics in greater detail and with greater precision. By capturing fine-grained data at both very high spatial and temporal resolution, these tools enable us to analyze changes in the landscape over time with unprecedented precision.

The aim of this study is to monitor the spatial and temporal evolution of dune systems at high planimetric and altimetric accuracy in order to develop sustainable management strategies for the coastal zone.

2 Material and Methods

2.1 Study Site

The *Anse Du Guesclin* (48° 41′ 31 N; 1° 53′ 49 W) is located in the commune of Saint-Coulomb in Ille-et-Vilaine (Brittany; Fig. 1). A dune mass-oriented east–west is a natural formation created by the accumulation of wind-blown sand along the

Fig. 1 The *Anse Du Guesclin* study site. In orange, a road crossing the dune

coastline. The dunes separate the fine sandy beach to the north from the rear dunes to the south.

The *Anse Du Guesclin* dune has been subjected to human activity over the past seven decades. The construction of a road and associated parking lots in the 1960s prevented both the natural movement of sediment and the establishment of dune vegetation, despite restoration attempts in the 1980s.

Since then, the dune mass has continued to move naturally in response to the weather and sea, until it covered the coastal road with sand, requiring intervention by the responsible authorities.

2.2 UAV Survey

Four LiDAR drone surveys have been carried out since 2022: May 3, 2022; September 1, 2022; January 19, 2023; and April 17, 2023.

A multi-rotor aerial drone, DJI Matrice 300, equipped with a LiDAR L1 sensor, was used for topographic surveys. In addition to the L1 sensor, this drone was featured with a 20 M-pixel red–green–blue camera.

The study area was covered in a single flight and at a height of 70 m (AED, i.e. above ground level, with an average topography of 3 m according to the national vertical datum, IGN 69).

The flight parameters applied were (Fig. 2):

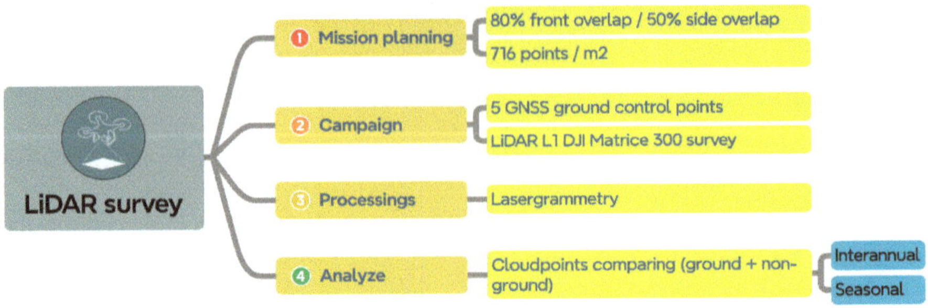

Fig. 2 Methodology workflow

- Flight altitude: 70 m AED (Above elevation),
- Estimated flight time: 26 min,
- Point density: 716 points/m^2,
- Blue-green–red image resolution (DJI CMOS 1″ sensor, 20 Mpixels): 0.019 m,
- Average speed: 4 m/s,
- Lateral/frontal image overlap: 50/80%.

A simultaneous ground campaign was carried out in conjunction with the drone surveys. A series of 5 ground control points were placed on the study site and measured using GNSS RTK.

2.3 *Lasergrammetry and Points Cloud Analysis*

DJI Terra software, known for its advanced lasergrammetric capabilities, played a crucial role in the reconstruction process (Fig. 3). This software not only has the ability to read the proprietary format of the survey files, but also processes the RTK positioning files associated with LiDAR surveying. Based on these files, DJI Terra ensures a centimetric reconstruction of the point cloud.

CloudCompare is a powerful open-source software tool commonly used for point cloud analysis (https://www.cloudcompare.org). It offers a range of features that go beyond point cloud display and management. CloudCompare enables advanced point cloud processing and supports pairwise point cloud comparison.

Volume calculations were performed using CloudCompare's 'volume calculation' tool, based on the reconstructed point clouds. The evaluations were conducted in pairs on an inter-annual and seasonal scale.

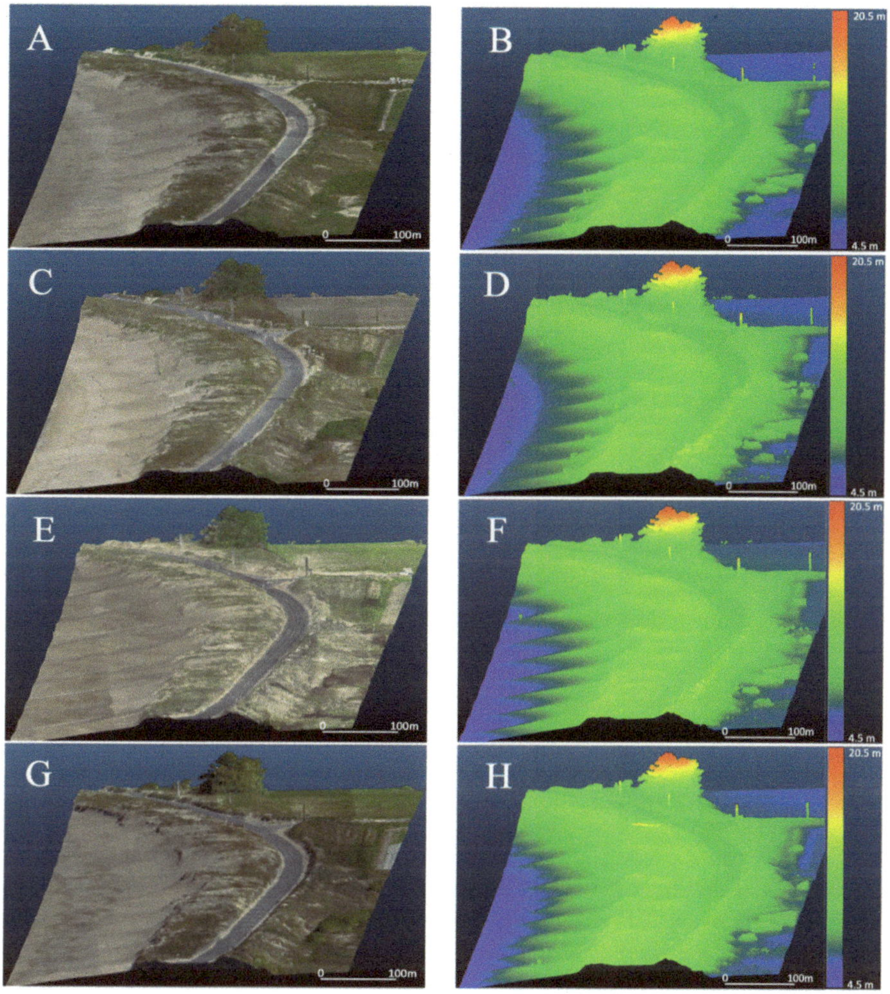

Fig. 3 Results from LiDAR drone surveys: point clouds in natural color and colorized according to altitude for May 3, 2022 (**a** and **b**), September 1, 2022 (**c** and **d**), January 19, 2023 (**e** and **f**) and April 17, 2023 (**g** and **h**)

3 Results and Discussion

3.1 Seasonal Analysis

Further analysis of the factors contributing to these volume differences on a seasonal scale provides us with valuable insights into the processes shaping the landscape.

During the summer accretion period, which is characterized by favorable conditions for sediment accumulation and vegetation growth, there was a significant gain

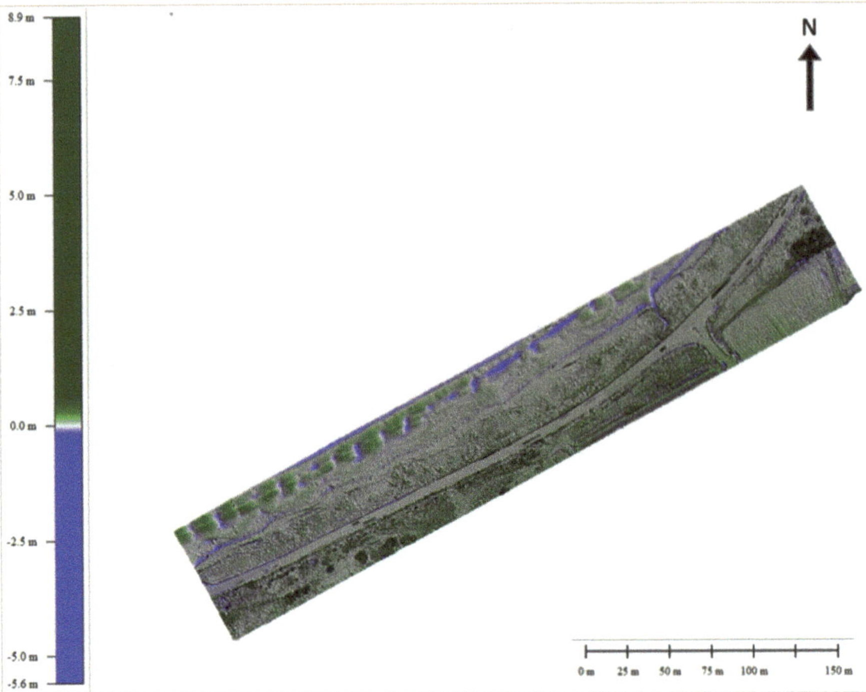

Fig. 4 Summer mapping differences (May vs. September 2023)

of + 2359 m^3 (Fig. 4). This indicates that the site experienced a significant increase in sediment deposition during this period on the foredune. However, there was also a slight loss of 884 m^3 on a seasonal scale, suggesting that some erosion occurred during specific events or periods during the summer season due to a difference in vegetation cover between the two periods.

Conversely, the winter season, known for its erosive storms, showed a substantial gain of + 8848 m^3 (Fig. 5). This indicates that, despite the erosive nature of winter storms, the site experienced a significant net increase in sediment deposition during this season. However, a loss of – 1 034 m^3 was also estimated, suggesting that erosion processes were still active during certain events or periods of the winter season. An increasing mobility of the dune mass as observed, suggesting a significant positive difference between the point clouds.

3.2 *Inter-Annual Analysis*

The results of the analysis revealed a significant inter-annual difference in volume between the scatterplots (Fig. 6). Specifically, the volume increased by 7292 m^3 in

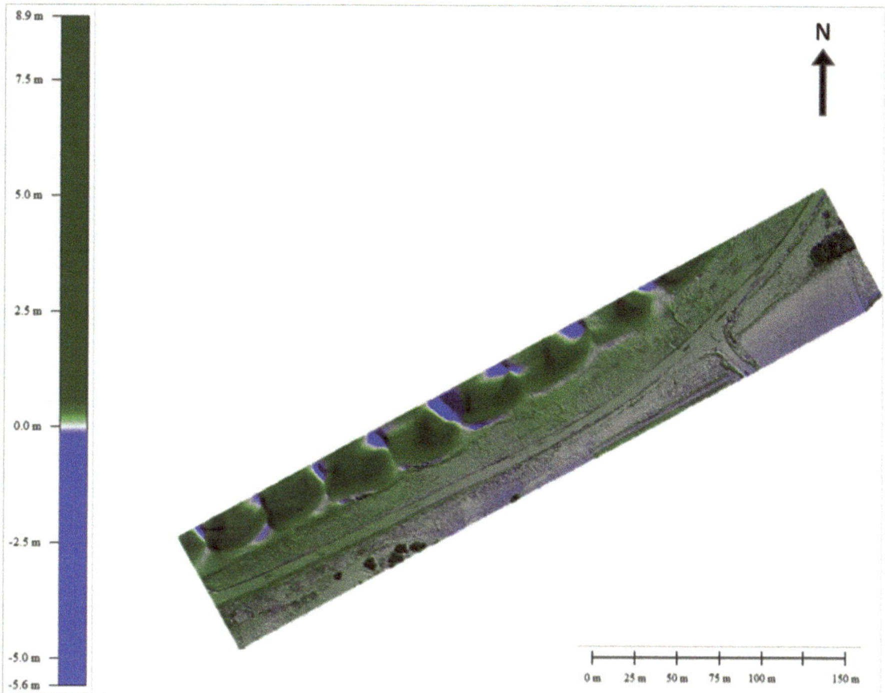

Fig. 5 Winter mapping differences (January vs. April 2022)

one year, indicating substantial growth or accumulation of material. On the other hand, the volume decreased by – 2637 m^3 in another year, suggesting a reduction or loss of material.

The analysis of the volumes of the entire point cloud, which includes both ground and overburden components, on an inter-annual or seasonal scale, revealed the positive geo-ecological morphodynamism of the site. This means that, overall, the site has shown a favorable trend in sediment accumulation and ecological dynamics over time. This analysis provides valuable information on the site's geomorphological processes.

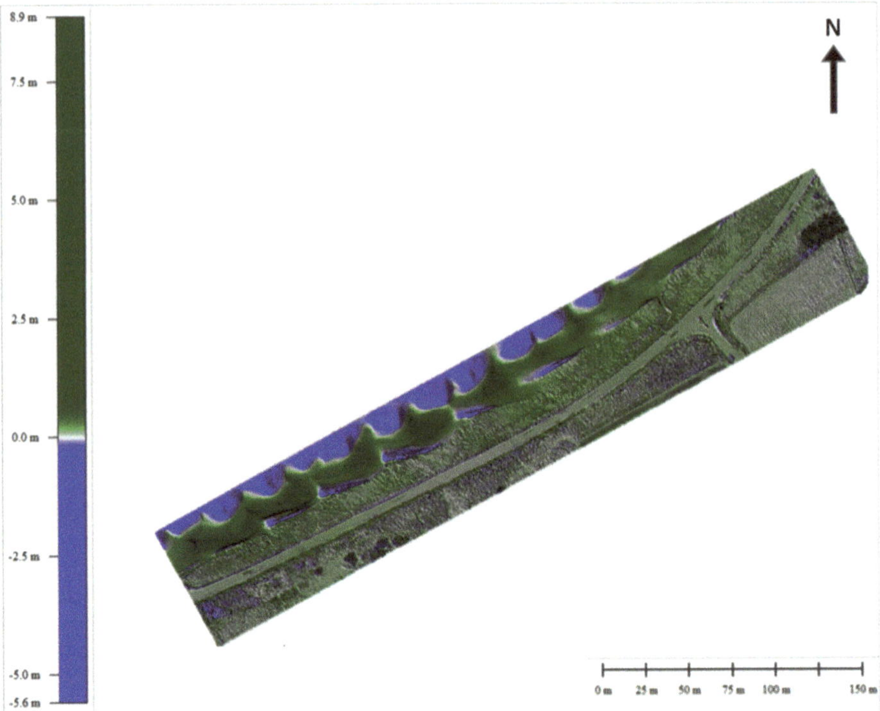

Fig. 6 Inter-annual difference mapping 2022 versus 2023

4 Conclusion

To address the impact of climate change on terrestrial landscapes, it is crucial to have precise and accurate techniques for understanding the geo-ecological dynamics of an environment. This technology allows for increased precision and accuracy in terms of spatial, spectral, and temporal resolution.

Drone-mounted LiDAR (Light Detection and Ranging) provides a solution that is both easy to deploy, thanks to the agility offered by the drone, and provides unrivaled spatial precision and accuracy.

LiDAR drone technology has made it possible to analyze the volumes of sites like the *Anse Du Guesclin*. The point cloud has revealed the site's significant geo-ecological spatial dynamism, with an estimated inter-annual difference in volume between the point clouds of $+ 7\ 292\ m^3$ and $- 2\ 637\ m^3$.

A study is currently underway on a broader time scale to enable more detailed analysis and a better understanding of geo-ecological processes. To complete this study, LiDAR infrared indicators of vegetation vitality could be included (Collin et al. 2023).

References

Collin A, James D, Gallon R, Poizot E, Feunteun E (2023) The use of ultra high resolution UAV Lidar infrared intensity for enhancing coastal cover classification. In: Presented at the 5th international electronic conference on remote sensing, vol 7, p 21

IPCC (2023) Climate change 2023: synthesis report. A report of the intergovernmental panel on climate change. Contribution of working groups I, II and III to the sixth assessment report of the intergovernmental panel on climate change. IPCC, Geneva

James D, Collin A (2023) Photogrammétrie vs. lasergrammétrie par drone aérien sur la Côte d'Émeraude. J Drones Cap

James D, Collin A, Houet T, Mury A, Gloria H, Le Poulain N (2020) Towards better mapping of seagrass meadows using UAV multispectral and topographic data. J Coast Res 95(SI):1117–1121

Molino GD, Kenney MA, Sutton-Grier AE (2020) Stakeholder-defined scientific needs for coastal resilience decisions in the Northeast US. Mar Policy 118:103987

Mousavi ME, Irish JL, Frey AE, Olivera F, Edge BL (2011) Global warming and hurricanes: the potential impact of hurricane intensification and sea level rise on coastal flooding. Clim Change 104(3–4):575–597

Pérez T, Garrabou J, Bonhomme D, Bensoussan N, Boury-Esnault N, Chevaldonné P, Zibrowius H (2002) Dynamique de la biodiversité marine: Perturbations récentes des paysages sous-marins méditerranéens, signal d'un changement climatique. Act J Inst Franç Biodiv Tours

Robert S, Quercy A, Schleyer-Lindenmann A (2023) Territorial inertia versus adaptation to climate change. When local authorities discuss coastal management in a French Mediterranean region. Glob Environ Change 81:102702

Monitoring the Submersion Using Monthly Time-Series at High Spatial Resolution: PlanetScope's Five-Year-Old Beaussais' Marsh Restoration

A. Collin, D. James, N. Lamontagne, E. Feunteun, T. Durozier, and G. Hervouet

1 Introduction

1.1 Coastal Risks

Addressing today's planetary crisis necessitates to adopt a proactive emergency approach (IPCC 2021). Emergency is the product of the urgency and the risk: while the first factor equals to the ratio of reaction and intervention time, the second factor is itself a combination of the hazard, the exposure and the vulnerability (Fig. 1).

On the coastal area, the hazard can be the acute cyclone/storm-driven wave and surge (Lochin et al. 2020), or the chronic sea-level (Jeuffrard et al. 2023) and temperature rise (Bitoun et al. 2018). The exposure of the humans with their assets, but also other living species with their habitats, results from the concomitance of the distance and elevation to the shoreline. The vulnerability reflects both the degree of the stakes' sensitivity to cope with hazards, and the stakes' adaptive capacity to respond to hazards (Mury et al. 2020).

A. Collin (✉) · D. James · N. Lamontagne · E. Feunteun
Centre de GéoEcologie Littorale, EPHE-PSL Université, Dinard, France
e-mail: antoine.collin@ephe.psl.eu

T. Durozier
Conservatoire du Littoral et des Rivages Lacustres—Délégation Normandie, Hérouville-Saint-Clair, France

G. Hervouet
Conservatoire du Littoral et des Rivages Lacustres—Délégation Bretagne, Plérin, France

A.-C. Bennis et al. (eds.), *Constraints and Adaptations to Global Change at the Land-Sea Interface: For a Shared Ecological and Energy Transition*,
https://doi.org/10.1007/978-3-031-90050-1_3

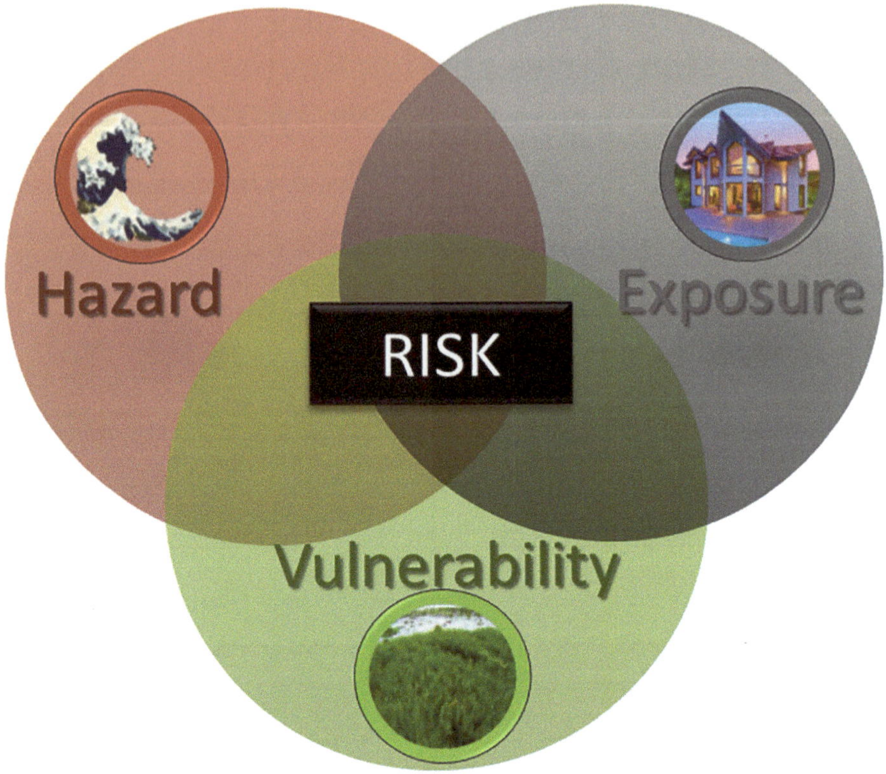

Fig. 1 The risk intersects the hazard, the exposure and the vulnerability

1.2 Coastal Remote Sensing

The coastal risks occur at various spatial and temporal scales, sometimes in remote areas, exacerbating their sensitivity. As nowadays' acute and chronic hazards are increasing at unrecorded rates, their reliable monitoring at the global scale is required. Since it exists a trade-off between the spatial extent and resolution, the scale of interest strongly constrains the platform to be spaceborne, and the pixel size to be medium (> 10 m, such as Sentinel-2, Collin et al. 2016, or Landsat-9, Tieng et al. 2019) or coarse (> 100 m, such as Sentinel-3 or MODIS, Collin et al. 2021a, b). Conversely, the coarser the spatial resolution is, the higher the spatial temporal is. Sentinel-3 therefore leverages a daily revisit, while Sentinel-2 may capture a same scene every ten days. However, erosion or submersion hazards, that can potentially lead to risks, mostly happen at both the meter and the day scale.

1.3 PlanetScope Remote Sensing

The advent of the nanosatellites has recently changed the traditional remote sensing trade-off between the spatial and temporal resolution. The three PlanetScope constellations, composed of more than 500 CubeSats, enable to capture a regional area (several km^2) at high spatial resolution (3-m pixel size) and very high temporal resolution (one-day revisit). This paradigm's shift of the spaceborne platform has been used to derive bathymetry (Collin et al. 2023a), to classify coastal area through a fusion with the 10-band Sentinel-2 (Collin et al. 2022a), and to derive the Normalized Difference Vegetation Index (NDVI) at ultra high spatial resolution (0.03 m) through a fusion with a consumer-grade airborne drone (Collin et al. 2022b).

1.4 Issue

The agility of the PlanetScope imagery has offered the opportunity to explore the monitoring of a coastal phenomenon of interest, namely the salt marsh restoration of Beaussais' marsh in Brittany, France (Fig. 2). The monitoring capacity of the PlanetScope imagery was evaluated using the first constellation, called Dove-C (PS2), that provided five cloudless multispectral (blue-green–red-infrared) imageries of October's climatology from 2016 to 2020 (Collin et al. 2021a, b). We propose here to deepen both the spatial–temporal monitoring and processing of that site by filtering, selecting, and mosaicking PlanetScope imageries at the month lag, to create, analyze (by derivation) and discuss a time-series of the NDVI along 59 months, from February 2017 to April 2022 (as preliminary introduced in the related symposium's poster, Collin et al. 2023b).

2 Methodology

2.1 Study Site

The original PlanetScope time-series has been applied to Beaussais' marsh in Brittany (France). The entire salt marsh (Fig. 2a) is located along the upper bay of Beaussais (48° 34′ 53″ N, 2°09′ 50″ W) and is fragmented by a dyke, separating a seaward nature-driven salt marsh (Collin et al. 2018) and a landward polderized fodder crop field or pastureland (Fig. 2b).

The dyke was built in the XIXth century in order to (re-)claim land over sea for agricultural purposes. The polderized part of the marsh, spanning over 0.6 km^2, is

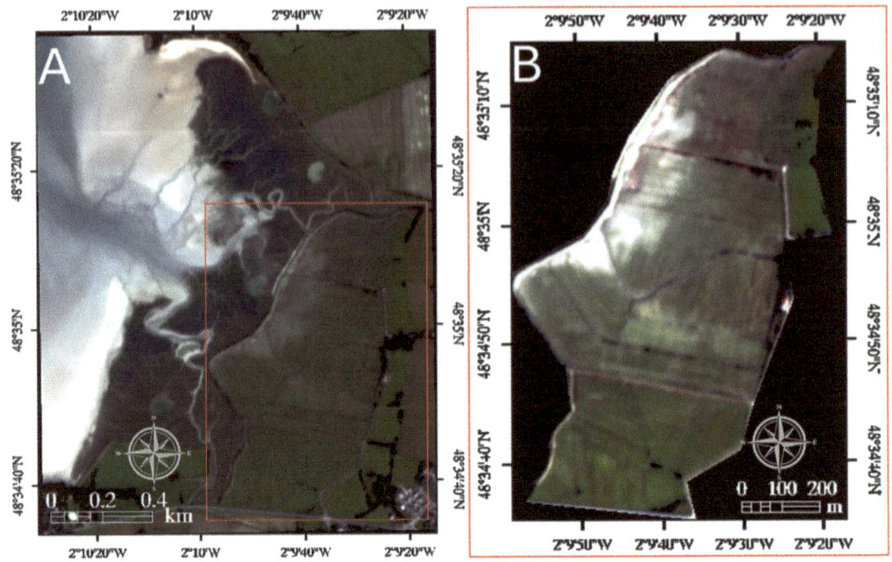

Fig. 2 The study site of **a** entire, and **b** (de)polderized part of Beaussais' marsh

featured by an average altitude of 5 m, that is say, almost 2 m below the highest astronomical tide level. The natural part of the salt marsh, above the tidal mud flat, is characterized by lower, middle and upper schorre communities, as well as meandering channels. It also hosts five hunter's ponds (see bean-like pattern in Fig. 2a).

Beaussais' marsh is an interesting site in transformation given its specific management and its cultural value.

Belonging to the *Conservatoire du littoral*, it was chosen as a pilot site for the European Life program "Adapto", aiming at experimenting nature-based adaptation solutions to the sea-level rise (https://www.lifeadapto.eu/home.html). Following the polder's acquisition between 1994 and 2018, the landowner the landowner supported farmers in adopting a practice resilient in the face of occasional marine intrusion, transforming cultivated fields into wetlands maintained by extensive grazing. In concertation with local farmers, policy makers and managers, was raised the issue of recurrent marine inundation and associated costs of dyke's maintenance and repair. It was then decided not to repair the next breach and to let this natural area evolve freely, what happened in March 2020, during the high spring tide. Following that event, at each spring tide (coefficients above 100–110), the sea has entered the polder and enlarged the breach, whose width has increased from 3 m in June 2020 to 23 m in December 2020. This was accompanied by regressive erosion and the digging of a channel on the foreshore. This led, at the spring tide of April 28, 2021, to a clear reconnection of the polder to the sea. After that, the sea flows unimpeded into the polder at full tide (even outside spring tide periods, see Le Roy and Brivois (2022), for further chronologic illustrations).

The process of the salt marsh "passive" restoration was tracked, first, by the preliminary use of PlanetScope-originated October's climatology, from 2016 to 2020 (Collin et al. 2021a, b), and then, by centimeter-scale and season-scale drone imagery from 2021 to 2023 (James et al. 2022).

The marsh reconnection to the sea has rapidly brought the marine sediments and salt marsh propagules, leading to the colonization of the lower salt marsh vegetation, namely the glassworts and the saltworts. That landscape transformation has triggered the artistic inspiration of Nancy Lamontagne to build a dedicated project. "*Relier: à l'abri dans la marée*" (https://nancy-lamontagne.wixsite.com/artiste/a-labri-dans-la-maree, Fig. 3) was selected by the "*Mondes nouveaux*" program (https://mondes-nouveaux.culture.gouv.fr/fr), then funded by the French Ministry of Culture, between 2021 and 2023. A flood-prone installation, 20-m long and 5-m high, was created as a wooden, wave-shaped pattern evoking the metamorphosis of a wall (dyke) into a bridge (walkway) (Fig. 4). It was inspired by the tide, dyke, breach, agriculture posts, and nature-based solution. From that artwork's process has emerged, in 2022, one media library exhibition (Lamontagne and Collin 2022a), one symposium conference (Lamontagne and Collin 2022b), and one research paper (Lamontagne and Collin 2022c). The artwork also provided incentives for emphasizing the role of the culture as an ecotone between economy and nature (Lamontagne et al. 2023, 2024).

Fig. 3 The breach, flooded marsh and artwork «*Relier: à l'abri dans la marée*» from Nancy Lamontagne, lying on Beaussais' marsh at high spring tide

Fig. 4 The close up on the artwork «*Relier: à l'abri dans la marée*» from Nancy Lamontagne, lying on Beaussais' marsh at high spring tide

2.2 PlanetScope Imagery

2.2.1 Source

Every nanosatellite (CubeSat) of the three PlanetScope constellations measures 10 cm × 10 cm × 30 cm), weighs 5.8 kg with a designed 2/3-year-old lifetime. With at least 130 nanosatellites orbiting simultaneously, PlanetScope can cover Earth's terrestrial surface, from 70° to − 70° latitude every day with 3-m pixel size. The constellations have succeeded each other with temporal overlapping, from 2014 to now.

The first constellation, called Dove-Classic or Dove-C, was composed of the CubeSat bearing a PS2 telescope, collecting blue-green–red-infrared imagery (Table 1). The typical scene captured an area of 25 km × 11.5 km. Dove-C flocks were operational from 2014 to 2022.

The second constellation, called Dove-R, orbited PS2-based CubeSats but provided with Bayer pattern and pass-band filters, thus the telescope's name of PS2.SD (Table 1). The acquired scene extended over a surface of 25 km × 23 km. Dove-R flocks were operational between 2019 and 2022.

The third constellation, called SuperDove, are constituted of a PSB.SD telescope, that captures eight spectral bands: coastal blue (purple), blue, green 1, green 2,

Table 1 Spectral and spatial specificities of the CubeSat's PlanetScope constellations

Band names	Lower wavelength (nm)	Upper wavelength (nm)	Pixel size (m)
Dove-C (PS2)			
Blue	455	515	3
Green	500	590	3
Red	590	670	3
Near-infrared	780	860	3
Dove-R (PS2.SD)			
Blue	464	517	3
Green	547	585	3
Red	650	682	3
Near-infrared	846	888	3
SuperDove (PSB.SD)			
Blue	465	515	3
Green 2	547	583	3
Red	650	680	3
Near-infrared	845	885	3

yellow, red, red edge, infrared (Table 1). The scene spans area of 32.5 km × 19.6 km. SuperDove flocks have started to be operational from March 2020 until now.

A total amount of 535 nanosatellites[1] were launched between 2014 and 2023.

2.2.2 Processing

The time-series of the PlanetScope imagery was created using a workflow applied to the Planet Explorer's geoportal.

First, the scene of Beaussais' bay was drawn. Second, the filtering of the blue-green–red-infrared PlanetScope imagery, the scene coverage at 100%, and the cloud coverage of the scene less than 10% was fixed. Third, the clearest imagery of every month, between February 2017 and April 2022 (the change date in radiometry), was manually validated. Fourth, the resulting imageries were exported as surface reflectance (ratio between scene radiance and sun irradiance, both units in $W\ m^{-2}\ sr^{-1}\ nm^{-1}$), in order to be comparable across space and time. A total of 59 screened imageries constituted the time-series to be further analyzed. For each of the 59 dates, a 3D matrix contained 4 images (blue-green–red-infrared) dimensioned by 520 pixels × 660 pixels.

Since the phenomenon of interest was the submersion process, a spatial subset was realized over Beaussais' ancient polder (Fig. 5). As a valued eco-indicator sensitive to water content and chlorophyll-laden vegetation, the pixel-wise NDVI was computed

[1] https://www.eoportal.org/satellite-missions/planet#launches.

as the ratio between the difference of infrared and red surface reflectance values with their sum (Collin et al. 2022b). The spatial–temporal variance of the NDVI was then implemented to distinguish areas where drastic changes (toward red in Fig. 5) occurred from stable areas (toward blue in Fig. 5). The NDVI is worldwide recognized as the mostly used vegetation remote sensing indicator, owing to its easiness of computation, and its robustness, time-tested since its creation in 1974 (Rouse et al. 1974) following the launch of the Landsat-1 in 1972. The NDVI continues to be daily produced from the spaceborne platforms with no need of ground-truthing.

A profile of eight virtual stations (in white in Fig. 5) was created to study in-depth changes, from the breach (northwesternmost station in Fig. 5) to marsh's back (southeasternmost station in Fig. 5). The NDVI time-series of the eight virtual stations were line-plotted to evaluate the temporal behavior of that NDVI profile. Finally, the first derivatives of the eight lines were calculated so as to identify rhythmic patterns and tipping points.

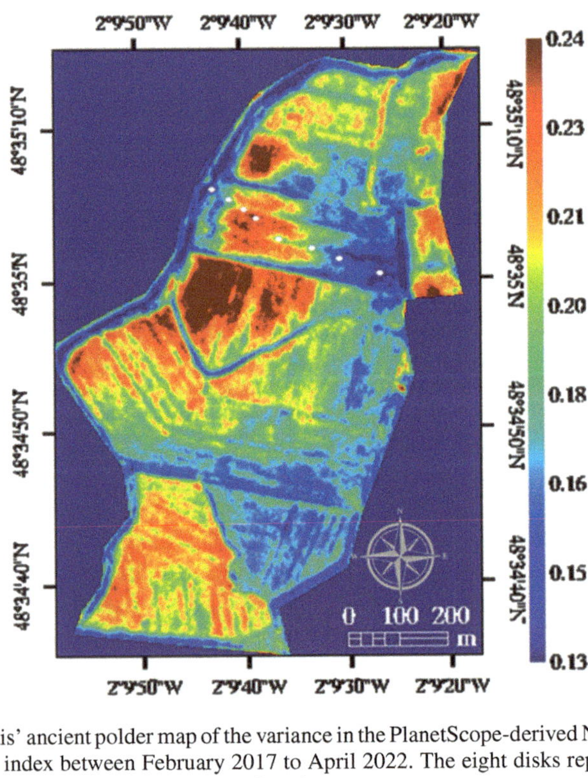

Fig. 5 Beaussais' ancient polder map of the variance in the PlanetScope-derived Normalized difference vegetation index between February 2017 to April 2022. The eight disks represent the virtual stations, whose time-series were further analyzed

3 Results and Discussion

3.1 *Mapping the Submersion*

Out of the hundreds of partial or complete scenes captured over Beaussais' bay from February 2017 to April 2022, an easy-to-select set of 59 multispectral imageries was found out at the month lag. This cloudless, dense time-series, provided with four bands at 3-m pixel size, equals the capabilities of a temporally-very high but spatially-coarse sensor, such as Sentinel-3 or MODIS. This unique opportunity enabled to map the variance in the surface reflectance-derived NDVI values, focused on Beaussais' ancient polder.

The northwesternmost virtual station symbolizes the dyke's breach (see Fig. 5). From that point, the closer the polder's area is, the more variant is. The closest parts, first, underwent seawater tidal flood which "burnt" the fodder crops, then, received seawater-laden mud sediment, and finally, were colonized by lower salt marsh vegetation. The change gradient between the breach and marsh's back might be very well correlated with fine-scale topography: the lower is, the more frequent the tidal flood happen. This likely hypothesis can be strengthened by the low variations in NDVI values observed for farmer-dug channels, since they were topographically low and thus water-saturated, most of the time. The airborne, be it manned (Collin et al. 2022c) or unmanned (James et al. 2022), lasergrammetry or photogrammetry digital models will be good assets to corroborate the hypothesis.

Both northeastern and southwestern parts of the polder also showed great variations in the NDVI values despite their remote distance from the breach.

For the first part, the changes could be logically explained by the presence of hedges, apple's orchards and birch trees. Trees with underlying vegetation were flooded at lower rates but the salt impacted those arborescent and arbustive strata for longer times than the herbaceous stratum closer to the breach. That longer process might be correlated with the size of the vegetation: the taller the vegetation is, the longer the devitalization is.

About the second part, the changes took place south to an internal smaller dyke, behind which the agricultural use continued over a longer period than other places. That use was essentially focused on the pasture and the fodder crop, implying seasonal changes of the cover, thus the detected higher variance.

3.2 *Analyzing the Time-Series*

The first derivative NDVI analysis across 59 months for the eight stations revealed two distinct patterns: the pre-breach and post-breach behaviors, with an obvious breach-stemmed shift between March–May 2020 (Fig. 6).

Prior to the breach, the location of the stations satisfactorily translated the expected NDVI in a polderized area. The more remote from the breach the station is, the higher

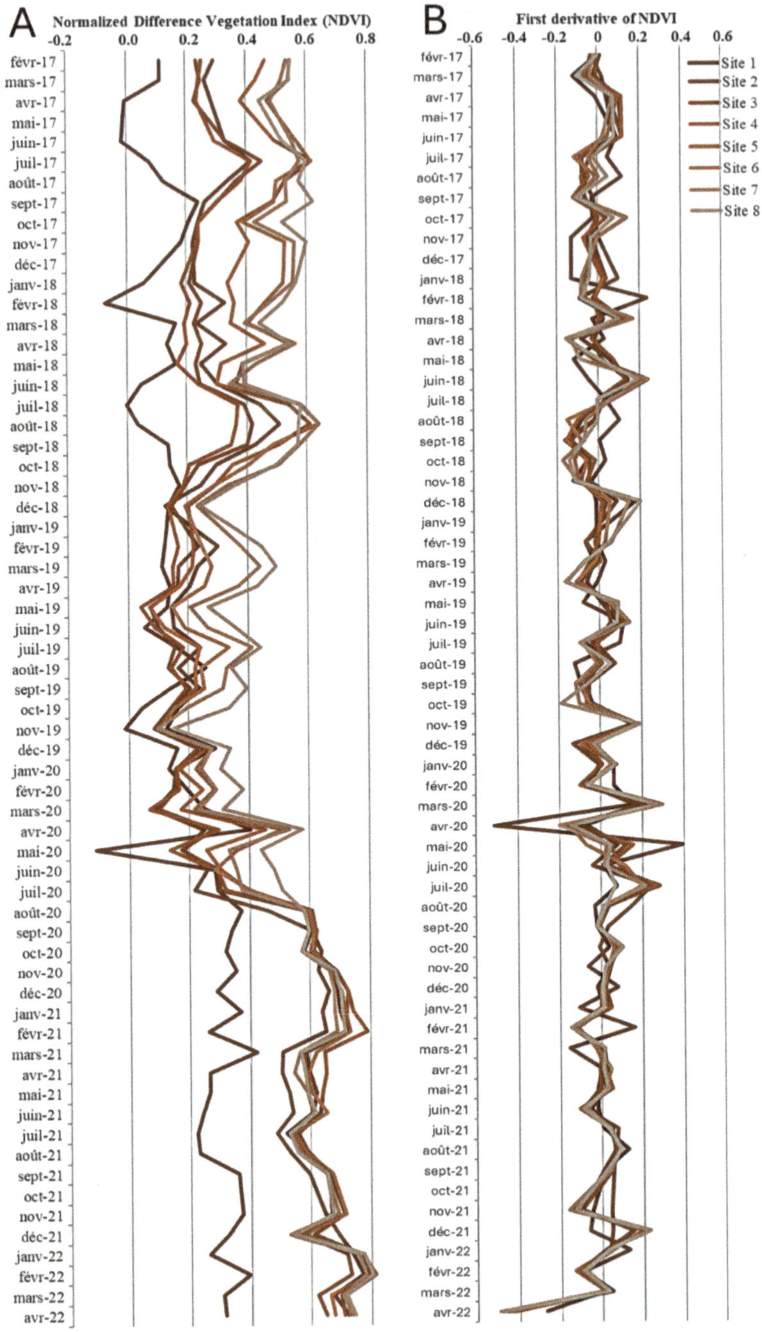

Fig. 6 Time-series of the **a** normalized difference vegetation index, and **b** its first derivative, derived from PlanetScope CubeSats from Feb. 2017 to Apr. 2022 for the eight virtual stations

the NDVI is. This might be explained by (1) the presence of the water in the channel just behind the dyke, (2) in addition to the lower rate of underground salinization with the distance to dyke, increasingly letting the fodder crop gain in chlorophyll concentration. It is besides relevant to notice a kind of decrease in the NDVI for the seven more remote stations from the dyke between July 2018–March 2020. We could assume that gradual decline might be the consequences of more frequent marine intrusions in the groundwater, whose effects were also gradually visible in the diminution in the chlorophyll amount at the surface. Those potential underground processes could therefore be detected from the surface reflectance, what could be considered as early warnings.

The reconnection of the polder to the sea domain was clearly quantified by the abrupt NDVI changes in all stations in March 2020. That deep signal extended over three months for the closest station, and four months for the other stations.

Following the breach's event, the NDVI of the closest station reached a slightly higher plateau and all other stations dramatically increased their NDVI to more similar values than before the breach. The channel station became the breach's pond station, which linked its water level to sea's one, letting more sediment reflecting the sun irradiance, thus a higher NDVI. As for the other stations, their pre-breach NDVI differences rapidly shrank. That might be elucidated by the successive mud deposit and lower marsh vegetation growth, whose surface was large enough to encompass the seven stations. We could also refine that post-breach description by three sub-periods: the first one, from March 2020 to February 2021, featured by a rapid increase; the second one, from February to March 2021, by a marked decline; and the third one, from February 2021 to April 2022, by a constant, gradual augmentation. The first quick gain in NDVI might correspond to the pioneering colonization of glassworts and saltworts following the breach's aperture (3 m) in March 2020. The second episode evoke the plants' fully coverage by mud after the widening of the breach in March 2021 (23.5 m). Finally, the plants since kept growing and densifying with no massive load of mud due to a stabilization of the breach's width (26.75 m) (Collin and James 2022).

3.3 Perspectives

As perspectives, it would be insightful to test the capabilities of other four-band PlanetScope's spectral indices to detect early warnings, such as the Blue Normalized Difference Vegetation Index, the Green Normalized Difference Vegetation Index or the Structure Intensive Pigment Index 2. Furthermore, by using the eight-band SuperDove, new indices, such as the Normalized Difference Red Edge, or the Leaf Chlorophyll Index, could also be meaningful.

The fusion of the 3-m PlanetScope surface reflectance bands with those from higher spatial resolution, such as the spaceborne (Collin et al. 2018) and airborne (James et al. 2022) imagery, holds great promise to refine the spatial scale of time-series. This spatial enhancement has proven to detect more subtle changes, such as the colonization of small patches (< 1 m^2) of salt marsh vegetation (Collin et al. 2022b).

4 Conclusions

The submersion process of Beaussais' marsh, that is in restoration' progress, has been successfully monitored using a monthly time-series of 59 PlanetScope imageries, from February 2017 to April 2022. Every multispectral imagery is composed of four bands, namely, blue, green, red and infrared, collected at 3-m pixel size.

Separating the seaward nature-driven salt marsh from the landward polderized wetland, the dyke breached in March 2020. The calculation of the NDVI across the 59 spatial–temporal data enabled to (1) map the variance, and (2) analyze the time-series, through the first derivative, of eight virtual stations, drawing a profile from the breach to marsh's back.

About the mapping in the NDVI variance, the closer to the breach the station was located, the more changes they have undergone for the 2017–2022 period. For the pre-breach period (from February 2017 to March 2020), the results showed a constant decrease in the NDVI for the stations out of the channel (just behind the dyke), which might reflect the marine intrusions in the groundwater. We therefore advocate to use the NDVI as a surface proxy for the underground salinization, conferring it the status of early warnings. The breach's aperture (in March 2020) provoked an abrupt augmentation, like a tipping point, of all NDVI's stations. The post-breach period (from March 2020 to April 2022) reflected the successive (1) first colonization of lower marsh vegetation, (2) large marine sediment deposit due the breach's widening, and (3) gradual growth of lower marsh vegetation due to the breach's stabilization.

We recommend to adopt this transferable PlanetScope-derived approach to create and analyze time-series of other coastal areas in submersion and erosion.

References

Bitoun R, Collin A, James D, Pioch S (2018) Modelling saltmarsh spatial evolution at high resolution from 1948 to 2100 (Beaussais's Bay, France). J Coastal Res 85:676–680

Collin A, James D (2022) Suivi par orthomosaïque multispectrale et modèle numérique de surface de la reconnexion à la mer du polder de Beaussais sur Mer. Scientific report, Conservatoire du Littoral, France, 113p

Collin A, Laporte J, Koetz B, Martin-Lauzer FR, Desnos YL (2016) Mapping bathymetry, habitat, and potential bleaching of coral reefs using Sentinel-2. In: 13th international coral reef symposium (ICRS 2016), pp 405–420

Collin A, Lambert N, Etienne S (2018) Satellite-based salt marsh elevation, vegetation height, and species composition mapping using the superspectral WorldView-3 imagery. Int J Rem Sens 39(17):5619–5637

Collin A, Durozier T, Hervouet G, Costa S (2021a) Détecter le processus de dépoldérisation par l'observation satellite à très haute résolution temporelle. In: 16th SAGEO, 5–7 May 2021, pp 250–252

Collin A, Prot P, Bouazzaoui H, Bnikkou S (2021b) Spatiotemporal patterns of the sea surface temperature using statistical anomaly detection across the 18-year-old global scale. In: OCEANS 2021: San Diego–Porto. IEEE, pp 1–7

Collin A, James D, Feunteun E (2022a) Towards better coastal mapping using fusion of high temporal SENTINEL-2 and PLANETSCOPE-2 imageries: 12 bands at 3 M through neural network modelling. Int Arch Photogramm Remote Sens Spat Inf Sci 43:479–484

Collin A, James D, Mury A, Letard M, Houet T, Gloria H, Feunteun E (2022b) Multiscale spatiotemporal NDVI mapping of salt marshes using sentinel-2, Dove, and UAV imagery in the bay of Mont-Saint-Michel, France. In: European spatial data for coastal and marine remote sensing: proceedings of international conference EUCOMARE 2022-Saint Malo, France. Springer International Publishing, Cham, pp 17–38

Collin A, Pastol Y, Letard M, Le Goff L, Guillaudeau J, James D, Feunteun E (2022c) Increasing the nature-based coastal protection using Bathymetric Lidar, Terrain classification, network modelling: reefs of Saint-Malo's Lagoon? In: European spatial data for coastal and marine remote sensing: proceedings of international conference EUCOMARE 2022-Saint Malo, France. Springer International Publishing, Cham, pp 235–241

Collin A, James D, Lamontagne N, Feunteun E, Durozier T, Hervouet G (2023a) Before/after the submersion process using planetscope monthly time-series at high spatial resolution: the case study of Beaussais' Marsh. Coast, Caen, 23–27th Oct 2023

Collin A, Palola P, James D, Pastol Y, Monpert C, Loyer S et al (2023b) Superdove-modelled bathymetry using neural networks along a turbidity gradient: Brehat, Saint-Barthelemy and Tetiaroa Islands. Int Arch Photogramm Remote Sens Spat Inf Sci 48:1351–1356

James D, Collin A, Mury A, Letard M, Legal O, Le Quilleuc A (2022) Very high-resolution monitoring and evaluation of tidal and ecological restoration in Beaussais' Bay. In: European spatial data for coastal and marine remote sensing: proceedings of international conference EUCOMARE 2022-Saint Malo, France. Springer International Publishing, Cham, pp 207–211

Jeuffrard I, Collin A, Duthion G (2023) Suivi pluri-décennal rétrospectif et prospectif du trait de côte dunaire par imagerie multispectrale. 4th merIGéo, 14–16 Mar 2023, pp 147–149

Lamontagne N, Collin A (2022a) Inoculation. C. ART Cartographies artistiques, Exarmas#19, Bordeaux, 9th Nov 2022

Lamontagne N, Collin A (2022b) Evolution of the sensitive connectivity in a salt marsh under metamorphosis. In: International geographic union, Paris, 18–22 July 2022

Lamontagne N, Collin A (2022c) Inoculation. L'Inf Géograph 86(4):107–109

Lamontagne N, Collin A, James D, Feunteun E (2023) The cultural interface between human needs and nature's integrity: the case study of Beaussais' Marsh. Coast, Caen, 23–27th Oct 2023

Lamontagne N, Collin A, James D, Feunteun E, Wedding L (2024) The cultural interface between human's economy and ecosphere's integrity. In: Proceedings of the 19th Franco-Japanese oceanography symposium: submitted. Springer International Publishing, Cham

Le Roy S, Brivois O (2022) Projet adapto—Baie de Lancieux (22): submersion marine en contexte de changement climatique sur les polders de Lancieux et Ploubalay. Rapport Final, p 217

Lochin P, Collin A, James D, Pastol Y, Costa S (2020) Blue carbon attenuation of future coastal risks related to extreme stream- and sea-water levels: the high-resolution 2D simulation as a management tool. Rev Invest Mar 27(1):47–50

Masson-Delmotte VP, Zhai P, Pirani SL, Connors C, Péan S, Berger N, Scheel Monteiro PM et al (2021) IPCC 2021: summary for policymakers. In: Climate change 2021: the physical science basis. contribution of working group I to the sixth assessment report of the intergovernmental panel on climate change

Mury A, Collin A, Jeanson M, James D, Gloria H, Pastol Y, Etienne S (2020) Mapping nature-based marine flooding risk using VHR wave, airborne LiDAR and satellite imagery: the case study of the Dol Marsh (Bay of Mont-Saint-Michel, France). J Coast Res 95(SI):743–747

Rouse JW, Haas RH, Scheel JA, Deering DW (1974) Monitoring vegetation systems in the great plains with ERTS. In: Proceedings of the 3rd earth resource technology satellite (ERTS) symposium, vol 1, pp 48–62

Tieng T, Sharma S, Mackenzie RA, Venkattappa M, Sasaki NK, Collin A (2019) Mapping mangrove forest cover using Landsat-8 imagery, Sentinel-2, very high-resolution images and Google Earth Engine algorithm for entire Cambodia. IOP Conf Ser Earth Environ Sci 266(1):012010

Spatial Variation of Sea-Air CO_2 Fluxes in the Seto Inland Sea and Analysis of Contributed Parameters

Masahiro Fujita, Mitsuru Hayashi, Eiji Yamashita, and Soichi Hirokawa

1 Introduction

Carbon dioxide (CO_2) is exchanged between the air and sea. According to Nellemann et al. (2009), the ocean absorbs 30% of the total anthropogenic CO_2 emissions, while Takahashi et al. (2009) revealed that the open ocean serves as a net sink for CO_2. In contrast to the open ocean, coastal seas are recognized as a significant source of CO_2 due to the inflow of organic matter and organic decomposition (Borges and Abril 2011; Chen et al., 2013). However, recent studies revealed that in certain coastal seas, such as Tokyo Bay and Ise Bay, CO_2 is absorbed due to the reduction in organic matter flowing into the ocean and high primary production (Kubo et al., 2017; Tokoro et al., 2021). Compared to that in the open ocean, the behaviour of CO_2 in coastal seas, such as the variation in CO_2 and amount of sea-air CO_2 exchange, has not been studied well.

Ocean acidification occurs when seawater becomes more acidic due to the increase in CO_2 in seawater. According to the Japan Meteorological Agency (2023), ocean acidification is progressing across open oceans around Japan due to the increase in anthropogenic CO_2. This could cause serious damage to ecosystems, especially in

M. Fujita (✉) · S. Hirokawa
Graduate School of Maritime Sciences, Kobe University, Kobe, Japan
e-mail: 230w325w@stu.kobe-u.ac.jp

S. Hirokawa
e-mail: hirokawa@maritime.kobe-u.ac.jp

M. Hayashi
Research Center for Inland Seas, Kobe University, Kobe, Japan
e-mail: mitsuru@maritime.kobe-u.ac.jp

E. Yamashita
Okayama University of Science, Okayama, Japan
e-mail: eiyama-mirai@lagoon.ocn.ne.jp

A.-C. Bennis et al. (eds.), *Constraints and Adaptations to Global Change at the Land-Sea Interface: For a Shared Ecological and Energy Transition*,
https://doi.org/10.1007/978-3-031-90050-1_4

coastal seas, where various marine organisms exist. Thus, the behaviour of CO_2 in coastal seas should be studied in depth.

The Seto Inland Sea, the largest enclosed Sea of Japan, is one of the regions affected by ocean acidification. CO_2 in seawater in the Seto Inland Sea has been reported for Osaka Bay by Fujii and Fujiwara (2012), Endo et al. (2017), and others. Considering the whole region, Hayashi and Yamashita (2020) analysed the spatial variation in the partial pressure of CO_2 in seawater (pCO_2) in spring and summer covering seven areas of the Seto Inland Sea. They reported that the pCO_2 distribution varies regionally. They also clarified that from the difference in the partial pressure of CO_2 between seawater and the atmosphere, the direction of sea-air CO_2 exchange changed spatially and seasonally. However, the amount of CO_2 exchange and the parameters that determine this exchange remain unclear. In this study, based on in situ data obtained in the Seto Inland Sea, we explained the spatial distribution of pCO_2, focusing on regional characteristics. In addition, we analysed the difference in the parameters contributing to the CO_2 flux.

2 Methods

2.1 *Observation and Measurements*

Measured data we obtained during research cruises involving TS Fukaemaru of the Graduate School of Maritime Sciences, Kobe University. Water samples were collected via a water intake at a depth of 3 m below the water surface. We measured the partial pressure of CO_2 in seawater (pCO_2, μatm), seawater temperature (T, °C), salinity (S), pH and dissolved oxygen (DO, mg/L) content of the samples.

pCO_2 was measured every 15 min using an automatic CO_2 measurement system of the Okayama type (Ohtaki et al. 1994 revised version; Yamashita et al. 1993), as shown in Fig. 1. The measurements were conducted with an equilibrator, a water bath, a nondispersive infrared (NDIR) gas analyser, and a recorder. The equilibrator and seawater sample were stored in a water bath whose temperature was controlled by a regulator under field conditions. A known concentration of CO_2 gas was continuously passed through a reference cell of the NDIR analyser. Other CO_2 standard gases were passed through the measuring cell of the NDIR analyser as carrier gases. The carrier gas was flushed through a drying column with $Mg(ClO_4)_2$ at the final stage of the carrier gas circuit. The series of procedures were repeated 5 times with different concentrations. A calibration curve was created from 5 types of differences in the concentrations and concentrations of carrier gases. We determined that the pCO_2 concentration corresponded to the difference in the concentration at 0 in the calibration curve.

T and S were measured with a DS-15 conductivity meter (HORIBA), and the pH was measured with a F-52 pH Meter (HORIBA) at 5-min intervals. DO was measured with a DO-25A meter (TOA) every minute. Air samples were collected from an inlet

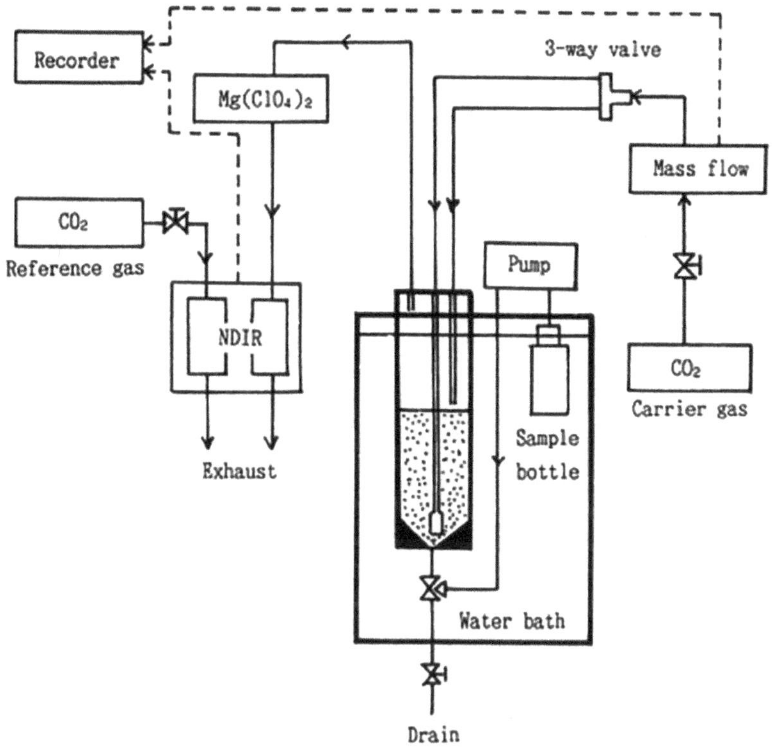

Fig. 1 Schematic of the measuring instrument setup for pCO_2 (Yamashita et al. 1993)

at a height of approximately 10 m. The partial pressure of CO_2 in the atmosphere (PCO_2, μatm) was measured every minute using a nondispersive infrared (NDIR) analyzer (LI6262, LI-COR), while the wind speed (U, m s^{-1}) was also measured. All parameters were averaged over 15 min.

The cruises were conducted in spring (March) and summer (July, August, September) from 1994 to 2010, as listed in Table 1. The data were measured around the Seto Inland Sea and adjacent regions (Fig. 2). The area comprises the Bungo Channel and Kii Channel, which are connected to the open ocean at the ends, and a series of straits and Bay-Nada areas. The straits are narrow with strong tides, while the Bay-Nada areas are wide with relatively weak tides. The channels are strongly affected by the open ocean. To explain how the variations in pCO_2 are affected by regional characteristics, we divided the Seto Inland Sea into 13 regions: 5 straits (Houyo Strait, Kurushima Strait, Bisan Seto, Akashi Strait and Kitan Strait) and 8 non-strait areas (Pacific, Bungo Channel, Aki Nada, Iyo · Suo Nada, Hiuchi Nada, Harima Nada, Osaka Bay and Kii Channel). We treated Iyo Nada and Suo Nada as one area. We studied the spatial distribution using seasonally divided data and averaged the data from each cruise.

Table 1 PCO_2 measurement period during the cruise

Year	Spring	Summer
1994	16–18, March	–
1995	–	34–31, July
1996	18–21, March	–
1997	18–21, March	26–29, September
2000	7–10, March	6–14, September
2001	12–16, March	2–10, August
2002	18–22, March	1–8, August
2003	–	1–8, August
2004	18–22, March	–
2005	14–18, March	10–16, September
2006	–	9–15, September
2007	8–13, March	4–10, September
2008	3–11, March	4–10, September
2009	6–11, March	3–9, September
2010	5–10, March	–

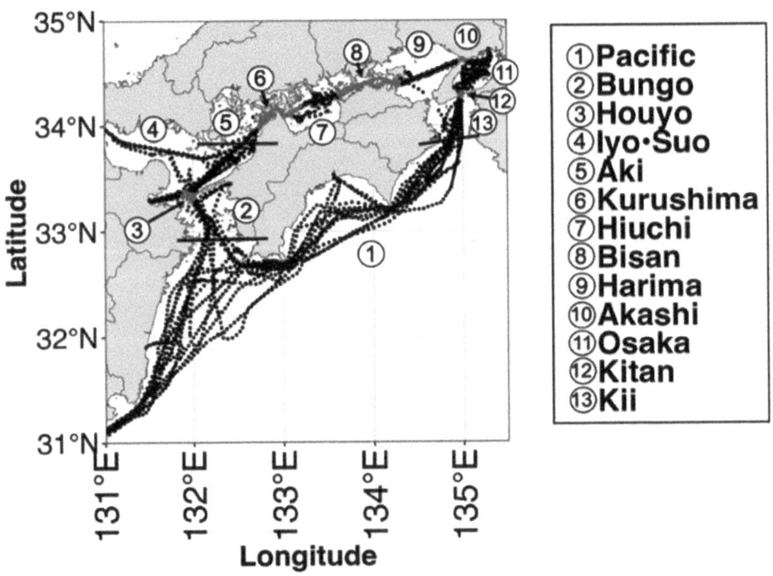

Fig. 2 Cruise tracks of Fukaemaru, divided into 13 regions

2.2 Calculation of the CO_2 Flux

The CO_2 flux (the amount of sea-air CO_2 exchange) can be calculated as follows (Andrié et al. 1986):

$$CO_2 flux = 0.24\alpha k \Delta pCO_2 \quad (1)$$

where α (mol kg^{-1} atm^{-1}) is the solubility of CO_2, which can be calculated as follows (Weiss 1974):

$$\alpha = \exp\left\{A + B\left(\frac{100}{T'}\right) + C \times \ln\left(\frac{T'}{100}\right) + S\left(D + E\left(\frac{T'}{100}\right) + F\left(\frac{T'}{100}\right)^2\right)\right\}$$
$$A = -60.2409 \quad B = 93.4517 \quad C = 23.3585$$
$$D = 0.023517 \quad E = -0.023656 \quad F = 0.047036 \quad (2)$$

where, T′ is the absolute temperature. Moreover, k (cm h^{-1}) in Eq. (1) indicates the gas exchange coefficient, which can be obtained as follows (Wanninkhof 1992):

$$k = 0.31U^2\left(\frac{Sc}{660}\right)^{-\frac{1}{2}} \quad (3)$$

The Schmidt number in Eq. (3) can be calculated as follows (Wanninkhof 1992):

$$Sc = G - HT + IT^2 - JT^3$$
$$G = 2073.1 \quad H = 125.62$$
$$I = 3.6276 \quad J = 0.043219 \quad (4)$$

ΔpCO_2 (μatm) is the difference between the partial pressure of CO_2 in seawater and that in the atmosphere, which can be obtained as follows:

$$\Delta pCO_2 = pCO_2 - PCO_2 \quad (5)$$

2.3 Quantification of the Contribution to the CO_2 Flux

CO_2 fluxes were calculated by 5 parameters, namely, T, S, U, pCO_2 and PCO_2. We clarified the contribution of each parameter to the CO_2 flux by season and area. Since U generally amplifies the CO_2 flux, the effect of U was excluded. We used k′, which was substituted for U (1.0 m s^{-1}) in Eq. (3), instead of k. We defined the flux’ using k′ instead of k, as follows:

$$CO_2 \mathit{flux} = 0.24\alpha k' \Delta pCO_2 \tag{6}$$

We quantified the contributions of the parameters to the flux' via three methods. The first method is the correlation coefficient, which can be calculated between the flux' obtained from the measured parameters and each parameter. The second method is the root mean square error (RMSE), which is calculated from the flux' based on the measured parameters and the theoretical line. We defined the theoretical line as one specific parameter ranging from its minimum to maximum values while keeping the other parameters at their average values. Figure 3 shows the relationships between the flux' and pCO_2. With increasing contribution to the flux', the measured data increasingly followed the theoretical line. Since T was used to calculate α and k, T_α and T_k could be distinguished. The third factor is the contribution rate. The contribution rates indicate the proportion of the variation in each parameter relative to the flux' accounting for the variation in the flux'. For example, the contribution rate of T_α can be obtained as follows:

$$Contribution\ rate_{T_\alpha} = \frac{Variation_{T_\alpha}}{Variation_{T_\alpha} + Variation_S} \times Contribution\ rate_\alpha \tag{7}$$

$$Contribution\ rate_\alpha = \frac{Variation_\alpha}{Variation_\alpha + Variation_k + Variation_{\Delta pCO_2}} \tag{8}$$

The contribution rate of a specific parameter can be calculated by the variation in the flux' divided by the sum. Since T_α, S, T_k, pCO_2 and PCO_2 were used to calculate the flux' through α, k′ and ΔpCO_2, the contributions of α, k′ and ΔpCO_2 to the flux' were first calculated, after which the contributions of T_α and S to α, of T_k to k′, and of pCO_2 and PCO_2 to ΔpCO_2 were multiplied. The parameter *variation* can be obtained by the slope, which can be derived from the regression equation between the flux' and the considered parameter.

$$Variation_p = \sigma_p \times |slope|_p \tag{9}$$

where p denotes a given parameter, and σ_p is the standard deviation of p. Since these parameters exhibit different scales, we standardized each parameter through multiplication by the standard deviation of the measured parameter.

After we quantified the contributions of each parameter to the flux' with the above three methods, we compared them qualitatively with the criteria of these methods. Regarding the correlation coefficient, values of ± 0.5 to ± 0.7 indicated a slight correlation, while values of ± 0.7 to ± 1.0 indicated a strong correlation. Regarding the RMSE, it should be smaller than the standard deviation of the flux'. Regarding the contribution rate, a value greater than 20% indicated a large contribution, while a value less than 10% indicated a small contribution. The data of regions accounting for less than 1% of the total data were excluded.

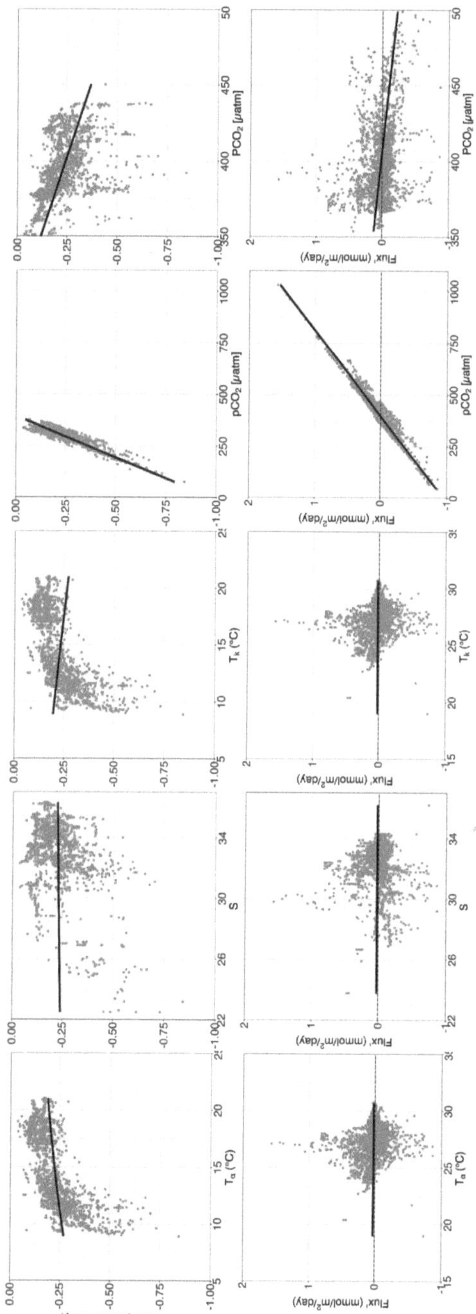

Fig. 3 Relationship between the flux' calculated from the measured parameters and the theoretical line. Upper row indicates spring, lower row indicates summer. The points indicate the flux', and the line is the theoretical line

3 Results

Figure 4 shows the spatial distributions of pCO_2 and PCO_2 in spring and summer. The variation in pCO_2 was higher than that in PCO_2. It is suggested that seawater is not advective or diffusive as air. PCO_2 was higher in the central part of the Seto Inland Sea probably due to the notable influence of factories around the region.

pCO_2 was lower in spring than in summer. pCO_2 was lowest in the Osaka Bay during both seasons, highest in the Houyo Strait in spring, and highest in the Bisan Seto in summer. pCO_2 in the straits was higher than that in the adjacent non-strait areas in summer, except in the Akashi Strait. In summer, the average pCO_2 was greater than that in spring. The differences were larger than 100 μatm in the non-strait areas and larger than 150 μatm in the strait areas.

Figure 5 shows the spatial distribution of the pH in spring and summer, while Fig. 6 shows the spatial distribution of the DO in spring and summer. The pH and DO were greater in spring than in summer. For both parameters, the Osaka Bay exhibited the highest values during both seasons, while Bisan Seto exhibited the lowest values in summer. In summer, the pH was greater and the DO concentration was lower in the straits, except for the Akashi Strait, than in the non-strait areas. Compared to the variations in pCO_2, those in the pH and DO were reversed.

Figure 7 shows the spatial variations in the CO_2 fluxes in spring and summer. The CO_2 fluxes were negative in spring. This suggests that there was an absorbing area in spring. In summer, CO_2 was emitted in the Seto Inland Sea, except of Bungo Channel, Kii Channel and Osaka Bay. CO_2 was also absorbed in the Pacific in summer.

Table 2 provides the correlation coefficients between the flux' and each parameter, i.e., T, S, pCO_2 and PCO_2, in each area in spring and summer. There were strong

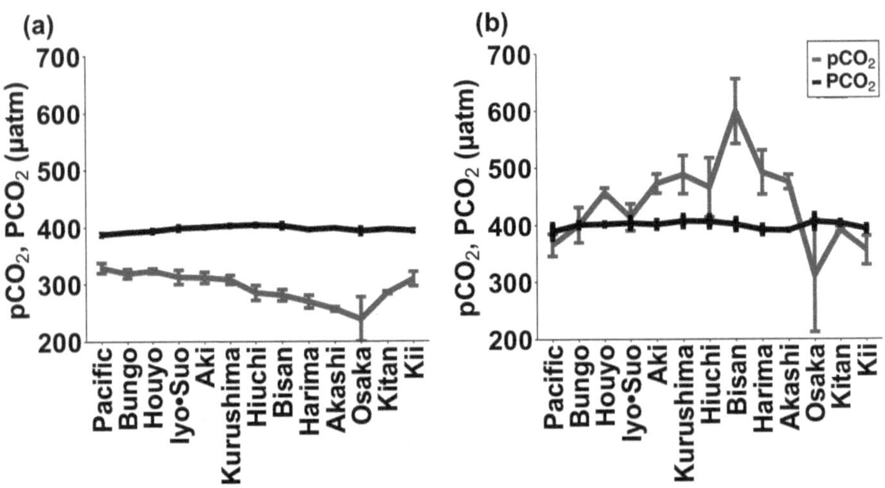

Fig. 4 Spatial distributions of pCO_2 and PCO_2 **a** in spring and **b** in summer. The grey lines indicate pCO_2 and the black lines indicate PCO_2, with the error bars indicating the standard deviation

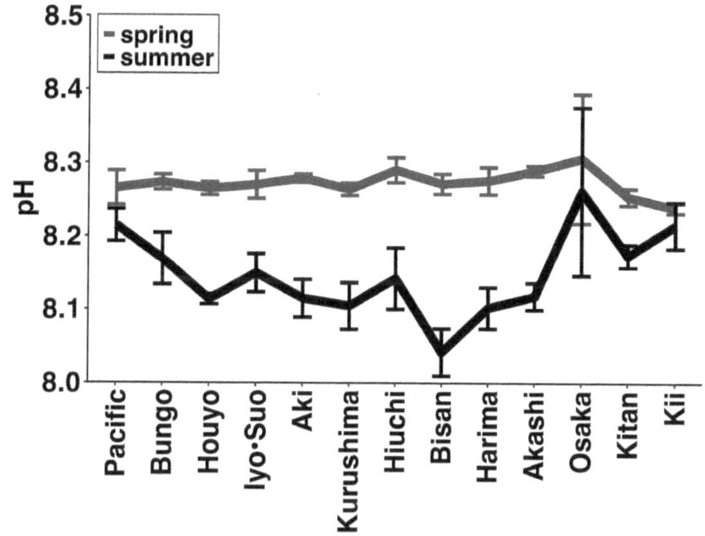

Fig. 5 Spatial distributions of the pH in spring (grey line) and in summer (black line). The lines indicate average values, and the error bars indicate the standard deviation

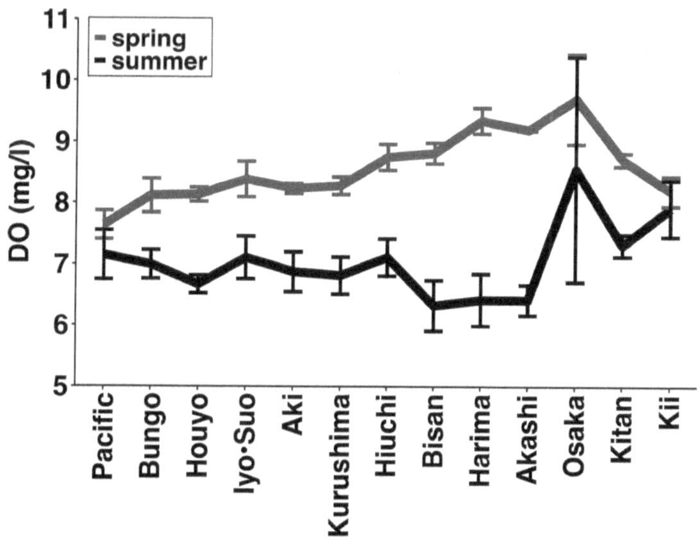

Fig. 6 Spatial distributions of DO in spring (grey line) and in summer (black line). The lines indicate average values, and the error bars indicate the standard deviation

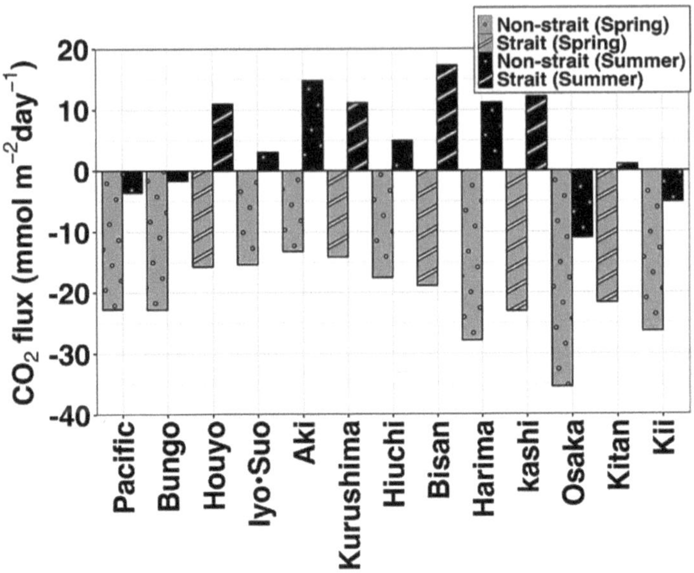

Fig. 7 Spatial variation in the CO_2 fluxes

positive correlations between pCO_2 and the flux' in almost all areas in spring. While there were strong negative correlations between PCO_2 and the flux' in the remaining areas, where there were no correlations with pCO_2, in spring. There were also strong positive correlations between pCO_2 and flux' in entire area in summer. However, there were few correlations between the other parameters and the flux' across the entire area in summer. Table 3 provides the RMSE between the flux' based on the measured parameters and the theoretical line in each area. The RMSE of pCO_2 was the smaller, especially in summer. Figure 8 shows the contribution rates in each area. During both seasons, the contribution rates of pCO_2 and PCO_2 accounted for the majority of the total contribution. The contribution rates of T_α, S, T_k were lower than 20% in all areas, and higher than 10% in a few areas.

As shown in Fig. 8, pCO_2 and PCO_2 contributed greatly to the flux'. In particular, in summer, the contributions were greater than 90% in most areas, suggesting that the relationship between pCO_2 and PCO_2 mostly determines the flux'. In spring, pCO_2 was considered to greatly contribute to the flux' because the correlation coefficient, RMSE and contribution rate exceeded the respective criteria in almost all areas. PCO_2 was also considered to generate a notable contribution because two or more criteria were exceeded in many areas. In the Pacific, Bungo Channel, and Houyo Strait, the indicators of PCO_2 exceeded the criteria, but only one criterion of pCO_2 was exceeded, suggesting that the contribution of PCO_2 was largest. The indicators of T_α, S, and T_k did not exceed the corresponding criteria in most of areas, suggesting their contributions were considered small. In the Kii Channel, the three criteria of T_α were also exceeded, and it could be considered a significant contributor to pCO_2. These results also suggest relatively limited effects of pCO_2 in spring.

Table 2 Correlation coefficients between the flux' and each parameter (a) in spring and (b) summer

(a) spring

	Pacific	Bungo	Houyo	IyoSuo	Aki	Kurushima	Hiuchi	Bisan	Harima	Akashi	Osaka	Kitan	Kii
T	0.08	0.08	-0.01	0.23	0.15	0.45	**0.52**	**0.59**	0.24	-0.11	0.34	0.46	**0.59**
S	-0.28	-0.05	0.14	0.22	0.04	-0.13	-0.25	0.01	0.30	**0.62**	0.41	0.45	0.14
pCO_2	0.34	0.15	0.01	**0.71**	**0.60**	**0.71**	**0.71**	**0.84**	**0.81**	**0.88**	**0.91**	**0.64**	**0.61**
PCO_2	**-0.70**	**-0.63**	**-0.76**	**-0.58**	-0.49	**-0.57**	**-0.71**	-0.42	-0.22	**-0.52**	-0.30	-0.49	-0.44

(b) summer

	Pacific	Bungo	Houyo	IyoSuo	Aki	Kurushima	Hiuchi	Bisan	Harima	Akashi	Osaka	Kitan	Kii
T	0.16	-0.36	0.18	0.04	-0.27	-0.02	-0.35	-0.03	-0.03	×	0.38	×	0.10
S	0.24	0.09	-0.28	-0.05	-0.29	-0.25	**-0.56**	-0.20	-0.21	×	0.18	×	**0.58**
pCO_2	**0.73**	**0.59**	**0.92**	**0.77**	**0.89**	**0.80**	**0.93**	**0.97**	**0.95**	×	**0.96**	×	**0.75**
PCO_2	**-0.60**	-0.45	-0.04	-0.36	-0.02	-0.25	-0.23	-0.34	-0.18	×	0.14	×	-0.21

Black letters indicate a slight correlation, black letters with grey indicate a strong correlation, and cross indicates insufficient data points for the calculation

Table 3 RMSE between the flux' calculated from the measured parameters and the theoretical line (a) in spring and (b) summer

(a) spring

	Pacific	Bungo	Houyo	IyoSuo	Aki	Kurushima	Hiuchi	Bisan	Harima	Akashi	Osaka	Kitan	Kii
T_a	4.7	4.4	4.1	5.9	5.0	5.8	6.2	7.7	9.1	13.7	13.7	6.2	4.9
S	4.8	4.4	4.1	5.9	5.0	6.0	6.5	8.2	9.1	13.2	14.0	6.5	5.5
T_k	4.6	4.5	4.1	6.0	4.9	5.9	6.2	8.0	8.5	12.6	14.3	7.2	6.3
pCO_2	4.8	5.8	5.4	4.3	4.7	4.5	5.0	5.4	7.0	7.4	7.4	5.6	4.9
PCO_2	3.7	4.5	3.5	5.0	5.0	5.1	5.0	7.7	9.8	11.7	14.0	6.4	5.6

× 10^2 (mmol/m^2/day)

(b) summer

	Pacific	Bungo	Houyo	IyoSuo	Aki	Kurushima	Hiuchi	Bisan	Harima	Akashi	Osaka	Kitan	Kii
T_a	7.9	5.0	7.0	6.3	9.5	11.0	18.4	32.0	14.5	×	26.3	×	11.3
S	7.9	5.0	7.0	6.3	9.6	11.0	18.4	31.9	14.5	×	26.6	×	11.2
T_k	7.6	4.6	6.4	5.9	8.8	9.9	18.1	28.9	12.7	×	25.9	×	10.6
pCO_2	5.8	4.2	3.8	4.4	5.7	7.9	8.7	17.9	8.5	×	11.8	×	8.1
PCO_2	6.6	4.6	7.3	6.0	10.1	11.1	18.0	31.1	14.3	×	28.0	×	11.7

Grey indicates that the RMSE is smaller than the standard deviation of the parameter, and cross indicates insufficient data points for the calculation

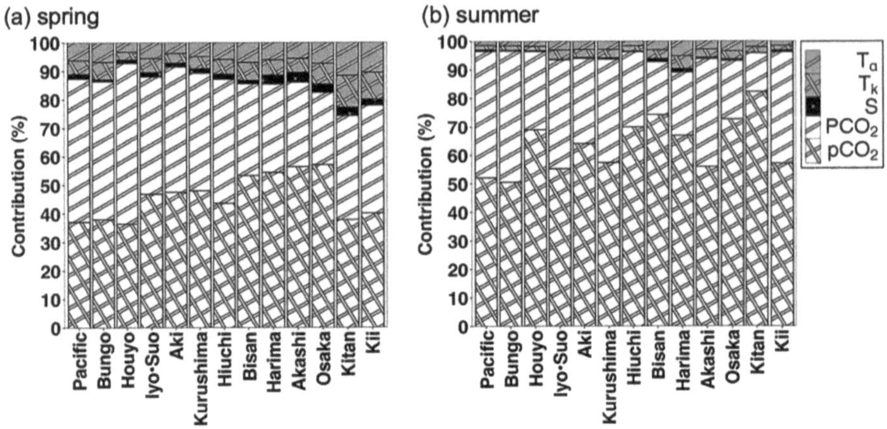

Fig. 8 Contribution rates in the different areas **a** in spring and **b** in summer

4 Discussion

In summer, the average pCO_2 was greater than that in spring. Previous researches conducted in the Seto Inland Sea are very little. Endo et al. (2017) reported that the surface pCO_2 level ranged from approximately 50–200 μatm in spring to approximately 200 to 900 μatm in summer in the inner part of Osaka Bay. Yamashita (2020) reported that pCO_2 was lower from January to March and was the higher from August to October in Ushimado, located in northern part of Harima Nada. Overall, the trend of higher pCO_2 levels in summer was consistent.

According to Takahashi et al. (1993), pCO_2 is mainly a function of temperature, total CO_2 concentration, alkalinity and salinity of that water, and the effects of temperature is one of the most important ones in surface ocean water. They indicated that pCO_2 changes about 4.2% per 1-°C. Similarly, Yamashita et al. (1993) confirmed that, without considering biological activity and changes in the total carbon dioxide concentration, pCO_2 increased by approximately 4.5% per 1-°C increase in the water temperature using seawater in the Seto Inland Sea. Thus, the pCO_2 concentration was higher in summer since the water temperature was higher in summer than in spring.

CO_2 is consumed by photosynthesis to produce oxygen and, conversely, by respiration and decomposition of organic matter. With the use of DO as an indicator of the primary production, DO was the highest in Osaka Bay during both seasons. Hashimoto et al. (1997) reported that the primary production in there was highest in the Seto Inland Sea because of the large amount of nutrients supplied from land. Therefore, the pCO_2 concentration in Osaka Bay was lower than that in the other areas because CO_2 in seawater was consumed during photosynthesis.

The pH and pCO_2 are inversely correlated due to the equilibrium relationship of the carbonic acid system. pCO_2 attained a strong negative correlation with the pH, and exhibited the opposite spatial distribution. The pCO_2 level was the highest, while the pH was the lowest in the Bisan Seto. According to Kobayashi and Fujiwara

(2006), in the surface layer of the Bisan Seto, density flow towards Hiuchi Nada and Harima Nada. In the lower layer, flow towards the Bisan Seto to compensate for the outflow. It is suggested that the currents are mixed in the Bisan Seto after flowing into the lower layer. Taguchi and Fujiwara (2010) confirmed that the pH and DO were low in the lower layers of Hiuchi Nada and Harima Nada. This was supported by the decrease in the oxygen supply due to stratification and the sinking and accumulation of organic matter, which is decomposed by microorganisms. Based on the pH and DIC values in the lower layer of Harima Nada reported in Taguchi and Fujiwara (2010), we calculated the pCO_2 level in the lower layer using the R package seacarb (Gattuso et al. 2021), and we found that the value was approximately 570 μatm. This value is approximately 100 μatm higher than that in the surface layer of Harima Nada in this study, suggesting the existence of high-CO_2 water masses in the lower layer. For the same reasons, CO_2 could be considered to accumulate in the lower layer of other Bay-Nada areas, which are hypoxic due to stratification in summer.

The pH and DO were lower in the straits than in the non-strait areas in summer, suggesting that the lower water with a high pCO_2 level, produced by organic matter decomposition in the lower part of the stratified area, flowed horizontally into the adjacent straits and was then transported to the surface by vertical mixing. This could be the result of the higher pCO_2 levels in the straits than in the adjacent Bay-Nada areas in summer.

In addition, regarding Bisan Seto, where the pCO_2 level was highest in summer, this area is strongly influenced by river water (Takeoka 1985). Nishimura et al. (1998) measured pCO_2 in river water flowing into these waters and reported that it was highly concentrated in summer. It was suggested that the inflow of a large amount of river water may also contribute to the high pCO_2 level in the Bisan Seto. However, in contrast to other straits, the Akashi Strait exhibited a lower pCO_2 level than the adjacent Nada areas, due to the notably influence of Osaka Bay.

According to Takahashi et al. (2009), the mid-latitudes of the North Pacific region served as a sink in spring. However, they comprised a slight source in summer. In this study, pCO_2 in the Pacific region was almost equilibrium with the atmosphere. Therefore, there were no significant difference.

The Bungo Channel and Kii Channel were considered sinks due to the notable influence from the Pacific region. Osaka Bay served as a CO_2 sink because of the high primary production, in contrast to the other areas. Almost all the straits released more CO_2 than the adjacent non-strait areas in summer. Although the effect of U was excluded in the following analysis, it was taken into account in here.

Table 4 lists qualitative comparison of the contributions of the parameters to the flux'. The contribution of pCO_2 was greatest in most areas in spring but dominated in all areas in summer. The contributions of PCO_2 and T_α were larger in spring, while they were smaller in summer. The contributions of S were the smallest during both seasons.

Considering the low temporal and spatial variations in PCO_2, pCO_2 contributed the most to the CO_2 flux in terms of the direction of emission or absorption, and U contributed the most to the magnitude of the CO_2 flux. The contributions of T and S were considered very small. However, higher T values could result in higher pCO_2

Table 4 Correlation coefficients, RMSE and contribution rates of the parameters in each area, the left table shows in spring, and the right table shows summer

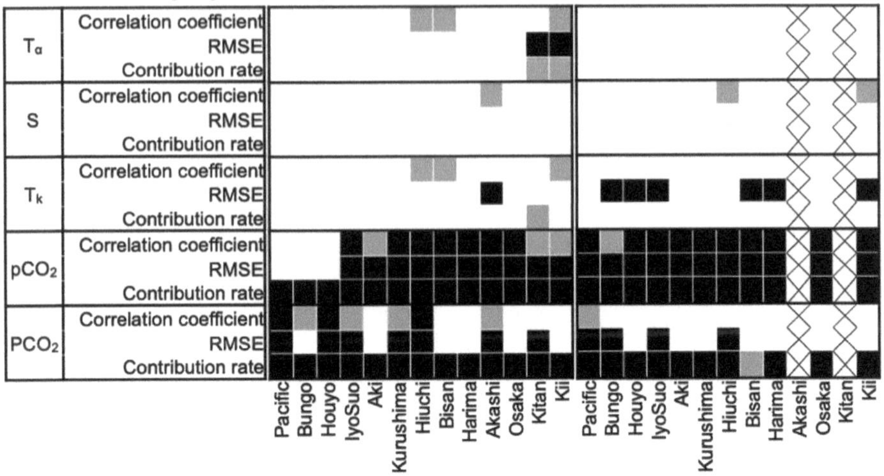

levels (Takahashi et al. 1993), while lower S values could result in higher pCO_2 levels due to the notable influence of river water and the decomposition of organic matter derived from river water (Kubo et al. 2017).

According to Takahashi et al. (2002), pCO_2 can be standardized by T as following equation:

$$(pCO_2 \text{ at } T_{mean}) = (pCO_2)_{obs} \times \exp[0.0423(T_{mean} - T_{obs})] \quad (10)$$

where the subscripts "mean" indicate the average values and "obs" indicate observed values. Adapting this equation into Eq. (5) and calculating the contribution rate, we considered the indirectly effect of T to the flux'. As shown in Fig. 9, the contribution of T_p, which indicate T affecting pCO_2, was greater than 10% in spring, as much as that of PCO_2 in summer. Even in the case of spring, considering T_α and T_k, the contribution the sum of T was as much as PCO_2. This is suggesting that the contribution of T to the flux' is significant as much as PCO_2. Although T and S hardly contribute directly to the amount of sea-air CO_2 exchange, they indirectly contribute to the direction of CO_2 fluxes because their variability significantly impacts pCO_2.

5 Conclusions

In this study, we determined the spatial distributions of pCO_2 and CO_2 fluxes and analysed the parameters contributing to the flux'. In Osaka Bay, pCO_2 was the lowest and was a sink of CO_2 during both seasons because of high CO_2 consumption due to the high primary production. In summer, pCO_2 was greater in the straits than

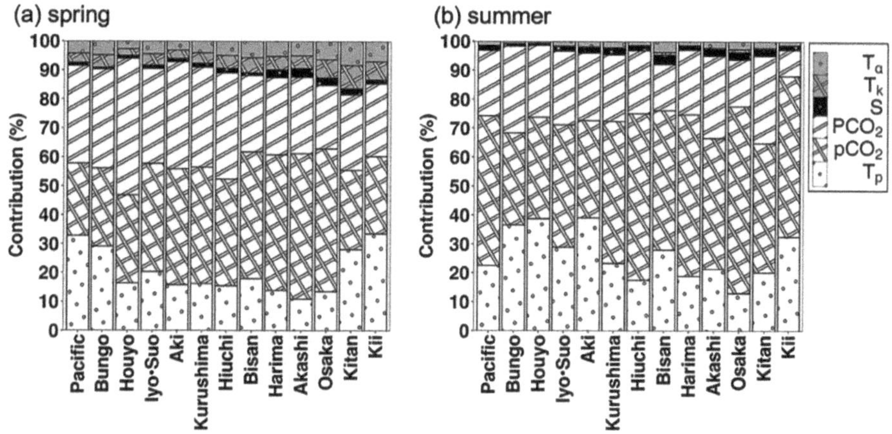

Fig. 9 Contribution rates in the different areas **a** in spring and **b** in summer, taken account of the contribution of T_p, which is T affecting pCO_2

in the adjacent non-strait areas. In the straits in summer, high CO_2 water masses accumulated in the bottom layer of the Bay-Nada areas, flowed to the bottom layer of the straits, and then were transported to the surface layer by vertical mixing. These processes could directly affect CO_2 release. Regarding CO_2 fluxes, pCO_2 contributed the most to the direction of release or absorption, and U contributed the most to the magnitude of CO_2 fluxes. T and S hardly contributed directly to sea-air CO_2 exchange. However, they significantly affected pCO_2. Thus, they indirectly contributed to the direction of the CO_2 fluxes.

References

Andrié C, Oudot C, Genthon C, Marlivat L (1986) CO_2 fluxes in the tropical atlantic during FOCAL cruises. J Geophys Res 91:741–755

Borges AV, Abril G (2011) Carbon dioxide and methane dynamics in estuaries. Treat Estuar Coast Sci 5:119–161

Chen C et al (2013) Air-sea exchanges of CO_2 in the world's coastal seas. Biogeosciences 10:6509–6544

Endo T, Shimano J, Sakai D, Fujiwara R (2017) Relationship between vertical distribution of pCO_2 in sea and air-sea CO_2 exchange in inner part of Osaka Bay. J Jpn Soc Civ Eng B2 73: I_1231–I_1236 (in Japanese)

Fujii T, Fujiwara T (2012) Continuous measurement of carbon dioxide concentration in the surface Seawater of Osaka Bay. J Jpn Soc Civil Eng B2 68:I_1061–I_1065 (in Japanese)

Gattuso J-P, Epitalon J-M, Lavigne H, Orr J (2021) Seacarb: seawater carbonate chemistry. R package version 3.3.0. http://CRAN.R-project.org/package=seacarb

Hashimoto M, Yamamoto T, Tada K, Matsuda O, Nagasue T (1997) Primary production and physical structure of the Seto Inland Sea, Japan. Bull Coast Ocean 35:109–114 (in Japanese)

Hayashi M, Yamashita E (2020) Spatial variation in pCO_2 based on 16 years of in situ measurements in the Seto Inland Sea, Japan. In: Evolution of Marine coastal ecosystems under the pressure of global changes. Springer, Cham. https://doi.org/10.1007/978-3-030-43484-7_6

Japan Meteorological Agency (2023) Ocean acidification around Japan. Viewed in 22 Feb 2024. https://www.data.jma.go.jp/gmd/kaiyou/english/oa_jpn/pHtrend_jpn_en.html

Kobayashi S, Fujiwara T (2006) Seasonal variation in intrusion depths from straits to adjoining basins in the Seto Inland Sea. Umi to Sora 82:1–11 (in Japanese)

Kubo A, Maeda Y, Kanda J (2017) A significant net sink for CO_2 in Tokyo Bay. Sci Rep 7:44355. https://doi.org/10.1038/srep44355

Nellemann C, Corcoran E, Duarte C M, Valdés L, De Young C, Fonseca L, Grimsditch G (eds) (2009) Blue carbon. A rapid response assessment. United Nations Environment Programme, GRID-Arendal, www.grida.no

Nishimura N, Watanabe K, Kishida T, Iwata T, Ohtaki E (1998) Variation of partial pressure of carbon dioxide (pCO_2) in river water. Environ Sci Jpn 11:297–304 (in Japanese)

Ohtaki E, Fujiwara F, Yamashita E, Ito H, Liu X (1994) Development of measuring instrument of pCO_2 in seawater. Oceanogr Jpn 3:413–418 (in Japanese)

Taguchi F, Fujiwara T (2010) Carbon dioxide stored and acidified low oxygen bottom waters in coastal seas, Japan. Estuar Coast Shelf Sci 86:429–433

Takahashi T et al (2002) Global sea-air CO_2 flux based on climatological surface ocean pCO_2, and seasonal biological and temperature effects. Deep-Sea Res II 49:1601–1622

Takahashi T et al (2009) Climatological mean and decadal change in surface ocean pCO_2, and net sea-air CO_2 flux over the global oceans. Deep-Sea Res II 56:554–577

Takahashi T, Olafsson J, Goddard J, Chipman D, Sutherland S (1993) Seasonal variation of CO_2 and nutrients in the high-latitude surface oceans: A comparative study. Global Biogeochemical Cycles 7:https://doi.org/10.1029/93GB02263

Takeoka H (1985) Density stratification in the Seto Inland Sea. Umi Sora 60:145–152 (in Japanese)

Tokoro T, Nakaoka S, Takao S, Kuwae T, Kubo A, Endo T, Nojiri Y (2021) Contribution of biological effects to carbonate-system variations and the air–water CO_2 flux in urbanized bays in Japan. J Geophys Res Oceans 126:e2020JC016974

Wanninkhof R (1992) Relationship between wind speed and gas exchange over the ocean. J Geophys Res 97(7373):7382

Weiss RF (1974) Carbon dioxide in water and seawater: the solubility of a non-ideal gas. Mar Chem 2:203–215

Yamashita E (2020) Carbon dioxide in the sea water seasonal variation at coastal site, Sibukawa and Ushimado in the Seto Inland Sea, Japan. Bull Inst Front Sci Tech, Okayama Univ Sci 2:59–64 (in Japanese)

Yamashita E, Fujiwara F, Liu X, Ohtaki E (1993) Measurements of carbon dioxide in the Seto Inland sea of Japan. J Oceanogr 49:559–569

Multidisciplinary Study of the Population Dynamics of the Non-indigenous Manila Clam *Ruditapes philippinarum* Along the Western Coast of Cotentin, France

O. Basuyaux, A.-C. Bennis, S. Ajmi, G. Croguennec, M. Boutet, V. Lefebvre, Y. Basuyaux, and J.-C. Dauvin

1 Introduction

There are an estimated 2.75 million recreational shellfish harvesters in France (France Agri Mer 2018). More than a third of these only harvest one to three times a year, but 17% land more than 15 catches per year. Cockle *Cerastoderma edule* (Linnaeus, 1758) is the main species targeted (36% of the harvesters), closely followed by the European clam *Ruditapes decussatus* (Linnaeus, 1758), the Japanese clam *R. philippinarum* (Adams and Reeve, 1850) (32% of the harvesters), then mussels, *Mytilus edulis* (Linnaeus, 1758) and *M. galloprovincialis* (Lamarck, 1819) (30% of the fishermen). The European clam *R. decussatus* is found across a large part of the European coastline, including the northern and southwestern Mediterranean Sea, the seaboards of the East Atlantic Ocean as well as the English Channel and the North Sea (Breber 1985; Global Biodiversity Information Facility 2023). The Japanese clam was introduced in France and Great Britain in order to make up for shortages of indigenous clams during the early 1970s, and spat has since be intentionally transferred to other European countries (Italy, Spain, Norway) (Utting and Spencer 1992; Flassch and Leborgne 1992; Cordero et al. 2017). Fifty years after its introduction, the Japanese clam has settled in all the same areas as the native European species. In 2015, Basuyaux et al. (2017) mapped clam occurrence (both

O. Basuyaux (✉) · Y. Basuyaux
Mer Expertise Environnement Etudes et Conseils, La Boivinerie, Blainville-sur-Mer, France
e-mail: obasuyaux@m2e-ec.fr

O. Basuyaux · V. Lefebvre
Centre Expérimental, ZAC Blainville-sur-Mer, SMEL, Blainville-sur-Mer, France

A.-C. Bennis · S. Ajmi · G. Croguennec · M. Boutet · J.-C. Dauvin
Morphodynamique Continentale et Côtière, CNRS UMR 6143 M2C, Normandie Université, UNICAEN, UNIROUEN, Caen, France

A.-C. Bennis et al. (eds.), *Constraints and Adaptations to Global Change at the Land-Sea Interface: For a Shared Ecological and Energy Transition*,
https://doi.org/10.1007/978-3-031-90050-1_5

species) along the western coast of Cotentin, and found the presence of clams on about 10% of the foreshore, distributed in five sectors of high abundance. However, abundances were relatively low, with overall mean densities of 5.1 clams/m^2. Based on examination of external morphology, 98% of the at the Blainville-sur-Mer site clams were Japanese clams.

While the two clam species differ morphologically with the Japanese clam being rounder with a well-marked lunula while the European species is sharper with well-separated siphons–most fishermen do not distinguish between the species, and they are fished the same way under the same quota that varies according to region (150 clams by fisherman per day in Brittany, 100 per day in Normandy). Moreover, Hurtado et al. (2011) have demonstrated cross-breeding between the two species. The wide foreshores of the western coast of Cotentin are important for recreational shellfish harvest: more than 6000 fishers of crustaceans and shellfish visit the foreshores annually, particularly during spring and summer spring tides. Basuyaux et al. (2017) estimate 120 t of clams collected by recreational fishers and 8 t by professional fishermen in this sector during 2015. Despite this high fishing pressure on an easily-accessible stock, the density of clams seems to have increased, allowing professionals to intensify their harvest to almost 20 t in 2021 and 2022 (Comité Régional des Pêches et des Elevages Marins, pers.com).

In France, the first attempts at breeding Japanese clams were made in the early 1970s, but the appearance of brown ring disease (*Vibrio tapetis*) in Brittany led shellfish farmers to find new virgin production sites. On the Chausey Islands in Normandy (France), Japanese clam farming was introduced in the late 1980s. In order to limit the risk of pathology, breeding densities and surface areas are limited. Today, 30 ha of the Chausey Islands foreshore are dedicated to shellfish farming (Comité Régional de la Conchyliculture, pers. Comm.) with a seeding of approximately 300 shells per square meter. Do the Japanese clams farms of the Chausey Islands supply larvae to the western coast of Cotentin, enabling renewal of its highly exploited populations? (Fig. 1; Table 1).

In the Chausey Islands farms, Japanese clam spat from hatcheries are scattered at 1 year old when they are mean size of 10 mm. They are farmed at a density of 200 calms per m^2 for three years, until they reach about 40 mm. The size and age of sexual maturity in *Ruditapes philippinarum* is highly variable and depends on environmental conditions; Moura et al. (2017) reports a mean size of 29.4 mm at 1 year old in the Tage estuary of Portugal) while Sanchez et al. (2022) report that the size at which *R. philippinarum* achieves sexual maturity is 24 mm (22.3–25.9 mm) for females and 21.6 mm (19.2–23.0 mm) for males in Arcachon Bay of France. By contrast, Sugiura and Kikuya (2022) report a size of 21.1 mm for females and 18.5 mm for males in Japan. In the Chausey Islands clam farms, the size of first sexual maturity is not known but it may be reached after a year, enabling the clams to reproduce several times during the three-year farming cycle. While the number of eggs emitted by a female varies with size (Ee-Yung et al. 2001), with up to 8 million per season in some cases, each female generally produces 2–3 million eggs per season (Helm et al. 2006). With a sex ratio of 1:1 (Sanchez et al. 2022), the potential egg production on a 30-ha farm is considerable, and fertilization with be

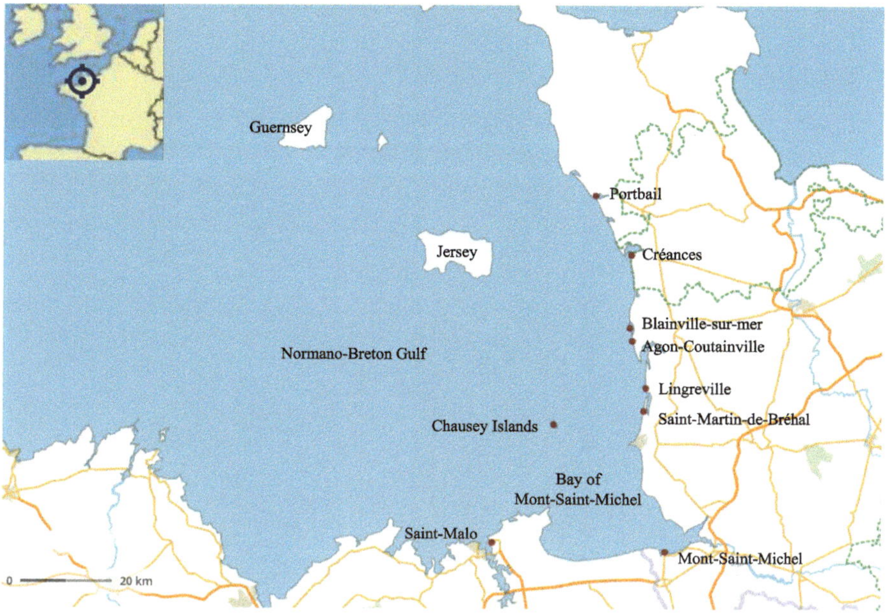

Fig. 1 Location of sites covered in this study

boosted by the high population density. After fertilization, clam larvae are dispersed by marine currents for a period of 15–30 days depending on the water temperature and trophic conditions, before settling on the foreshore.

The marine currents in the Western English Channel are very complex and the 14 m tidal ranges of the Norman-Breton Gulf during spring tides are among the largest in the world. The presence of the numerous islands (Chausey, Les Minquiers, Jersey, Guernesey) as well as shallow water create a current system with many complex gyres (Salomon and Breton 1991a, b, 1993). Previous observations (Ménesguen and Gohin 2006) have suggested the existence of two gyre types in the Channel, each with distinct behaviours: (1) gyres induced by cape effects that act like tracer accumulation points and, (2) gyres that develop around the islands, for example around Jersey, and that act as smaller accumulation zones or even as dissemination zones. Previous modelling of *Sabellaria alveolata* (Linnaeus, 1767) larval dispersion in the Western English Channel have indicated that the gyres off the coast of Cancale's Pointe du Grouin play a role in this species' larvae retention in the Mont-Saint-Michel Bay (Ayata et al. 2009), thus facilitating larval exchange between the permanent reefs of the bay. In addition, through modelling of larval dispersion over six weeks (the length of the planktonic larval stage in *S. alveolata*), Ayata et al. (2009) has shown that there is also a dispersal of larvae northwards along the Cotentin coast due to south-westerly winds. Due to the tidal currents and the presence of gyres around the Channel Islands (Salomon and Breton 1993), it is unlikely that the *S. alveolata* populations north of Granville contribute to the recruitment of juveniles along the

Table 1 Total area and number of clams per m^2 (> 40 mm and > 20 mm) in five areas of high density of the Cotentin coast in 2015; values with the same superscript letters do not differ significantly (Tukey-test, $p < 0.001$)

	Saint-Martin-de-Bréhal	Lingreville	Agon-Coutainville	Blainville-sur-mer	Chausey Islands
Total area (km^2)	3.1	0.5	3.1	3.7	0.3
Number of clams > 40 mm (/m^2)	2.3 ± 1.0^{a}	$1.4 \pm 0.8^{a,c}$	0.6 ± 0.23^{c}	$1.1 \pm 0.4^{b,c}$	67
Number of clams > 20 mm (/m^2)	5.81	3.54	1.52	2.78	133
Estimated number of clams > 40 mm (in millions)	7.3 ± 3.2	0.7 ± 0.5	1.7 ± 0.7	4.2 ± 1.5	20
Estimated number of clams > 20 mm (in millions)	18	2	4	11	30

According to Basuyaux et al. (2017) and CRC data

Cotentin coast. If *R. phillipinarum* follow the same pattern as *S. alveolata*, then the supply of larvae along the western coast of Cotentin depends on the southern populations in the Mont-Saint-Michel Bay (Lecornu et al. 2016). The objective of this study is to understand the dispersal dynamics of clam larvae on the western coast of Manche (France). For this purpose, it is necessary to know precisely the spawning period of the clams and to use numeric modelling to identify the factors that could affect this dispersion.

2 Materials and Methods

2.1 Identification of the Spawning Period

Gamete emission results in a strong decrease (up to 50%) in clam weight. The condition index (Ic) of Walne and Mann (1975) is commonly used to quantify this change in bivalves over time.

$$Ic = \frac{Dry\,flesh\,weight}{Shell\,dry\,weight} \times 100 \qquad (1)$$

A sample of 30 clams of at least 40 mm in length was collected every month between 2015 and 2022 (except 2017) in Blainville-sur-Mer on the western coast of Cotentin (France). A sample of 30 clams was collected every two weeks of the summer from the Chausey Islands farms between 2019 and 2022. The clams' fresh weights were measured (± 0.01 g) after draining during 2 min on a paper towel along with width, length and shell height using callipers (± 0.1 mm). The shells were opened by heating them in a microwave for 30 s at 750 W (for six clams). After 2 min draining on a paper towel, the weight of the drained flesh and the shell valves were measured. After 48 h in the oven at 70 °C, the flesh and the shells were weighed again, permitting the calculation of Ic [see Eq. (1)].

2.2 Modelling

Larval dispersal was simulated with a Lagrangian transport numeric model of particles moving passively with the currents, coupled with a numeric model of regional current circulation using the bathymetry of the area and integrating the effects of tide, wind, marine turbulence and temperature (Lett et al. 2008; Ayata et al. 2009; Nicolle et al. 2013).

2.2.1 Hydrodynamic Model

The dispersal of the clam larvae in the gulf was simulated with a 3D hydrodynamic model, MARS [Model for Applications at Regional Scale, Ifremer (Lazure and Dumas 2008)]. This model resolves the primitive equations using the method of finite differences under the Boussinesq and hydrostatic hypothesis (Boussinesq 1877). The MARS numeric model of regional circulation uses the terrain coordinate sigma as vertical coordinates defined as:

$$\sigma = \frac{z + H}{\xi + H} \tag{2}$$

where σ is the vertical coordinate, H(x, y) is water depth in reference to hydrographic zero (lowest sea), and ξ (x, y, t) is the elevation of the sea surface. H(x; y) + ξ(x, t) represents water column height D(x, y, t). The scales are expressed in a cartesian coordinate system (x, y) and which fluctuate over time (t).

To simplify the advection term of any variable, A, the notation L(A) as follows is used:

$$L(A) = u\frac{\partial A}{\partial x} + v\frac{\partial A}{\partial y} + w\frac{\partial A}{\partial \sigma} \tag{3}$$

where u is zonal velocity, v is meridional velocity, w is vertical velocity, and σ is a following-terrain vertical coordinate, within the scope of the coordinate systems (x, y, σ), given by:

The conventional Boussinesq and hydrostatic assumptions are used to simplify the Navier–Stokes equations. The resulting equations are called primitive equations of the ocean and are written as:

$$\frac{1}{D}\frac{\partial \rho}{\partial \sigma} = -\rho g \tag{4}$$

$$\frac{\partial \xi}{\partial t} + L(D) = 0 \tag{5}$$

$$\frac{\partial u}{\partial t} + L(u) - fv = -g\frac{\partial \xi}{\partial x} - \frac{1}{\rho_0}\frac{\partial P_a}{\partial x} + \pi_x + \frac{1}{D}\frac{\partial[(n_z/D)(\partial u/\partial \sigma)]}{\partial \sigma} + F_x \tag{6}$$

$$\frac{\partial u}{\partial t} + L(v) - fu = -g\frac{\partial \xi}{\partial y} - \frac{1}{\rho_0}\frac{\partial p_a}{\partial y} + \pi_y + \frac{1}{D}\frac{\partial[(n_z/D)(\partial v/\partial \sigma)]}{\partial \sigma} + F_y \tag{7}$$

where ρ is the water density that is function of the salinity, the temperature et the pressure,

g is gravitational acceleration, f is the Coriolis parameter, p_0 is the density of seawater, P_a is the atmospheric pressure at sea level, π_x and π_y are the zonal and meridional components of the barocline pressure gradient, n_z is the vertical viscosity associated with vertical turbulence, calculated from a RANS model of two equations following the k − ε formulation and adapted to the coastal domain (Luyten et al. 1996).

$$w = \frac{1}{D}\left[\omega - \sigma\frac{\partial\xi}{\partial t} - u\left[\sigma\frac{\partial\xi}{\partial x} + (\sigma - 1)\frac{\partial H}{\partial x}\right] - v\left(\sigma\frac{\partial\xi}{\partial y} + (\sigma - 1)\frac{\partial H}{\partial y}\right)\right] \quad (8)$$

F_x and F_y are the horizontal friction terms.

This creates a specific barotropic model by integrating the Eqs. (4) and (6) on the vertical axis. The MARS 3D model has previously been used to study dissolved and particulate matter transport along the continental shelf of Northwest Europe (Lazure and Jegou 1998, Delhez et al. 2004; Allain et al. 2007; Xie et al. 2009).

2.2.2 Particulate Transport Model

The particulate transport model, ICHTHYOP, was used to simulate the dispersion of the clam larvae (ICHTHYOP, IRD: Lett et al. 2008). ICHTHYOP was originally developed to study the physical (oceanic currents, temperature) and biological (growth, mortality) factors affecting ichthyoplankton movement.

This model uses water velocity, temperature and salinity computed by a hydrodynamic model akin to MARS 3D to reproduce water mass movement, enabling us to virtually follow larval drift. This model, based on advection–diffusion processes, solves the continuity equation in its conservative form, describing the transport of a scalar quantity in two dimensions, given by:

$$\frac{\partial(HC)}{\partial t} + \frac{\partial(HUC)}{\partial x} + \frac{\partial(HVC)}{\partial y} - \frac{\partial^2\left(KH\frac{\partial C}{\partial x}\right)}{\partial x} - \frac{\partial^2\left(KH\frac{\partial C}{\partial y}\right)}{\partial y} - mC = 0 \quad (9)$$

where U and V are the components of the hydrodynamic speeds along the directions x and y, m is the larval mortality rate, C is the larval concentration, H is the mean water height in a tidal cycle and K is the turbulent diffusion coefficient.

2.2.3 Numerical Configurations

In this study, the MARS 3D model was used within the domain from 48° to 50° N in latitude and from 4° W to 1° E in longitude and incorporating actual bathymetry (SHOM-HOMONIM at 100 m resolution), at two different resolutions: (i) the parent grid at a resolution of 300 m and (ii) daughter grids at a resolution of 100 m. Twenty uniformly spread sigma levels were used for vertical discretization. Meteorological conditions essential to particle movement and heat exchange, such as wind, are integrated into MARS 3D model based on conditions (e.g. wind speed) at 10 m over the sea surface, taken from replays of the METEO-FRANCE ALADIN model

(Daniel et al. 2019; Nabat et al. 2020). These calculations are conducted for both parent and daughter grids. The CST FRANCE 2012 tidal forecast of the SHOM (Le Roy and Simon 2003), containing 114 tidal components, was applied to the boundaries of the parent grid in order to generate tidal dynamics within each grid. The temperature was not explicitly calculated, but instead was assumed to be uniform across the Normano-Breton Gulf, changing with season: 15.4 °C on the 15th of June and 18.5 °C on the 15th of August. The salinity was fixed at 35 ppt in these two model configurations.

2.2.4 Projected Scenarios

From the MARS 3D hydrodynamic model, the 3D currents and different-temperature water masses impacting the clam larvae dispersion were calculated over time according to the oceanic and weather conditions. ICHTHYOP creates a virtual population of individuals that is released into a simulated physical environment, as dictated by the hydrodynamic model. This population increases in age with each time step. Each passive particle (larva) is exposed to its own environmental conditions, determined by its position, and follows its own path, determined by local hydrodynamics. Each particle has a memory of the place and the time where it has been released (location, date, depth). In each simulation, the number of released particles, release date, release depth, lethal temperature, monitoring period, saving frequency, time step and spawning zones are defined by the user. By coupling these two models, several simulations were run in order to test different dispersion scenarios depending on: (i) the release time of gametes according to tidal phase (mid-rising tide, mid-ebb tide) and tidal coefficients (spring tide, neap tide), (ii) the larvae source sites (Chausey Islands, Saint-Martin-de-Bréhal, Lingreville, Agon-Coutainville, Blainville-sur-Mer), and (iii) wind direction (East Winds, West Winds, variable).

3 Results

3.1 Spawning Season

There was little difference between the condition index of clams from Chausey Islands farms and wild clams of the western coast of Cotentin (Blainville-sur-mer) (Fig. 2). During the seven seasons observed, the condition index varied from 4 in autumn and winter and increased to 8 at the beginning of the summer. A strong drop in this index was observed at the end of June or at the beginning of July. This drop is associated with spawning, resulting in a strong decrease in dry weight. A second, smaller peak was sometimes (2020, 2021) observed in August, suggesting a second spawning phase (Fig. 2). These results suggested that larvae were released between the 15th of June and the 15th of July.

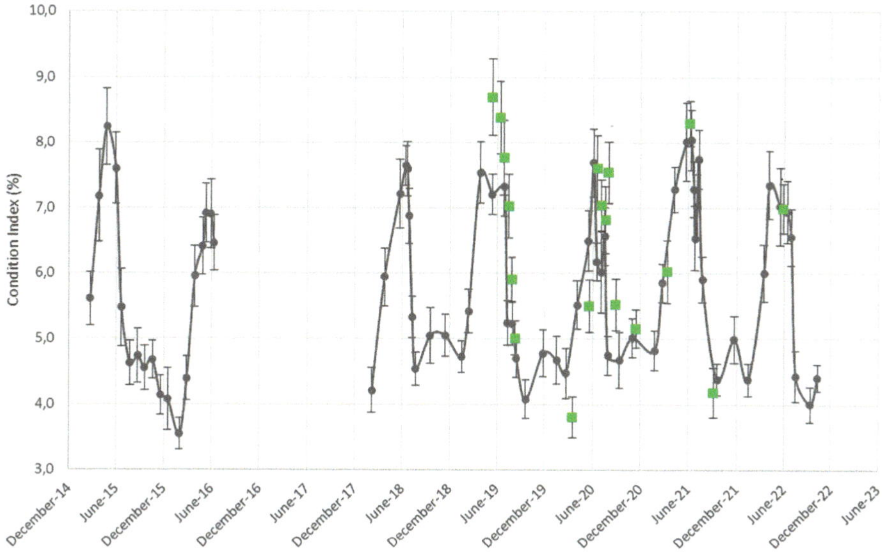

Fig. 2 Mean condition index (Ic) of Walne and Mann (1975) (and 95% confidence interval) for the clams sampled from Blainville-sur-mer (France) (grey) and from the Chausey Islands (green)

3.2 Dispersal According to Larval Stage Duration in 2019

Supposing a release of 10,000 identical larvae per site (Chausey, Saint-Martin-de-Bréhal, Lingreville, Agon-Coutainville, Blainville-sur-Mer), the simulation shows increasing dispersal with time from Southwest to North and then around northern Jersey. After 30 days, the majority of the larvae have been lost due to a lack of suitable substrate (Fig. 3).

3.2.1 Influence of the Tidal Coefficient

The simulations of larvae dispersal according to varying tidal coefficients show few differences between emission during spring tides or neap tides. Emission from the Chausey Islands during spring tides results in 7% of the larvae reaching the western coast of Cotentin after 30 days of simulation, versus 8% during neap tides. During spring or neap tides, only 1% of larvae stays in the Islands. During spring tides, the rate increases to 11% when the wind data from 2019 is incorporated into the simulations.

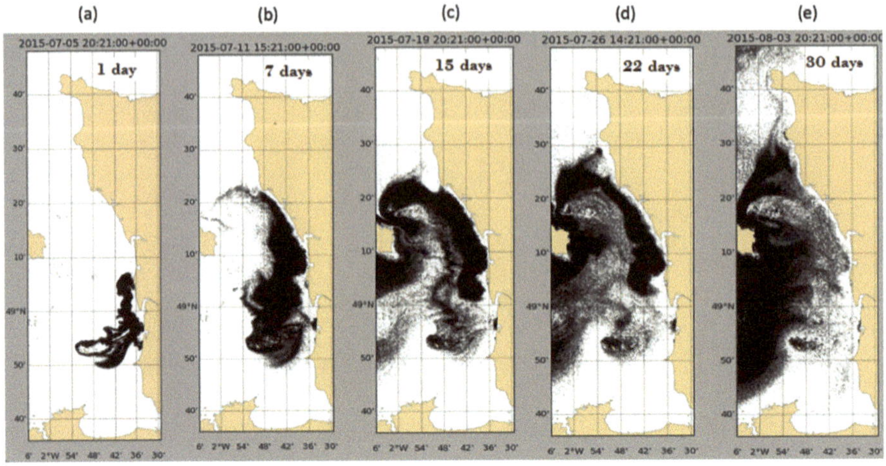

Fig. 3 Dispersion of larvae from five sources accounting for the effect of tide and wind after 1 day (**a**), 7 days (**b**), 15 days (**c**), 22 days (**d**) and 30 days (**e**) since larval emission

3.2.2 Influence of Wind Direction

The Norman-Breton Gulf is known to endure strong easterly winds. Simulations incorporated estimated or real wind data from Météo-France ALADIN from the 15th of June to the 15th of July 2019.

These simulations suggest that a majority of the larvae is retained to the east of Jersey Island, and that some of the larvae is directed towards the Mont-Saint-Michel Bay when actual wind data, rather than an idealized unidirectional wind, are used (Fig. 3).

3.2.3 Discrimination of Fertilization Sites

The simulations enable visualisation of larvae origins (fertilization sites), and clearly show that the larvae emitted from the coastal sites on the west coast of Cotentin are directed towards the Northwest of thir site origine and that few are recruited to the wester coast of cotentin. By contrast, the majority of larvae originating from the Chausey Islands seem to reach the western coast of Cotentin as well as the Mont-Saint-Michel Bay (Fig. 4).

Other origin sites were also identified from these simulations. For instance, larvae from Saint-Malo do not recruit to the west coast of Cotentin. However, populations with other origin sites might contribute to recruitment in the West coast of Cotentin, especially the populations of the Mont Saint-Michel Bay (Fig. 5).

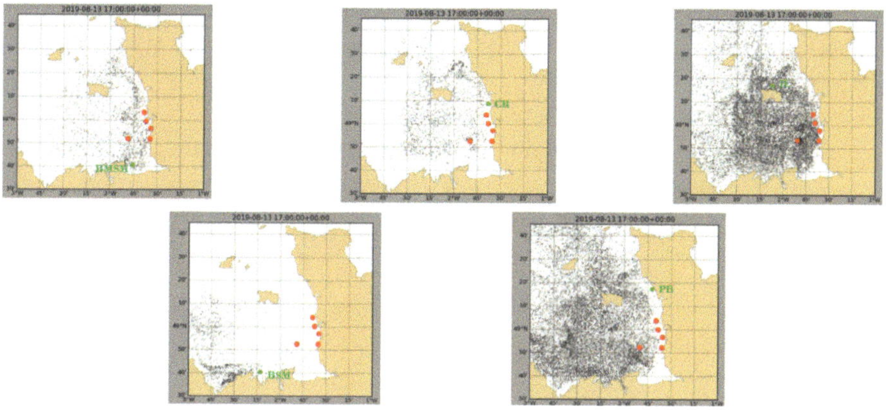

Fig. 4 Dispersion of clam larvae 30 days after egg fertilization at each source site, marked with a green dot (CH: Chausey; SM: Saint-Marin-de-Bréhal; LI: Lingreville; AC: Agon-Coutainville; BL: Blainville-sur-mer)

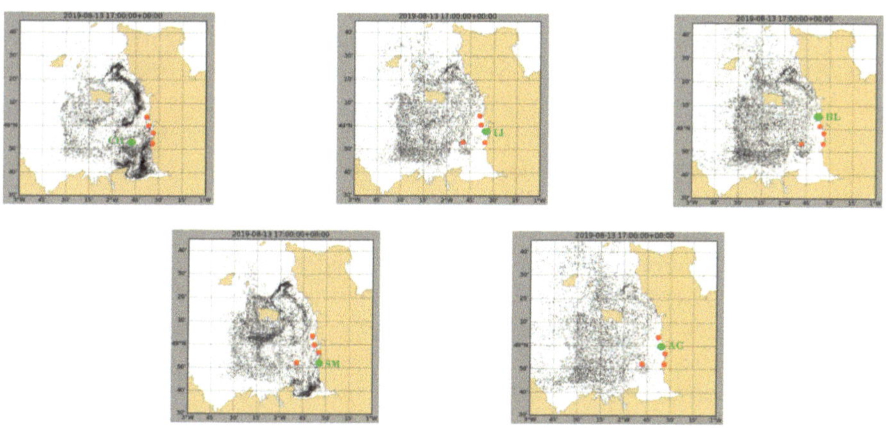

Fig. 5 Dispersion of clam larvae 30 days after egg fertilization at Mont Saint-Michel Bay (BMSM), Saint-Malo Bay (BSM), Créance (CR), Portbail (PB) and Jersey (JE) (green dots). Red dots are the reference points of our study

3.3 Connectivity of the Sites

These simulations enable prediction of clams' settlement site. Clam density at each site is imprecisely know, and therefore larval emission rate cannot be determined. The main contributors of larvae to the west coast of Cotentin are likely the Chausey Islands and the Mont-Saint-Michel Bay (Table 2), as self-recruitment in this sector seems negligible. Conversely, it seems that other populations in the Mont-Saint-Michel Bay do experience a large amount of self-recruitment but also receive Chausey Islands larvae.

Table 2 Matrix of the connectivity (in % of larvae settled) between Mont-Saint-Michel sites (BMSM), the Chausey Islands site, the sites from Saint-Martin-de-Bréhal to Blainville-sur-mer (CW), the Portbail site and Jersey sites

		Origine of the larvae				
		BMSM	Chausey	CW	Portbail	Jersey
Recrutement	BMSM	81	43	0	2	2
	Chausey	10	14	20	51	56
	CW	7	9	0	0	2
	Portbail	1	29	50	1	1
	Jersey	1	5	30	46	39

4 Discussion

In the Norman-Breton Gulf, the Japanese clam *Ruditapes philippinarum,* introduced for aquaculture in the 1980s, has nearly replaced the European clam *R. decussatus.* This has also occurred at numerous other European sites, including the Arcachon basin (France), the Venice lagoon (Italy) and the Tage estuary (Portugal) (Caill-Milly et al. 2006; Dias et al. 2019). However, there are some sites where the two species coexist without either dominating (Santander's Bay, Spain) (Juanes et al. 2012), so this non-native species is not solely responsible for decreases in the native European clam. The other causes of its decline are unknown, but several hypotheses, including predation, parasitism, diet, disease, anthropogenic modification of the habitat, cross-breeding and competition, have been suggested. However, Bidegain and Juanes (2013) have demonstrated that in Santander's Bay (Spain), the increase in the population density of *R. philippinarum* does not affect the growth and the mortality of *R. decussatus* but that predation might play an essential part in the regulation of the populations. Bidegain et al. (2015) identified depth, current speed and salinity as addition potential reasons for one species' supremacy over the other.

On the west coast of Cotentin, clam population densities are relatively low–generally less than 5 clams/m^2 (Basuyaux et al. 2017)—whereas the mean is almost 40/m^2 in the Arcachon basin (France) (Sanchez et al. 2018), and higher than 150/m^2 in Marennes Oléron (France) (Prieur et al. 2018) where the native population has practically disappeared. Despite the fact that brown ring disease, engendered by *Vibrio tapetis* (Borrego et al. 1996), is present on the west coast of Cotentin, it does not seem to be a major cause of mortality (Basuyaux et al. 2023). However, predation of clams by Muricidae has been found to be an important cause of mortality on the west coast of Cotentin (Pezy et al. 2021), but this predation affects both species equally, so doesn't explain the overall decline in European clam reserves.

Instead, the relative decline of native clams and increase in Manila clams can be explained by differences in reproductive efficiency coupled with a large increase in shore fishing. Fishing pressure on the west coast of Cotentin is increasing. Harvesters collect the two clam species without distinguishing between them, and while the reproduction cycle of the two species is roughly the same, *R. philippinarum* seems to have a longer reproductive period and more frequent spawning than *R. decussatus* in the Morbihan Gulf (France) (Laruelle et al. 1994). The ecological distribution of the Japanese clam is also much greater than that of the European clam: it extends higher on the foreshore and colonizes a greater diversity of habitats, such as mud, sand and gravel floors (Dauvin, personal observation).

The simulations performed in this study show that the self-recruitment in the clam population of the west coast of Cotentin is very low. The larvae from this area develop in Jersey, the Chausey Islands, and Portbail if they find favourable substrate in the foreshore zone. Otherwise, they are lost, especially if they settle on subtidal sediments. The larvae that settle on the west coast may come from *R. philippinarum* farms in the Chausey Islands or Mont-Saint-Michel Bay, as the high population density of these two sites could supply the massive amount of Japanese clam larvae that permits it to dominate over the European species in this area.

The Norman-Breton Gulf hydrodynamic system is highly complex, with a central gyre around Jersey and the Chausey Islands, as well as a nested gyre system to the south of the Mont-Saint-Michel Bay. This double gyre system promotes rapid larval circulation depending on the meteorological conditions (Ayata et al. 2009). The matrix of connectivity shows that it is a metapopulation, with the Chausey Islands driving most of the larvae supply, although the populations of the Mont-Saint-Michel Bay can also self-recruit, permitting very high population densities in some areas. This supports intensive shellfish harvest in the south-west of the Mont-Saint-Michel Bay, where 33 fishermen are each permitted a quota of 100 kg/day (Comité Régional des Pêches Maritimes et des Elevages Marins, Ille-et-Vilaine) operate.

Climate change is increasing seawater temperatures, which may alter the temporal dynamics of clam spawning, including inducing earlier and multiple spawns, which may be favourable to the species if microalgae food resources are sufficient and if it does not result in an increase in pathogens. However, climate change also modifies wind patterns, which could disrupt larval dispersion. Our simulations suggest that the inputs of larvae from the Chausey Islands are largely unaffected by the wind, while the larval supply to the west coast of Cotentin from Mont-Saint-Michel Bay depend an easterly wind orientation. Any increase in Northeasterly winds may trap larvae within the Mont-Saint-Michel gyre, thus depriving the west coast of Cotentin of clams.

Acknowledgements This study was carried out under the Framework of the RS2S program (Reconstitution d'un Stock de bivalves Indicateur de Stocks et Vigie des Havres) supported by the Seine Normandy Water Agency (AESN). The organizations responsible for this work are the SMEL (Synergie Mer Et Littoral), the GEMEL (Groupe d'Etudes des Milieux Estuariens et Littoraux), Normandie, Avril Association, the UMR-CNRS laboratory "Morphodynamique continentale et côtière" of Caen Normandy University, and the UMR-CNRS laboratory LEMAR of Occidental Bretagne University. The authors thank Naida Laisney and the APP2R association for their assistance during the sampling and the CRC and CRP for the provision of fishing and shellfish data. We are grateful to C.E. O'Brien for editing the English style and grammar.

References

Allain G, Petitgas P, Lazure P, Grellier P (2007) Biophysical modelling of larval drift, growth and survival for the prediction of anchovy (*Engraulis encrasicolus*) recruitment in the Bay of Biscay (NE Atlantic). Fish Oceanogr 16:489–505

Ayata SD, Ellien C, Dumas F, Dubois S, Thiébaut É (2009) Modelling larval dispersion and settlement of the reef-building polychaete *Sabellaria alveolata:* role of hydroclimatic processes on the sustainability of biogenic reefs. Cont Shelf Res 29:1605–1623

Basuyaux O, Beck F, JP Pezy, Baffreau A, Joncourt Y, Tétard X, Dauvin JC (2017) Evaluation of the *Ruditapes* spp. clam stock on the western coast of Cotentin (English Channel). J Mar Biol Oceanogr 7(1)

Basuyaux O, Angelo M, Arrenoux G, Baffreau A, Bennis AC, Boutet M, Brieau L, Champolivier L, Croguennec G, Dauvin J-C, Charrier, Desfeux-Mezière V, Garcia A, Haitz F, Huet M, Jolly G, Joncourt Y, Laisney N, Laplace Dolonde A, Lefebvre J, Lefebvre V, Legrand A, Lehuen A, Lemoine A, Michel J, Moal S, Ollivier E, Paillard C, Sahnoun-Ajmi S, Spagnol M, Timsit O, Vieira P, Vigoureux P, Zablocki Thomas L (2023) Reconstitution d'un Stock de bivalves, Indicateur de Stocks et Vigie des havres. Rapport de synthèse final, AESN 1074142-2017

Bidegain G, Juanes JA (2013) Does expansion of the introduced Manila clam *Ruditapes philippinarum* cause competitive displacement of the European native clam *Ruditapes decussatus*? J Exp Mar Biol Ecol 445:44–52

Bidegain G, Bárcena JF, García A, Juanes JA (2015) Predicting coexistence and predominance patterns between the introduced Manila clam (*Ruditapes philippinarum*) and the European native clam (*Ruditapes decussatus*). Estuar Coast Shelf Sci 152:162–172

Borrego JJ, Castro D, Luqué A, Paillard C, Maes P, Garcia MT, Ventosa A (1996) *Vibrio tapetis* sp. nov., the causative agent of the brown ring disease affecting cultured clams. J Int Bactériol Syst 46(2):480–484

Boussinesq VJ (1877) Essai sur la théorie des eaux courantes. Mém Présent Divers L'acad Sci Paris 23:1–680

Breber P (1985) On-growing of the carpet-shell clam (*Tapes decussatus* (L.)): two years' experience in Venice Lagoon. Aquaculture 44:51–56

Caill-Milly N, Duclercq MN, Morandeau G (2006) Campagne d'évaluation du stock de palourdes du bassin d'Arcachon. Rapport IFREMER. https://archimer.ifremer.fr/doc/00000/2218/

Cordero D, Delgado M, Liu B, Ruesink J, Saavedra C (2017) Population genetics of the Manila clam (*Ruditapes* philippinarum) introduced in North America and Europe. Sci Rep 7:39745

Daniel M, Lemonsu A, Déqué M, Somot S, Alias A, Masson V (2019) Benefits of explicit urban parameterization in regional climate modeling to study climate and city interactions. Clim Dyn 52(5–6):2745–2764

Delhez ÉJ, Damm P, De Goede E, De Kok J, Dumas F, Gerritsen H, Jones JE, Ozerg J, Pohlmann T, Rasch PS, Skogen M, Proctor R (2004) Variability of shelf-seas hydrodynamic models: lessons from the nomads2 project. J Mar Syst 45:39–53

Dias E, Chainho P, Barrocas-Dias C, Adao H (2019) Food sources of the non-indigenous bivalve *Ruditapes philippinarum* (Adams and Reeve, 1850) and trophic niche overlap with native species. Aquat Inv 14:638–655

Ee-Yung C, Bum HS, Young-Baek H, Sick LJ (2001) Gonadal Maturation and Artificial Spawning of the Manila Clam, *Ruditapes philippinarum* (Bivalvia: *Veneridae*), in Komso Bay, Korea. Fish Aquat Sci 4:208–218

Flassch J, Leborgne Y (1992) Introduction en Europe, de 1972 à 1980, de la palourde japonaise. CIEM Mar Sci Symp 194:92–96

France Agri Mer (2018) Étude sur l'évaluation de l'activité de pêche de loisirs en France métropolitaine (dont la Corse)—Synthèse. Les études de France Agri Mer. https://www.franceagrimer.fr/fam/content/download/59226/document/SYN-MER-PECHE-DE-LOISIRS-2017.pdf?version=4

Global Biodiversity Information Facility (2023) https://www.gbif.org/species/4372686

Helm MM, Bourne NM, Lovatelli A (2006) Ecloserie de bivalves. Un manuel pratique. FAO Document technique sur les pêches. FAO, Rome. https://www.fao.org/3/y5720f/y5720f00.htm

Hurtado NS, Pérez-García C, Morán P, Pasantes JJ (2011) Genetic and cytological evidence of hybridization between native *Ruditapes decussatus* and introduced *Ruditapes philippinarum* (Mollusca, Bivalvia, *Veneridae*) in NW Spain. Aquaculture 311:123–128

Juanes JA, Bidegain G, Echavarri-Erasun B, Puente A, García A, Garcia A, Barcena JF, Alvarez C, Garcia-Castillo G (2012) Differential distribution pattern of native *Ruditapes decussatus* and introduced *Ruditapes phillippinarum* clam populations in the Bay of Santander (Gulf of Biscay). Coast Manag 69:316–326

Laruelle F, Guillou J, Paulet YM (1994) Reproductive pattern of the clams, *Ruditapes decussatus* and *R. philippinarum* on intertidal flats in Brittany. J Mar Biol Ass UK 74:351–366

Lazure P, Dumas F (2008) An external–internal mode coupling for a 3d hydrodynamical model for applications at regional scale (mars). Advan Wat Res 31:233–250

Lazure P, Jégou AM (1998) 3D modelling of seasonal evolution of Loire and Gironde plumes on Biscay Bay continental shelf. Oceanol Acta 21:165–177

Le Roy R, Simon B (2003) Réalisation et validation d'un modèle de marée en Manche et dans le Golfe de Gascogne. Application à la réalisation d'un nouveau programme de réduction des sondages bathymétriques. Rapport SHOM n°002/03

Lecornu B, Schlund E, Basuyaux O, Dauvin JC (2016) Dynamics (2010–2014) of the *Sabellaria alveolata* reefs on the west coast of Cotentin (France). Reg Stud Mar Sci 8:157–169

Lett C, Verley P, Mullon C, Parada C, Brochier T, Penven P, Blanke B (2008) A Lagrangian tool for modelling ichthyoplankton dynamics. Environ Model Softw 23:1210–1214

Luyten PJ, Deleersnijder E, Ozer J, Ruddick KG (1996) Presentation of a family of turbulence closure models for stratified shallow water flows and preliminary application to the Rhine outflow region. Cont Shelf Res 16:101–130

Ménesguen A, Gohin F (2006) Observation and modelling of natural retention structures in the English Channel. J Mar Syst 63:244–256

Moura P, Vasconcelos P, Pereira F, Chainho P, Costa J, Gaspar M (2017) Reproductive cycle of the Manila clam (*Ruditapes philippinarum*): an intensively harvested invasive species in the Tagus Estuary (Portugal). J Mar Biol Ass UK 98:1–13

Nabat P, Somot S, Cassou C, Mallet M, Michou M, Bouniol D, Decharme B, Drugé T, Roehrig R, Saint-Martin D (2020) Modulation of radiative aerosols effects by atmospheric circulation over the Euro-Mediterranean region. Atmos Chem Phys 20:8315–8349

Nicolle A, Dumas F, Foveau A, Foucher E, Thiébaut E (2013) Modelling larval dispersal of the king scallop (Pecten maximus) in the English channel: examples from the bay of Saint-Brieuc and the bay of Seine. Ocean Dyn 63:661–678

Pezy JP, Raoux A, Dauvin JC, Schwanka T, Michelet C, Basuyaux O (2021) Prey-predation relationships between bivalves and predatory gastropods: experiments on English Channel populations. J Mar Biol Oceanog 4:172–180

Prieur S, Costales L, Gerard S, Jeune N, Morel M (2018) Bilan des trois années d'études : suivi participatif des effets de la pêche à pied de loisir sur la ressource de palourdes européennes et japonaises (*Ruditapes decussatus* et *Ruditapes philippinarum*). Rapport de l'action B6 du projet Life+ Pêche à Pied de Loisir. Fédération Nationale de la Plaisance et des Pêches en mer. https://www.pecheapied-loisir.fr/wp-content/uploads/2023/06/C4-2017-FNPPSF-RAPPORT-FINAL-2015-C3A0-2017-FNPPSF-VSC.pdf

Salomon JC, Breton M (1991a) Courants résiduels de marée dans la Manche. Oceanol Acta 11:47–53

Salomon JC, Breton M (1991b) Courants résiduels de marée dans la Manche. Oceanol Acta 1:47–53

Salomon JC, Breton M (1993) An atlas of long-term currents in the channel. Oceanol Acta 16:439–448

Sanchez F, Caill-Milly N, Lissardy M (2018) Campagne d'évaluation du stock de palourdes du bassin d'Arcachon. Année 2018. Rapport Ifremer R.ODE/LITTORAL/LER AR 18.015. https://archimer.ifremer.fr/doc/00477/58897/

Sanchez F, Caill-Milly N, Briaudeau T, Benito D, Ruiz P, Izagirre U (2022) Détermination de la taille de première maturité sexuelle de la palourde japonaise (*Ruditapes philippinarum*) du bassin d'Arcachon. Rapport Ifremer ODE/LITTORAL/LERAR 22.013. https://archimer.ifremer.fr/doc/00793/90464/

Sugiura D, Kikuya N (2022) Cycle of reproduction, oocyte atresia, and maturation size of the Manila clam *Ruditapes philippinarum* on a pebble beach in Mutsu Bay, Northern Japan. Invert Repr Dev 66(1):8–15

Utting SD, Spencer BE (1992) Introductions of marine Bivalvia Mollusca into the United Kingdom, International Council for Exploration of the Marine. Mar Sci Symp 194:84–91

Walne P, Mann R (1975) Growth and biochemical composition in *Ostrea edulis* and *Crassostrea gigas*. In: Barnes H (eds) Proceedings of the 9th European marine biology symposium. Aberdeen University Press, Aberdeen, pp 587–607

Xie H, Lazure P, Gentien P (2007) Small scale retentive structures and *Dinophysis*. J Mar Syst 64:173–188

Spatio-Temporal Distribution of Intertidal Shrimps, *P. serratus* and *P. elegans* from Two Western and Eastern Cotentin Sites (Normandy, France)

K. Costil, A. Serpentini, C. Guégan, R. Boisserie, C. Caplat, G. Guichard, J.-P. Pezy, A. Raoux, C. Roger, V. Lefebvre, R. Coulaud, X. Tétard, O. Basuyaux, and J.-C. Dauvin

1 Introduction

Palaemon serratus (Pennant, 1777) and *Palaemon elegans* (Rathke, 1837) are Caridea shrimps that belong to Palaemonid family and that are morphologically quite similar. These transparent shrimps show dark lines and yellowish dots, and mainly differ by the rostrum that is long, slender and upwardly curved in *P. serratus* whereas it is shorter and right in *P. elegans*. In this latest species, three teeth behind

K. Costil (✉) · A. Serpentini · C. Guégan · C. Caplat · C. Roger
Normandie Université, UNICAEN, UMR BOREA (MNHN, Sorbonne Université, UCN, CNRS-8067, IRD-207), Caen Cedex, France
e-mail: katherine.costil@unicaen.fr

Normandie Université, UNICAEN, Laboratoire MERSEA (Research on Marine Ecosystems and Organisms), Caen Cedex, France

R. Boisserie · J.-P. Pezy · A. Raoux · J.-C. Dauvin
Normandie Université, UNICAEN, UNIROUEN, Laboratoire Morphodynamique Continentale et Côtière, CNRS UMR 6143 M2C, Caen Cedex, France

G. Guichard · V. Lefebvre · O. Basuyaux
SMEL, Centre Expérimental, ZAC Blainville-sur-Mer, Blainville-sur-Mer, France

X. Tétard
CRPMEM (Comité Régional des Pêches Maritimes et des Elevages Marins de Normandie), Port-en-Bessin, France

O. Basuyaux
Mer Expertise Environnement Etudes et Conseils, Blainville-sur-Mer, France

R. Coulaud
Université Le Havre Normandie, UMR SEBIO (Stress Environnementaux et Biosurveillance des Milieux Aquatiques), Le Havre Cedex, France

A.-C. Bennis et al. (eds.), *Constraints and Adaptations to Global Change at the Land-Sea Interface: For a Shared Ecological and Energy Transition*,
https://doi.org/10.1007/978-3-031-90050-1_6

Fig. 1 Photographs of the rostrum of two *Palaemon* shrimps: *P. serratus* (top photo) and *P. elegans* (bottom photo). *Photo credit* Roberto Pillon, doris.ffessm.fr

the eye can be observed while only two post-orbital teeth are present in *P. serratus* (Fig. 1). These species also differ by the size as the maximum total length of *P. elegans* and *serratus* reaches, respectively, 63 and 110 mm (Hayward and Ryland 2017).

Palaemon serratus is a shrimp distributed along the coasts of the FAO area "Est Atlantic" from Norway to Mauritania and including the Baltic Sea, the English Channel, the Mediterranean Sea and the Black Sea (INPE MNHN 2024a). The geographical distribution of the close species *Palaemon elegans* is similar but more extended since this species can also be found in the Caspian Sea, the Aral Sea and in other locations such as the coasts of Namibia, Indonesia and New England, USA where it was introduced and sometimes considered as an invasive species (INPE MNHN 2024b). Both shrimp species live on rocky littoral habitats, under rocks and among algae and are also reported to inhabit ports and estuaries with low salinity, *Zostera* spp. meadows on intertidal muds and be particularly abundant in the intertidal pools explaining it is sometimes called "the rockpool prawn" (La Rivière et al. 2022; Marine Biological Association (MarLIN) 2023a, 2023b). The two species can be collected in the intertidal zone but *P. serratus* lives until 40 m depth and *P. elegans* until 5–10 m depth. Both species are euryhalin shrimps, *P. elegans* tolerating salinity range of 5–45 PSU (INPE MNHN 2024b) and *P. serratus* 1–40 PSU.

Because of this larger size compared to *P. elegans*, *P. serratus* is especially fished by both professional and recreational fishermen. In France, the shrimp size to be fished is regulated and set at a total length of 50 mm. The professional fishing is coastal by using traps or shrimp trawl. In Normandy, France, on the period of 2012–2020, 20–25 tons were fished per year by about 55 fishing vessels (Tétard, pers. com.). These landings can appear quite low but the shrimp commercial value is the highest one among all the seafood (about 60–80 € kg^{-1} in 2023–2024). There is an important traditional activity around *P. serratus* in some fishing communities such as Spain, Portugal, UK and Ireland where the value can reach 240 €/kg (Perina et al. 2018). Moreover, *P. serratus* is much in demand among recreational fishermen which was relatively easy to fish during the summer and the beginning of the autumn where the species is present and particularly in Normandy where the species can be qualified as an emblematic species. However, the biology and the ecology of this species had never been studied in Normandy and this is also the reason why a European project called "Gedubouq" was launched in 2020 for a period of three years.

Gedubouq project (Analyse préalable à une gestion durable des pêcheries de bouquet en Normandie, i.e. Analysis prior to sustainable management of the common shrimp fisheries in Normandy) was conducted on three topics: (1) biology and ecology of *P. serratus* in Normandy; (2) cumulated effects of anthropogenic activities on *P. serratus* and (3) place of *P. serratus* in Normandy in terms of socioeconomical aspects. In the first topic, field work was carried out to more especially investigate at Grandcamp-Maisy (GRE) and Blainville-sur-Mer (BLE) the shrimps' health and physiological status; their life cycle, growth and reproduction; the diet at four sites (GRE, BLE plus two other sites) and the population genetics at 23 sites. Besides these studies focused on *P. serratus*, it appeared us important from an ecological point of view to also consider another Palaemonid species, *P. elegans* that sympatric at many sites along the coasts of the English Channel. Therefore, four sampling campaigns were conducted to collect both species, and the aim of the present study was (1) to compare the distribution of *P. serratus* and *P. elegans* along transects and according to the hypsometric level in order to assess potential interspecific competition; (2) to evaluate the seasonal variation in spatial distribution of both species and (3) to investigate the size (total length) of *P. serratus* individuals at both sites by considering the potential variations intra-sites and across seasons.

2 Materials and Methods

2.1 Study Sites

The two study sites were located on the coast of the English Channel, Grandcamp-Maisy (GRE) on the eastern part of the Cotentin peninsula and Blainville-sur-Mer (BLE) on the western part (Fig. 2). The extreme geographical coordinates of the two study areas were, for latitudes (north) and longitudes (east), respectively: at GRE: 49° 23′ 29.4″–49° 23′ 47.6″ and 1° 03′ 10.7″–1° 03′ 21.8″, and at BLE: 49° 03′ 10.8″–49° 04′ 33.5″ and 1° 36′ 50.8″–1° 38′ 16.9″.

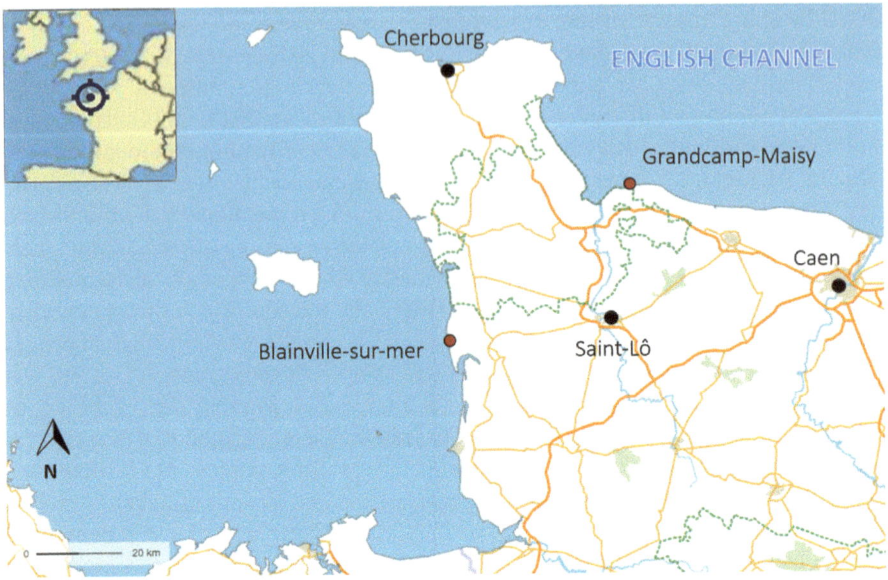

Fig. 2 Location of the two study sites on the coasts of the English Channel in Normandy (France). Background map: IGN. France

The water masses of these two sites differed since GRE is situated in the Eastern Basin of the English Channel (EBEC) and belongs to subdivision named "Baie de Seine" whereas BLE is located in the Western Basin of the English Channel (WBEC) and opens on the "Golfe Normano-Breton" (Dauvin 2012). GRE is quite far from the Seine estuary but can be still impacted by this large estuary, which is highly contaminated (e.g. Cachot et al. 2006; Chiffoleau et al. 2007), even if improvements are now being noted. It is noticeable that GRE is close to the Baie des Veys where the largest watersheds in Normandy leds (3500 km^2). By contrast, anthropogenic pressures are far lower at BLE.

The temperature of the two water masses differed with a higher range of sea temperature between winter and summer in EBEC compared to WBEC, as recorded in the present study that showed a differential of − 1 °C in winter and + 2 °C in summer (Table 1). The tidal regime of both sites is qualified as megatidal but maximal tidal (by a coefficient of 120) ranges higher at BLE (~ 14 m) by comparison with GRE (~ 6.8 m). At GRE, the upper part of the intertidal area is sandy, and the rocky substratum potentially populated by Palaemonid shrimps begins at half-tide. At this site belonging to the Parisian basin, rocky substratum is composed of Middle Jurassic sedimentary rocks (limestone) (Fily et al. 1989) while at BLE belonging to the Armorican massif, Precambrian sedimentary rocks (i.e. sandstone and grauwacke) were predominant (Dupret et al. 1989). Finally, it is noticeable that the study site of BLE is partially included in a 230-ha oyster farming area.

Table 1 Sampling date, tidal coefficient and surface water temperature during the sampling campaign to collect Palaemonid shrimps at two study sites (GRE and BLE)

	Season	Sampling date	Tidal coefficient	Water temperature (°C)
Grandcamp-Maisy (GRE)	Spring	2022-04-01	98	10
	Summer	2022-06-14	93	19
	Autumn	2022-10-10	101	17
	Winter	2023-01-23	105	9
Blainville-sur-Mer (BLE)	Spring	2022-03-31	95	11
	Summer	2022-06-15	96	21
	Autumn	2022-10-11	98	17
	Winter	2023-01-24	103	8

2.2 *Sampling Design*

2.2.1 Sampling Protocol and Sample Treatment

To study the spatio-temporal distribution of Palaemonid species, shrimps were sampled four times, one time per meteorological season (Table 1). These dates were chosen for the high tidal coefficients, and the minimum coefficient was 93 at GRE in summer. At each date, before the beginning of the sampling and about 3 h before the low tide, surface water temperature was measured by diving a thermometer at a depth of about 30 cm under the surface (Table 1).

At each date and sampling point, three persons fished shrimps with a traditional hand net (diameter of 50 cm and mesh size of 5 mm) for 20 min and strictly in a circle of 30 m from a central point. To minimize the variability between teams in terms of fishing experience, we ensured that the teams were as balanced as possible by mixing experienced fishermen with less experienced fishermen. The geographical coordinates of the central point were chosen before the beginning of the study. The fished shrimps were regularly transferred in a bucket filled with water (one bucket per person). A fourth person checked that the sampling protocol was well respected. After 20 min, sampling was stopped and all Palaemonid shrimps in each bucket were carefully counted and then shrimps were gathered in a same bucket (one bucket per sampling point). Then, each group of persons went down towards the next sampling point.

All collected shrimps were brought back to the laboratory where they were fixed into ethanol. Thereafter, these shrimps were identified and then the study focused on *P. serratus* because it was the species particularly studied in Gedubouq project. All males and females were distinguished. Thus, the abdomen of all female *P. serratus* was examined to determine the presence of eggs and then calculate the percentage of berried shrimps. Then, all the shrimps that were in a good state (e.g. rostrum not broken) were measured with a vernier caliper (to the nearest of 0.01 mm). It was chosen to measure the length of the cephalothorax (including the rostrum) (Fig. 3),

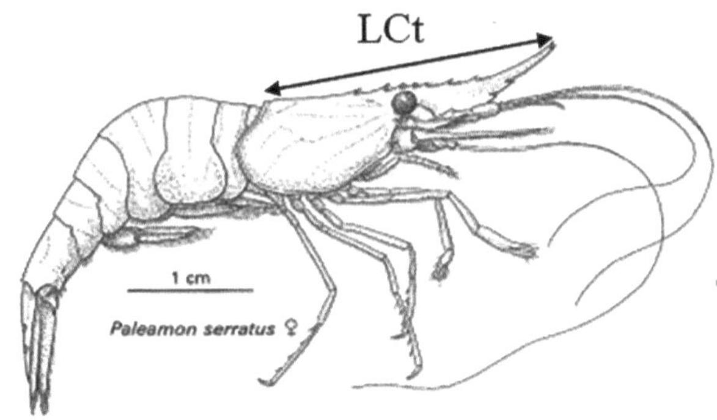

Fig. 3 Measurement of the length of *Palaemon serratus* cephalothorax, LCt. From Hayward and Ryland (2017)

Table 2 Results about the linear regression model between the length of cephalothorax (including the rostrum) of shrimps and their total length

Gender	Linear regression equation	Pearson's correlation	Number
Females + Males	LT = 2.16(LCt) + 0.398	0.975	1867
Females	LT = 2.13(LCt) + 5.30	0.977	1066
Males	LT = 2.06(LCt) + 6.01	0.977	801

LCt that appeared to be the most reliable measurement, and because of the possibility to link this parameter to age classes determined for female and male shrimps by Campillo (1979). The distribution of data about the LCt and the total length (TL) revealed normal, and were positively correlated (Pearson's correlation, $p < 0.001$) (Table 2). Therefore, the regulatory size of 50 mm corresponds to individuals showing a mean LCt of 21.63 mm (i.e. 21.39 mm for female shrimps and 21.70 mm for male ones).

2.2.2 Characterization of Sampling Areas

At each site, three transects were prospected for shrimps but it is noticeable that they were far separated at BLE compared to GRE (Fig. 4). At GRE, we could distinguish an eastern, a central and a western transect (Fig. 4a) while at BLE, there were a southern, a central and a northern transects (Fig. 4b). For each transect, five hypsometric levels separated by a difference of 0.5 m were prospected for shrimps at GRE: level 1: ~ 3.5 m above chart datum; level 2: ~ 3 m; level 3: 2.5 m; level 4: 2 m and level 5: 1.5 m. At BLE where the intertidal area is very large, shrimps were sampled at only three levels: level 1: 3–3.5 m above chart datum; level 2: 2–2.5 m and level 3: 1–1.5 m.

During the study at GRE, the algal community changed over time, but it mainly included *Fucus serratus*, and *Sargassum muticum* and at a lesser extent *Ulva* ssp. and red algae such as *Chondrus crispus* or *Osmundea pinnatifida* (Fig. 5). At BLE, the algal community appeared us less developed but it was quite similar in terms of composition. However, *Fucus vesiculosus* was observed at the first level of

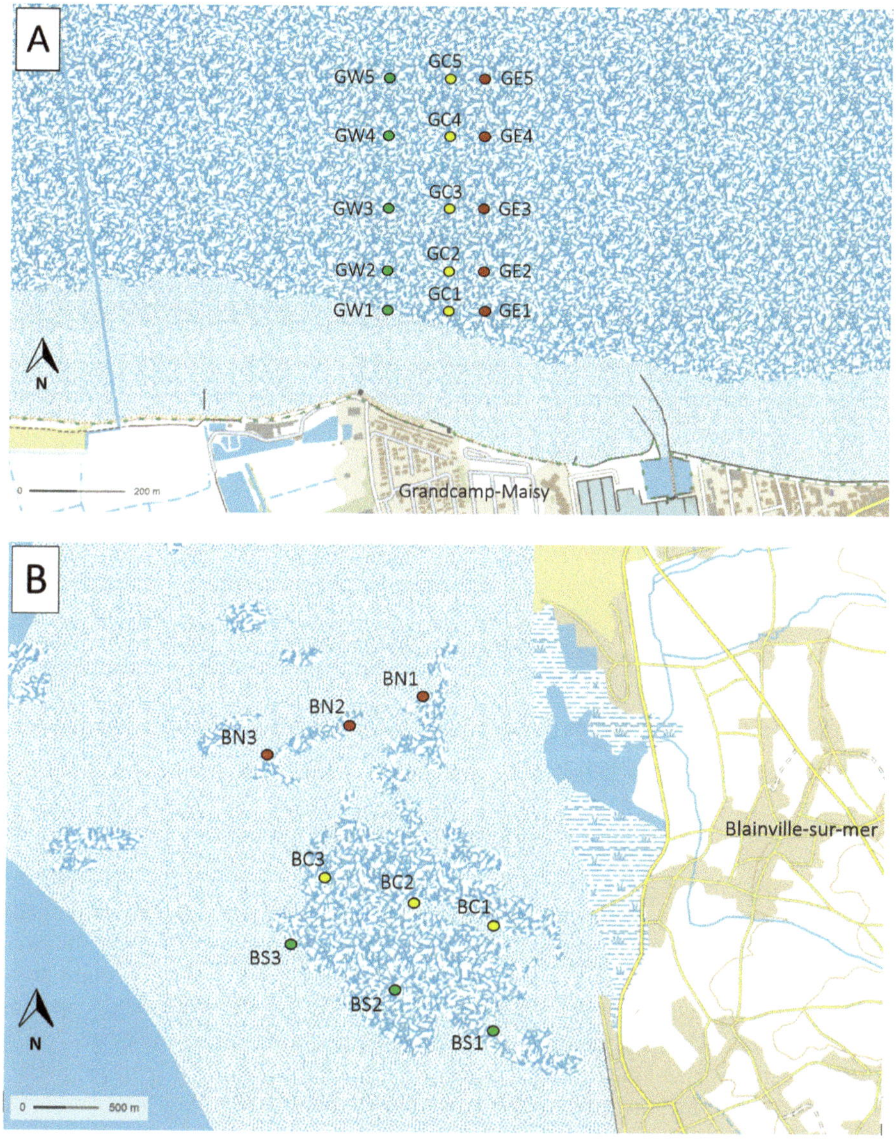

Fig. 4 Location of the shrimps' sampling stations along three transects at Grandcamp-Maisy (**a**) and at Blainville-sur-Mer (**b**). Rocky substrates where sampling points were positioned and distinguished from sandy substrates. For the nomenclature of the sampling stations at Grandcamp-Maisy (**a**) or Blainville-sur-Mer (**b**), E = East, C = Centre, W = West, N = North, and S = South. Background map: IGN. France

Fig. 5 Photographs of the intertidal area at Grandcamp-Maisy (GRE, left) with *Fucus serratus* and *Ulva* ssp and at Blainville-sur-Mer (BLE, right) with the eelgrass meadows and "oyster tables". *Photo credits* Claire Guegan (GRE) and Katherine Costil (BLE)

the central and south zones prospected for shrimps. Moreover, eelgrass meadows (*Zostera marina*) were present especially on the north zone (Fig. 5).

2.3 Data Treatment and Statistical Analyses

The standardized "niche breadth", B_A of each species at both sites was calculated by using Levins' index:

$$B_A = \frac{B - 1}{N - 1}$$

with $B = 1/\Sigma p_j^2$ and $p_j = n_j/N$ where, here, $N = \Sigma\, n_j$, the total number of individuals of species j collected in all sampling stations and n_i the number of individuals of species j collected in each sampling station (Levins 1968). N was the number of sampling stations (considered as a "resource"): 36 at BLE and 60 at GRE. A species is considered to have a wide niche breadth when $B_A \geq 0.5$ (Caldart et al. 2012). For each site, we also assessed the degree of overlap between both species "niche" (habitat) by using the "niche overlap" index O_{jk} (Pianka 1973):

$$O_{jk} = \frac{\sum p_{ij} p_{ik}}{\sqrt{\sum p_{ij}^2 \sum p_{ik}^2}}$$

where j and k are the two species compared (i.e. *P. serratus* and *P. elegans*); p_{ij} and p_{ik} are, respectively, the proportions of the individuals of species j and k were fished in sampling station i. The niche overlap index ranges from 0 (no overlap) to 1 (full overlap).

The number of shrimps belonging to the two species were compared between sites, transects, seasons and hypsometric levels by using Chi^2 tests. As data did not meet the conditions required by parametric tests (even with various transformations), Kruskal–Wallis tests followed by Dunn tests were used to assess the significance of inter-season differences in shrimps' numbers, and to determine the homogenous groups whereas inter-sex differences were tested by Mann–Whitney tests. Data about the size of length of *Palaemon serratus* shrimps' cephalothorax (LCt) met the assumption of both normality and homoscedasticity, and differences in hypsometric levels were thus tested by using ANOVAs followed by Tukey tests. Finally, the significance of differences in "niche breadth" and "niche overlap" between species or sites were tested by t-tests. The statistical analyses were performed by using STATISTICA 13.0 software (Statsoft®, Tulsa, OK, USA).

3 Results and Discussion

3.1 *Number of Collected Shrimps According to Areas and Transects*

Besides Palaemonid species, shrimps belonging to other families such as Alpheidae (i.e. *Athanas nitescens*), Crangonidae (i.e. *Crangon crangon*) and Hippolytidae (i.e. *Hippolyte varians*) were also collected in few abundances in our records and these common species in Normandy coasts were not counted.

Only two of the five Palaemonid species present in the Bay of Seine were collected: the littoral species, *P. elegans* and *P. serratus* while the estuarine species, *P. longirostris* and *Palaemonetes varians* were not logically sampled in marine waters. Furthermore, the alien estuarine species *P. macrodactylus* was not found in the study sites. With a sampling effort strictly comparable, a total of 3786 Palaemonid shrimps (2287 *P. serratus* and 1499 *P. elegans*) were collected in the intertidal zone of Grandcamp-Maisy (GRE) whereas 4665 Palaemonid shrimps (1598 *P. serratus* and 3067 *P. elegans*) were sampled at Blainville-sur-Mer (BLE). The environment at GRE revealed significantly more "favorable" to *P. serratus* while *P. elegans* were comparatively more abundant at BLE (Chi^2, $p < 0.001$). At both sites, the hypsometric level reached at the end of each sampling date was similar (1.5 m at GRE and 1–1.5 m at BLE) but at BLE it could not be ruled out that it would have been necessary to fish even lower on the foreshore. Finally, these total number conceal important and significant differences between transects at GRE and above at BLE (Fig. 6).

At GRE, while *P. serratus* were almost equally collected along the three transects (Fig. 6a), the East transect presented a significantly lower number *of P. elegans* (Chi^2, $p < 0.001$) compared to the two other transects that did not differ (Chi^2, $p = 0.44$) (Fig. 6b); this relative scarcity of *P. elegans* was probably due to the proximity of the ship channel leading to the port, and above all to a silting of rocks towards

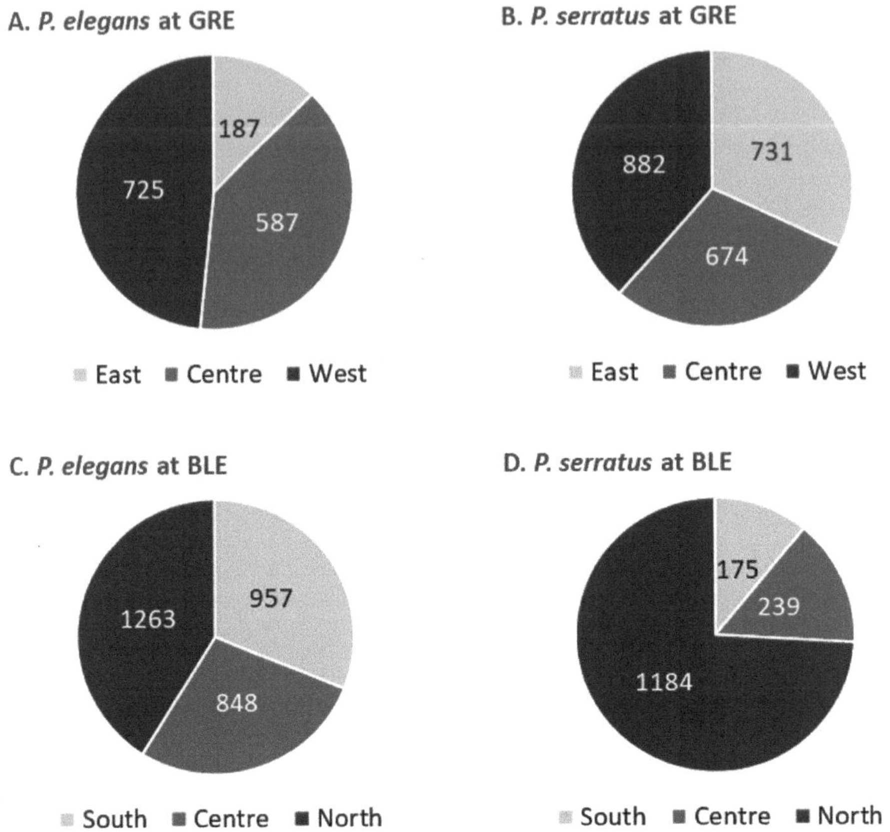

Fig. 6 Spatial origin of the total number of Palaemonid shrimps collected along the three transects at Grandcamp-Maisy (East. Centre and West): *Palaemon elegans* (**a**) and *Palaemon serratus* (**b**) and in the three zones at Blainville-sur-Mer (South, Centre and North): *Palaemon elegans* (**c**) and *Palaemon serratus* (**d**)

the upper part of the foreshore (silting that is not visible in Fig. 4a). Our results highlighted the importance of rocky substratum for Palaemonid species as already reported by Başçınar et al. (2002) and, for example, in the MarLIN internet site (2023a, 2023b: https://www.marlin.ac.uk/species/detail/2035 and https://www.marlin.ac.uk/species/detail/2033). In shallow rock pools, *P. serratus* often could be found in large numbers hiding underneath the seaweed canopy and along the boulders on the bottom (Joint Nature Conservation Committee (JNCC), 2015). At BLE compared to GRE, the three transects were far more distant from each other (Fig. 4b), and the number of *P. serratus* and *P. elegans* significantly differed between the three transects (Chi^2, $p < 0.001$); in the Palaemonid community, *P. serratus* represented only 15.5% in the South zone, 22.0% in the central zone and 48.4% in the North zone (Fig. 6c, d). The particularity of the North area was to be richer than the two other ones in *Fucus* algae and higher number of rock pools retaining water during low tide. The differences of number of shrimps belonging to both species had to be searched by also considering the potential seasonal variations and hypsometric differences.

3.2 Spatio-Temporal Distribution of **Palaemon serratus** *and* **P. elegans** *in the Intertidal Zone of Both Sites*

At GRE, the number of *P. serratus* highly fluctuated and ranged from 234 (16.5% of the maximum value) to 1415 while the minimum and maximum values for *P. elegans* were 271 (47.4%) and 572, respectively (Fig. 7a).

Significant differences in the proportion of both species were observed between seasons ($Chi^2 = 650.68$; $p < 0.001$) and each season differed from the others (Chi^2, $p < 0.001$). At BLE, the seasonal variations in shrimps' numbers were significant ($Chi^2 = 1063.00$; $p < 0.001$) and particularly high both for *P. serratus*: from 48 (4.6% of the maximum value) to 1033, and for *P. elegans*: 452 (40.5% of the maximum value) to 1117 (Fig. 7b). With different shrimp abundances and different proportions of both species, the two study sites however showed similar temporal changes: in samples from the two sites, *P. serratus* was more abundant in autumn while more *P. elegans* were mainly collected in winter (Fig. 7). The temporal differences can be explained by the reproduction and recruitment periods as well as potential migrations between the intertidal and subtidal zones. In the southern Black Sea, ovigerous female of *P. elegans*, were observed from March to August and juveniles appeared in the benthic population in August (Bilgin et al., 2009). For *P. serratus*, a long period of berried females was observed at both sites with a maximum in summer at GRE and spring–summer at BLE (see Sect. 3.3). The new recruits appeared in autumn (October) and swelled the population numbers. Nevertheless, the use of hand nets with a mesh size of less than 5 mm would have made it possible to collect smaller individuals which appear on the foreshore in summer.

Palaemon elegans and *P. serratus* could be fished together in the same stations but the spatial distribution of these sympatric species rather differed (Figs. 8 and 9). Indeed, at each season and at both sites, the spatial distribution of shrimps revealed quite different, with *P. elegans* that was more abundant at the highest hypsometric levels (levels 1 and 2 at GRE and BLE, corresponding to + 3 m-3.5 m above chart datum at GRE and + 3.5–2 m at BLE). By contrast, *P. serratus* dominated the Palaemonid community at the lowest hypsometric levels: from level 3 (i.e. about + 2.5 m at GRE and at + 1.5 m at BLE). This apparent difference in the shrimp location regarding the hypsometric levels can be probably attributed to the geomorphological characteristics. Indeed, the intertidal zone at BLE is wider and flatter in comparison to GRE where more hypsometric levels could be prospected (5 vs. only 3 at BLE).

For *P. elegans* at BLE, the higher number of individuals were recorded at level 1 (+ 3–3.5 m) in spring, autumn and winter whereas numbers were similar at levels 1 and 2 (+ 2–2.5 m) in summer (N = 318 and N = 339, respectively) (Fig. 6). For the same species at GRE, the higher numbers of shrimps were fished at level 1 (+ 3.5 m),

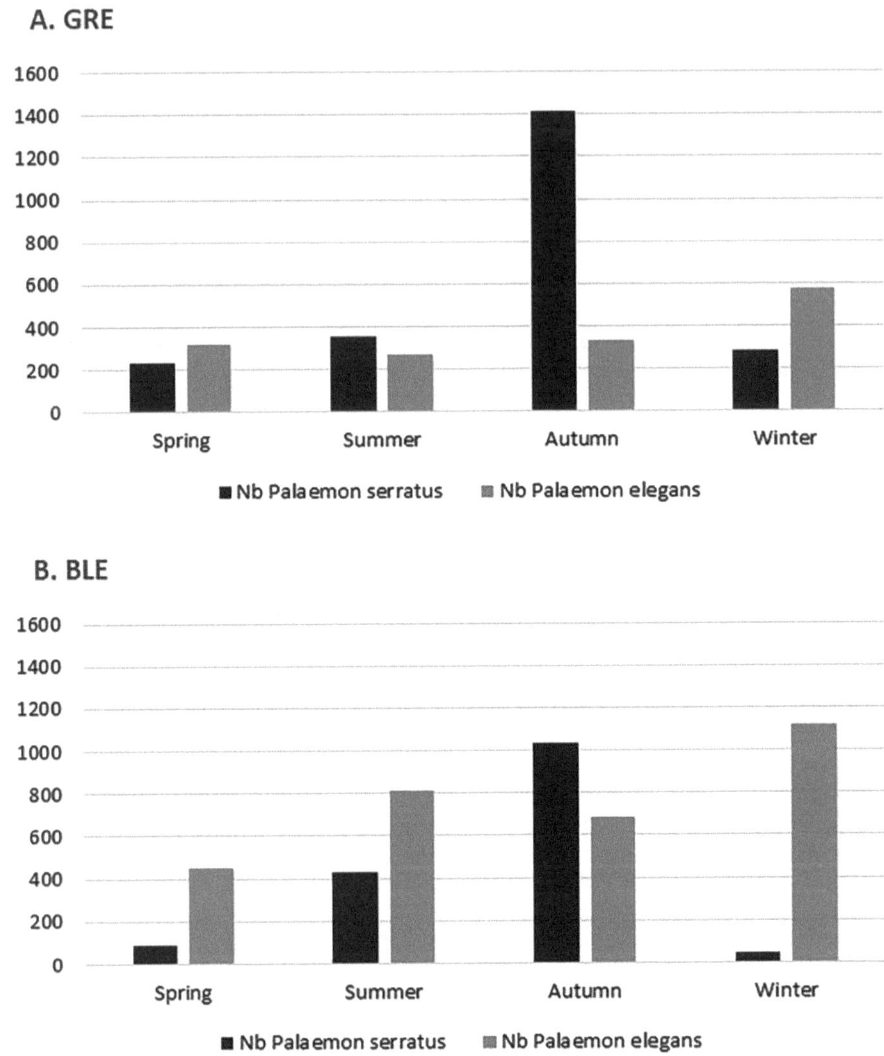

Fig. 7 Number (Nb) of shrimps of both species (*Palaemon serratus* and *P. elegans*) collected in the intertidal zone of **a** Grandcamp Maisy at four seasons (spring: 2022/04/01; summer: 2022/06/14; autumn: 2022/10/10 and winter: 2023/01/23) and **b** Blainville-sur-Mer at four seasons (spring: 2022/03/31; summer: 2022/06/15; autumn: 2022/10/11 and winter: 2023/01/24)

in autumn and winter or level 2 (+ 3 m), in spring and summer. Moreover, it is noticeable that no individuals at all were sampled at level 1 in summer. MarLIN (2023a) has reported an intertidal distribution of *P. elegans* around mid-tide but sometimes higher, and it has been specified that the species moves offshore in winter in northern latitudes. Both laboratory and field experiments suggested a winter migration of large individuals of *P. elegans* from intertidal pools to deeper waters in relation to

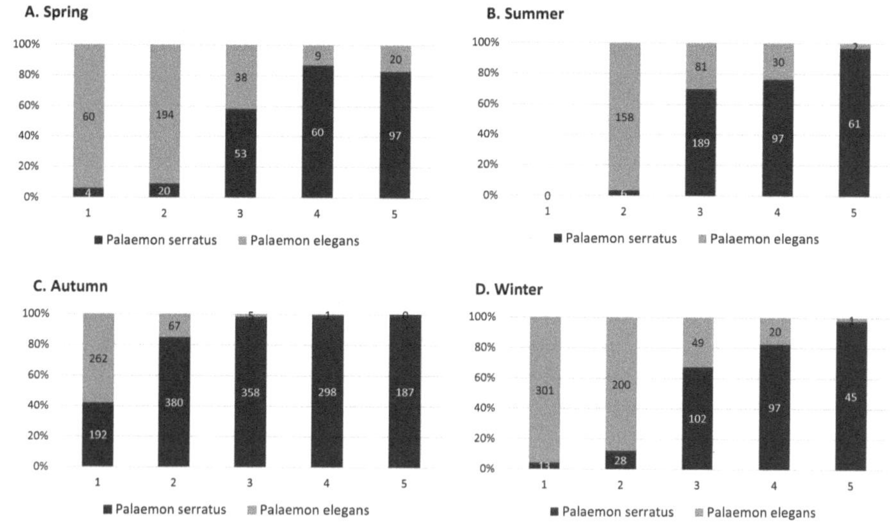

Fig. 8 Percentages (histograms) and number (added into histograms) of shrimps of both species (*Palaemon serratus* and *P. elegans*) collected at five levels of the intertidal zone of Grandcamp Maisy (see Table 1): **a** Spring (2022/04/01); **b** Summer (2022/06/14); **c** Autumn (2022/10/10) and **d** Winter (2023/01/23)

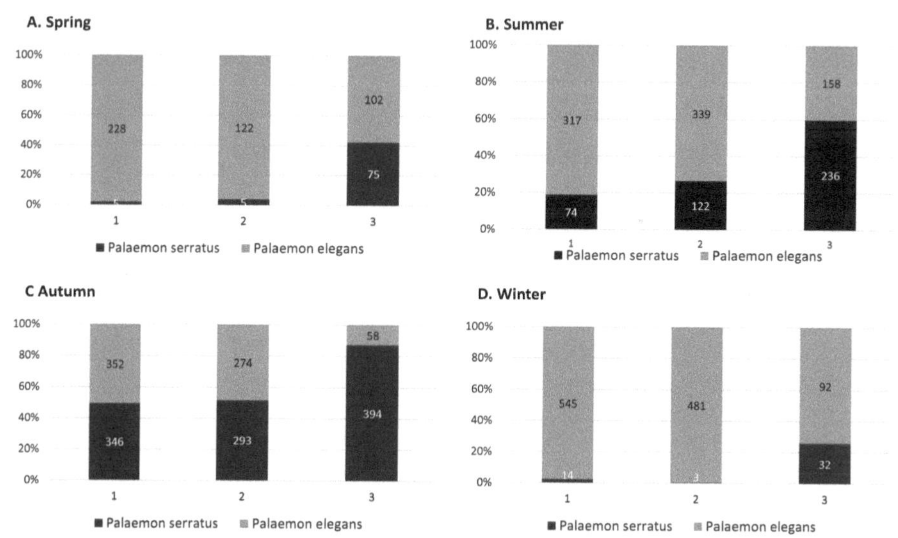

Fig. 9 Percentages (histograms) and number (added into histograms) of shrimps of both species (*Palaemon serratus* and *P. elegans*) collected at three levels of the intertidal zone of Blainville (see Table 1): **a** Spring (2022/03/31); **b** Summer (2022/06/15); **c** Autumn (2022/10/11) and **d** Winter (2023/01/24)

less osmoregulation ability at low temperature (2 °C) (Janas and Spicer 2008). In our study, no obvious offshore migrations seemed occurring during winter; the water temperature recorded in this season was 8 °C at BLE and 9 °C at GRE, and it is possible that such temperatures did not induce offshore migration in this species.

At both sites and at each season except in summer, *P. serratus* showed a minimum number of individuals at levels 1 and 2 (Figs. 5 and 6) indicating a distribution of this species at lower hypsometric levels by comparison with *P. elegans*. In literature, *P. serratus* is indicated to inhabit from intertidal pools to the shallow subtidal (depths of 40 m) (Hayward and Ryland, 2017; MarLIN 2023b). *Palaemon elegans* can be found under a few cm of water and at depths of up to 2 m in Mediterranean Sea but it can go down to 10 m in northern Europe (i.e. Baltic Sea) (DORIS 2023).

The spatiotemporal differences in *P. serratus* numbers illustrated in Figs. 4, 5 and 6 can be explained by potential migrations between the shoreline and offshore. Indeed, it appeared at both sites that in spring (end of March) the shrimps were more numerous at the lowest levels corresponding to the hypsometric level of + 1–1.5 m above chart datum (level 5 at GRE and 3 at BLE). Such a result was also obtained in summer at BLE (236 individuals at level 3) but shrimps were quite abundant at the intermediate level (122 individuals at level 2 *vs* only 5 in spring). In summer (end of June) at GRE, the highest number of shrimps was collected at the intermediate level 3 (+ 2.5 m) suggesting a move from offshore towards inshore. In autumn (October) at both sites, shrimps were dispatched across all the intertidal area, and were more numerous at the highest levels (+ 3 m): 192 and 380, and 346 individuals at, respectively, GRE and BLE. In winter (January), shrimps went back to the offshore with maximum number at level 3 at GRE (+ 2.5 m) and BLE (+ 1.5–1 m). Some slight differences were observed between the two sites due to differences in geomorphological and hypsometric features but the general pattern was similar: shrimps at the lowest levels in spring and at the highest ones in autumn with inverse migrations in summer and winter. Such a result is consistent but slightly different from the migration pattern reported by Campillo (1979) for individuals from Britany (France) and Foster (1951) for a population fished in Paignton (Southern England). Briefly, for these authors, the shrimps moved away from the coast and/or stayed the subtidal zone, from November to May; it was not until May that the individuals reached the higher hypsometric levels and colonize the rocky intertidal zone. The migration pattern in our two study sites should be confirmed and refined by monthly samplings.

Regarding the "niche breadth", that of *P. serratus* did not significantly differ between sites (test t, $p = 0.11$) while *P. elegans* showed a narrower "niche breadth" at GRE compared to BLE (test t, $p < 0.05$) (Table 3). This result suggested a more favorable environment for *P. elegans* at BLE, as already mentioned by considering the total number of collected shrimps. The main other difference between both species was the higher variation in the "niche breadth" (B_A index) in *P. serratus* (by a factor of 2.28 at GRE and as high as 5.66 at BLE) compared to *P. elegans* (by factors of 1.11 and 1.79, respectively, at GRE and BLE). Indeed, *Palaemon elegans* showed low temporal variations in the widest "niche" (from 0.338 to 0.392) at GRE; by contrast, these variations were higher at BLE where the maximum values were

Table 3 Standardized Levins' index (B_A) illustrating the "niche breadth" of *Palaemon serratus* and *P. elegans* collected at four seasons and two sites (GRE and BLE) and Pianka's index reflecting the "niche overlap" of these two shrimp species

Site	Index	Spring	Summer	Autumn	Winter	t-tests
GRE	B_A *P. serratus*	0.636	0.370	0.841	0.649	$p = 0.035$
	B_A *P. elegans*	0.385	0.338	0.392	0.353	(n = 8)
	Niche overlap	0.074	0.132	0.109	0.056	
BLE	B_A *P. serratus*	0.100	0.332	0.564	0.469	$P = 0.187$
	B_A *P. elegans*	0.631	0.684	0.382	0.493	(NS)
	Niche overlap	0.035	0.145	0.246	0.146	(n = 8)

p-values of t-tests indicated here were related to the "niche breadth" comparison between the two species at both sites. *NS* non-significant

recorded in spring and summer: respectively, 0.631 and 0.684. This inter-species difference could be partly explained by the migrations already described in *P. serratus* while the B_A fluctuations in *P. elegans* had to be more linked with the life cycle including reproduction and recruitment. At GRE and BLE, the widest "niche breadth" of *P. serratus* was calculated in autumn (respectively, 0.841 and 0.564) when the shrimps were the most numerous, and this parameter was especially low at BLE in spring (0.100). Finally, the "niche overlap" between the two species did not differ significantly between sites (test t, $p = 0.32$), and it was limited throughout the year (values lower than 0.5), with the highest "niche overlap" in summer at GRE and in autumn at BLE. These results suggested quite a low competition between the two sympatric shrimp species throughout the year.

3.3 *Reproduction Period and Size of the Individuals of* Palaemon serratus *Collected at Both Sites*

Besides the migration mentioned above, it could be relevant to consider the period of reproduction and recruitment as well as the size of the shrimps *P. serratus* to explain the abundance and spatiotemporal differences in *P. serratus*. Regarding the reproduction, berried females were observed at GRE in spring (14.2% of female individuals), summer (26.0%) and winter (only 8.8%); at BLE, the percentages of berried females were: 23.6% (spring), 20.2% (summer) and finally 3.0% (winter). The reproduction pattern of *P. serratus* was similar at both sites but showed slight differences as the main reproduction period seemed to occur a little bit earlier at BLE (spring) by comparison to GRE (summer). These results should be refined by monthly samplings but it appeared that shrimps reproduced when they moved from offshore to the shoreline and were at the proximity of the shore. In Wales (UK), a peak of berried female shrimps (i.e. spawning) was recorded during spring (Emmerson et al. 2017). However, the reproduction pattern is rather complicated in *P. serratus* as several reproduction periods can succeed depending on the size and age of the individuals (Foster 1951; Campillo 1979).

Table 4 shows the length of shrimps' cephalothorax including the rostrum (LCt) by considering female and male shrimps separately and together.

It appeared that the mean size of female individuals was always larger than that of male ones, except at BLE in winter. Thus, overall, the study period, female shrimps were significantly larger than male ones at GRE (Mann–Whitney test, $Z = 12.60$; $p < 0.001$) and BLE ($Z = 11.35$; $p < 0.001$). Such a difference in size between genus had already been reported by different authors (e.g. Campillo 1979; Huxley 2011; Emmerson et al. 2017) and also in *P. elegans* (Bilgin et al. 2009). According

Table 4 Mean (±SD), minimum and maximum cephalothorax length (Ct L) of female and male individuals of *Palaemon serratus* collected at 4 seasons and 2 sites (GRE and BLE)

Site	Sex	Seasons (number of shrimps measured)	Ct L Mean ± SD	Minimum–maximum Ct L
GRE	All shrimps	Spring (227)	19.73 ± 5.36^{a}	8.65–33.16
		Summer (383)	24.00 ± 4.13^{b}	10.35–40.54
		Autumn (1415)	20.59 ± 4.73^{a}	7.01–40.96
		Winter (237)	23.02 ± 4.62^{c}	10.36–38.09
	Female	Spring (148)	20.76 ± 5.73^{a}	8.65–33.16
		Summer (246)	25.34 ± 3.67^{b}	14.56–40.54
		Autumn (632)	21.81 ± 5.52^{a}	7.53–40.96
		Winter (136)	24.06 ± 5.13^{c}	10.36–38.09
	Male	Spring (79)	17.80 ± 3.94^{a}	9.53–28.68
		Summer (137)	21.58 ± 3.81^{b}	10.35–29.34
		Autumn (783)	19.61 ± 3.71^{c}	7.01–33.6
		Winter (101)	21.47 ± 3.25^{b}	10.83–30.01
BLE	All shrimps	Spring (84)	21.34 ± 6.58^{a}	9.00–40.60
		Summer (629)	22.95 ± 4.26^{b}	7.15–40.42
		Autumn (964)	19.01 ± 5.19^{c}	8.17–41.56
		Winter (58)	17.23 ± 3.89^{d}	11.18–34.23
	Female	Spring (55)	21.57 ± 7.15^{a}	9.00–40.60
		Summer (420)	23.52 ± 4.26^{b}	14.30–40.42
		Autumn (495)	20.46 ± 5.78^{a}	8.87–37.95
		Winter (62)	17.15 ± 4.57^{c}	11.70–34.23
	Male	Spring (29)	21.02 ± 5.46^{a}	12.91–32.76
		Summer (209)	21.83 ± 4.05^{a}	7.15–33.60
		Autumn (469)	17.49 ± 3.95^{b}	8.17–41.56
		Winter (26)	17.33 ± 2.93^{b}	11.18–25.48

Groups that share a common letter did not significantly differ (Kruskal–Wallis tests followed by Dunn tests to Campillo (1979), except for the age class 0 (individuals of 10–11 months), mean LCt of female shrimps sampled in Roscoff (Brittany, France) was always higher than that of male individuals: class 0 (10–11 months): 23.5 versus 24.2 mm; class 1 (12–14 months): 30.6 versus 30.0 mm; class 2 (22–26 months): 38.6 versus 36 mm and class 3 (34–38 months): 48 versus 40.6. Moreover, female shrimps could live one year more: class 4 (3–4 years): mean LCt of 50.5 mm.

When considering the temporal evolution of the mean size, there were significant differences between the four seasons and also between genus (Table 4). At GRE, the highest mean LCt for both female and male shrimps was observed in summer (mid-June) when the largest individuals ready to reproduce had moved from the lowest hypsometric levels to the shoreline. The decrease of LCt recorded in autumn (October) was due to the recruitment of small individuals after the reproduction of the most precocious females as suggested by the appearance of individuals as little as 7.01 mm or 7.53 (respectively, a male and a female). These low values of LCt were lower than that of age class of 0 (10–11 months) as described by Campillo (1979), and thus corresponded to new hatched individuals. The increase of mean LCt from autumn to winter (January) was probably linked with shrimps' growth and still the presence of quite large individuals in the intertidal area. Finally, the lowest mean value of LCt was recorded for both sexes in spring (March) when the largest individuals were still offshore. At BLE similarly to GRE, a decrease of LCt and appearance of small individuals was recorded between summer and autumn in relation to the recruitment. By contrast, the differences were (i) a higher mean LCt in spring suggesting a less extent of migration towards the lowest hypsometric levels (in relation with the very large intertidal area) and/or an earlier moving up the highest levels; and (ii) the absence of increase in LCt between autumn and winter possibly due to an earlier migration of the largest individuals toward offshore.

To investigate more deeply these results, the LCt averages at each season were calculated for three comparable hypsometric levels at both GRE and BLE (Fig. 10). For that, at GRE, data of levels 1 and 2 (+ 3–3.5 m above chart datum) on one hand, and 3 and 4 (+ 2–2.5 m), on the other hand, were pooled. This also allowed to increase the numbers which were sometimes low, especially at levels 1 and 2. Nevertheless, because of very low numbers, caution was needed in the interpretation of the highest levels at GRE in summer (Fig. 10b) and at BLE in spring (Fig. 9e). Figure 9 shows that at GRE, the lower the hypsometric level, the bigger were the shrimps. Indeed, the size of *P. serratus* individuals was always significantly smaller at the highest levels (+ 3–3.5 m above chart datum) compared to the two other levels (from 2.5 m) (ANOVAs, $p < 0.001$), except in winter when no significant difference was revealed (ANOVA, $p = 0.12$). These results confirmed the reproduction occurring at the highest levels, and suggest a relative mixture of individuals during migration in winter and a migration to a lesser extent for smallest individuals (that were well represented at + 3–3.5 m in spring). At BLE, the spatiotemporal differences in the shrimps' size showed a different pattern. Indeed, except in autumn when *P. serratus* recruitment took place at the highest hypsometric level with thus small individuals in abundance at this level (ANOVA, $p < 0.001$), no differences in LCt were significant between the three hypsometric levels in the three other seasons (ANOVAs, $0.07 < p < 0.71$).

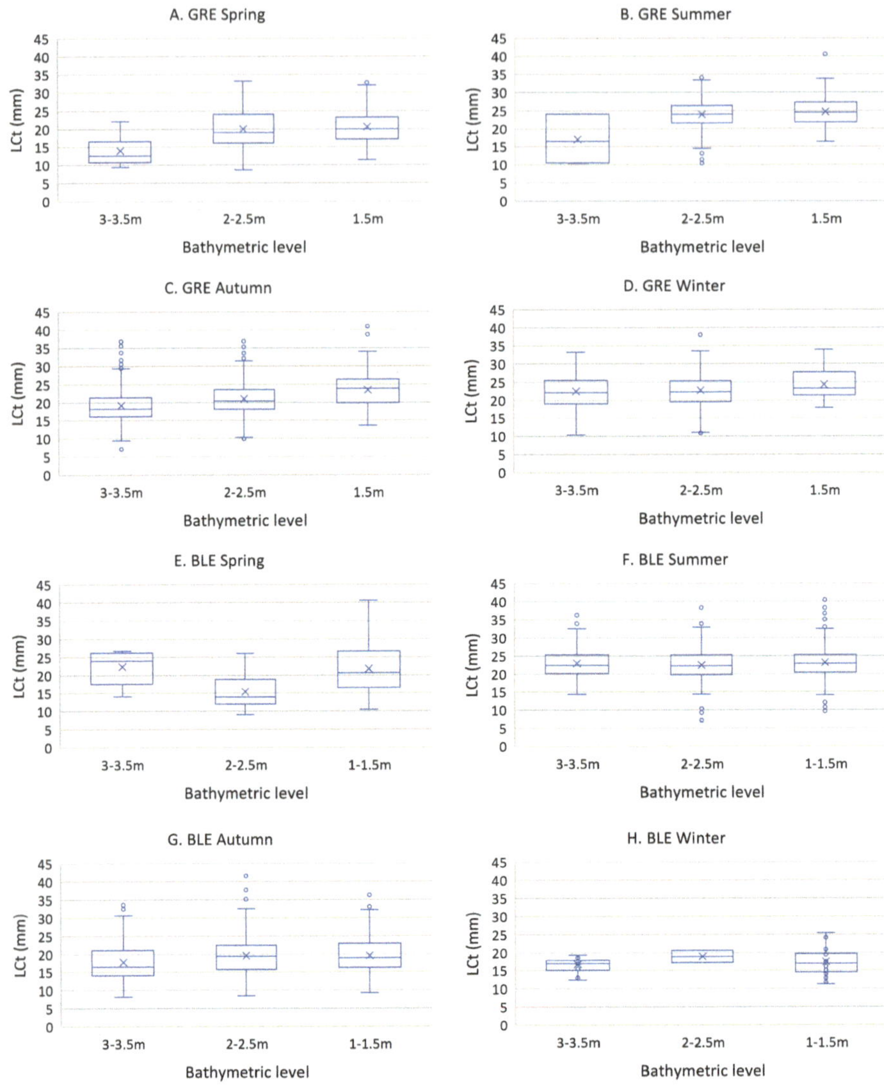

Fig. 10 Box plots illustrating the cephalothorax length (CtL in mm) of individuals of *Palaemon serratus* (males and females cofounded) collected at three hypsometric levels (+ 3–3.5 m above chart datum; + 2–2.5 m; + 1.5 m at GRE and 1–1.5 m at BLE) in 4 seasons and 2 sites: GRE (**a**, **b**, **c** and **d**) and BLE (**e**, **f**, **g** and **h**)

4 Conclusion

Our results showed that *Palaemon serratus* and *P. elegans* were the only Palaemonid shrimps collected in the intertidal areas of both Normandy sites along the French side of the English Channel. Their spatial distribution (inter-transects differences) depended on particularities at small scale such as the presence of algae, crevices and microrelief in rocky substrates allowing the shrimps to hide.

The general spatial distribution of the two species differed as *P. elegans* was most frequent and/or abundant at the highest hypsometric levels whereas *P. serratus* is more prevalent at the lowest levels of the intertidal zone. This result was in lines with other studies.

At the two study sites, *P. elegans* showed quite slight seasonal variations in number which however the largest abundances in winter and thus no obvious migration pattern. By contrast, *P. serratus* showed a marked migration pattern in relation notably to reproduction that occurred at the highest hypsometric levels.

Population number and spatial distribution of each species were determined by reproduction and recruitment and for *P. serratus* as well as shoreline-offshore migrations.

The two species had quite similar biological and ecological features but at Grandcamp-Maisy, *P. serratus* showed a wider "ecological niche" than *P. elegans* that seemed favored at Blainville-sur-Mer where the foreshore is particularly large and flat. From the quite low values of "niche overlap", it could be deducted that the interspecies competition was limited, except in autumn when *P. serratus* shrimps were numerous at the highest hypsometric levels.

Considering the smaller size of *P. elegans* and regulation allowing the fishing of only individuals $\geq$ 50 mm and the professional fishing occurring only in the subtidal zone, this species is probably little impacted by fishing. However, *P. elegans* inhabits the closest areas from shoreline, and is submitted to various environmental constraints including temperature and salinity changes and possibly contamination from runoffs and watersheds. By contrast, *P. serratus* is probably more impacted by professional and above all recreational fishing in our region. Therefore, more data from field study are needed for the sustainable management of the English Channel populations in the future.

Acknowledgements Our warmest thanks go to all the students who contributed so effectively to the fieldwork. We would also like to thank Dr. Patrick Prouzet for his proofreading of the first draft of our manuscript and for his very useful suggestions.

This work was supported by the EMFF Gedubouq program of the European Union. Project «Gedubouq»: Analyse préalable à une gestion durable des pêcheries de bouquet en Normandie.

References

Başçınar NS, Düzgüneş E, Başçınar N, Sağlam HE (2002) A Preliminary study on reproductive biology of *Palaemon elegans* Rathke, 1837 along the South-eastern Black Sea coast. Turkish J Fish Aquat Sci 2:109–116

Bilgin S, Ozen O, Samsun O (2009) Sexual seasonal growth variation and reproduction biology of the rock pool prawn, *Palaemon elegans* (Decapoda: Palaemonidae) in the southern Black Sea. Sci Mar 73:239–247. https://doi.org/10.3989/scimar.2009.73n2239

Cachot J, Geffard O, Augagneur S, Lacroix S, Le Menach K, Peluhet L, Couteau J, Denier X, Devier MH, Pottier D, Budzinski H (2006) Evidence of genotoxicity related to high PAH content of sediments in the upper part of the Seine estuary (Normandy, France). Aqua Toxic 79:257–267. https://doi.org/10.1016/j.aquatox.2006.06.014

Caldart VM, Iop S, Nunes Bertaso TR, Zanini Cechin S (2012) Feeding ecology of *Crossodactylus schmidti* (Anura: Hylodidae) in southern Brazil Zool Stud 51:484–493

Campillo A (1979) Contribution à l'étude de la crevette rose *Palaemon serratus* (Pennant): exploitation, biologie, élevage, épidémiologie. Thèse des Sciences et Techniques du Languedoc. Nantes, 247pp

Chiffoleau JF, Claisse D, Cossa D, Ficht A, Gonzalez JL, Guyot T, Michel P, Miramand P, Oger C, Petit F (2007) Metal contamination. The Seine-Aval scientific programme. Editions Ifremer and QUAE, 39pp

Dauvin JC (2012) Are the eastern and western basins of the English channel two separate ecosystems? Mar Pollut Bull 64:463–471. https://doi.org/10.1016/j.marpolbul.2011.12.010

DORIS (Données d'Observations pour la Reconnaissance et l'Identification del a faune et de la flore Subaquatiques) (2023) Petite crevette rose: *Palaemon elegans*. https://doris.ffessm.fr/Especes/Palaemon-elegans-Petite-crevette-rose-337. Accessed in Mar 2024

Dupret L, Poncet J, Lautridou JP, Hommeril P (1989) Notice explicative, Carte géol. France (1/50000), feuille Coutances (142). Bureau de recherches géologiques et minières, Orléans, 58pp

Emmerson JA, Haig JA, Robson G, Hinz H, Le Vay L, Kaiser MJ (2017) Size-selective fishing of *Palaemon serratus* (Decapoda, Palaemonidae) in Wales, UK: implications of sexual dimorphism and reproductive biology for fisheries management and conservation. J Mar Biol Assoc UK, 1223–1232. https://doi.org/10.1017/S0025315416000722

Fily G, Coutard JP, Rioult M, Auffret JP, Larsonneur C, De La Querrière P (1989) Notice explicative, Carte géol. France (1/50000), feuille Grandcamp-Maisy (95). Bureau de recherches géologiques et minières, Orléans, 55pp

Foster GR (1951) The biology of the common prawn, *Leander serratus* Pennant. J Mar Biol Assoc UK 30:333–360

Hayward PJ, Ryland JS (2017) Handbook of the marine fauna of north-west Europe. Oxford University Press, Oxford, UK, 785pp

Huxley R (2011) Population structure and morphology of the prawn *Palaemon serratus* (Pennant, 1777) in Welsh coastal waters with a consideration of two options for regulating the fishery. Ph.D. thesis. Inst Biol Environ Rural Sci, 283pp

INPN MNHN (2024a) Bouquet commun. https://inpn.mnhn.fr/espece/cd_nom/18410/tab/carte. Accessed in Mar 2024

INPN MNHN (2024b) Bouquet flaque. Petite crevette rose. Bouquet chevrette. https://inpn.mnhn.fr/espece/cd_nom/18412/tab/carte. Accessed in Mar 2024

Janas U, Spicer JI (2008) Does the effect of low temperature on osmoregulation by the prawn *Palaemon elegans* Rathke, 1837 explain winter migration offshore? Mar Biol 153:937–943. https://doi.org/10.1007/s00227-007-0865-z

JNCC (Joint Nature Conservation Committee) (2015) LR.FLR.Rkp.FK.Sar *Sargassum muticum* in eulittoral rockpools. https://mhc.jncc.gov.uk/biotopes/jnccmncr00000764. Accessed in Mar 2024

La Rivière M, Delavenne J, Janson AL, Andres S, de Bettignies T, Blanchet H, Decaris FX, Derrien R, Derrien-Courtel S, Grall J, Houbin C, Latry L, Le Gal A, Lutrand A, Menot L, Percevault L, Tauran A, Thiébaut E (2022) Fiches descriptives des habitats marins benthiques de la Manche. de la Mer du Nord et de l'Atlantique. PatriNat (OFB-CNRS-MNHN). Paris, 578p. https://inpn.mnhn.fr/docs-web/docs/download/399329

Levins R (1968) Evolution in changing environments. Princeton Univ Press (Princeton),120pp

MarLIN (Marine Biological Association) (2023a) The Marine life information network (information on the biology of species and the ecology of habitats found around the coasts and seas of the British Isles): rock prawn (*Palaemon elegans*). https://www.marlin.ac.uk/species/detail/2035. Accessed in Mar 2024

MarLIN (Marine Biological Association) (2023b) The Marine life information network (information on the biology of species and the ecology of habitats found around the coasts and seas of the British Isles): common prawn (*Palaemon serratus*). https://www.marlin.ac.uk/species/detail/2033. Accessed in Mar 2024

Perina A, Mari-Mena N, Torrecilla Z, Gonzalez-Tizon AM, Gonzalez-Castellano I, Gonzales-Ortegon E, Martinez-Lage A (2018) Assessment of genetic diversity and population structure of the common littoral shrimp *Palaemon serratus* (Pennant, 1777) by microsatellites: towards a sustainable management. Aquatic Cons Mar Fresh Ecosys 29:528–536. https://doi.org/10.1002/aqc.3011

Pianka E (1973) The structure of lizard communities. Annu Rev Syst 4:53–74

Current Status, Causes and Countermeasures for the Change from Seaweed Beds to Barren Grounds in Japan Based on the Reports of the Project Promoting the Multifunctional Roles of Fisheries

T. Komatsu

1 Introduction

Seaweed beds are photosynthetic ecosystem engineers and provide habitats for many marine organisms. For example, they are habitats for attached animals, sea urchins, abalone and turban shells, as well as spawning grounds for bigfin reef squid (*Sepioteuthis lessoniana*), flying fish species, Japanese halfbeak (*Hyporhamphus sajori*) and nursery grounds for rockfish (e.g. *Sebastes inermis*). Seaweed beds exert a number of effects on the physical environment (e.g. Komatsu et al. 1982; Komatsu and Murakami 1994). Seaweeds carry out photosynthesis, absorb carbon dioxide, accordingly change pH of seawater, generate oxygen, absorb nutrients and purify seawater (e.g. Komatsu and Kawai 1986). In recent years, attention has also focused on the roles that seaweeds play as blue carbon, producing persistent organic matter and transporting organic matter to the deep sea (Kuwae et al. 2016).

Nowadays, declines in seaweed beds have been reported in various parts of the world (Filbee-Dexter and Wernber 2018; Martins et al. 2019). In the Mediterranean Sea, the decline of seaweed beds has been reported due to feeding damage to seaweed beds by fish invading from the Suez Canal (Boudouresque et al. 2023). In Japan, the decline of seaweed beds is called '*îsoyaké*' of which "îso" and "yaké" mean coastal rocky beds and fire, respectively. That is to say, canopy-forming seaweeds on the rocky and boulder beds disappear like forests and bushes burnt by fire. Fujita (2002) defined '*îsoyaké*' as a phenomenon in which seaweed beds in rocky and boulder beds in shallow waters decline or disappear significantly beyond the range of seasonal

T. Komatsu (✉)
Japan Fisheries Resource Conservation Association, Chuo-ku, Tokyo, Japan
e-mail: komatsu@fish-jfrca.jp

A.-C. Bennis et al. (eds.), *Constraints and Adaptations to Global Change at the Land-Sea Interface: For a Shared Ecological and Energy Transition*,
https://doi.org/10.1007/978-3-031-90050-1_7

and interannual changes, resulting in a poor vegetation condition. The word *isoyake* is used in this meaning hereafter. Japan Fisheries Agency has started the Project Promoting the Multifunctional Roles of Fisheries since 2013 to maintain aquatic ecosystems healthy and support activities of fishermen's groups to restore seaweed beds (Japan Fisheries Agency 2023). This article summarises the actual situation, causes and countermeasures for *isoyake* in Japan, based on reports of activities in the Fisheries Agency's Project Promoting the Multifunctional Roles of Fisheries. In addition, the results of interviews with officials, researchers and fishermen in several prefectures in Japan will be used as a reference.

2 Materials and Methods

2.1 Distribution of Seaweeds in Japan

The main factor defining the distribution of seaweeds is coastal water temperature, which is strongly influenced by the ocean current system.

In the long north–south Japanese archipelago, a cold current of the Oyashio Current affects the area from eastern Hokkaido to Sanriku coast facing the Pacific Ocean, a warm current of the Kuroshio Current affects the Pacific coast south of Tohoku region in Honshu Island and the Tsushima Warm Current affects the coast from the northwest of Kyushu Island to western Hokkaido Island through Honshu Island facing the Sea of Japan (Fig. 1). The effect of Tsushima Warm Current is limited in the northern part of the Sea of Japan, with the northern west coast of the Tohoku Region and the west coast of Hokkaido receiving only a weak influence of the Tsushima Warm Current in summer and autumn. As a result, the coast influenced by the Tsushima Warm Current is dominated by *Sargassum* species, the coast influenced by the Kuroshio Current is dominated by *Ecklonia cava* Kjellman and *Eisenia bicyclis* (Kjellman) Setchell, and the coast around Hokkaido and the northeastern Tohoku region is dominated by *Laminaria* and *Saccharina* species such as *Saccharina japonica* (Areschoug) C. E. Lane, C. Mayes, Druehl et G. W. Saunders (Fig. 2). However, *Sargassum* species are distributed along all coasts. *Ec. Cava* is distributed on the east coast of Honshu Island from Boso Peninsula to southern Sanriku, and *Ec. cava* and *Ei. bicyclis* are also distributed west of Niigata on the coast facing the Sea of Japan (Fig. 2).

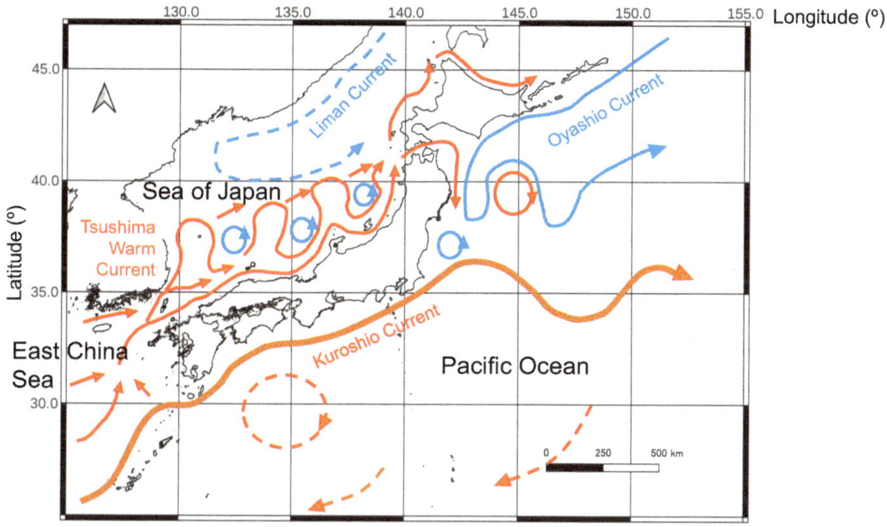

Fig. 1 Current systems around Japanese Archipelago. Warm and cold currents are drawn with red and blue colours, respectively. Names of ocean or sea are indicated with black characters

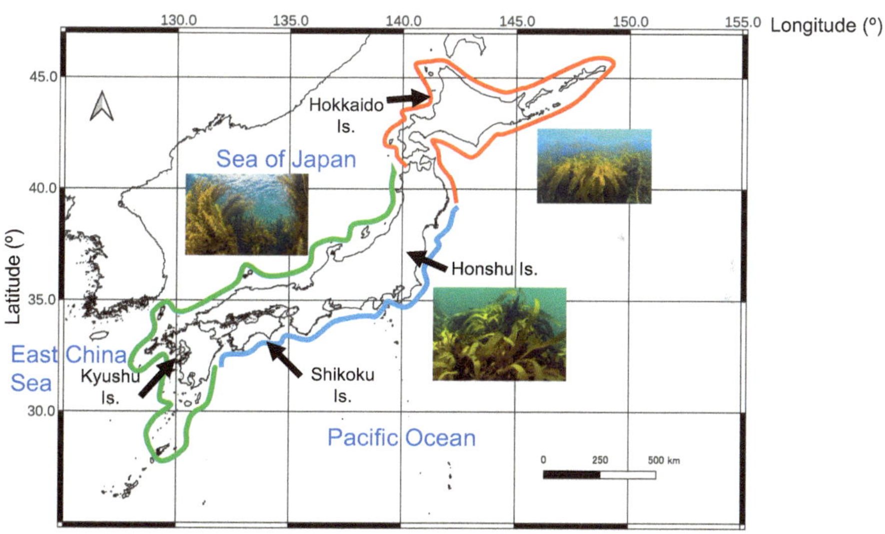

Fig. 2 Distribution of main canopy-forming seaweeds around Japanese Archipelago with red, blue and green lines corresponding *Laminaria* and *Saccharina* species, *Eisenia* and *Ecklonia* species and *Sargassum* species, respectively

2.2 *Extraction of Restoration Activities of Seaweed Beds from* Îsoyaké *Under the Fisheries Agency's Project Promoting the Multifunctional Roles of Fisheries*

Restoration activities of seaweed beds from *îsoyaké* were extracted based on reports from organisations that are working under the Fisheries Agency's Project Promoting the Multifunctional Roles of Fisheries.

2.2.1 Brief Description of the Project

Fisheries and fishing villages have long played the roles of providing various multifunctional roles to local people and other people, including the stable supply of safe and fresh marine food products to the public, the preservation of people's lives and property through border surveillance and maritime rescue, and the provision of places for health, recreation, exchange and education (Japan Fisheries Agency 2023). However, as the problems facing fisheries and fishing villages become more serious due to the ageing of the fishing industry and the decline in the fishing village population, these multifunctional roles are being hampered. The multifunctional roles of fisheries and fishing villages in the future will bring wide-ranging benefits to the entire nation. Thus, it is necessary to promote activities by fishermen and others to enhance and keep the multifunctional roles of fisheries (Japan Fisheries Agency 2023). Since 2013, Japan Fisheries Agency has provided grants for activities that contribute to the fulfilment of the multifunctional roles of fisheries and fishing villages carried out by groups of fishermen and others.

2.2.2 Support Menu of the Project

The support menu of the project can be divided into two main categories: one is environmental and ecosystem conservation, and the other is marine security (Japan Fisheries Agency 2023). The activities must involve efforts to understand and promote multifunctional roles of fisheries and fishing villages. Education and learning are essential to understand and promote multifunctional roles of fisheries and fishing villages. The first category includes the conservation of seaweed/seagrass beds, coral reefs and the release of fish and shellfish, while activities for the conservation of waterfronts include the conservation of tidal flats, reed zone conservation, cleaning or disposal of drifting and stranded matters and sediment, and the maintenance and conservation of inland water ecosystems. The other category covers ensuring maritime safety covers monitoring of borders and waters, strengthening of maritime surveillance networks, and maritime rescue training.

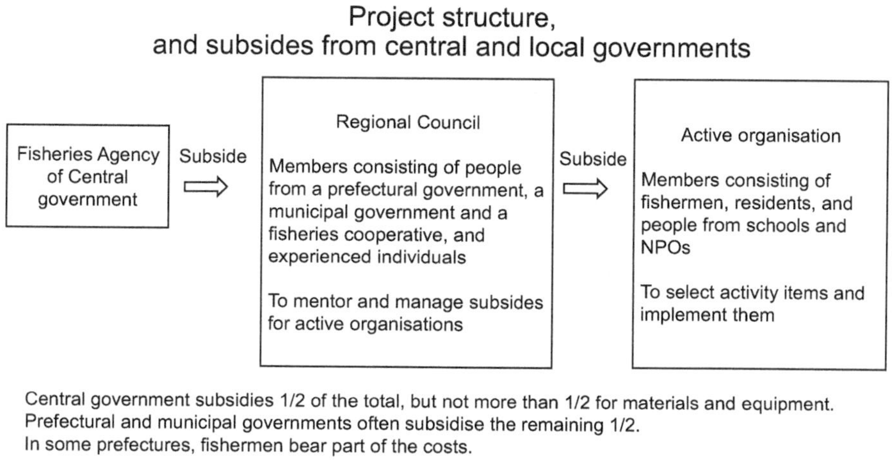

Fig. 3 Structure of project promoting multifunctional roles of fisheries (Japan Fisheries Agency 2023)

2.2.3 Structure of the Project

The project structure is shown in Fig. 3.

Japan Fisheries Agency grants subsidies to the regional councils in each province, which in turn subsidise the organisations under their jurisdiction. Japan Fisheries Agency subsidises 1/2 of the costs required by each organisation for its activities (Japan Fisheries Agency 2023). However, the cost of materials and equipment is limited to 1/2 of the total cost. In most cases, the remaining 1/2 of the costs are covered by the prefecture and municipality. Otherwise, in addition to the prefectural and municipal subsidies, the fisheries cooperatives bear a smaller share of the costs.

2.2.4 The Third Phase of the Project Promoting the Multifunctional Roles of Fisheries

The third phase of the Project Promoting the Multifunctional Roles of Fisheries was launched in 2021 fiscal year and ends in 2025 fiscal year. Records of these activities are uploaded to the web on the project as information on nationwide initiatives by the Fisheries Policy Department of the Japan National Federation of Fisheries Co-operative Associations (2024). The information is classified into the following categories: conservation of seaweed/seagrass beds, conservation of tidal flats, conservation of reed beds, conservation of coral reefs, conservation of rivers and lakes, beach cleaning, release of fish or shellfish seedlings, utilisation of waste related to marine product processing, monitoring activities and marine accident rescue training. In the report on the conservation of seaweed beds, areas where seaweed beds have been reported to have been *îsoyaké* and the causes of such *îsoyaké* were extracted and captured as data on the geographical information system. The geographical information system used was QGIS 2.3.4.

2.3 Interview Research

Researchers of fisheries experimental stations in Hokkaido, Aomori, Miyagi, Kanagawa, Shizuoka, Ishikawa, Toyama, Tottori and Nagasaki prefectures, as well as administrative officials, were interviewed in person from 2020 to 2024 to ask about the situation, causes and countermeasures for *îsoyaké*. From 2022 to 2024, the author also met the heads or councillors of fisheries cooperatives in Hokkaido, Nagasaki, Tottori, Yamaguchi, Oita and Mie asked them about the same issues and owners of diving shops organising volunteer divers to remove sea urchins by scuba diving. The reason for the five year period of the survey was that there was a period of suspension of interviewing activities due to the Japanese Government's request for the public to refrain from taking action as a result of COVID-19. In particular, the self-restraint was strictly observed in rural areas, making it difficult for people from outside the region to conduct interviews with fisheries cooperatives.

2.4 Information on *Îsoyaké* and *Îsoyaké* Countermeasures

Information on *îsoyaké* and *îsoyaké* control was collected based on information on the internet. Regarding the current situation, 'Seagrass/seaweed bed vision' was compiled by each prefecture at the request of the Japan Fisheries Agency and was used where applicable information was available. The study collected information on local activities on *îsoyaké* presented at the National Conference for the Control of *Îsoyaké*, which was held as part of the Japan Fisheries Agency's Project Promoting the Multifunctional Roles of Fisheries (Japan Fisheries Agency 2021).

2.5 Temporal Change in the Number of Entities Engaged in Shellfish and Seaweed Harvesting Fisheries

Japan Fisheries Agency conducts a census of fisheries in Japan every five years (Ministry of Agriculture, Forestry and Fisheries of Japan 2024). Data for 2003, 2008, 2013 and 2018, which can be obtained on the web, were therefore used. A fisheries census was also conducted in 2023, but the data have not yet been compiled and are not available on the web. The fisheries census compiles the number of shellfish and seaweed harvesting fishery entities per municipality. This fishery is conducted by one or two fishermen boarding on the small boat with an outboard motor, which

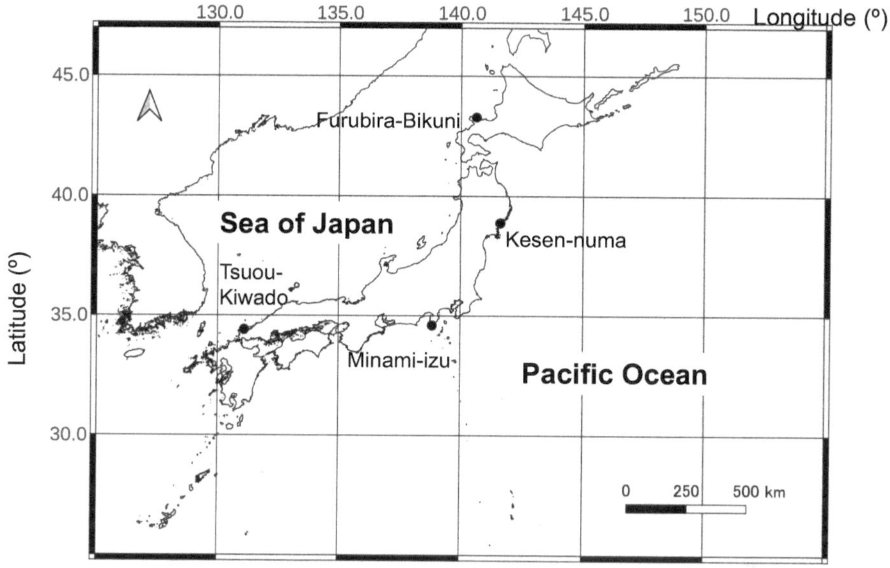

Fig. 4 Locations indicated with closed circles of communities of which the number of entities engaged in shellfish and seaweed harvesting fisheries has been investigated by the Ministry of Agriculture, Forestry and Fisheries of Japan since 2003 to 2018 at five years intervals as Japanese national fisheries census

permits the boat to look for target species in very shallow waters. The methods used to catch shellfish include divers catching shellfish with or without scuba, fishermen catching shellfish with dip nets or hooks attached to the end of long poles and searching for shellfish with underwater glasses, and fishermen deploying traps with baits attached. Those used to harvest seaweeds involve using an L-shaped tree or similar attachments fixed to the end of a long stick with a handle to wrap around the seaweeds and hooking the seaweed with a hook at the end of a rope from the boat. Four locations were selected to draw temporal changes of the entities. They were Furubira-Bikuni in Shakotan Peninsula belonging to Shiribeshi region in Hokkaido facing the Sea of Japan, Kesen-numa on the Sanriku coast facing the Pacific Ocean, Minami-izu in the top of Izu Peninsula and Tsuou-Kiwado west of Honshu Island facing the Sea of Japan (Fig. 4).

2.6 Sea Surface Temperature Information

The Japan Meteorological Agency (JMA) publishes daily mean coastal sea surface temperature data over a narrower area than the prefectural level on the web (Japan Meteorological Agency 2024). In order to confirm whether an interannual increase in sea surface temperature is occurring, daily mean sea surface temperature data

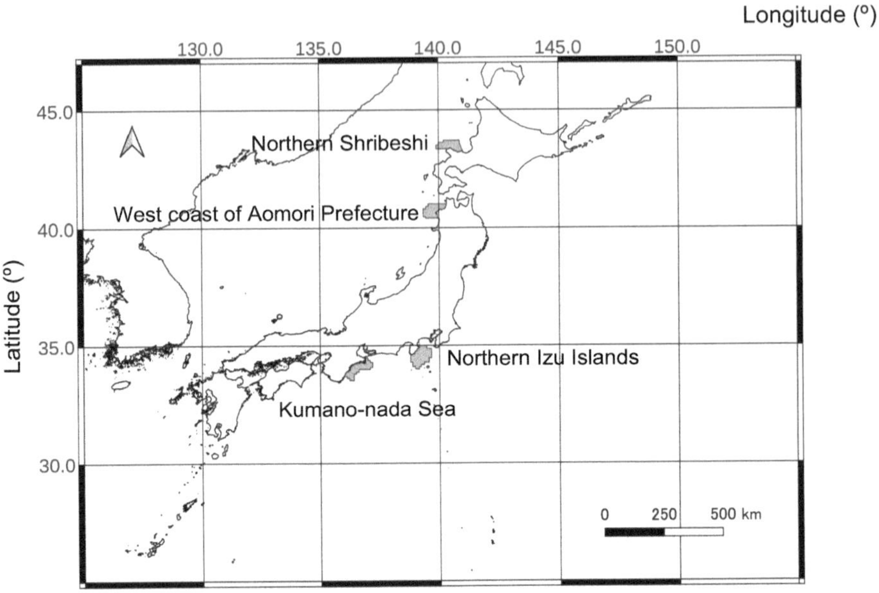

Fig. 5 Map showing the three areas of Aomori coast of the Sea of Japan, Northern Izu Archipelago and Kumano-nada Sea drawn with grey colour of which daily mean sea surface temperature data were obtained from Japan Meteorological Agency (2024)

for 1991–2020 and 2018–2022 were obtained for the coastal waters of Northern Shiribesh, the waters of West Coast of Aomori Prefecture, the coastal waters of the Northern Izu Islands and the coastal waters of the Kumano-nada Sea shown in Fig. 5.

3 Results

3.1 Isoyaké *by Sea Urchins*

In 2023, the organisations reporting activities against *îsoyaké* caused by sea urchins due to Project Promoting Multifunctional Roles of Fisheries were distributed in Hokkaido, Honshu, Shikoku and Kyushu (Fig. 6). The activities of fishermen's groups to restore *îsoyaké* by sea urchins are introduced by species.

Isoyaké caused by northern sea urchin *Strongylocentrotus nudus* (Agassiz 1863) occurred in southern Hokkaido and along the Sanriku coast in east coast of the Tohoku region (Fig. 7). *Isoyaké* by purple sea urchin *Anthocidaris crassispina* (Agassiz 1863) was found from the Noto Peninsula central of Honshu Island to Kyushu Island facing the Sea of Japan and from Miura Peninsula near Tokyo in central Honshu Island to Kyushu Island through Shikoku Island facing the Pacific Ocean (Fig. 7). In other

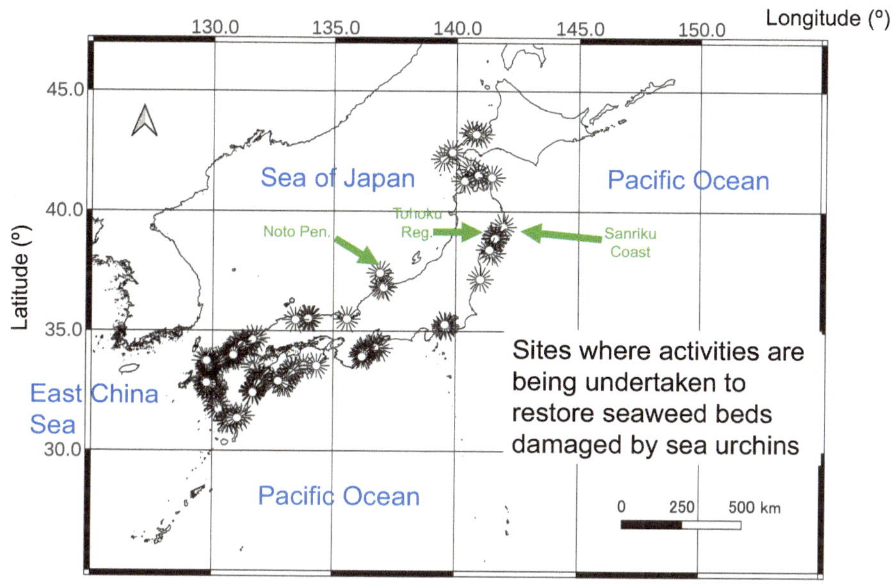

Fig. 6 Sites where groups of fishermen and others have undertaken activities to restore seaweed beds damaged by sea urchins under the project promoting the multifunctional roles of fisheries

words, *îsoyaké* by northern sea urchins occurred along the coasts with lower water temperatures, while *îsoyaké* by purple sea urchins occurred along the coasts with higher water temperatures.

Two long-spined urchins of *Diadema* species were considered to be distributed in Japan, the *Diadema setosum* (Leske 1778) and *Diadema savignyi* (Audouin 1829). However, a genetic and morphological re-examination led to the recognition of *Diadema clarki* Ikeda, 1939 as a separate species, bringing the total to three species (Chow et al. 2016). As the species of long-spined urchins is not specified in the report of the Project Promoting Multifunctional Roles of Fisheries, it was referred to as a long-spined urchin. An area of *îsoyaké* by long-spined urchins is found on the Kii Peninsula in central Honshu Island and Kyushu Island (Fig. 8).

Isoyaké by subtropical sea urchins, such as the white spin sea urchin*Tripneustes gratilla* (Linnaeus 1758), was also observed along the west coast of Kyushu Island and south-west Shikoku Island (Fig. 9). These coasts are under the influence of the Kuroshio Current and the Tsushima Warm Current, which branches off from the Kuroshio Current and flows along the west coast of Kyushu Island.

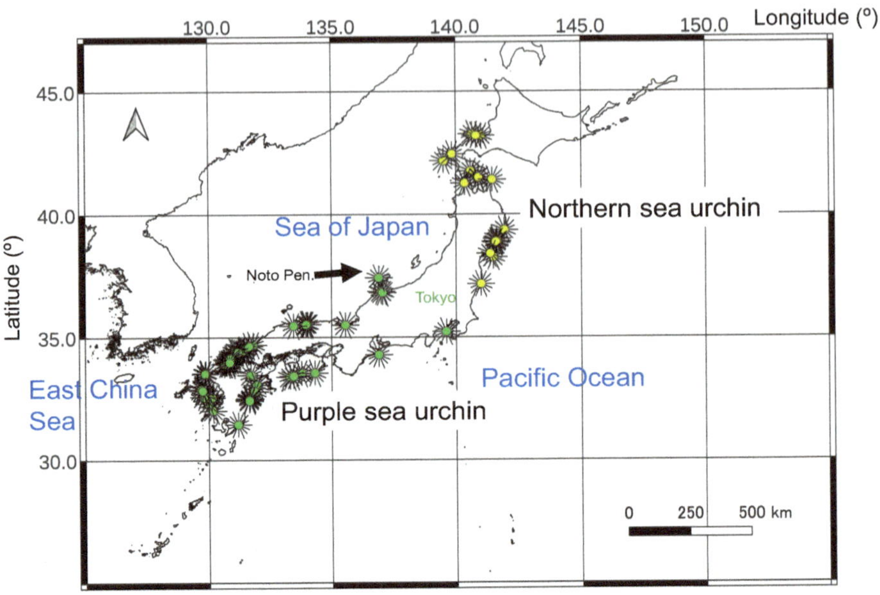

Fig. 7 Sites where groups of fishermen and others have undertaken activities to restore seaweed beds damaged by northern sea urchins (yellow) and purple sea urchins (green) under the project promoting the multifunctional roles of fisheries

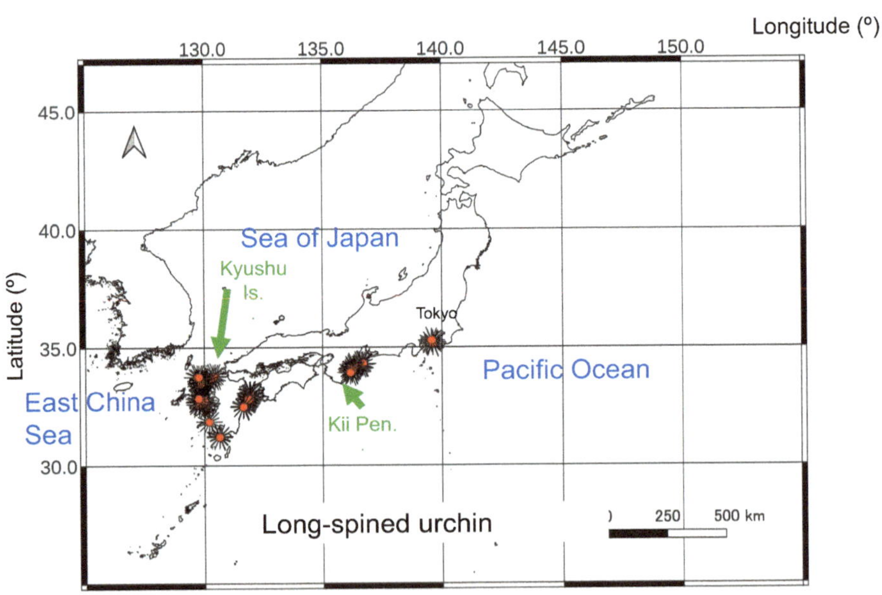

Fig. 8 Sites where groups of fishermen and others have undertaken activities to restore seaweed beds damaged by long-spined urchins (red) under the project promoting the multifunctional roles of fisheries

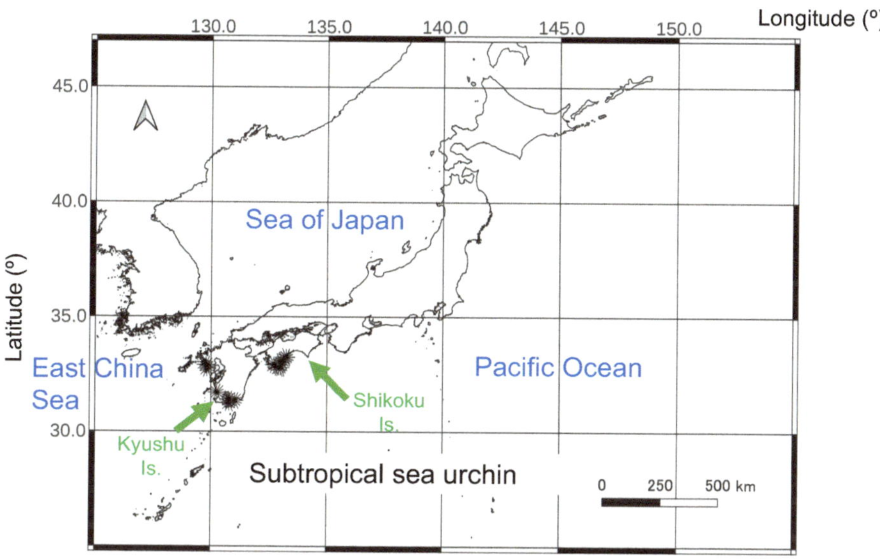

Fig. 9 Sites where groups of fishermen and others have undertaken activities to restore seaweed beds damaged by subtropical sea urchins such as *Tripneustes gratilla* (black) under the project promoting the multifunctional roles of fisheries

3.2 Isoyaké *by Herbivorous Fish*

Isoyaké by herbivorous fish was mainly distributed along the southern coasts of Kyushu Island, Shikoku Island and Honshu Island along the Kuroshio Current, and along the coasts along the Tsushima Warm Current branching off the Kuroshio Current northwards along the western coast of Kyushu Island, where herbivorous fish was found to cause *îsoyaké*.

Activities of fishermen's groups have tried to restore *îsoyaké* by herbivorous fish. Hereafter the activities are introduced by fish species. The activities against mottled spinefoot (*Siganus fuscescens*) were distributed along the coast from Miura Peninsula near Tokyo to Kyushu Island facing the Pacific Ocean (Fig. 10).

Those against Japanese parrotfish (*Calotomus japonicus*) were distributed at one site on the Izu Peninsula of Honshu Island, and in Shikoku Island and southern Kyushu Island (Fig. 11).

Four species of the genus *Kyphosus Kyphosus bigibbus* (Lacepède 1801), *K. cinerascens* (Forsskål 1775), *K. pacificus* (Sakai and Nakabo 2004) and *K. vaigiensis* (Quoy and Gaimard 1825), are abundant along the coasts of Honshu, Shikoku and Kyushu islands and on southern small islands in Japan (Honda et al. 2022; Sakai 1991). Since most of reports mentioned Japanese name of the species *K. bigibbus*, its English standard name southern drummer is used hereafter. The activities against southern drummer were restricted to the north-west and east coasts of Kyushu Island

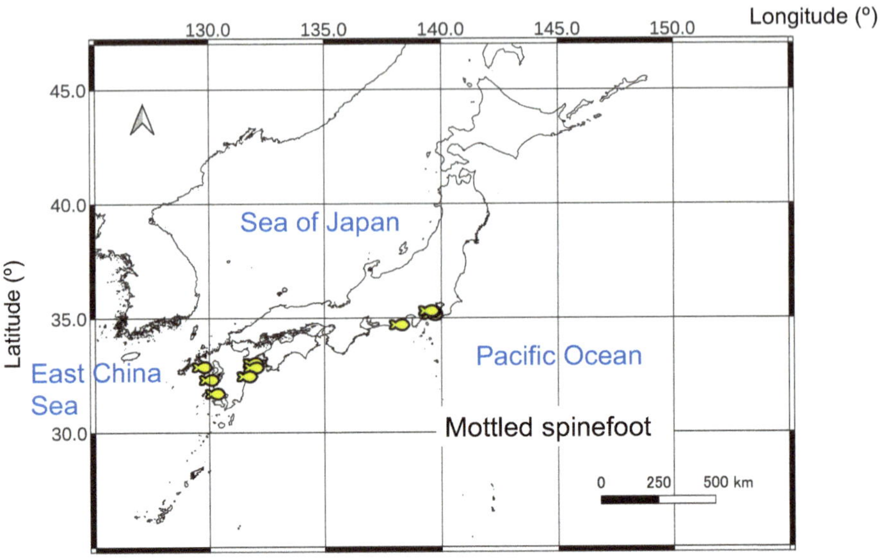

Fig. 10 Sites where groups of fishermen and others have undertaken activities to restore seaweed beds damaged by mottled spinefoot (yellow) under the project promoting the multifunctional roles of fisheries

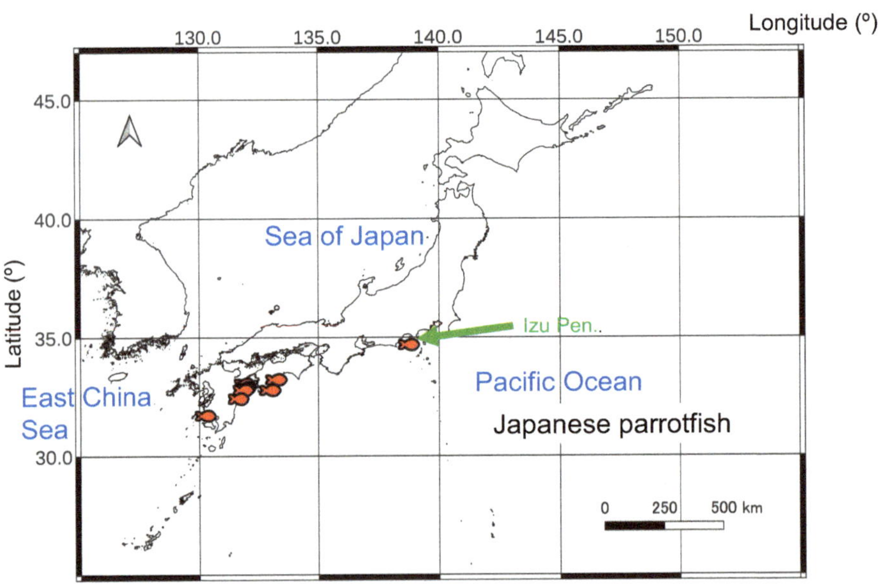

Fig. 11 Sites where groups of fishermen and others have undertaken activities to restore seaweed beds damaged by Japanese parrotfish (red) under the project promoting the multifunctional roles of fisheries

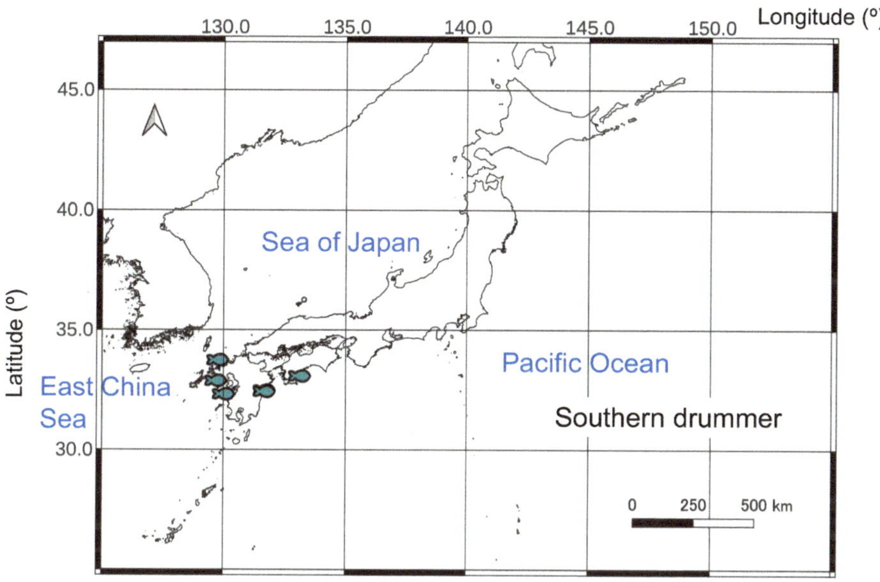

Fig. 12 Sites where groups of fishermen and others have undertaken activities to restore seaweed beds damaged by southern drummer (blue) under the project promoting the multifunctional roles of fisheries

and Shikoku Island, and its distribution was the narrowest of the three fish species (Figs. 10, 11, and 12).

3.3 Temporal Changes in the Number of Entities Engaged in Shellfish or Seaweed Harvesting Fisheries

In Furubira-Bikuni, located on the top of Shakotan Peninsula in Shribeshi region of Hokkaido Prefecture facing the Sea of Japan side of Hokkaido Island, the number of entities engaged in shellfish and seaweed harvesting by one or two fishermen, which had remained unchanged between 2003 and 2008 at 74–75 operators, consistently decreased at a rate of 3.8 entities per year in 2013 and 2018. In the same way, in Kesen-numa on the Pacific coast of the Tohoku region, the number of entities engaged in shellfish or seaweed harvesting fisheries declined from 951 in 2003 to 306 in 2008.The coastal fishing industry in Kesen-numa was also severely damaged by the huge tsunami that hit the Sanriku coast in 2011. However, the abalone or seaweed harvesting fishery recovered quickly as it does not require expensive equipment other than boats with an outboard motor. The number of fisheries had already recovered to 163 in 2013, compared to 190 in 2018. In Minami Izu, the number of fishery operators was 77 in 2003, halved to 36 in 2008, and then further halved to 19 in 2013

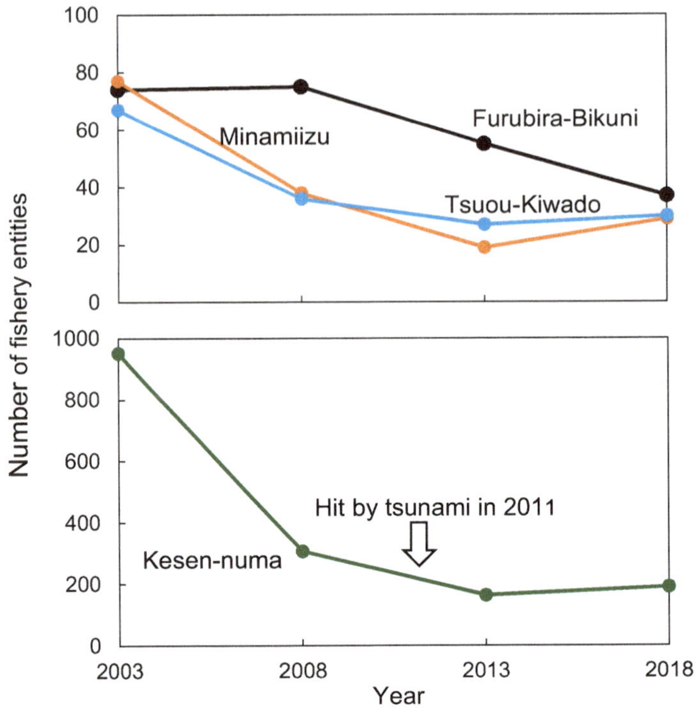

Fig. 13 Upper panel showing the number of entities engaged in shellfish or seaweed harvesting fisheries in Furubira-Bikuni (blue), Minami-izu (red) and Tsuou-Kiwado (green) and lower panel showing that in Kesen-numa for each five-year period 2003 to 2018

and 29 in 2018, a significant decrease compared to 2003. In Tsuoh-Kiwado, located in Yamaguchi prefecture in the western part of Honshu facing the Sea of Japan, the number of entities decreased from 67 in 2003 to 38 in 2008, about half the number in 2003. Thereafter, from 2013 to 2018, the number of the entities remained low and stable at approximately 30 (Fig. 13).

3.4 Interannual Changes in Daily Mean Sea Surface Temperature in the Sea Area

The daily mean sea surface temperature (SST) in the coastal waters of the Northern Izu Islands from 1991 to 2020 reached a minimum of 16.22 °C in late February and a maximum of 27.1 °C in mid-August.

In contrast, the daily mean SST from 2018 to 2022 reached a low of 17.05 °C in late February and a high of 28.16 °C in early September. An increase in water temperature of about 1 °C was observed during the minimum and maximum water temperature periods.

The daily mean SST in the coastal waters of the Kumano-nada Sea between 1991 and 2020 reached a minimum of 16.49 °C in late February/early March and a maximum of 27.53 °C in mid-August. In contrast, the daily mean sea SST from 2018 to 2022 reached a minimum of 17.04 °C in mid-March and a maximum of 28.39 °C in late August to early September. Increases in water temperature of about 1 and 1.5 °C were observed during the minimum and maximum water temperature periods.

The daily mean SST in the coastal waters of the West Coast of Aomori Prefecture from 1991 to 2020 was a minimum of 8.48 °C in mid-March and a maximum of 24.9 °C in mid-August. In contrast, the daily mean water temperature from 2018 to 2022 was a minimum of 8.81 °C in early and mid-March and a maximum of 25.69 °C in early August, with a water temperature of 25.64 °C in early September, almost identical to the maximum water temperature (Fig. 14).

The daily mean minimum and maximum water temperature values were 0.4 °C and 0.7 °C higher in 2018–2022 than in 1990–2020. Furthermore, for the daily mean SSTs in 2022 and 2023, the annual minimum and maximum water temperatures were 7.96 °C and 8.94 °C in early and mid-March, respectively, and 25.94 °C and 29.38 °C in mid and late August, respectively (Fig. 15). On the other hand, the daily mean SSTs in the coastal waters of the Northern Shiribeshi in summer from 1991 to 2020 and from 2018 to 2022 were also lower than 25 °C (Fig. 14).

4 Discussion

Three species of sea urchin, namely the northern sea urchin, the purple sea urchin and the long-spined urchin, are mainly responsible for *isoyaké*. The northern sea urchin is found from the southwest coast of Hokkaido Island to the Pacific side of the Tohoku region. It is omnivorous, but feeds mainly on seaweed (Kawamura 1964). The northern sea urchin has a strong dietary selectivity (preference). It prefers *Laminaria* and *Saccharina* species (Machiguchi et al. 1994). In experiments with *Laminaria* and *Saccharina* species, food intake varied with water temperature. It increased almost linearly with increasing water temperature from 2 to 22 °C, reaching a maximum at 22 °C and decreasing at 26 °C (Machiguchi 1997). The recent increase in water temperature along the coast affected by the Tsushima Warm Current in western Hokkaido suggests that the feeding activity of northern sea urchins is increasing. Increased feeding activity of northern sea urchins may also be due to rising coastal water temperatures along the Pacific coast of the Tohoku region.

In fishing villages along the west coast of Hokkaido, the number of entities engaged in shellfish or seaweed harvesting fisheries has declined since 2003 due to the ageing of the fishing population and a decrease in the number of new workers. In Sanriku, there has been a similar decline in the number of entities. This is illustrated by the example of Kesen-numa. Therefore, in Hokkaido and along the Sanriku coast, it is likely that northern sea urchin catches have also declined. Along the Sanriku

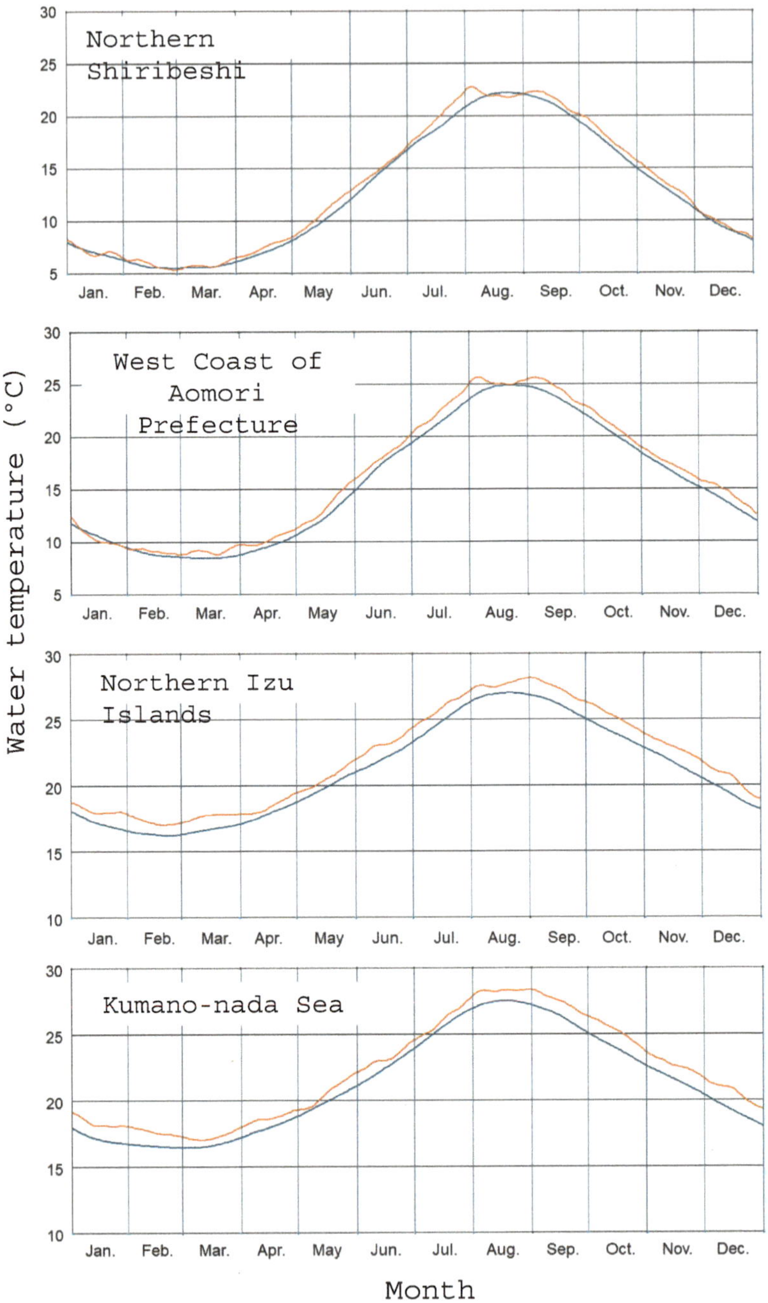

Fig. 14 Daily mean sea surface temperatures in a year (from top to bottom) in the coastal waters off the northern Shiribeshi in Hokkaido Island, of the northern Izu Isalnds, of Kumano-nada Sea and off the west coast of Aomori Prefecture from 1991 to 2020 (blue line) and from 2018 to 2022 (red line)

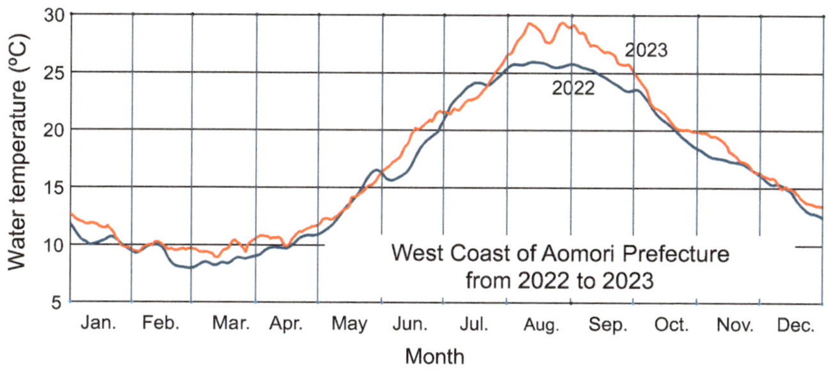

Fig. 15 Daily mean sea surface temperatures in a year in the coastal waters off the west coast of Aomori Prefecture in 2022 (blue line) and 2023 (red line)

coast, fishermen believed that sea urchins may be feeding on the corpses of Sanriku tsunami victims on the bottom of the seabed. Therefore, fishermen suspended sea urchin fishery (Masataka Kurosawa, personal communication 2011). The suspension of sea urchin fishery lasted around three years depending on individual's thought (Komatsu et al. 2019) and decreased fishing mortality of sea urchins. While fishing mortality is decreasing, the life expectancy of the northern sea urchin is about 15 years (Aomori Industrial Technology Centre 2023). Therefore, the population density of northern sea urchins is increasing and they are devastating seaweed beds due to increased predation. Kuwahara et al. (2006) report that *îsoyaké* occurs when the population density of sea urchins reaches 5 to 10 inds/m^2 or more. The percentage of the gonadal weight of the northern sea urchin where *îsoyaké* is low and such sea urchins cannot be sold as a product (Kuwahara and Miyata 2012). Fishermen therefore do not fish for northern sea urchins from areas of *îsoyaké*, and *îsoyaké* will continue (Komatsu et al. 2019). Appropriate fisheries targeting sea urchins play an important role to maintain seaweed beds by controlling a population density of sea urchins. The Project Promoting the Multifunctional Roles of Fisheries focuses on restauration of seaweed beds in areas of *îsoya*ké by controlling population density of sea urchins instead of commercial sea urchin fisheries.

Water temperatures in waters of the West Coast of Aomori Prefecture have been on an upward trend, with a minimum water temperature of 8.98 °C and a maximum water temperature of 29.38 °C in 2023. According to the Aomori Industrial Technology Centre (2023), northern sea urchins die at high water temperatures above 26–30 °C. The highest daily mean SSTs in 2018–2020 and 2023 were almost equal to or higher than 26 °C. The high-water temperatures during the summer months may have reduced the northern sea urchin population, resulting in the lack of *îsoyaké*. In the coastal waters of the Northern Shiribeshi, the daily mean SST in summer 2018–2022 was also lower than 25 °C, suggesting that *îsoyaké* caused by northern sea urchins is still present in the area due to summer water temperature lower than higher lethal temperature for the northern sea urchin.

It is known that the quantities of seaweeds fed on per day per individual increase with water temperature at 10–30 °C for the purple sea urchin and the long-spined urchin (Marine Ecology Research Institute 2012), which are persistent factors of *îsoyaké* by sea urchins in warm waters. Therefore, the influence of sea urchins on seaweed beds may have become more significant under the recent high water temperature trend. In the western coast of Honshu Island facing the Sea of Japan, the number of entities engaged in shellfish and seaweed harvesting halved from 2003 to 2008 and has remained stable at a low level since then. This has led to the occurrence of *îsoyaké* due to a decrease in fishing mortality of purple sea urchins. A decrease in fishing mortality of sea urchins would lead to an increase in population density and cause *îsoyaké*.

Long-spined urchins are not targeted by fishermen because their gonads have an astringent taste and are not suitable for food (Kaneko et al. 2009). They are only valuable as fishing bait for rockfish. Their toxic spines can cause serious injuries if they sting people. Since the lethal water temperature (on the cold side) in laboratory tests of adult long-spined urchin is 9 °C (Marine Ecology Research Institute 2012), they are not distributed in waters where winter water temperatures are below 9 °C. However, the recent increase of the water temperature promotes to expand their habitable coastal area due to the trend towards higher winter water temperatures, leading to successful reproduction and the spread of *îsoyaké* caused by long-spined urchins from central and western Honshu Island to Kyushu Island. In this way, other tropical sea urchins, rising winter water temperatures would also have led to an expansion of their distribution area and increased winter-feeding activity, causing *îsoyaké*.

The most common methods of preventing *îsoyaké* caused by sea urchins are collection from boats with box glasses and poles, and by divers. Collected sea urchins are used in some places as a fertiliser for seaweed as well as for terrestrial plants. In Hokkaido, individuals with a high gonad weight [e.g. 1 kg with shell costs about 3000 yen (Uenoya 2024)], so they are often collected from *îsoyaké* areas and released into luxuriant seaweed beds. The released individuals feed seaweeds, resulting in the further expansion of *îsoyaké* (Hiroshi Hoshikawa personal communication 2019). In Aomori Prefecture, collection is done using traps, which is less efficient but less expensive. When professional divers are hired to collect, the efficiency is very high, but the cost is high rather than traps (Dotsu et al. 2012), as hiring one diver costs several tens of thousands of yen. Since long-spined urchins haven't important commercial value except bait for rockfish, they are generally crushed in the sea with a hammer. Crushing sea urchins with a hammer damages the wrist. An air gun called a sea urchin buster may also be used (Fisheries Infrastructure Development Center 2024).

Three species of herbivorous fish cause feeding damage to seaweed, which, along with sea urchins, is one of the two main causes of *îsoyaké* (Noda 2001; Japan Fisheries Agency 2021). The control of herbivorous fish is more difficult than the control of sea urchins because they are more mobile. The *îsoyaké* areas by mottled spinefoot and the Japanese parrotfish that fishermen's groups tried to remove under the Project Promoting Multifunctional Roles of Fisheries were conducted are distributed from

Kyushu Island to central Honshu Island, whereas those by the southern drummer were distributed in Shikoku Island and Kyushu Island. According to Noda et al. (2016) who studied the feeding habits of the southern drummer, it prefers *Sargassum* species to *Eisenia* and *Ecklonia* species. The fact that *Ei. bicyclis* and *Ec. cava* beds are dominant around the Izu Peninsula may be one of the reasons why the southern drummer has not been targeted for eradication of herbivorous fish species in the activities of the Project Promoting Multifunctional Roles of Fisheries. The lower limit of winter water temperature for the emergence of the species is reported to be 11 °C for mottles spinefoot, and 14.5 °C for Japanese parrotfish and southern drummer (Japan Fisheries Agency and Japan Fisheries Research and Education Agency 2023). It is presumed that the eradication activities targeting the mottled spinefoot were carried out around the Miura Peninsula, where winter water temperatures are lower than in coastal waters north of the Izu Peninsula, whereas those targeting the southern drummer were carried out up to the Izu Peninsula.

The mottled spinefoot has poisonous spines on its dorsal, ventral and anal fins, so it must be handled with care. Its flesh has an ammonia-like odour and is not eaten in eastern Japan, but is eaten in western Japan. The flesh is chewy and is cooked as sashimi, boiled or simmered with milt, or sun-dried. In summer, the Japanese parrotfish feeds on calcareous algae, crustaceans and benthic animals, while in winter it feeds mainly on seaweed. In winter, the Japanese parrotfish has no odour, is fatty and firm, and is delicious as sashimi or in stewed dishes (https://gourmet-note.jp/posts/11630). The various ways of cooking Japanese parrotfish include boiled, souped, steamed, deep-fried, sashimi, grilled and sautéed (https://www.zukan-bouz.com/syu/%E3%83%96%E3%83%80%E3%82%A4). Southern drummer is considered to be not tasty and smells of the sea, and is unpopular as a food source, so it is rarely found in shops. However, in a very limited area such as Kushimoto in Wakayama Prefecture, southern drummer is sold in supermarkets and other local fish shops, and popularly eaten as a standard foodstuff (https://kurashi-no.jp/I0020411). Therefore, in areas where it is not customary to eat it, it is not a target for fishing and should be actively exterminated. Social movements have been developed by local people and fishery authorities to encourage the consumption of herbivorous fish in areas where it is not customary to cook herbivorous fish, in order to encourage fishing of herbivorous fish, reduce their population and restore seaweed beds (e.g. Fisheries Civil Engineering and Construction Technology Centre 2013).

In general, it is efficient to set up fishing gear at time periods when the target fish are active (Nakajima et al. 2015). Therefore, gill nets are often cast at night and lifted the next morning, targeting the feeding time in the morning and in the evening when fish are more active. For mottled spinefoot, gill nets can be set efficiently to catch adult fish that are active at night during the breeding season in early summer and those that come to feed on seaweed in the daytime during the non-breeding season (Japan Fisheries Agency 2021). Japanese parrotfish are diurnal and are less active at night. However, for Japanese parrotfish, the more efficient method is to cast early in the morning (or the day before), lift the net after noon (Nakajima et al. 2015). For this species, removal methods combining longlines are effective when seaweeds are available as their prefered food, and gill nets when no seaweeds are available

(Kadota et al. 2021). Biotelemetry studies of the Japanese parrotfish have shown that the behavioural range is as narrow as 2 km and the species is highly sedentary, so a local fisheries cooperative may be sufficient to remove them (Kadota et al. 2021). On the other hand, Yamaguchi et al. (2006), who investigated the behaviour of the mottled spinefoot and the southern drummer using biotelemetry, reported that the southern drummer was more mobile than the mottled spinefoot. Southern drummer tends to congregate on blocks of wave-dissipating structures during winter and spring (Kadota et al. 2017). This habit is efficient when fished with gill nets and traps (Kuwahara 2015). Traps using seaweeds as bait may reduce removal efficiency in areas where seaweed remains. As the behavioural range of the southern drummer is several tens of kilometres, in such areas it is desirable to conduct removal operations by gill nets without bait in collaboration with several sites (Japan Fisheries Agency 2021). In Iki City, Nagasaki Prefecture, a buy-back system was established to encourage fishermen to actively fish for southern drummer (Japan Fisheries Agency 2021), with a subsidy of 150 yen/fish for individuals caught in fixed nets and 200 yen/fish for those caught for extermination purposes. This policy has proved effective. In addition, 50 spear divers, known as "spear hunters", have been employed to eliminate southern drummer and have succeeded in restoring seaweed beds (Nagao 2024).

A circular obstacle called a sea urchin fence has been devised to prevent or slow the entry of sea urchins into seaweed beds (Nanri 2015) (Fig. 16a). Box-like structures made of wire mesh or a net surrounding seaweed bed have also been used to protect seaweed from feeding damage by herbivorous fish, allowing it to mature and supply embryo to the surrounding area (Japan Fisheries Agency 2021) (Fig. 16b). Although effective, such devices are difficult and costly to maintain, as they are washed away during typhoons and heavy storms, and require regular cleaning by diving.

Reasons for and countermeasures against *îsoyaké* in Japan are reviewed. The main reasons are the rising water temperatures and the ageing of the fishing workforce. It is difficult to resolve the first problem. In response to the second problem, volunteer divers have recently been recruited through dive shops to work with local fishermen to remove sea urchins in place of professional divers (Katsumi Suzuki personal

Fig. 16 A circular obstacle called a sea urchin fence to prevent or slow the entry of sea urchins into seaweed beds (left picture: **a**) and a box-like structure made of wire mesh or a net surrounding seaweed bed to prevent herbivorous fish from their predation (right picture: **b**) (Japan Fisheries Agency 2021)

communication 2024). The development of a catching method known as the southern drummer trap has also been effective in targeting highly migratory, herbivorous fish species, particularly this species. Furthermore, seaweed beds are recovering thanks to support from local and central governments, such as a buy-back system for herbivorous fish. In the future, it is hoped that above-mentioned efforts will be expanded throughout Japan to restore seaweed beds that have been scorched by herbivorous animals, thereby restoring coastal fisheries resources in Japan, increasing biodiversity and promoting carbon sequestration through blue carbon.

Acknowledgements The author is deeply grateful to all members of SFJO France for giving me the opportunity to present my research on *îsoyaké* at International Conference COAST CAEN 2023, the 19th Japanese-French Oceanography Symposium, at Caen, France in October 2023. Their support and encouragement, as well as the discussions at the symposium, were extremely helpful in the preparing of this article.

References

Aomori Industrial Technology Centre (2023) Sea urchin *Strongylocentros nudus*. Aomori Industrial Technology Centre, Aomori, 1pp (in Japanese). https://www.aomori-itc.or.jp/_files/00126717/19-2012mirai_kitamurasakiuni.pdf. Accessed on 12 Aug 2023

Boudouresque C-F, Blanfune A, Changeux T, Pergent G, Perret-Boudouresque M, Ruitton S, Thibaut T (2023) Marine biodiversity in the Mediterranean in the era of global warming. La Mer 61:189–231. https://www.sfjo-lamer.org/la_mer/lamer.61.3-4/lamer.61.3-4_189.pdf

Chow S, Konishi K, Mekuchi M, Tamaki Y et al (2016) DNA barcoding and morphological analyses revealed validity of *Diadema clarki* Ikeda, 1939 (Echinodermata, Echinoidea, Diadematidae). ZooKeys 585:1–16

Dotsu M, Nomura H, Ohta M, Saitoh J (2012) Preliminary comparison of catching performances of the sea urchin Strongylocentrotus nudus by divers and cage fishing. Rep Mar Ecol Res Inst 15:63–69 (in Japanese with English abstract). https://kaiseiken.or.jp/publish/reports/lib/2012_15_04.pdf

Filbee-Dexter K, Wernberg T (2018) Rise of turfs: a new battlefront for globally declining kelp forests. Bioscience 68:64–76

Fisheries Civil Engineering and Construction Technology Centre (2013) Countermeasures against isoyake by eating. Fisheries Civil Engineering and Construction Technology Centre, Tokyo, 3pp (in Japanese). https://www.fidec.or.jp/library/58d8822a490358483545e577/5d131a4b03c0839e34a38c41.pdf

Fisheries Infrastructure Development Center (2024) Sea-urchin buster. Fisheries Infrastructure Development Center, Tokyo (in Japaese). https://www.fidec.or.jp/research3. Accessed on 1 Apr 2024

Fisheries Policy Department, Japan National Federation of Fishery Cooperatives (2024) Information on national initiatives. Japan National Federation of Fishery Cooperatives, Tokyo (in Japanese). https://hitoumi.jp/torikumi/. Accessed on 1 May 2024

Fujita D (2002) Isoyake. In: Hori T, Ohono M, Horiguchi T (eds) Current status of phycology in the early 21st century. Japanese Society of Phycology, Yamagata, pp 102–105 (in Japanese). http://sourui.org/publications/phycology21/materials/file_list_21_pdf/31Isoyake.pdf

Honda K, Wada H, Seno H (2022) Diagnosis and ontogenetic changes in morphological characters and coloration in young stages of four Japanese species of *Kyphosus* (Perciformes: Kyphosidae). Ichthy Nat Hist Fish Jpn 21:13–26 (in Japanese with English abstract and figure captions)

Japan Fisheries Agency (2021) Guidelines for isoyake control, third edition. Japan Fisheries Agency, Tokyo, 250p (in Japanese)

Japan Fisheries Agency (2023) Project promoting the multifunctional roles of fisheries, 3rd edn. Fisheries Agency of Japan, Tokyo (in Japanese). https://www.jfa.maff.go.jp/j/gyoko_gyozyo/g_gideline/ https://www.jfa.maff.go.jp/j/gyoko_gyozyo/g_thema/sub391.html. Accessed on 10 Aug 2023

Japan Fisheries Agency and Japan Fisheries Research and Education Agency (2023) Marine environment in the waters around our country. Fisheries Agency of Japan and Japan Fisheries Research and Education Agency, Tokyo and Yokohama (in Japanese). https://abchan.fra.go.jp/pr/kaiyou/. Accessed on 10 Sept 2023

Japan Meteorological Agency (2024) Sea surface temperature information for coastal areas of Japan. Japan Meteorological Agency, Tokyo (in Japanese). https://www.data.jma.go.jp/kaiyou/data/db/kaikyo/series/engan/engan.html. Accessed on 1 May 2024

Kadota T, Kiyomoto S, Fukuda H, Yoshimura T (2017) Temporal patterns of aggregation formation around wave-dissipating blocks in the grey sea chub *Kyphosus bigibbus*. Nippon Suisan Gakkaishi 83:74–76 (in Japanese)

Kadota T, Kiyomoto S, Yamada H et al (2021) Aiming for the recovery of Ecklonia cava ssp. kurome beds through the removal of parrotfish—a new challenge in the Nagoya area, Oita Prefecture! In: National council for the prevention of barren ground in 2020 fiscal year. Fisheries Agency of Japan, Tokyo (in Japanese). https://www.jstage.jst.go.jp/article/pamjsfe/2015/0/2015_83/_pdf/-char/ja

Kaneko K, Shirai T, Tanaka M et al (2009) Taste characteristics of the gonad of longspine black urchin Diadema setosum. Nippon Suisan Gakkaishi 75(4):689–694 (in Japanese with English abstract and figure captions)

Kawamura K (1964) Sea urchin *Strongylocentros nudus* off Toei of Urakawa Town. J Hokkaido Fish Exp Sta 21:8–22 (in Japanese)

Komatsu T, Kawai H (1986) Diurnal changes of pH distribution and the cascading of shore water in a *Sargassum* forest. J Oceanogr 42:447–458

Komatsu T, Murakami SI (1994) Influence of a *Sargassum* forest on the spatial distribution of water flow. Fish Oceanogr 3:256–266

Komatsu T, Ariyama H, Nakahara H, Sakamoto W (1982) Spatial and temporal distributions of water temperature in a *Sargassum* forest. J Oceanogr Soc Jpn 38:63–72

Komatsu T, Sasa S, Murata H, Sawayama S et al (2019) Trophic cascade in seaweed beds in Sanriku Coast hit by the huge tsunami on 11 March 2011: sea urchin fishery as a satoumi activity serving for increase in marine productivity and biodiversity. In: Komatsu T, Ceccaldi H-J,·Yoshida J, Prouzet P (eds) Oceanography challenges to future earth: human and natural impacts on our seas. Springer International Publishing, Cham, pp 95–104

Kuwae T, Kanda J, Kubo A, Nakajima F, Ogawa H, Sohma A, Suzumura M (2016) Blue carbon in human-dominated estuarine and shallow coastal systems. Ambio 45:290–301

Kuwahara H (2015) The evaluation of current status for grazing control techniques of herbivorous fish. Fish Eng (Jpn) 51:253–257 (in Japanese with English abstract)

Kuwahara H, Miyata T (2012) Maintaining seaweed beds and sea urchin fisheries. Res J Food Agr 35(3):21–25 (in Japanese)

Kuwahara H, Watanuki A, Aota T, Yokoyama J, Fujita D (2006) Results of questionnaire on 'isoyake' in Japan. Fish Eng Jpn 43:99–107 (in Japanese with English abstract and figure captions)

Machiguchi Y (1997) Effects of seaweed preference and feeding pressure of echinoderms on isoyake. In Iizumi H (ed) Comprehensive development of technology for predicting the occurrence of isoyake. Council of Agricultural, Forestry and Fisheries Technology, Tokyo, pp 49–59 (in Japanese)

Machiguchi Y, Mizutori S, Sanbonsuga Y (1994) Food preference of sea urchin *Strongylocentros nudus* in laboratory. Bull Natl Fish Res Inst 58:35–43 (in Japanese with English abstract and figure captions)

Marine Ecology Research Institute (2012) Report of the environmental review study on thermal and nuclear power (Investigation of the effects of thermal effluent on biological communities). Marine Ecology Research Institute, Tokyo, 191p (in Japanese)

Martins GM, Harley CDG, Faria J, Vale M, Hawkins SJ, Neto AI, Arenas F (2019) Direct and indirect effects of climate change squeeze the local distribution of a habitat-forming seaweed. Mar Ecol Prog Ser 626:43–52. https://doi.org/10.3354/meps13080

Ministry of Agriculture, Forestry and Fisheries of Japan (2024) Fisheries census. Ministry of Agriculture, Forestry and Fisheries of Japan, Tokyo (in Japanese). https://www.maff.go.jp/j/tokei/census/fc/. Accessed on 1 May 2024

Nagao Y (2024) Initiatives for seaweed bed restoration in Iki City—efforts to control herbivorous fishes and their effects. In: National Council for the Prevention of Barren Ground in 2023 fiscal year. Fisheries Agency of Japan, Tokyo (in Japanese). https://www.jfa.maff.go.jp/j/seibi/attach/pdf/R5_isoyake_kyogikai-7.pdf

Nakajima Y, Kuwahara H, Nishizaki T, Watanabe K (2015) Methods to remove herbivorous fish Japanese parrotfish. In: Proceedings of Scientific Conference of Japanese Society of Fisheries Engineering in 2015. Japanese Society of Fisheries Engineering, Tokyo, pp 83–84 (in Japanese)

Nanri K (2015) Technologies and problems for control of sea urchin population. Fish Eng (Jpn) 51:213–219 (in Japanese with English abstract and figure captions)

Noda M (2001) Current status of Isoyake. In: Proceedings of scientific conference of Japanese society of fisheries engineering in 2001, Japanese Society of Fisheries Engineering, Tokyo, pp 227–228 (in Japanese)

Noda M, Kawano M, Okamoto K, Murase N (2016) Estimation of daily feeding rate and diel feeding rhythm of herbivorous fish *Kyphosus bigibbus* in captivity. J Nat Fish Univ 64(4):219–225 (in Japanese with English abstract and figure captions)

Sakai K (1991) There are four species in the genus *Kyphosus*. I.O.P. Diving News 2(8):2–5 (in Japanese)

Uenoya (2024) Raw sea urchins with shells, caught in the morning in Matsumae, Hokkaido. Uenoya, Matsumae, Hokkaido (in Japanese). https://sakura-maguro.com/pages/nama-uni. Accessed on 12 Jan 2024

Yamaguchi A, Inoue K, Furumitsu K, Kiriyama T, Yoshimura T, Koido T, Nakata H (2006) Behavior and migration of rabbitfish *Siganus fuscescens* and grey seachub *Kyphosus bigibbus* off Nomozaki, Kyushu, tracked by biotelemetry method. Nippon Suisan Gakkaishi 72:1046–1056 (in Japanese with English abstract and figure captions)

Resilience of Marine Benthic Communities in an Area Highly Impacted by Human Activities: The Case of the English Channel

J.-C. Dauvin

1 Introduction

The English Channel (EC) is a sea shared between the United Kingdom to the north and France to the south; it is a shallow epicontinental sea with a maximum depth of 174 m (Dauvin 2019). Its western basin is influenced by Atlantic waters and the presence of a summer thermocline off Cornwall and Devon (Plymouth, south-west England), while its eastern basin is mainly affected by the River Seine. The tidal regime in the EC is dominated by semidiurnal tides with a period of 12 h 25 min and tidal ranges reaching a maximum of 12 m along the French coast mainly in the Mont-Saint-Michel Bay compared with an average of 4–5 m along the English coast. Tidal currents dominate the sedimentation processes: extensive pebbly sediments are predominant in areas of strong tidal currents located offshore and to the east of gyres; coarse sands and pebbles cover > 80% of the EC surface (Larsonneur et al. 1982). On the other hand, fine sands and muddy fine sands occur in areas of weak tidal currents in bays and estuaries. Fine-grained sediments are more widespread on the English side of the EC than on the French side, and only extend over a significant area (400 km^2) in the eastern part of the Bay of Seine (Larsonneur et al. 1982).

The EC is strongly impacted by human activities (Dauvin and Lozachmeur 2006; Amara 2010; Dauvin 2012, 2019). Halpern et al. (2008) have highlighted that the North Sea combined with the EC and China Sea represent the three coastal marine zones where cumulative human impacts have the greatest influence in the worldwide Ocean. Moreover, in European waters, the eastern part of the EC is more impacted

J.-C. Dauvin (✉)
Laboratoire Morphodynamique Continentale et Côtière M2C, Normandie Univ., UNICAEN, UMR 6143 M2C, F-14000, Caen, France
e-mail: jean-claude.dauvin@unicaen.fr

A.-C. Bennis et al. (eds.), *Constraints and Adaptations to Global Change at the Land-Sea Interface: For a Shared Ecological and Energy Transition*,
https://doi.org/10.1007/978-3-031-90050-1_8

than the western part (Dauvin 2012, 2019), due to the density of traditional activities and the emergence of new activities such as granulate extraction, dredging and deposition of spoil sediment and the installation of offshore wind farms.

Global warming, ocean acidification, and cliff collapse due to coastal erosion, impacted also the ecosystems of the English Channel (Dauvin 2019), and their evolution were due to a mix of both natural and human disturbances. The more important impact of temperature increase in the English Channel is the distribution changes of fish in the English Channel. As an example, with the availability of a series of 32 years of trawl surveys in the Bay of Somme, Auber et al. (2017) showed the existence of tropicalization, i.e. the presence of species of warm temperate affinities, of the community of fish in the middle of the 1990s, when an exceptional rise in temperatures occurred in the North-East Atlantic.

In spite of the cumulative human activities as seen in the eastern part of the Bay of Seine, benthic communities show long-term stability (Bacouillard et al. 2020) and even display high resilience following a perturbation such as in the case of sediment deposition (Pezy et al. 2017a, b, 2018; Dauvin et al. 2022). In this context, resilience corresponds to the ability of a habitat, a community, or an ecosystem to recover its structure and function after a disturbance (stress) (Elliott et al. 2016; O'Leary et al. 2017; Farr et al. 2021; Dauvin 2023).

This paper aims to illustrate the resilience of ecosystems by comparison with the state of neighbouring areas observed prior to perturbations in coastal and offshore areas. For this purpose, some examples are chosen from soft-bottom benthic communities of the EC (Bay of Morlaix, Rance Basin, eastern part of the Bay of Seine and eastern part of the EC; Fig. 1).

2 Resilience of *Ampelisca* Populations in the Bay of Morlaix After the Amoco Cadiz Oil Spill

The Amoco Cadiz oil spill remains one of the most important pollution accidents in the north-eastern Atlantic and one of the most documented in the world (Anonymous 1981). The wreck occurred during the night of 16–17 March 1978 off the North-Brittany coast in the western approaches to the EC; 223000 t of crude oil were discharged into the sea in two weeks, polluting 350 km of the Brittany coastline including the seabed down to depths of 50 m (Cabioch et al. 1978). Subsequently, subtidal sediments were polluted in the Bay of Morlaix at a distance of 60 km to the east the wreck site (Beslier et al. 1980).

Acute lethal toxicity of stranded Amoco Cadiz oil had been investigated by Vandermeulen et al. (1978). Invertebrate mortalities in the intertidal zone were highly variable, some species experiencing near total mortality while others, notably burrowing annelids and nemertean's, appeared highly resistant to oiling. Moreover, hydrocarbon concentrations and aryl hydrocarbon hydroxylase activity of oiled birds and oiled fish were found to be markedly higher than those in non-oiled species.

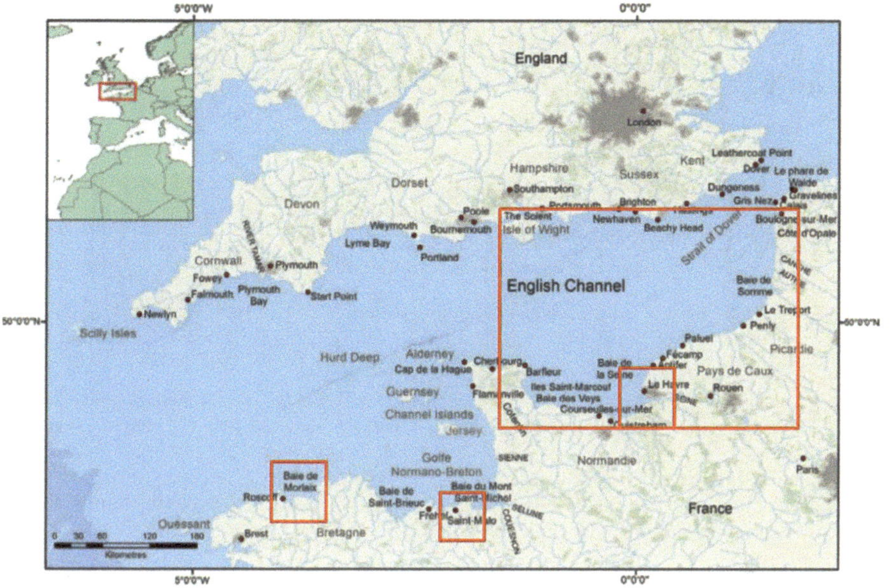

Fig. 1 Map of the English Channel showing the zones where studies have been carried out (from Dauvin 1999)

The presence of hydrocarbons was observed in sediments of the Pierre Noire fine sand community on 3 April 1978 (Cabioch et al. 1978). The seasonal changes in this benthic community were observed in April 1977 one year before the Amoco Cadiz spill. During the one-year assessment 1977–1978, three amphipod species belonging to the genus *Ampelisca* were dominant: *A. armoricana* Bellan-Santini and Dauvin 1981; *A. sarsi* Chevreux 1888, and *A. tenuicornis* Liljeborg, 1856 (up to 40,000 individuals per m^2 in September 1977), representing 90% of the abundance (Fig. 2), 38% of the biomass and 50% of the secondary production of the community (Dauvin 1998). The *Ampelisca* species are very sensitive to hydrocarbon pollution, which explains why they disappeared in the fine sand community as a consequence of the Amoco Cadiz oil spill, showing that this community was not resistant to the chemical perturbation (Dauvin 1998). A similar population collapse was also observed following the Aegean oil spill to the north of Galicia, Spain (Gomez Gesteira and Dauvin 2005).

In 1990, 12 years after the Amoco Cadiz oil spill, the *Ampelisca* abundances returned to the same order of magnitude as those observed before the incident (Fig. 2) (Dauvin 1998; Dauvin et al. 2021a). At this time, *Ampelisca sarsi* dominated the community, replacing *A. armoricana* which became the second most abundant species, while *A. tenuicornis* remained the third most abundant (Fig. 2).

The long-term monitoring of the Pierre Noire community illustrates its low level of resistance, reflected by the sudden disappearance of the *Ampelisca* population just after the arrival of hydrocarbons on the sea bottom (Fig. 2). However, this same

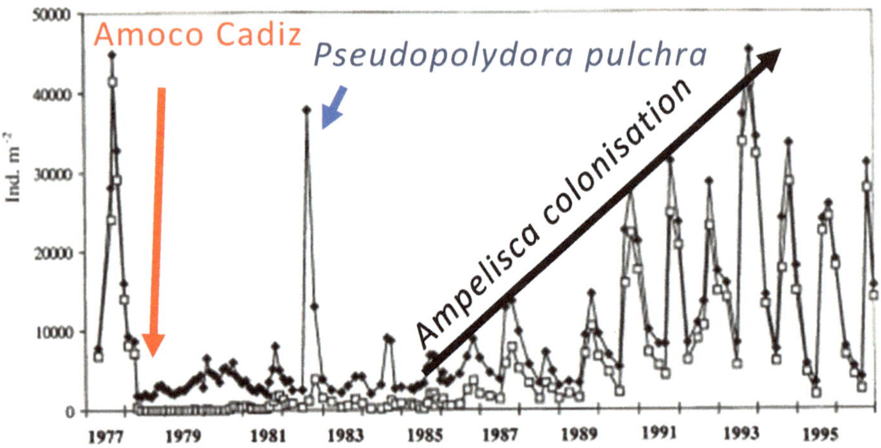

Fig. 2 Annual changes (1977–1996) in total number of individuals of all species per m^2 (black squares) and abundance of *Ampelisca* populations per m^2 (white squares) at Pierre Noire Station in the Bay of Morlaix (from Poggiale and Dauvin 2001)

community shows a high resilience with the reconstitution of the *Ampelisca* populations after a decade. These holobenthic species are without pelagic larvae, but have a high capacity to re-form their populations (Dauvin et al. 1998). In this case, there was a massive and brutal disturbance provoked by the Amoco Cadiz oil spill followed by a fairly rapid decontamination of hydrocarbons from the sediments (Beslier et al. 1980). Nonetheless, given the low fecundity of the *Ampelisca* amphipods and the distance of dispersal from their source areas to allow recolonization of the bay of Morlaix (Cabioch 1968), the reconstitution of populations was slow and gradual but finally effective.

3 Resilience of Benthic Communities in the Rance Estuary After Construction of the Tidal Power Plant

The only existing tidal power plant in France was built between 1963 and 1966 on the Rance estuary near the Bay of Mont-Saint Michel, where the tidal range exceeds 12 m during spring tides. During the period of dam construction, the estuary was isolated from the sea and the marine and estuarine communities completely disappeared, with the ecosystem becoming totally occupied by freshwater draining from the Rance river watershed. Surveys were conducted in 1966, 1971, 1975, 1995, 2010 and 2020 to study the recolonization of this habitat by macrobenthic fauna (Desroy and Retière 2010; Trigui et al. 2011; Brébant 2024). Although the number of stations and their location upstream from the dam varied over a period of 54 years (1966–2020), the available data remains sufficient to identify some general patterns in the long-term changes of the macrobenthic communities (Brébant, 2024).

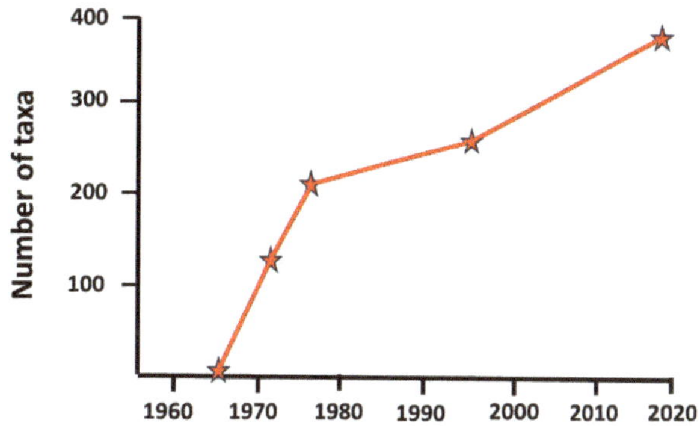

Fig. 3 Evolution of the number of taxa recorded on the soft-bottom communities of the Rance Basin from 1966 to 2025 (from Trigui et al. 2011; Brébant 2024)

A rapid recolonization of the marine fauna was observed after the opening to marine intrusion: the number of species was 200 in 1986, and then stabilized to around 240 species in 1995 (Fig. 3). By 2010, 45 years after the end of the works, the benthic communities appeared to be stabilized and their faunal compositions became similar to other estuaries in northern Brittany (Trigui et al. 2011).

However, between 1976 and 1995, an increase of fine particles was observed mainly in the upper part of the Rance Basin, resulting from a fall in the low mean water level due to human control in relation to electricity production. This led to a proliferation of the gastropod mollusc *Crepidula fornicata* (Linnaeus 1758) as well as a reduction of the taxonomic richness of the macrofauna (Trigui et al. 2011; Desroy and Retière 2010). Recently, Brébant (2024) compared the sediment data from 1994 to 2020, which show an overall stability of the surface cover. However, in the upstream part of the estuary, coarser sediments (medium to coarse sands), fine and medium sands and silty fine and medium sands have reappeared since 2010 in the central channel. In both upstream and downstream areas, the upper parts of the intertidal mudflats have become richer in particles finer than 100 μm, showing the long-term extent of the sedimentation processes. Moreover, the taxonomic richness highlights a significant change in the upstream brackish area, with a deeper penetration of marine species suggesting a marinization of this part of the Rance Basin; the total taxonomic richness reached 392 in 2020. Mud content, hydrodynamic conditions and salinity variations can explain these changes. Moreover, despite significant human activities in the Rance Basin and substantial modifications in hydrological conditions, the macrobenthic communities have maintained good to very good ecological conditions over time (Brébant 2024).

In this case, after three years of isolation of the Rance Basin from marine influence, the marine colonization was rapid and most of the marine and estuarine characteristics were restored in one decade, thus illustrating the high resilience of this system. This was followed by long-term fluctuations of the soft-bottom benthic ecosystem. The changes observed here have been related to human activities mainly due to the operation of the tidal power plant, i.e., an increase in fine sediments as well as the biomass of Non-Indigenous Species populations such as *Crepidula fornicata*. Hydrological changes have occurred in relation to Climatic Change, mainly in the upper part of the estuary where there is a marinization of the system with an increased presence of marine species.

Similarly, in the south Korea, the Sihwa Tidal Power Plant (the second tidal plant in the world) had played a big role in restoring the Lake Sihwa ecosystem and water quality through the continuing circulation of sea water. The plant is to open the existing dam in 2012 to allow the circulation and exchange of water between the Lake Sihwa and the sea and to improve not only the water quality but also introduced marine species in the freshwater originated lake (Park 2007).

4 Resilience of Benthic Communities of the Seine Estuary After 150 years of Anthropogenic Work

Owing to the importance of the two main French harbours (Le Havre and Rouen) at the national and European level since the middle of the nineteenth century, the successive major engineering projects (i.e., dams, land reclamation, management and dredging of navigational channels and harbour basins) have disturbed the natural environment of the lower part of the Seine estuary (Dauvin and Desroy 2005; Dauvin et al. 2006). As a result, the surface-area of the intertidal zone has been reduced from 130 km^2 in 1850 to less than 20 km^2 at the beginning of the years 2000.

The latest development of significant infrastructure in the lower part the Seine estuary has been the Port 2000 project, which aimed to extend the Le Havre harbour. Port 2000 has impacted water circulation in the North Channel and in the navigational channel of the estuary, and has affected the local flora and fauna of the estuarine habitats (Dauvin et al. 2006).

Dauvin and Desroy (2005) recognized two main macrobenthic communities on the intertidal flats: a *Nephtys cirrosa* community on fine and medium sand substrates dominated by amphipods in the marine part of the estuary and a *Limecola balthica* community occupying the mudflats of both the North and South Channels as well as the border of the Navigational Channel. Similarly, these latter authors have described two main subtidal communities in the lower part of the Seine estuary: (1) an *Abra alba-Pectinaria koreni* muddy sand community occupying the outer part of the estuary and the entrance to the North and South Channels and (2) a *Limecola balthica* community in the inner subtidal muddy sediments of the North, South and Navigational Channels. In spite of a reduction in the areas of the intertidal and subtidal zones due to the Havre harbour extension and the construction of the Normandy bridge, these macrobenthos communities are permanently established and continue to be important sources of preys for birds at low tide and juveniles of fish at high tide (Dauvin and Desroy 2005). Moreover, based on a series of beam trawls (1995–2019) Cariou et al. (2020) have highlighted the importance of this part of the eastern part of the bay of Seine and Seine estuary for the fish populations of the EC. Dauvin and Desroy (2005) also emphasised that most of the fish individuals harvested in the eastern part of the bay and the Seine estuary belong to the G0 age group, which corresponds to fish that settle in the estuary during the year of the survey.

Two main suprabenthic communities have been identified across the salinity gradient in the lower part of the Seine estuary (Dauvin and Desroy 2005): a marine community located in the downstream part of the estuary, with a diversified fauna dominated by amphipods and mysids *Mesopodopsis slabberi* (Van Beneden 1861), *Schistomysis* spp. and *Gastrosaccus* spp., and an estuarine assemblage essentially dominated by the mysid *Neomysis integer* (Leach 1814). These communities were not affected by the Port 2000 construction (Dauvin et al. 2006), but appear sensitive to salinity changes due to the decreased input of freshwater from the River Seine and the marinization of the lower part of the Seine estuary (Pezy et al. 2017a). The changes in suprabenthos are mainly characterized by an increase in species richness in the upper part of the North Channel and a decrease in species density and biomass of the dominant mysid species over time (Pezy et al. 2017a) (Fig. 4).

The studies cited above concluded that benthic and suprabenthic communities have persisted in the lower part of the Seine estuary in spite of successive significant works since the middle of the nineteenth century (Dauvin and Desroy 2005), and have confirmed the awareness of their importance as underlined by Elliott and McLusky (2002). Estuaries are habitats that humans have exploited and often destroyed, but remain among the most resilient habitats on earth, maintaining their attractiveness for wildlife, despite the industrialization and land reclamation.

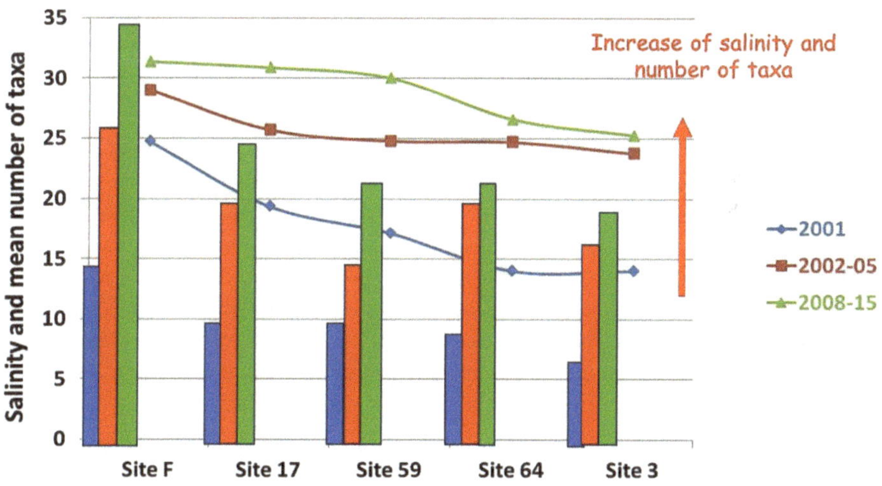

Fig. 4 Evolution of the suprabenthos taxa and salinity recorded in the North Channel of the Seine estuary from 1991 to 2012 (from Pezy et al. 2017a)

5 Resilience of Benthic Communities of the Eastern Bay of Seine After Dredge Spoil Deposition

The eastern part of the Bay of Seine and the Seine estuary are among the most studied sites on the EC coast due to the accumulation of anthropogenic activities (maritime transport, dredging and deposition of harbour sediment, aggregate extraction, and fishing). This has generated studies of the impact of such activities on benthic ecosystems and the need to protect the habitats and functionalities of natural heritage areas (Dauvin 2015; Dauvin et al. 2021b).

Based on summer sampling (41 stations in September 2008 and September 2009) Dauvin et al. (2017) have used sedimentary, hydrodynamic and morphometric data from the eastern Bay of Seine and estuary to study the spatial distribution of benthic species and assemblages in relation to environmental variables. Redundancy Analyses allow us to distinguish six assemblages according to the presence of several granulometric classes in sediments at most of the stations studied (mainly mud, sand and coarse sand) as well current speed, bathymetry and salinity. Generalized Linear Models explain between 30 and 89% of the variance within the chosen environmental factors. The dominant species with a wide distribution and high fecundity (i.e., *Owenia fusiformis* Delle Chiaje, 1844, *Aphelochaeta marioni* (Saint-Joseph 1894), *Abra alba* (W. Wood 1802), *Magelona johnstoni* Fiege, Licher & Mackie 2000, *Chaetozone gibber* Woodham and Chambers 1994 and *Ampharete baltica* Eliason 1955) were those showing the lowest percentage of explained variance, while those species located in few stations [i.e. *Thyone fusus* (O. F. Müller, 1776), *Nephtys cirrosa* Ehlers, 1868, and *Limecola balthica* (Linnaeus, 1758)] showed the highest percentage of explained variance.

To study the long-term temporal changes of benthic communities, a network of 60 stations was sampled in winter during seven sampling campaigns spread over 28 years (1988, 1991, 1996, 2001, 2006, 2011 and 2016) (Bacouillard et al. 2020). The main results show that, despite exposure to multiple human activities, plus the forcing due to climate change, the muddy fine sands community of the eastern Bay of Seine exhibited resilience in terms of species composition and the occurrence of patches of abundance and biomass of the macrofauna. The relative densities of dominant species as well as the spatial distribution of the community remained stable over the 30 years of the study. The community was dominated by highly resistant species with bentho-pelagic cycles (planktonic larvae), capable of rapidly rebuilding their populations after exposure to multiple stressors and occupying a wide area in the eastern part of the bay at the mouth of the Seine estuary (Dauvin et al. 2017, 2021b).

To find an alternative zone for the dumping of dredge spoil from Rouen harbour, experimental operations were carried out at two adjacent sites (MASED and MABIO) at the entrance of the navigational channel offshore from the Seine estuary, where one million cubic metres were deposited at each site on a medium sand community at a depth of about 10 m over the period 2012–2013 (Marmin et al. 2016). A 'Before After Control Impact' approach was adopted to assess the bio-sedimentary changes brought

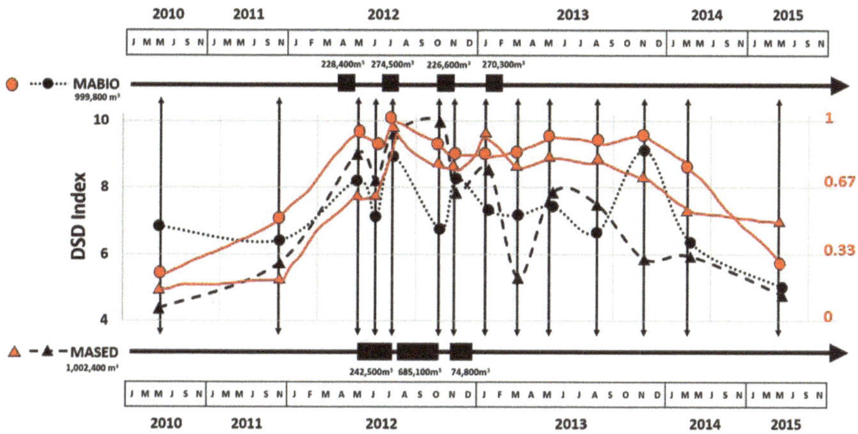

Fig. 5 Schedule of dumping operations on the two dredge spoil disposal sites MABIO and MASED in the eastern part of the Bay of Seine, with indication of the corresponding accumulated dumped volumes at each site and dates of macrofauna sampling, with values of the DSD Index and D2SI for both sites (from Dauvin et al. 2018; Baux et al. 2020)

about by these experimental deposits: macrofauna and sediments were sampled using a 0.1m^2 Van Veen grab carried out 14 times from 2010 to 2015 (Fig. 5): twice before dumping (April 2010 and November 2011), seven times during the overall dumping period (May, June, July, October and November 2012, and January and March 2013) and five times after dumping operations (May, August and November 2013, March 2014 and March 2015).

To study the macrofauna, the sediment was sieved on 1-mm mesh circular size. Monitoring showed a very local effect limited to the vicinity of the deposition areas, with significant but not total mortality of fauna in the most affected areas, and rapid recolonization just after a period of massive deposition (Pezy et al. 2017b). Moreover, during the 2012–2013 deposition phase, three zones were identified: a deposition zone, a deposition-influenced zone and an unaffected zone (Pezy et al. 2018). The Quality Benthic Ecological Status indices used here reveal that the system was highly resilient (Pezy et al. 2018). Using this classic BACI case, Dauvin et al. (2018) proposed a Dredge Spoil Disposal (DSD) index based on the response of selected species to spoil disposal and a ratio between the abundance of each of these species at impacted and unimpacted stations. The DSD index varied from 10, when there were no selected species at the impacted stations, to 0 when there were no selected species at the unimpacted station. In the Bay of Seine, the DSD index varied between 4 and 10 in October 2012 just after the end of a massive dumping operation (Fig. 5). For both sites, the values increased just after the beginning of the dumping, remained high (impacted conditions) throughout the dumping operation, then decreased rapidly just after the cessation of spoil disposal; similar values were obtained at the end of the observations in March 2015 compared with values measured at the two dates before the dumping.

Along the same lines, Baux et al. (2020) later developed the Dredge Disposal Sediment Index (D2SI) incorporating six sub-indices taking into account the variation in four biological and two physical parameters: (i) presence of species and abundances of benthic macrofauna; (ii) presence of species and abundances of ichthyofauna; (iii) ability of species to resist deposition of dredge spoil during disposal of harbour sediment; (iv) ability of drifting individuals of a given species to recolonize an area after disturbance near the sea bed; (v) percentage of modification of sediment between its natural state before dredge spoil disposal and the deposited sediment after dumping operations, and (vi) thickness of deposit accumulated per year (measured with bathymetric differentials) and considering the initial bathymetry before operations at the dumping site. As in the case of DSD, the D2SI considers two types of station: impacted and unimpacted stations. The D2SI varies from 0 for an unimpacted state to 1 for an impacted state. As observed for the DSD, the D2SI increased during the dumping operations and decreased afterwards to recover similar values calculated before the experiments (Fig. 5).

All these results show that, after the cessation of a perturbation of the benthic habitat, the ecosystem recovers rapidly to the state observed before the stress, thus illustrating the high resilience of the eastern part of the bay of Seine (Dauvin et al. 2021b).

6 Resilience of Benthic Communities After Aggregate Extraction

In the EC, the marine reserves of granulates are important, being mainly located in paleo-valleys formed by ancient rivers during successive Quaternary regressions and transgressions (Dauvin 2019). The exploitation of granulates should eventually replace the extraction of alluvial sediments on land. The extraction of marine sand, gravel and pebbles has long been carried out, particularly in the eastern basin of the EC where concessions are exploited on both the English and French sides. The volumes extracted reach 182 million t for the United Kingdom at 25 sites over an area of 232 km^2 and 149 million m^3 for France at 14 sites over an area of 113 km^2 (Dauvin 2019). Extraction is mainly carried out from suction dredgers that remove the sediment using an articulated arm of the dredge. The impacts of dredging are temporary but cause changes to the marine environment with various physical and biological impacts related to the passage of the articulated arm on the bottom and overflow of the finer sediments (Desprez et al. 2000, 2010, Duclos 2012; Lozach 2011). Dredges suck up the sediment as they move and create furrows about 2–3 m wide and about 30 cm deep, inducing a 30–70% decrease in the number of species, 40 to 95% in their abundance, and roughly the same percentage reduction in macrobenthos biomass (Lozach and Dauvin 2012).

In France, the maximum depths of extraction currently do not exceed 2 m, using a strategy that recommends leaving fallow strips without exploitation in order to facilitate benthic reconstitution after extraction operations. In the past, some operations were carried out in restricted areas in water depths of 5–10 m, which resulted in the trapping of fine particles in the depressions so formed associated with faunal changes leading to the replacement of species characteristic of coarse sediments by sandy sediment species (Desprez 2000). Nowadays, for each zone of granulate extraction, a 'Before After Control Impact' approach has been developed aimed at assessing the physical recovery times of the sea bed and the biological recovery of the benthic communities, i.e., their resilience.

Physical recovery is considered to be complete when the presence of furrows is no longer detectable by imaging techniques and when the sediment composition is similar to that prior to pre-dredging operations. Biological recovery corresponds to the recovery of a community to a state close to that found at non-impacted reference sites, comprising 80% of the species richness and biomass observed before extraction operations (Foden et al. 2009; Newell et al. 1998, 2004; Barrio-Froján et al. 2011).

Lozach (2011) has reviewed the factors influencing the physical and biological recovery times of exploited areas in the eastern part of the EC and the southern North Sea (Fig. 6). This latter author pointed out that estuaries are among the habitats least sensitive to extraction since they are naturally stressed environments with the shortest physical and biological recovery times: between 1.5 and 5 years. In the case of coarse sand with moderately energetic hydrodynamics, the biological recovery time would appear to be much longer (9 years), and the physical recovery even longer still (20 years). Very highly energetic hydrodynamic sites show shorter recolonization times (3–4 years). However, the physical and biological impacts vary greatly from one site to another and are frequently site-specific. This limits the possibilities of extrapolating the measured values of recovery time from one site to another on a case-by-case basis; even if the general ecological changes are the same, the processes controlling the recovery are specific to each site (Lozach and Dauvin 2012). In all cases, the cessation of extraction operations results in a recolonization of the site by benthic invertebrates, showing again the resilience of benthic communities in the EC following a disturbance. Moreover, there are several stages in the colonization process: the first stage corresponds to a rapid settlement of opportunist species and some mobile species, followed by a second stage which can last several years, associated with the colonization by species characteristic of gravelly habitats and a less marked increase of the biomass (Newell et al. 1998, 2004).

In the case of extraction operations lasting from a few dozen to a few hundred hours per year, the recovery time is from 2 to 20 years and there is a no permanent disturbance in the long-term (in France, concessions have been given for 20–30 years). At all the studied sites, similar sediments and communities are re-established comparable to those observed before the exploitation, provided there has been no change of sediment granulometry. The resilience is low and gradual but finally effective.

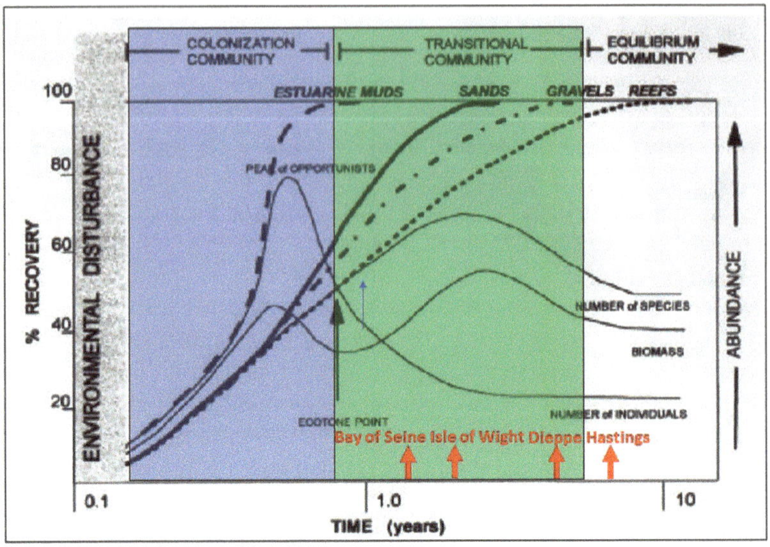

Fig. 6 Biological colonization times after the cessation of granulate extraction in the eastern part of the English Channel (from Newell et al. 1998, 2004; Lozach 2011)

7 Conclusions

The examples from the EC presented here show that the soft-bottom communities exhibit high resilience. This is due to several factors: in relation to highly energetic hydrodynamics, the benthic communities are adapted to continuous natural stress; most of the dominant species have a bentho-pelagic life cycle and are able to produce numerous larvae which are dispersed by tidal currents and can rapidly colonize the benthic habitats after the cessation of a disturbance (Dauvin et al. 2021b). Moreover, the *Ampelisca* populations of the fine sand community of the Bay of Morlaix also show a high resilience, with the ability to reconstitute their populations in a decade in spite of the absence of a pelagic larval phase.

This resilience appears to be an intrinsic property of marine ecosystems (Hughes et al. 2005). These authors (op. cit.) also suggested that high species richness should confer greater resilience to marine ecosystems. Gattuso and Duarte (2021) have pointed out that the global ocean is surprisingly resilient, provided that suitable conditions are created along with the cessation of disturbances. Fortunately, there are grounds for optimism in view of our improved understanding of the processes that drive the resilience of marine ecosystems (Hughes et al. 2005).

Finally, Gattuso and Duarte (2021) have questioned the choice to pass on to future generations either a functioning ocean full of life and resources, or an irretrievably dysfunctional ocean: this challenge to substantially replenish marine life is realistic, ethically just and economically profitable, as well as being scientifically sound.

Acknowledgements The author thanks Pr. Kenji Okoshi for his useful comments on the first version of the typescript.

References

Amara R (2010) Impact de l'anthropisation sur la biodiversité et le fonctionnement des écosystèmes marins. Exemple de la Manche-mer du Nord. VertigO la revue électronique en sciences de l'environnement Hors-série 8. https://journals.openedition.org/vertigo/10129

Anonymous (1981) Amoco Cadiz: Conséquence d'une pollution accidentelle par les hydrocarbures. Actes du Colloque International—Centre océanologique de Bretagne 19–22 novembre 1979, Centre National pour l'Exploitation des Océans, Paris

Auber A, Gohin F, Goascoz N, Schlaich I (2017) Decline of cold-water fish species in the Bay of Somme (English Channel, France) in response to ocean warming. Estuar Coast Shelf Sci 189:189–202

Bacouillard L, Baux N, Dauvin JC, Desroy N, Geiger KJ, Gentil F, Thiébaut E (2020) Long term spatio-temporal changes in the species diversity of the fine sand benthic community of the Bay of Seine (eastern English Channel). Mar Env Res 161:105062

Barrio-Froján CRS, Cooper KM, Bremner J, Defew EC, Wan Hussin WMR, Paterson DM (2011) Assessing the recovery of functional diversity after sustained sediment screening at an aggregate dredging site in the North Sea. Estuar Coast Shelf Sci 92:358–366

Baux N, Chouquet B, Martinez M, Pezy JP, Raoux A, Balay P, Dancie C, Baffreau A, Dauvin JC (2020) The Dredge disposal sediment index (D^2SI): a new specific muticriteria index to assess the impact of harbour sediment dumping. Ecol Ind 112:106109

Beslier A, Birrien JL, Cabioch L, Larsonneur C, Leborgne L (1980) La pollution des Baies de Morlaix et de Lannion par les hydrocarbures de l'"Amoco Cadiz": répartition sur les fonds et évolution. Helg Meeresunt 33:209–224

Brébant T (2024) Les communautés benthiques du bassin maritime de la Rance et de la baie de Saint-Malo: structure, fonctionnement et organisation trophique. Dissertation, Université de Bretagne Occidentale, Brest, France

Cabioch L (1968) Contribution à la connaissance des peuplements benthiques de la Manche occidentale. Cah Biol Mar 9:488–720

Cariou T, Dubroca L, Vogel C (2020) Long term survey of the fish community and associated benthic fauna of the Seine estuary nursery grounds. Sci Data 7:229

Dauvin JC (1998) The fine sand *Abra alba* community in the Bay of Morlaix twenty years after the Amoco Cadiz oil spill. Mar Poll Bull 36:669–676

Dauvin JC (2012) Are the eastern and western basins of the English Channel two separate Ecosystems? Mar Poll Bull 54:463–471

Dauvin JC (2015) History of benthic research in the English Channel: from general patterns of communities to habitat mosaic description. J Sea Res 100:32–45

Dauvin JC (2019) English Channel: La Manche. In: Sheppard C (ed) World of the Seas: an environmental evaluation, vol I, 2nd edn. Europe, the Americas and West Africa. Elsevier Ltd., Amsterdam, pp 153–188

Dauvin JC (2023) Érosion de la biodiversité marine, méprise ou réalité: le cas des invertébrés marins benthiques en Manche (Atlantique Nord-Est). VertigO la revue électronique en sciences de l'environnement 23. http://journals.openedition.org/vertigo/39835

Dauvin JC, Desroy N (2005) The food web in the lower part of the Seine estuary. Hydrobiologia 540:13–27

Dauvin JC, Lozachmeur O (2006) Mer côtière à forte pression anthropique propice au développement d'une gestion intégrée: exemple du bassin oriental de la Manche (Atlantique Nord- Est).

VertigO la revue électronique en sciences de l'environnement 7. https://journals.openedition.org/vertigo/1914
Dauvin JC, Desroy N, Janson AL, Vallet C, Duhamel S (2006) Recent changes of estuarine benthic and suprabenthic communities in response to harbour infrastructural development. Mar Poll Bull 53:80–90
Dauvin JC, Lucas S, Navon M, Lesourd S, Mear Y, Poizot E, Alizier S (2017) Does the hydrodynamic, morphometric and sedimentary environment explain the structure of the soft-bottom benthic assemblages in the Eastern Bay of Seine (English Channel)? Estuar Coast Shelf Sci 189:156–172
Dauvin JC, Baux N, Pezy JP (2018) Dredge spoil disposal (DSD) index. Mar Poll Bull 129:102–105
Dauvin JC, Sampaio L, Rodrigues AM, Quintino V (2021a) Taxonomy, and ecology of sympatric Ampelisca species (Amphipoda, Crustacea) coast from the strait of Gibraltar to the Strait of Dover, North-Eastern Atlantic. Front Mar Sci 8:643078
Dauvin JC, Raoux A, Pezy JP, Baux N, Niquil N (2021b) The Bay of Seine: a resilient socio ecosystem under cumulative pressures. In: Hénoque Y, Komatsu T, Prouzet P, Sautour B, Yoshida J (eds) Ceccaldi HJ. Evolution of marine coastal ecosystems under the pressure of global changes, Springer Nature Switzerland, pp 95–109
Dauvin JC, Baux N, Lesourd S (2022) Benthic impact assessment of a dredge spoil disposal in a dynamic resilient environment. Mar Poll Bull 179:113668
Desprez M (2000) Physical and biological impact of marine aggregate extraction along the French coast of the Eastern English Channel: short and long-term post-dredging restoration. ICES J Mar Sci 57:1428–1438
Desprez M, Pearce B, Le Bot S (2010) The biological impact of overflowing sands around a marine aggregate extraction site: Dieppe (eastern English Channel). ICES J Mar Sci 67:270–277
Desroy N, Retière C (2010) Using benthos as a tool for coastal management: the impact of the tidal power station on benthic communities of the Rance basin. Aquat Eco Health Manag 7:59–72
Duclos A (2012) Impacts morpho-sédimentaires de l'extraction de granulats marins—application au bassin oriental de la Manche. Dissertation. Université de Rouen Normandie, France
Elliott M, McLusky DS (2002) The need for definitions in understanding estuaries. Estuar Coast Shelf Sci 55:815–827
Elliott M, Mander L, Mazik K, Simenstad C, Valesini F, Whitfield A, Wolanski E (2016) Ecoengineering with ecohydrology: successes and failures in estuarine restoration. Estuar Coast Shelf Sci 176:12–35
Farr ER, Johnson MR, Nelson MW, Hare JA, Morrison WE, Lettrich MD, Vogt B, Meaney C, Howson UA, Auster PJ, Borsuk FA, Brady DC, Cashman MJ, Colarusso P, Grabowski JH, Hawkes JP, Mercado-Allen R, Packer DB, Stevenson DK (2021) An assessment of marine, estuarine, and riverine habitat vulnerability to climate change in the Northeast US. PLoS ONE 16:e0260654
Foden J, Rogers SI, Jones AP (2009) Recovery rates of UK seabed habitats after cessation of aggregate extraction. Mar Ecol Progr Ser 390:15–26
Gattuso JP, Duarte C (2021) Érosion de la biodiversité marine: Quelles solutions ? Questions internationales. La Documentation Française, ffhal-03366434f
Gomez Gesteira JL, Dauvin JC (2005) Impact of the Aegean Sea oil spill on the Ría of Ares and Betanzos fine sand community (Northwest Spain). Mar Envir Res 60:289–316
Halpern BS, Walbridge S, Selkoe KA, Kappel CV, Micheli F, d'Agrosa C, Brunokenneth JF et al (2008) A global map of human impact on marine ecosystems. Science 319:948–952
Hughes TP, Bellwood DR, Folke C, Steneck RS, Wilson J (2005) New paradigms for supporting the resilience of marine ecosystems. Trends Ecol Evol 20:380–386
Larsonneur C, Bouysse P, Auffret JP (1982) The superficial sediments of the English Channel and its western approaches. Sedimentology 29:851–864
Lozach S (2011) Habitats benthiques marins du bassin oriental de la Manche: enjeux écologiques dans le contexte d'extraction de granulats marins. Dissertation Université de Lille1, France

Lozach S, Dauvin JC (2012) Temporal stability of a coarse sediment community from the central eastern English Channel palleovalleys. J Sea Res 71:14–24

Marmin S, Lesueur P, Dauvin JC, Samson S, Tournier P, Gallicher Lavanne A, Dubrulle Brunaud C, Thouroude C (2016) An experimental study on dredge spoil of estuarine sediments in the Bay of Seine (France): a morpho-sedimentary assessment. Cont Shelf Res 116:89–102

Newell RC, Seiderer LJ, Hitchcock DR (1998) The impact of dredging works in coastal waters: a review of the sensitivity to disturbance and subsequent recovery of biological resources on the sea bed. Oceanogr Mar Biol an Ann Rev 36:127–178

Newell RC, Seiderer LJ, Simpson NM, Robinson JE (2004) Impacts of marine aggregate dredging on benthic macrofauna off the South Coast of the United Kingdom. J Coast Res 20:115–125

O'Leary JK, Micheli F, Airoldi L, Boch C, De Leo G, Elahi R, Ferretti F, Graham NAJ, Litvin SY, Low NH, Lummis S, Nickols KJ, Wong J (2017) The resilience of marine ecosystems to climatic disturbances. BioSci 67:208–220

Park (2007) https://eer.wustl.edu/mcdonnellmayworkshop/Presentation_files/Saturday/Saturday/Park.pdf

Pezy JP, Baffreau A, Dauvin JC (2017a) What are the factors driving long-term changes of the suprabenthos in the Seine estuary? Mar Poll Bull 118:307–318

Pezy JP, Raoux A, Marmin S, Niquil N, Dauvin JC (2017b) Before-After analysis of the trophic network of a dumping site in the Bay of Seine (English Channel). Mar Poll Bull 118:101–111

Pezy JP, Raoux A, Marmin S, Balay P, Dauvin JC (2018) What are the most suitable indices to detect the structural and functional changes of a natural physical perturbed community after a local and short-term disturbance? Ecol Ind 91:232–240

Poggiale JP, Dauvin JC (2001) Long term dynamics of benthic *Ampelisca* (Crustacea-Amphipoda) populations from the Bay of Morlaix (western English Channel) related after the Amoco Cadiz oil spill. Mar Ecol Prog Ser 214:201–209

Trigui RJ, Desroy N, Le Mao P, Thiébaut E (2011) Preliminary results on long-term changes of estuarine benthic communnities 45 years after the implementation of a tidal power sation in the Rance basin (northern Brittany, France). Colloque Scientifique du golfe normand-breton 'Biodiversité, écosytèmes et usage du milieu marin: quelles connaissances pour une gestion intégrée du golfe normand-breton, 2 et 3 mars 2011, Saint-Malo. https://archimer.ifremer.fr/doc/00156/26692/

Vandermeulen JH, Buckley DE, Levy EM, Long B, McLaren P, Wells PG (1978) Immediate impact of Amoco Cadiz environmental oiling: oil behavior and burial, and biological aspects. Publications du Centre National pour l'Exploitation des Océans (CNEXO). Actes Colloq CNEXO 6:159–174

The Eel Paradox. Holistic Management of Migratory Fishes. Impossible Mission?

E. Feunteun, T. Higuchi, and P. Prouzet

1 Introduction

Anguillid eels have appeared c.a. 70 million years ago during the Cretacean. They were contemporary to dinosaurs and survived to the 5th mass extinction. Since then, they have experienced a number of environmental changes such as glaciations, marine transgressions and regressions that have certainly deeply modified environmental conditions both in the oceans and in the inland aquatic ecosystems.

Their longevity and resilience are founded by their extraordinary biological cycle. Freshwater eels are diadromous catadromous fish that spawn in marine offshore areas. The eggs produce leptocephalus leaf like larvae that are adapted to pelagic marine habitats. Leptocephali are oriented by marine currents towards continental areas. The marine larval migration lasts for several months to several years. When they reach the continental edges, they metamorphose into elongate snake like transparent yound eels: the glass eels. They enter the estuaries and lagoons where they settle and/ or they colonise upstream areas of river networks. They finally settle into brackish or freshwater, and practically any kind of habitats. They grow during 3 to more than 30 years according to species and geographical areas. All their life style is

E. Feunteun (✉) · T. Higuchi
Laboratoire de Biologie des Organismes et Ecosystèmes Aquatiques (UMR BOREA, MNHN, CNRS, SU, IRD, UCN, UA), MNHN Station Marine de Dinard, Dinard, France
e-mail: eric.feunteun@mnhn.fr

E. Feunteun
Centre de Géologie Littorale (EPHE-PSL), Dinard, France

T. Higuchi
Atmosphere and Ocean Research Institute, The University of Tokyo, Akahama, Otsuchi, Iwate, Japan

P. Prouzet
SFJO France, Sant-Jean de Luz, France

A.-C. Bennis et al. (eds.), *Constraints and Adaptations to Global Change at the Land-Sea Interface: For a Shared Ecological and Energy Transition*,
https://doi.org/10.1007/978-3-031-90050-1_9

devoted to growth and storage of fat. Once a target size is attained (variable among sex—males are always smaller than females—and species), eels start fasting and metamorphose into silver eels that are adapted to marine life where they will achieve their ultimate migration to spawn. They have a remarkable ecological valence and have an extraordinary plasticity that enable them to face a high range of environmental changes.

The genus *Anguilla* likely appeared in the indo-pacific and expanded to nearly every ocean (except Antarctic) and to nearly all continents (except Antarctic and south America) from tropical to temperate and even boreal latitudes (Aoyama and Tsukamoto 1997). If the total number of species that have occurred is not well known, the genus is highly conservative compared to other families and there are today only 16 species belonging to the genus *Anguilla* sp. Of these, five are found in temperate zones located in the North Atlantic or Pacific zones. Four species: A. *anguilla*, A. *japonica*, A. *rostrata* and A. *dieffenbachi* are considered endangered. Only A. *australis* is considered to be a species that is not greatly affected by human factors.

For eels found in tropical areas of the Indo-Pacific, only A. *borneensis* is considered vulnerable, while the others are considered to suffer only slight impacts or cannot be assessed for lack of sufficient data.

After millions of years of resilience, eels are currently declining to a point that IUCN has listed all species on the red list of threatened species.

It is interesting to note that eels from temperate areas, with growth areas located in highly anthropized and modified ecosystems are classified as endangered or critically endangered. These include *A. anguilla*, *A. rostrata* and *A. japonica*, which are fished at all the continental stages of their life cycles. The recruitment strength of glass eels simultaneously fell sharply in the second half of the twentieth century. While fishing mortality is an important anthropogenic factor that needs to be regulated, we will see that, as with purely marine species, this activity cannot explain alone the sharp decline in their abundance, which is closely linked to the productivity of the environments they colonise, a production that has been severely affected since at least the second half of the twentieth century.

Nevertheless, eels are resilient and adaptable species that can occupy a wide range of ecosystems that are subject to significant variations in temperature and salinity. This enables them to reside in a wide range of habitats from coastal areas, transitional areas, rivers lakes and backwaters, to the sources of river networks, but at the condition that the longitudinal and lateral ecological continuity between the juvenile and subadult essential habitats is maintained. Clearly, as numerous reports have pointed out, this far from being the case (Adam et al. 2008; Miller et al. 2015b; Righton et al. 2021; Drouineau et al. 2022).

Despite a large amount of work carried out on the physiology, ethology and fishing of certain species (*Anguilla*, *rostrata* or *japonica*), there are still many knowledge gaps (i.e. Righton et al. 2021; Drouineau et al. 2022). Some of these concerns the precise location of spawning grounds, migratory behaviour and numerous gaps in the assessment of the abundance and recruitment of all species of the *Anguilla* genus.

This species is part of the mythology of many cultures and at least characterises the fishing cultures of many communities living in their colonised areas.

The purpose of this article is to summarise some of the characteristics of these populations, and in particular to highlight recent findings about the ecology and conservation of European eel (*Anguilla anguilla*). In particular, we will discuss why a simple approach classically mainly centred on regulating fishing activity cannot efficiently restore the depleted species, without an integrated ecosystemic management that takes account of all the causes of the degradation.

2 Origin and Evolution of the Anguillid Eels

Minegeshi et al. (2005), analysed 15,187 mitochondrial DNA sites of specimens from 15 species of eels. From this, they suggested that *Anguilla mossambica* and *A. borneensis* are the most primitive Anguillid species. The genus *Anguilla* would then have spread from these species in the present-day Pacific Ocean.

Aoyama and Tsukamoto (1997) suggest that the genus *Anguilla* spread to the present-day Atlantic Ocean before the closure of the Tethys corridor 30 million years ago. The *Anguilla* genus is thought to have appeared between 50 and 70 million years ago. Another dispersal route is suggested by Lin et al. (2001), completely opposite to that suggested by Aoyama (2009). The migration towards the Atlantic Ocean would have taken place via the Isthmus of Panama, which closed 3 million years ago. These authors also believe that the Anguilla genus is more recent, having appeared 20 million years ago.

What has so far been established is that the *Anguilla* genus appeared in the Pacific Ocean and then diversified into three geographical clades: Atlantic (2 species), Oceania (3 species) and Indo-Pacific (11 species) (Tsukamoto et al. 2020).

In the Atlantic Ocean, as the separation between Europe and America became more and more pronounced due to continental drift from the mid-Atlantic ridge, two species developed, one colonising the west coast of North America (*Anguilla rostrata*) and the other, the east coast of Europe (*Anguilla anguilla*). Hybrids of both species have been found on the Icelandic coast (Albert et al. 2006).

Figure 1 adapted from Aoyama and Tsukamoto (1997) gives a schematic view of the diffusion pattern from an original centre in the Pacific Ocean. The spread of the eel into the current Atlantic Ocean would have occurred as the inland sea within Pangea expanded from the Triassic (250 to 200 million years ago) to the end of the Cretaceous (−145 to −66 million years ago) when the American continents were already well separated from Eurasia and Africa.

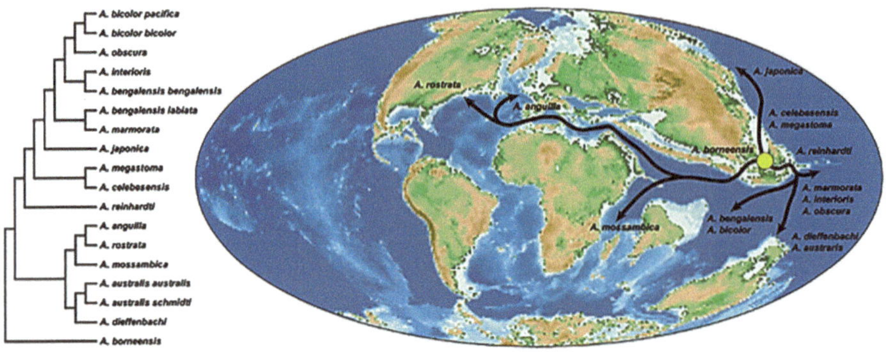

Fig. 1 Expansion of the genus *Anguilla* at the end of the Cretaceous, 70 million years ago (Aoyama 2009). The paleogeographic data at 70 million years ago were taken from Scotese and Wright (2018)

3 The Mystery of the Reproduction of Eels

The mystery of the reproduction of eels has intrigued researchers since the antiquity, mainly because they apparently seemed to have no eggs and no gonads, so how do they reproduce? More than 2400 years ago, Aristotle, believed that the eels were born from worms that formed spontaneously in mud and marshes. It is in the sea and rivers when putrefaction is at its peak that these worms form, probably by self-generation.

Some believed that the eels were born from the rotting of a dead horse and other beasts, echoing the thought of Aristotle (Byl and Schouls 1990). Pliny took up Aristotle's self-generation hypothesis and proposed another way of multiplying by the friction of eels on rocks, with the detached skins giving rise to eels. The eel has long been regarded as a kind of water snake. In the eighteenth-century texts, we find quotations such as: "Naturalists today agree with Homer, and distinguish the eel from fish, which they class as a water snake" (Scaccuto 2022).

In 1896, two Italian researchers, Grassi and Calandruccino, showed that the leptocephalus fish called *Leptocephalus brevirostris* was in fact the larva of the European eel (Grassi 1896). Grassi concluded that eels spawned in the Mediterranean Sea. In May 1904, the young Danish scientist Johannes Schmidt caught a leptocephali larvae while he was sampling herring larvae next to the Faroes Islands, more than 4000 km from the Mediterranean. This discovery marked the start of a series of oceanographic campaigns that gradually enabled Johannes Schmidt to locate the eel's spawning grounds in an area corresponding to the Sargasso Sea, the only non-coastal sea bounded by the great currents of the North Atlantic (Miller et al. 2015a).

Leptocephali, are laterally flattened and unpigmented and have strong swimming ability (Fig. 2). Leptocephali of anguilliform species exhibit a wide variety of shapes and sizes (Smith 1989; Miller and Tsukamoto 2004). Like most fish larvae, they are unpigmented and have a simplified digestive tract (Fig. 2).

Tropical eel larvae reach 50 to 55 mm before metamorphosis, as shown by the observations of Jespersen (1942) and further research cruises in the Indo-Pacific.

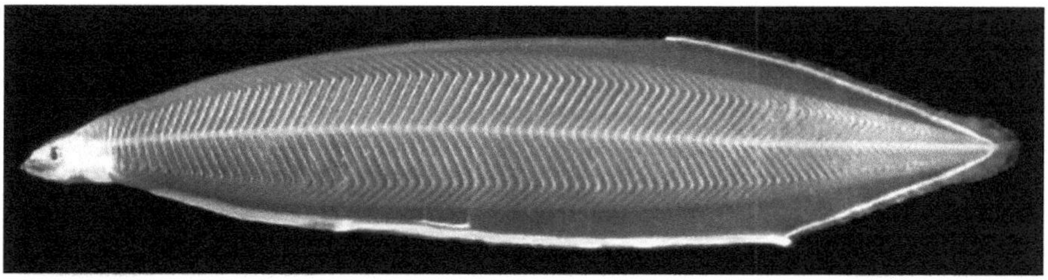

Fig. 2 Leptocephali of *Anguilla anguilla*. Note the particular shape that looks like a willow leaf and the body's transparency showing the myomeres (copyright E. Feunteun)

Leptocephalus eels from temperate zones are larger. *Anguilla rostrata* reaches 70 mm (Kleckner and McCleave 1985), while *Anguilla anguilla* is even larger: 88 mm (Boëtius and Harding 1985).

Kuroki et al. (2014, Fig. 8a and b) showed that age at metamorphosis and larval growth rate are significantly correlated. Leptocephali from species with a high larval growth rate metamorphose the fastest.

A number of studies using the increasing power of computing technologies have enabled to simulate marine migrations through the use of models combining biology and physics to study larval dispersal through the North-Atlantic oceanic circulation (Bonhommeau et al. 2009, 2010; Chang et al. 2020). These models indicate that the duration of the migration reaches up to several years if the transport is passive, but can be shortened to 18–24 months if eel larvae are able to swim at speeds of c.a. one body length per second (Chang et al. 2020). However, the analysis of daily rings of otoliths indicates much faster migration that last about one year (i.e., Lecomte-Finiger 1992; Réveillac et al. 2008). This difference is badly understood but might be due to a lack of knowledge about how fast leptocephali larvae swim and how well they are able to orientate and use oceanic features like eddies (Righton et al. 2021). The discrepancy could also be caused by the choice of the start-point of the migration, that is shorter when virtual eel larvae are released at the north east of the Sargasso Sea than when it is released at the south west (Chang et al. 2020). This assumption is supported by the microchemical difference of the core of the otoliths of leptocephali caught in the Sargasso Sea compared to those of glass eels captured along the European coasts. This finding was thought to support the existence of different spawning areas within the Atlantic Ocean (Martin et al. 2010; Chang et al. 2020, 2022).

Consequently, this opens the discussion about the relative efficiency of the exact spawning locations within the North Convergence zone of the North Atlantic Ocean that includes the Sargasso Sea and extends over more than 6 million km^2, e.g., to be compared to the surface of Europe that accounts for 10.5 million km^2! According to the exact location of the birth of leptocephali, be it in the eastern boundaries close to the Caribbean archipelago or at the north west boundaries, close to the mid-Atlantic ridge and the Azores, how long and successful is the larval migration? And how is the larval transport modified by the oceanic regime shifts that are currently observed/ modelled under the influence of the global change?

However, Johannes Schmidt was never able to observe any adult eels, nor eel eggs in this area. The delimitation of the zone between 22°N and 30°N–65°W and 48°W is based on fishing for the smallest leptocephali found in the North Atlantic. The F. W. Tesch's work (Tesch 2003), one of the greatest eel specialists in Europe, did not change Schmidt's hypothesis.

In recent years, the European project EELIAD (European Eels in the Atlantic: Assessment of their decline) (Righton et al. 2012) enabled to identify the migration routes of silver eels from different locations in the European eel colonisation zone located between northern Mauritania and northern Norway. Interestingly, all eels seemed to converge towards the Azores whatever their departure point throughout Europe (Righton et al. 2016). However, none of the tagged eels were detected beyond the Azores, which represents about 1/3 of the distance to the centre of the Sargasso Sea. Consequently, Righton's team decide to perform a complementary tagging programme in 2018 and 2019 (Wright et al. 2022) using silver eels that were captured in the Azores' rivers. Two of the silver eels reached the borders of the Sargasso Sea giving another support to Schmidt's hypothesis. Interestingly, although Wright et al. (2022) did not give emphasis on this point, two migration routes seem to occur: the first is an apparently direct trace to the centre of the Sargasso Sea, the second seems to follow a more eastern route following the mid-Atlantic Ridge as proposed by Chang et al. (2020, 2022).

During their oceanic migrations, future spawners perform large-scale daily vertical migrations between 100 and 1000 m, rising close to the surface at night and descending to the depths during the day. This is the case for *Anguilla anguilla* (Aarestrup et al. 2009; Amilhat et al. 2016, Righton et al. 2016), but also for the Indo-Pacific species: *Anguilla australis* (Koster et al. 2021), *Anguilla japonica* (Higuchi et al. 2018), *Anguilla marmorata* and *Anguilla megastoma* (Schabetsberger et al. 2016; Chang et al. 2020) (Fig. 2). The reasons of these dial vertical migrations are not well understood and debated. Predator avoidance, synchrony of the sexual maturation or gametogenesis promotion (Righton et al. 2016) or orientation mechanisms (Chang et al. 2020, 2022; Trancart et al. 2014)?

To migrate from their growth areas in the west of the North Atlantic to the Sargasso Sea in the east, there are two possible and maybe complementary hypotheses. The first is based on using the geomagnetic field (Van Ginneken and Maes 2005; Durif et al. 2021) and the second on locating the Mid-Atlantic Ridge (Chang et al. 2020).

For the first hypothesis, Erdman et al. (2021) point out in their summary that most, if not all, animals, including humans, have magnetoreception mechanisms that enable them to use variations in the geomagnetic field to orient themselves in the ocean. This is the case for migratory fish such as freshwater eels and some salmonids, which migrate several thousand kilometres to link growth habitats to spawning grounds (Fig. 3).

Durif et al. (2021) suggest a general navigation mechanism based on the larvae imprinting of a magnetic intensity maximum corresponding to the spawning area and of an intensity gradient that they experience during their oceanic migration: increasing from the spawning area to the production areas and which would be used

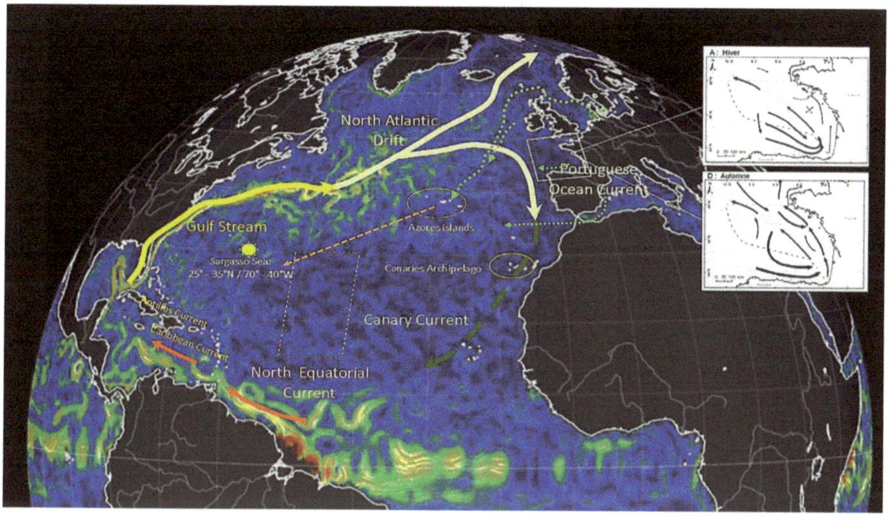

Fig. 3 Synthesis map based on publications by Schmidt (1922), Righton et al. (2016), Chang et al. (2020) on a background map of ocean currents taken from Earth. The green dotted lines show the general directions followed by the marked silver eels at different points in the range of the European eel. The yellow dotted parallelogram represents a possible eel spawning area further east than that originally defined by Schmidt at the beginning of the twentieth century. The orange dotted line shows the terminal migration route of silver eels as recorded by the latest tagging work carried out by Wright et al. (2022) from the Azores

in reverse by silver eels (decreasing from the production areas to the spawning areas) (Fig. 4).

According to Chang et al. (2020, 2022), European eels may likely spawn outside the eastward bounds of the Sargasso Sea following the Mid-Atlantic Ridge. This hypothesis proposes that possible olfactive and proprioceptive perception of particular current velocity, temperature and salinity gradients generated by the mid-Atlantic ridge may be perceived during vertical migration and help silver eels to orientate. The authors produced a larval transport model that supports this hypothesis by showing that virtual leptocephali released above the mid-Atlantic ridge had similar to higher probabilities of reaching European coasts compared to those released within the boundaries of the Sargasso Sea.

4 What Do Leptocephali Eat?

The intriguing question of "What do eel leptocephali eat?" has concentrated the attention of a number of scientists along decades. The digestive tracts of thousands of Anguilla leptocephali caught in the Ocean have been carefully examined by various scientists since 1935 (see review by Miller et al. 2020). Typically,

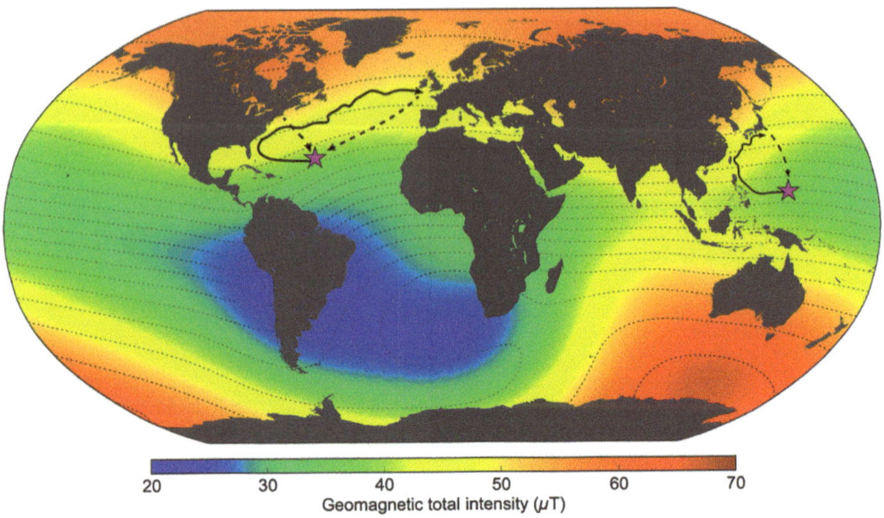

Fig. 4 Earth's magnetic field in 2024 calculated by the global geomagnetic field model, IGRF-13 (Alken et al. 2021). The colour and contours indicate the horizontal distributions of total intensity and inclination on sea surface layer. The purple stars, solid arrows, and dashed arrows indicate simplified spawning sites, larval migration routes, and spawning migration routes for anguillid eels in the northern hemisphere

zooplankton organisms have never been observed, only a few items have been identified while amorphous material is usually found. Studies based on isotope ratio analysis show that the leptocephali larvae have a trophic level, intermediate between most micro-zooplankton species and particulate organic matter (POM) (Otake et al. 1993; Mochioka and Iwamizu 1996; Feunteun et al. 2015; Miller et al. 2020). Recent DNA analysis suggested that the guts of small *A. Anguilla* leptocephali contain a variety of hydrozoans (Cnidaria) (Riemann et al. 2010; Ayala et al. 2018). However, the morphology of the mouth of leptocephali is adapted to ingest rather large and soft marine material but not to tear out parts of large living organisms as jellyfish (Miller et al. 2020). So, it is likely that these DNA materials are part of decaying organisms agglomerated in the marine snow (Miller et al. 2020; Righton et al. 2021).

Starvation of leptopcephali larvae may occur during ocean low productivity events that are might occur more frequently due to the oceanic regime shifts (Miller et al. 2015c). Considering that leptocephali likely depend on marine snow related particles, the high concentrations of microplastic that occurs in the oceans, especially in the North Atlantic convergence zone and the Sargasso Sea (Clark et al. 2016) is also a concern about the availability and pollution of food resources for eel leptocephali. Practically, nothing is known on this subject.

5 Migration of Glass Eels in Estuaries: A Steep Decline

Understanding the recruitment strength of glass eels in estuaries and mortality related to natural hydrodynamic conditions combined to anthropogenic pressures (fisheries, water quality including oxygen depletion) is a key to implement efficient management principles, especially in estuaries and lagoons that are often highly altered by anthropogenic activities.

Long recruitment series of European eels have been provided over more than 50 years in some estuaries and continental systems (see review in Righton et al. 2021; or Bornarel et al. 2018). They all converge to show a steep decline of the recruitment between the late seventies and 2000. The reasons of the decline are still debated and will be more detailed below.

Noteworthy, the relationship between spawning stock and recruitment strength is not well understood. Mainly because the overall spawning stock (i.e., yearly silver eel production from the growth habitats) and recruitment abundance (tons and numbers of glass eels recruiting each year at the distribution area's scale) are not properly assessed. However, the ICES working group on eels (ICES 2017) proposes that this is due to a reduction of the breeding stock abundance and quality and/or to the survival rate and quality of larvae. These two hypotheses are both due to habitat loss, degradation and fragmentation, contamination by pollutants, fisheries, … (see Adam et al. 2008; Righton et al. 2021 for review) while larval depletion could be due to oceanic changes (i.e., Miller et al. 2015c).

6 Assessing the Spawning Stock Biomass

Currently, the sampling design to assess the abundance of that population in the different aquatic ecosystems it colonises, is not well balanced and appropriate to define the current eel biomass. This is perfectly illustrated by the Fig. 8 from the ICES Working group Eel report (WGEEL report 2021).

Figure 5 shows that the series of observations come mainly from two countries: France and United Kingdom, and that they are mostly from scientific monitoring carried out in continental waters and by electrofishing, i.e. in shallow environments. Very few observations are made in other ecosystems: transitional water, coastal or open sea.

It is therefore very difficult to obtain reliable estimates of current eel biomasses, and even more difficult to obtain reliable estimates of pristine biomasses. This is more than detrimental for estimating the extent of silver eel escapement and even more so its evolution under the influence of the implementation of Eel Regulation 1100/2007.

Even the glass eel recruitment index defined by the ICES does not use the best available data, even though it is accurately collected by the French professional glass eel fisheries.

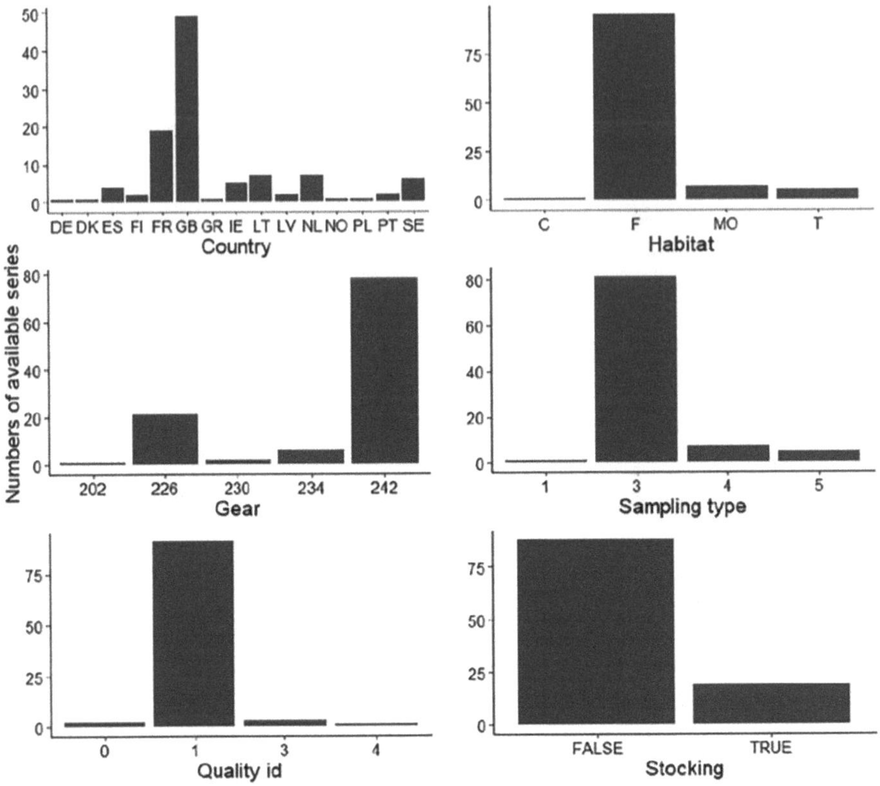

Fig. 5 Summary of available yellow eel series per country; habitat: C = coastal water, F = freshwater, MO = marine water (open sea), T = transitional water (according to WFD); gear: 202 = beach-seines, 226 = fykenets, 230 = traps, 234 = longlines; 242 = electric fishing; sampling type: 1 = commercial catch, 3 = scientific estimate, 4 = trapping all, 5 = trapping partial; quality id: 0 = missing data, 1 = good quality data, 3 = bad quality data, 4 = data used but with warnings; and stocking: FALSE = no impacts of stocking, TRUE = impacts of stocking. (From WGEEL report 2021, Fig. 3.9 page 30)

It is obvious that in this context, the ICES working group cannot assess the impact of numerous anthropogenic factors, including fishing, on the future of this resource (Biseau and Diaz 2021, Appendix 5). Under these conditions, and in accordance with the precautionary approach, the WGEEM should be more prudent when it recommends that all fishing be stopped. It points out that this advice relates only to fishing and that a more holistic approach should be taken, given that the species is subject to a number of anthropogenic pressures (Biseau and Diaz 2021).

However, although we are unable to assess silver eel escapement or bootlace eel recruitment at the scale of its distribution area, the national reports submitted to the ICES eel working group see national WGEEL report (2023) and numerous scientific publications make it possible to assess the impact of numerous anthropogenic factors on this species at various scales (see below).

7 A Panmictic Species with Significant Phenotypic Variability

Since Schmidt (1923) proposed that the single spawning area constituted a strong argument to the existence of one single panmictic stock. This assumption has been challenged by a number of studies based on genetic approaches (Baltazar-Soares et al. 2014; Maes and Volckaert 2007; Wirth and Bernatchez 2003). Other recent studies give support to Schmidt's original panmixia ecology (Als et al. 2011; Pujolar et al. 2014). However, given that spawning areas of both Atlantic eels overlap, hybridization between A. anguilla and A. rostrata was first demonstrated in 1990 (Avise et al. 1990) and mainly detected in Iceland. More recently, this was confirmed by a number of studies using various approaches (Albert et al. 2006; Gagnaire et al. 2011; Pujolar and Maes 2016). According to Wielgoss et al. (2014), the introgression is mostly found in Europe and declines from the North to the South of the distribution range. Interestingly, despite this strong propension of eels to hybridize when panmixia occurs, it seems that anguillid eel species are resilient (Barth et al. 2020). However, some variabilities of the selective pressures have been found in American eels, suggesting that different genomes might be selected according to latitude and other environmental parameters.

Even if the panmixia is generally admitted, the existence of different stocks based on life history traits variability (age, size, growth rate and fecundity of eels at maturation) has been considered in a number of studies (i.e. Bourillon et al. 2022; Teichert et al. 2023). There is a stunning variability of age at maturation which varies between 3 and more than 30 years. Total lengths of silver eels vary from 45 cm to more than a meter. There is a direct link between fecundity and total length of eels. Life history trait biogeographical pattern of female eels was first proposed by Vøllestad (1992). According to his study age and total length of females increased with latitude. This original pattern is apparently more complicated because local environmental conditions as salinity and pH of the water also control the growth rate and size at maturation (Teichert et al. 2023). More worryingly, these life history traits could be controlled by the contamination burdens of organic and mineral compounds produced by human activity. High contamination leading to reduced length and increased age at maturation (Bourillon et al. 2022).

Understanding panmixia and mechanisms controlling both genetic structure and life history traits of eels is the basement of management and conservation strategies of the endangered eel species (Jacobsen et al. 2018; Mateo et al. 2017; Pujolar and Maes 2016).

EIFAC-ICES Working group on eels has split the eel population in two distinct stocks: Northern Europe that includes the Baltic Sea and the North Sea, and Elsewhere Europe. It seems more relevant to consider three distinct stocks based on recruitment, life history and genetic particularities: Mediterranean stock (high growth rate, low age at maturation, reduced total length and high proportion of males), West European Atlantic Stock (high recruitment, age and size of female eels environmentally controlled, including density dependant processes), North Sea and Baltic Stock (low recruitment, low growth rate, large females, high proportion of females).

8 A Mysterious Fish that Has Cohabited with Humans for 6000 years

8.1 A Mythical Species for Many Civilisations

The eel is a mythical fish in the Indo-Pacific region. In Kanak culture, for example, it represents the link between land and sea: "At the dawn of the world, the moon placed its tooth on a rock emerging from the Ocean of Origins. Under the effect of the rays, the tooth decomposes. The first living creatures then appear. Those that remain on the rock become lizards, those that slip into the water become eels and snakes".[1]

In Polynesian culture, many legends refer to princess Hina who was forced to marry with the prince of eels. He was so ugly that she decided to run a way and found refuge with Hiro, the god of fishing. In order to protect the young beauty, he fished the prince of eels with a thread made of Hina's long hair. Once he was caught, the eel told Hina "Sooner or later you will kiss my lips, whatever happens". For that Hiro chopped the eel's head off and wrapped it in a banana leaf. Then he gave it to Hina and told her she could go back to her village. He clearly told her not to put the head on the ground before she returned to her village. But on her way, she decided to bathe in a River and she put the head on the floor. From the head grew as the first coconut tree of Polynesia. Years later, there was a terrible drought. And Hina was so thirsty that she drank the liquid from a fresh coconut. While she was drinking, the coconut changed in to an eel head. There were three holes, the mouth and the eyes of the eel that said: «I told you that one day you would kiss my lips «. In these legends eels play an important role for subsistence as they are at the origin of many exploited resources and sources of proteins (Feunteun and Robinet 2013) (Fig. 6).

The situation is different in Judeo-Christian cultures where eels are most often considered as unreliable or even evil. This probably is the consequence of its snake like appearance that refers to evilness: didn't the snake whisper to Adam and Eve that they should commit the original sin? (Feunteun and Robinet 2013).

Despite its mysterious origin and snake-like appearance, this fish is widely exploited and consumed by many communities in Europe. It is easy to transport alive and to smoke. It is found in a wide range of ecosystems, from mountain lakes to rivers, streams, estuaries and coastal areas (Gandolfi-Hornyold 1931).

---

[1] The birth of Tea Kawake. Tjibaou Cultural Centre—Nouméa—New Caledonia.

Fig. 6 Shortfin eel in New Caledonia (*Anguilla obscura*)—Copyright Patrick Prouzet

8.2 A Source of Proteins and a Commercial Fish

Eels are eaten and fished in a variety of ways: with a sieve to collect the fry known as glass eels, of which there can be considerable concentrations in late winter and early spring (Fig. 7); with a basket or a line, in barrier nets, some of which, known as "bordigues" or "capetchades", are typical fishing gear in Mediterranean lagoons (Fig. 8) to catch yellow or silver eels. It is also a popular fish for leisure anglers, some of whom use very special techniques such as "vermée" fishing.[2]

Japan is the world's leading consumer of eels where they have been consumed since the prehistoric times as in many countries. During the middle of the Edo period (1603–1868), eel cuisine acquired its full renown, notably with the preparation of "*Unagi no Kabayaki*": Japanese-style grilled eel. Since the Edo period, there has been an eel day, "*Doyo no Ushi no hi*", which celebrates the middle of summer (Anonymous 1992). In Japan, eel farming developed mainly from the 70s onwards (Kishida and Kanto 2013 cited by Shiraishi and Crook 2015). From 1969, following a sharp decline in the Japanese eel (*Anguilla japonica*), the country imported European

[2] Large earthworms are threaded onto wool yarns and then formed into a ball which is attached to the end of a rod-line. Then you wait for an eel to bite into the ball of earthworms. The line is then lifted and the eel is collected in an upturned umbrella by shaking the line over the umbrella.

Fig. 7 A professional fisherman landing glass-eel caught in one night with a single push-sieve on the Adour estuary just after the Second World War. (Collection Albert Lataillade)

eel from France. In Japan, farmed production reached 39,794 t in 1989, but has since declined.

The eel is currently classified as an endangered species in Japan, and its exploitation (particularly of glass eels) is highly regulated. Only around sixty tonnes of market eels are produced by fisheries, while Japan consumed 60,000 t in 2021 (mainly *A. japonica*), including 42,000 t imported from China and Taiwan (data from the French Ministry of Agriculture and Fisheries).

In Europe, eel production is far behind China, the main producer (233,222 t in 2018—FAO data) with 7663 t, representing 3% of world production by fishing and farming. However, Europe remains the main producer of European eel with 94% since the European Community banned export of glass eels outside the European Union in 2010. Production by fishing accounts for a third of European eel production, and around 30% of production by fishing for the Anguilla genus.

Fig. 8 Eel trap (Capetchade) in Thau Lagoon (Mediterranean Sea—France) copyright P. Prouzet

9 A Biological Cycle Subject to Multiple Pressures from Human Factors Others that Fishing

9.1 Dams and Ecological Continuity

In Europe, there are about 1,2 million dams of all kinds with heights ranging from a few cm to several tens of meters. The impacts of hydroelectric development and drinking water reservoirs or irrigation are identified by a number of eel-producing countries as a major cause of the depletion of the species in its range (WGEEL country report 2023). The dams create a fundamental modification of the aquatic landscapes and of their ecological functioning of inland aquatic ecosystems, in particular through the alteration of the hydrology. Moreover, they have strongly disturbed the river continuum and have led to a deep fragmentation of the aquatic habitats. Dams are thought to be one of the major causes of the decline of eels worldwide (Miller et al. 2015c; Righton et al. 2021).

Firstly, the construction of dams has drastically reduced the surface of habitats accessible for eels. Most dams are more or less impassable and they either reduce or stop the upstream migration. A significant number of dams have now been equipped with fish passage facilities that more or less efficiently restore upstream migration. But up to now, there is a strong lack of knowledge to mitigate downstream migration.

Indeed, the dams are responsible for direct mortality. Research has tested a number of technical solutions to deviate eels from turbines, but they all have proven their inefficiency. Recent work has been developed to predict the migration timing and to stop turbines for short periods that does not impede electric production too much (Teichert et al. 2020a, b).

Secondly, the presence of dams strongly disturbs the downstream migration behaviour. On the onset of their spawning migration, silver eels facing dams stop migrating for periods ranging from a few hours to a few months, or even renounce to complete their journey towards the sea (Besson et al. 2016; Trancart et al. 2019, 2022; Bourgeaux et al. 2022; Righton et al. 2021). This happens whatever the height of the dams, since the flow strength and orientation of the current are strongly disturbed by the dams.

9.1.1 Direct Mortality

Hydroelectric dams cause direct mortality to eels on their downstream migration. Mortality strongly varies according to the characteristics of the turbines, flow conditions and the presence of device to limit fish from passing through the turbines (Labedan and Sagnes 2021).

In Sweden, where the first hydroelectric power stations were built at the end of the nineteenth century, the mortality resulting since the mid-1980s is of the same order of magnitude as that caused by fisheries (WGEEL country report 2023—figure SE12 page 52). In England and Wales, 212 hydroelectric power stations are in use and affect 11,188 ha of habitats suitable for eel production. Data from the WGEEL country report (2023) (Table 4.4. page 365) indicate that impact on eel survival is half that caused by fishing and in a large number of management units represents the major pressure on the eel resource. In Germany, the loss of eels from hydroelectric power stations is estimated at 277 t in 2019, i.e. half of catches by professional and leisure fishers the same year: 485 t (WGEEL country report 2023). In Denmark, the country's largest hydroelectric power station (Tange power station), equipped with "Francis" turbines, is thought to be responsible for the mortality of 58% of the downstream stock (WGEEL country report 2023). In France, there are no estimates of the effect of hydropower production on eel mortality at this scale, but work carried out in certain catchment areas shows that the cumulative effects on certain water courses can significantly reduce silver eel production during the downstream migration. This is the case on the Gave de Pau, where 27 hydroelectric facilities have been built along the course of the river. The probability of a silver eel produced upstream to survive the downstream migration, even after the installation of more suitable facilities, is around 5% (Labedan and Sagnes 2021). The same authors estimate that eel survival after successive migrations through 7 structures from upstream is only 30%. They also point out that the presence of these obstacles also affects ecological continuity: preventing or delaying fish migrating upstream, modifying habitats and sediment transport, and causing the death of smaller yellow eels that are not stopped by the screens. On the Rhine, large dams such as those of the Fessenheim complex cause significant mortality, estimated at 60% of the downstream eel stock (Trancart et al. 2018b).

The Gave de Pau is obviously not the only river system affected in France, and other examples illustrate the negative impact of these facilities on the production of a species considered by the IUCN critically endangered of extinction.

9.1.2 Downstream Migration Delay or Failure

Because dams create hydrological disturbance, the orientation cues of eels on their downstream migration are disturbed and migration behaviour is deeply altered (Righton et al. 2021).

For example, the large floodplain river of Grand Lieu, France, a tributary of the Loire River, produces about 100,000 silver eels per year. It is a natural national reserve especially designed to protect water birds. The water levels and flow are controlled by a small sluice gate (1 m height and 5 m wide) following strict prefectural rules. These rules were mainly fixed for agricultural objectives. Consequently, the gates are not opened before the water increases at a threshold level after the summer when the first rain occur. Most often this occurs between October and November, well after the first downstream movements of male silver eels and during the floods of autumn. During dry years, the water level increases after the downstream runs and 70% to 90% of silver eels are blocked upstream the outlet dam and finally die in the lake (Trancart et al. 2018a). Simple rules were implemented and applied and enabled to restore the migration during the following years (Bourgeaux et al. 2022). On the Rance, the tidal power plant alone can block 70% silver eels that are therefore lost for reproduction (Trancart et al. 2022). In small river systems a number of reservoirs are implanted to provide drinking and irrigation water. In the very small Frémur River (60 km^2), a major dam (14 m high) slows the migration speed of silver eels during their migration that takes several days to swim 1 km as the cues are disrupted (no directional current, no temperature or natural gradients) and 30% fail to migrate downstream (Besson et al. 2016; Trancart et al. 2020).

9.2 *Loss of Habitats*

In order to develop its population from the glass eel recruits, the eel need so-called functional habitats. In other words, environments where it can normally live and flourish. The colonisation process may occur during all the life of the eels following individual behavioural skills and environmental characteristics (Feunteun et al. 2003). The eel is ubiquitous thanks to a strong phenotypic plasticity, which enables to adapt and live in a wide variety of aquatic environments from the coastal area, the estuaries, lagoons, stagnant backwaters and lakes, running rivers. They especially appreciate wetlands and marshes where their growth rate is favoured (Baisez et al. 2000; Elie et al. 1990; Feunteun et al. 1992, 1999; Laffaille et al. 2004; Lasne et al. 2008; Trancart et al. 2018a, b). But they also occur in large and small rivers both in lentic and lotic environments (Trancart et al. 2014; Lasne et al. 2008).

Wetlands, coastal areas and river bancks host 60% of the world's population and almost 4 billion people (Ducrotoy 2018). According to the IPBES (IPBES 2019), 85% of the world's wetlands disappeared between the eighteenth and twenty-first centuries. The 2018 RAMSAR (RAMSAR report 2018) report indicates that 35% of wetlands disappeared between 1970 and 2015, a loss three times faster than that recorded for forests. The causes include urbanisation, intensification of agriculture, forestry production, drainage, dams, extraction of materials, etc. (Perez 2021). France, like all European countries, is not an exception to the widespread degradation of these areas, considered by many of our decision-makers to useless and a source of multiple nuisances. The 1994 report on wetlands (Bernard 1994), indicates that France lost almost two-thirds of its wetlands during the twentieth century. As mentioned by Prouzet et al. (2023), many estuaries in France have been profoundly modified by urbanisation and industrialisation. This is particularly true for the Seine (Le Pape et al. 2003; Le Luherne et al. 2016) and the Loire (Lefeuvre et al. 2000). In Brittany (north-western France), 65% of salt marshes have disappeared due to diking, drainage or reclamation (Poncet 1984). These losses of functional habitats necessary for eel production have also been noted in many European countries (WGEEL country report 2023). For example, in southern Finland, only 38,000 ha of aquatic ecosystems out of the 23,408,500 available can be naturally colonised by eels. In Sweden, a large surface of habitats is available for eels, but the current low number of recruits and the inaccessibility of these areas severely limit natural production. In the Netherlands, the Ijsselmeer and Markermeer lakes reclaimed from the Wadden Sea are favourable habitats for eel production. Between 1932 and 1967, the total surface area of these lakes was reduced from 3470 km^2 to 1820 km^2 following the construction of a dam. In Lithuania, 75.8% of lakes and reservoirs favourable for eel production are located above hydroelectric dams with no possible connection to the sea. In Poland, only 24% of the lake surface area is freely connected to the sea. In Spain, Claveroso and Hermoso (2015) estimated that 80% of the eel's functional habitat had been lost due to the construction of dams. Work by Mateo et al. (2021) as part of the SUDOANG project confirms the extent of this loss. In the South of the Mediterranean Sea, a large number of dams were built, often at the mouths of Oueds, for agricultural purposes from the 1950s onwards, and then for tourism needs. There are 1200 large dams in the Mediterranean basin alone, with countries such as Spain and Turkey developing this type of facility rapidly (Ghiotti and Molle 2008), and more than 1,2 million across Europe. This has an impact on the eel's ability to colonise continental waters around the Mediterranean (Fig. 9). This restricts the species mainly to lagoon and coastal areas (Ciccotti and Morello 2023).

This considerable loss of habitat has a very strong impact on the recovery of the species, given the density-dependency phenomena (Minto et al. 2008). These have been observed, as for many fish species, in eels either in natural (Acou et al. 2011; Desprez et al. 2013; Machut et al. 2007) or in experimental environments.

Fig. 9 View of the Sidi Salem dam, built in 1981 on the Mejerda river and used for irrigation and hydroelectric power generation in Tunisia. (Copyright P. Prouzet)

9.3 Contaminants

Pollutant burdens in European and American eels have been quite well documented since a long time (for example, Buslé 1994; Belpaire et al. 2016, 2019; Freese et al. 2016). Every single eel sampled across Europe is contaminated by a large variety of contaminants, ranging from mineral, metallic and organic compounds (Bourillon et al. 2020). All of these compounds are produced by human industries, medicine and agriculture (Bourillon et al. 2020). The contamination burdens vary from moderate to much higher levels than the authorized thresholds fixed for human consumption (Bourillon et al. 2020; Castonguay et al. 1994). Locally, the contamination levels are strongly dependent upon the human activities (Bourillon et al. 2020; Holmqvist et al. 2006; Oliver et al. 2015).

Contamination is thought to represent one of the principal causes of the decline of eels (Bourillon et al. 2022; Feunteun 2002; Miller et al. 2015c; Robinet and Feunteun 2002). How the contaminants act on eels' fitness remains badly understood and relatively scarcely studied (Robinet and Feunteun 2002; Righton et al. 2021). Recent research using machine learning methods has shown that highly contaminated silver eels were smaller, older and had a lower growth rate than moderately contaminated eels (Bourillon et al. 2022). Highly contaminated female silver eels were consequently found to have a much lower fecundity that larger and younger moderatly contaminated females. This is thought to contribute to a recruitment deficit, estimated to c.a. 400 t of glass eels per year (Bourillon et al. 2022).

Up to now, little effort has been devoted to reduce pollutant inputs in the rivers as a measure to restore the eel population dynamics. It is still quite difficult to trace the origin of pollutants that are often produced and released diffusively (agriculture), remotely through the atmosphere (industry, transport) or the water (i.e. waste and sewage waters) (Bourillon et al. 2020). Contamination is most probably one of the key problems for eels and, more broadly, is probably the main causes of the current erosion of the biodiversity. Conservation ecology of eels, as a flagship and umbrella species for the conservation of biodiversity, should urgently focus on effects of contaminants on biological traits and fitness of species. In parallel, it is of uppermost priority to trace the origin of these pollutants and develop policies to simultaneously reduce their production and control their dispersion in the Nature, especially in aquatic environments. Nature based solutions should particularly be investigated.

Over the last 8000 years, the Earth has experienced a remarkably stable warm period, averaging around 15 °C. An equilibrium that should have persisted for a long time. But under the pressure of anthropogenic factors, global temperature has risen sharply since the pre-industrial period (1850–1899) (Delecluse 2014). In its latest report, IPCC notes a temperature rise of 1.1 °C compared with this period. It is estimated that global warming will reach 1.5 °C by the early 2030s.

While one of the most effective drivers of ocean movement is wind, the other is the exchange of heat, mass and chemical compounds between ocean and atmosphere (Delecluse 2014). Climate change has a major impact on general ocean circulation (Broecker 1979). Oceanic circulation used by migratory fish such as eels, whose leptocephali larvae need this dispersal vector.

Desaunay and Guérault (1997) suggested a possible link between the availability of food in the Atlantic Ocean and the survival and growth of leptocephali. Knights (2003) subsequently observed negative correlations between the glass eel abundance index observed at Den Oever (Netherlands) and temperature anomalies in the surface layer located between 100 and 250 m in the Sub-tropical gyre (STG) and the Sargasso Sea over the period from 1952 to 1995. This increase in temperature in the STG could result in a reduction in oceanic productivity, with an impact on leptocephalus mortality (Miller et al. 2015c). Bonhommeau et al (2008) noted a synchrony between the decline in the world's main commercial eel species: *Anguilla anguilla*, *A. rostrata* and *A. japonica*. This suggests the involvement of similar oceanic mechanisms that affect larval recruitment. They showed that the survival of larvae is strongly correlated with the availability of food during the early life stages. They noted that over the last 4 decades, changes in primary production linked to global warming may have led to a decline in these populations. Finally, they observed that in the Atlantic and Pacific oceans, changes in temperature have been followed by modifications in glass eel recruitment. Chang et al. (2018), using a 3D particle-type model (JCOP2: Japan Coastal Ocean Predictability Experiment 2), shows a decrease in particle transport from the spawning area during the last two decades, mainly towards Taiwan and Japan. Changes in ocean circulation in the western Pacific appear to be the cause of the weakening of tropical and subtropical winds over the last two decades. They suggest that these decadal changes in ocean circulation have affected the larval migration of Japanese eels towards their recruitment areas. In addition, climate change is influencing precipitation and temperature patterns across the continents. As a result, it is leading to changes in the hydrological and thermal conditions of river systems, which may modify the migratory behaviour of silver eels (Arevalo et al. 2021).

10 An Ecosystem Approach is Needed to Restore This Species

In 1998, the ices EIFAC working group on eels stated that eels "were outside safe biological limits" and recommended to limit fishery mortality as close a 0 as possible. Next year, WGeel confirmed the status of the European Eel but they recommend to reduce all anthropogenic mortality as close as 0 as possible. From 2003 to 2006, the ICES advice on eel management was to reduce all anthropogenic mortalities as far as possible until a restoration plan was put in place. After the restoration plan in 2007 was implemented and up to 2020, this advice changed little and insisted that all anthropogenic mortalities should be reduced to zero if possible.[3] Following the meeting of the working group on the wording of the advice in February 2021 (Biseau and Diaz 2021), this advice was modified by focusing on the impact of fishing, with other anthropogenic pressure factors being taken into account, but in an ancillary manner. Given the impossibility of assessing the eel resource within its distribution area, and in line with the ICES precautionary approach, the advice was not to authorise fishing, regardless of the use of catches (consumption or restocking).

However, this ICES sectoral advice, which only takes into account the effect of fishing on the resource, is not in line with the aims of Eel Regulation 1100/2007 formulated by the European Union to restore this species. This regulation clearly emphasises the need to implement a systemic approach that takes into account all the pressure factors affecting the species, and not just fishing.

In the text adopted by the European Parliament on November 21, 2023[4] on the implementation of measures concerning the restoration of the European eel known as the "Van Ruyssen report", the systemic approach taking into account all factors of anthropic pressure is strongly recommended.

In particular, a number of considerations were highlighted: (i) the eel regulation 1100/2007 is the most appropriate management instrument; (ii) the complexity of managing this stock, which cannot be focused solely on marine ecosystems; (iii) the need for a more ecosystem-based approach, rather than just fisheries management; (iv) too little attention on non-fisheries mortality.

[3] See document *ele.2737.nea.*

[4] (P9_TA(2023)0411).

10.1 A Sectoral Approach Based on Fisheries Regulation Alone won't Work

While it is impossible to estimate the effects of non-fishing anthropogenic impacts at the population scale, information summarized in paragraph 5 show that the non-extractive human activities such as pollution, dams and habitat destruction cause way more mortality, fecundity and fitness depletion than fisheries.

ICES and GFCM advices logically focus on controlling fisheries and appears to be inappropriate, alone, to efficiently manage this migratory species and target a sustainable exploitation. Controlling eel fisheries has proven to be efficient for for purely marine species such as bluefin tuna, cited as a management example by many NGOs. For diadromous species, it is necessary to minimize the ecological footprint of human activities other than fishing if we want to restore at least the European eel population to the levels of the late 70s.

However, it is interesting to note that overall assessment of fishing mortality of glass eels, yellow eels and silver eels is poorly documented. The recruit-stock relationship from glass eels to silver eels is not well understood, and the stock-recruit relationship from silver eels to glass eels remains a mystery. Indeed, the density dependant mortality, migration, growth rate and sex orientation processes are not well documented which reduces the efficiency of recruit- breeding stock models. While the survival of adult eels and larvae and the duration of the marine migration are still debated.

Moreover, the side effects of reducing fishing effort have never been assessed. Indeed, what really happens to a stock when nobody cares about it? Who will ring the alarm bell when hazards occur and impacting projects are planned?

Despite the Van Ruyssen report approved by the European Parliament in November 2023, and despite the achievement of the fishing mortality limitation targets imposed by EU regulation 1100/2007,[5] in view of the expert reports provided by ICES and GFCM, the European Commission at the last TACs and Quotas meeting in December 2023 tightened the restrictions on eel fishing, making this activity hardly profitable.[6]

One of the most representative examples of the decline in eel fishing effort is the French glass eel fishery (Fig. 10).

---

[5] eur-lex.europa.eu/legal-content/EN/TXT/PDF/?uri=CELEX:32007R1100—targets 13 and 14.

[6] "Finally, to protect eels, ministers agreed to a comprehensive package to protect the eel and help restore this unique stock in European waters. They agreed to prohibit recreational fisheries and to extend the closure for any commercial eel fishing activity to six months in marine and adjacent brackish waters in the north-east Atlantic (including the Baltic Sea) and in the Mediterranean (excluding the Black Sea) in a differentiated way to take into account different migration periods in different sea basins. Hence, member states will be able to adapt the closure period for different fishing areas, to take into account their specificities as well as the temporal and geographical migration patterns of eel at respectively the glass eel and silver eel life stage." Council approves fishing opportunities for 2023 in EU and non-EU waters—Consilium (europa.eu).

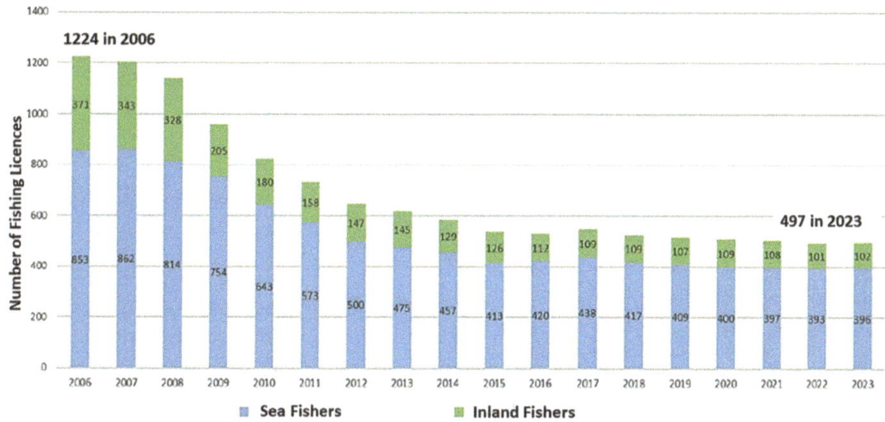

Fig. 10 Trend in the number of licences delivered for professional glass eel fishing. There was a sharp fall from 2008 to 2012, followed by a slower decline thereafter, according to CONAPPED and CNPMEM data

France currently supplies around 90% of the glass eel used either for direct consumption (mainly in Spain), for farming (mainly in northern European countries) or for restocking. France is the only country where a TAC is set annually by an independent scientific committee (SC) (Drouineau et al. 2022). It is then submitted to a Scientific and Economic Committee (SEC), on which professional fishers can express their opinion on the level of the overall TAC, taking into account social and economic criteria as well as their observations in the field. The total TAC is then split into a sub-TAC for consumption and a sub-TAC for restocking, according to the allocation key defined in regulation UE1100/2007 (40% consumption and 60% restocking).

Figure 11 shows the decline in the rate of exploitation index defined by the SC taking into account the two sub-TACs and that defined by the SEC from the sub-TAC consumption based on the justified assumption that restocking glass eels are destined for release alive in the European eel's distribution zone. These two curves don't take into account the decreasing trend of the fishing effort whereas a linear relationship has been shown between effective effort in terms of fishing days and exploitation rate (Susperregui 2007; Prouzet et al. 2008).

With regard to the French glass eel fishery, estimates of exploitation rates have been provided for major French rivers such as the Adour and Loire (Prouzet et al. 2008; Bru et al. 2009), based on sampling designs adapted to the glass eel's migratory behaviour (Prouzet et al. 2009; Ondulami and Vallet 2012). Estimates provided over a period of several fishing seasons show that seasonal exploitation rates ranged from 8 to 25%, with an average of 15.7% for the Adour between 1998 and 2005 and between 13.6 and 18.9% for the Loire from 2003 to 2006. Since this period, Fig. 10 shows that the exploitation rate has been significantly reduced. As a result, the exploitation rate is likely to be below 10% for this fishery, an order of magnitude which is also

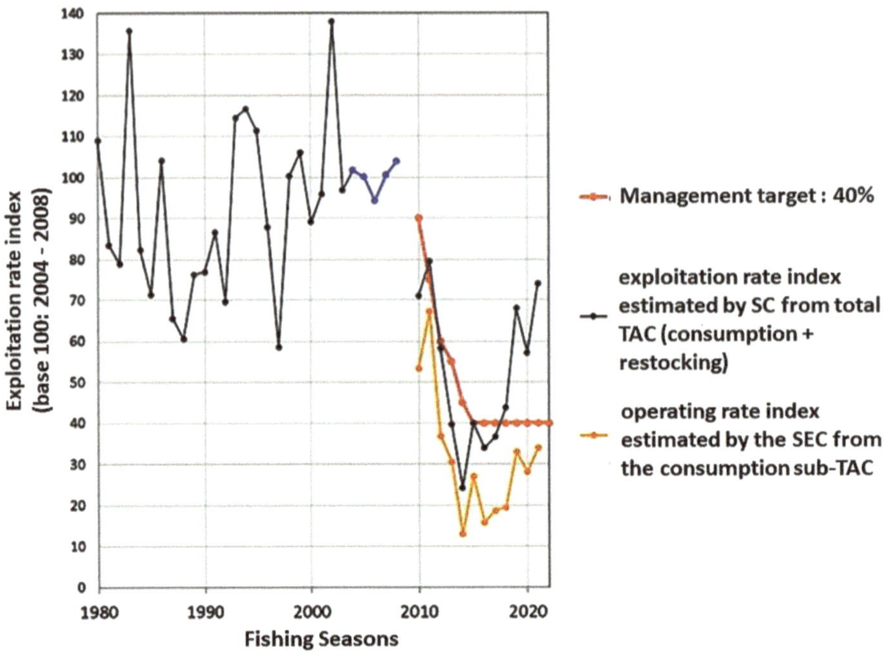

Fig. 11 Evolution of the exploitation rate index for the French glass eel fishery as estimated by SC and SEC from SC and SEC reports 2023

provided for the Severn River by Aprahamian and Wood (2021): 7.8%. Noteworthy, these values do not include the estuaries that are not subjected to glass eel fisheries.

For the various commercially exploited eel species: *A. anguilla*, *A. rostrata* and *A. japonica*, a number of authors have highlighted the significant effect of global change, or simply climate change, in the decline in recruitment of these species (Knights 2003; Machut et al. 2007; Kim et al. 2007; Bonhommeau et al. 2008; Bevacqua et al. 2015; Drouineau et al. 2018; Feunteun and Prouzet 2020; Kaifu et al. 2021), which means that a much more holistic approach is needed, taking into account the various pressures exerted on this population and not just its exploitation.

10.2 Active Rather Than Passive Restoration

Given the long life cycle of the species (on average between 8 and 15 years depending on the geographical area), the flow of glass eels essentially centred on the central Atlantic part of its distribution area, and the very reduced carrying capacity of this central part to support a large production of yellow eels likely to make a significant contribution to recruitment in peripheral areas to the north (North Sea and Baltic) or to the south (Mediterranean basin), it would seem that active restoration is more justified than a passive "renaturation" approach.

This active restoration is mainly based on transfers of individuals at various stages, as provided for in Eel Regulation 1100/2007. Restocking (or, more precisely, transferring glass eels within their natural range) is not a new practice. It has been widely used in many countries, including the Netherlands, Germany, Poland, Italy and France, since the end of the nineteenth century, using glass eels or young eels (Le Clerc 1930; Dekker and Beaulaton 2016). Since then, many countries have continued to release glass eels or young eels into various ecosystems, often to support inland water fisheries that would otherwise have been unable to continue due to a serious deterioration in ecological continuity (see § 5.1 and 5.2). This is the case in Sweden, Poland, Estonia, Lithuania and Finland (WGEEL country reports 2023). The Workshop on Eel Stocking (WKSTOCKEEL 2016) evaluated the results of eel stocking in Europe and concluded: "*it was clear from local studies that stocking had been beneficial by enhancing the yellow and silver eel stocks in a number of water bodies. These included several Danish, German, Swedish and Estonian Lakes, Lough Neagh in Northern Ireland as well as Danish streams and marine areas*". This working group also gave a definition of the net benefit of a restocking action: "*where the stocking results in a higher silver eel escapement biomass than would have occurred if the glass eel seed had not been removed from its natural (donor) habitat in the first place*" while admitting that the knowledge required to define this was very limited (natural mortality in the young stages; carrying capacity of the donor and recipient systems; density-dependency relationship).

In France (Fig. 12), since the implementation of the eel management plan, 14 t of glass eels, i.e. 45 million individuals, have been transferred to 6 eel management units between 2011 and 2021 (ARA France data). These operations are coordinated by ARA France (Association pour le Repeuplement de l'Anguille en France), which manages the funds allocated and organises the restocking in close collaboration with the professional fishermen organisations CNPMEM (Comité National des Pêches Maritimes et des Elevages Marins) and CONAPPED (Comité National des Pêches Professionnelles en Eau Douce).

For the French glass eel fishery, the mortality of glass eels due to fishing or to the storage and transport of glass eels has been greatly reduced in order to produce quality fish for European restocking. As a result, a code of good practice for fishing and restocking has been defined and implemented by professional maritime and continental fishing organisations in cooperation with the WWF and scientists. The batches are analysed before dispatching by the veterinary services of the donor and recipient countries. All or part of the transferred batches are marked in order to analyse the results obtained in terms of growth or survival according to the immersion environment chosen.

The report "ADRAF" (Feunteun et al. 2022) shows that the eel restocking programme is an effective measure for introducing eels into sections of rivers where they are not present. The growth rates measured increase with distance from the sea, particularly in large rivers, which suggests the value of selecting upstream sections of certain rivers. It complements and confirms the information obtained by other

Fig. 12 Glass eels transferred in aquatic ecosystems in France (Copyright N. Porcher—CONAPPED)

authors on site selection, density-dependency problems, variability in survival rates and growth (Kangur et al. 2002; Lin et al. 2007; Desprez et al. 2013; Simon and Dörner 2014; Pedersen and Rasmunssen 2015; Brämick et al. 2016; Matondo et al. 2021).

In the Mediterranean Sea, from the French Mediterranean lagoons, silver eels are released at the outlet of the lagoons in order to minimise downstream mortality. In 2023, just over 25 t were released and sub-samples were analysed (ARA France data) (Fig. 13). This type of transfer to minimise downstream mortality or facilitate migration to the sea is also carried out in many countries such as Ireland, Sweden, Greece and the Netherlands. In 2022, 250 t were transferred in Europe (WGEEL country report 2023).

If these transfers are to be as effective as possible, they must also be accompanied by strict control of the fishing effort and not simply constitute, as was the case in the twentieth century, a support for the maintenance of eel fisheries. It is also important to trace eels from the fishing point to the restocked area in order to prevent part of the stock from being used for other purposes that restocking. Moreover, a survey of the population from recruits to spawners (i.e. over several years and at the catchment level) is necessary to assess the efficiency of the restocking and to learn about recruit-stock relationships (including analysis of biological traits). Under these conditions, the net benefit of the operation, if any, would undoubtedly be greatly reduced.

Fig. 13 Release of silver eels in the port of Bouzigues (Hérault) on 12 January 2023 (Copyright Aguirrebarrena)

11 Conclusion

This paper aims to show that despite intensive research that has enabled key progress on eels' ecology, entire sections of their biological cycle remain unknown or badly understood thus preventing the design of efficient science-based restoration programmes and strategies.

Much more is known about marine migrations of leptocephali and adult but spawning locations are still not fully confirmed. The orientation mechanisms are still badly understood. The survival of larval and adult migrants at sea are still very poorly documented, although they are shown to be controlled by predation, and leptocephali are directly influenced by availability of food, speed and direction of marine currents. Filling these knowledge gaps is fundamental to understand and forecast possible effects of ocean regime shifts on the survival and migration of leptocephali larvae, followed in turn by the recruitment strength of glass eels.

Recruit to silver eel relation is better understood, thanks to a number of studies throughout Europe. The survival, sex determination, growth rate, fat content, age and size at silvering are highly variable and likely controlled by biogeographical gradients together with natural and anthropogenic parameters. However, quantitative estimation of the population size and characteristics at the distribution range of eels is still lacking. Thus, attempts to estimate outdated stock-recruitment relationships remain uncertain. This should lead to focus research on the complexity of recruit to stock relationships and to determine how this relationship is controlled by the various causes of decline of the eel population.

The genetic status of the population is still debated and controversial. However, whether there is one panmictic population or several populations will not challenge the management plan fundamentally. Indeed, it is more important to consider the existence of contrasted stocks that are characterized by their sex ratio, growth rates, age and size at silvering which induce adapted management strategies. Here we proposed the existence of at least 3 management stocks: North Sea and Baltic, Bay of Biscay and Atlantic, Mediterranean.

Fishing mortality is the most featured cause of the decline of eels. However, in our review we show that other anthropogenic pressures are much more involved than fisheries in the marked depletion of the stock. Although considerable efforts have been devoted to fishery reduction since 2007 (that is almost 2 generations of eels), the efficiency of the European eel management plans has not been demonstrated yet. A modest increase of the recruitment has been detected in 2011–14 and recruitment has recently increased. However, it is unclear whether this increase is significant given that the monitoring methods and their significance have changed since the management plan was implemented. The exact status of the yellow eels and silver eel stocks remains badly known.

Entire research programs have mobilized the scientific community to understand the effects of human activities on population decline. The conclusions of these studies are clear and unambiguous: eels are much more impacted and threatened by global change (climate change, ocean and rivers regime shifts, agriculture, industry and medication, habitat loss and fragmentation buy dams) than by fishery. Despite the accumulation of indisputable knowledge over nearly 20 years, management programs still emphasize fishing control and seem to elude actions on reducing pollution, restoring ecological continuity and aquatic habitats. At the same time, simple and fundamental questions about the comparative analysis of the links between the reduction of fishing mortality at all stages and the stock of spawning eels have still not been addressed. This type of approach is fundamental to step towards holistic approaches defining stock-to-stock relationships taking into account the effect of environmental variability and the pressures of human activities in the context of global change.

Easy to say. But this will only be possible if we rebuild trust between those involved in the management of eels, biodiversity and aquatic ecosystems, from the source of rivers to the deep ocean. Restoring trust and respect between fishermen, industrialists, farmers, environmental managers and scientists, an unrealistic utopia? What if the eel helped us achieve this?

All of these elements form the "eel paradox": a declining species with a wide distribution, very popular ultimately but whose restoration and management strategy remain inefficiently focused on one single cause: the fishery while the other impacts of human activity are merrily ignored.

References

Aarestrup K, Okland F, Hansen M M, Righton D, Gargan P, Castonguay M, Bernatchez L, Howey P, Sparholt H, Pedersen MI, McKinley RS (2009) Oceanic spawning migration of the European eel (*Anguilla anguilla*), Brevia, Science September, p 1660

Acou A, Rivot E, Van Gils JA, Legault A, Ysnel F, Feunteun E (2011) Habitat carrying capacity is reached for the European eel in a small coastal catchment: evidence and implications for managing eel stocks. Freshw Biol 56:952–968

Adam G, Feunteun E, Prouzet P, Rigaud C (2008) L'anguille européenne- Indicateurs d'abondance et de colonisation. Editions QUAE, Collection Savoir-faire, English version. https://www.researchgate.net/publication/266858339_The_European_Eel_abundance_and_colonization_indicators, p 393

Albert V, Jonsson B, Bernatchez L (2006) Natural hybrids in Atlantic eels (*Anguilla anguilla, A. rostrata*): evidence for successful reproduction and fluctuating abundance in space and time. Mol Ecol 15(7):1903–1916

Alken P, Thébault E, Beggan CD et al (2021) International geomagnetic reference field: the thirteenth generation. Earth Planets Space 73:49. https://doi.org/10.1186/s40623-020-01288-x

Als TD, Hansen MM, Maes GE, Castonguay M, Riemann L, Aarestrup KI, Munk P, Sparholt H, Hanel R, Bernatchez L (2011) All roads lead to home: panmixia of European eel in the Sargasso Sea. Mol Ecol 20:1333–1346. https://doi.org/10.1111/j.1365-294X.2011.05011.x

Amilhat E, Aarestrup K, Faliex E, Simon G, Westerberg H, Righton D (2016) First evidence of European eels exiting the Mediterranean Sea during spawning migration. Sci Rep 6:21817. https://doi.org/10.1038/srep21817

Anonymous (1992) Internationalization comes to the Japanese eel industry. Fish J 39:8

Aoyama J (2009) Life history and evolution of migration in catadromous eels (Genus Anguilla). Aqua-Biosci Monogr 2:1–42

Aoyama J, Tsukamoto K (1997) Evolution of the freshwater eels. Naturwissenschaften 84:17–21. https://doi.org/10.1007/s001140050340

Aprahamian M, Wood P (2021) Estimation of glass eel (*Anguilla anguilla*) exploitation in the Severn estuary, England. Fish Manage Ecol 28(1):65–75

Arevalo E, Drouineau H, Tétard S, Durif CMF, Diserud OH, Poole WR, Maire AL (2021) Joint temporal trends in river thermal and hydrological conditions can threaten the downstream migration of the critically endangered European eel. Sci Rep Nat Portfolio 11:16927. https://doi.org/10.1038/s41598-021-96302-x

Avise JC, Nelson WS, Arnold J, Koehn RK, Williams GC, Thorsteinsson V (1990) The evolutionary genetic status of Icelandic eels. Evolution 44:1254–1262. https://doi.org/10.1111/j.1558-5646.1990.tb05229.x

Ayala DJ, Munk P, Lundgreen RBC, Traving SJ, Jaspers C, Jørgensen TS, Hansen LH, Riemann L (2018) Gelatinous plankton is important in the diet of European eel (*Anguilla anguilla*) larvae in the Sargasso Sea. Sci Rep 8:6156

Baisez A, Rigaud C, Feunteun E (2000) Hétérogénéité de répartition de l'anguille européenne (*A. anguilla*) dans un marais endigué de la côte atlantique (marais breton). Cybium 34(3):23–32

Baltazar-Soares M, Biastoch A, Harrod C, Hanel R, Marohn L, Prigge E, Evans D, Bodles K, Behrens E, Böning CW, Eizaguirre C (2014) Recruitment collapse and population structure of the European eel shaped by local ocean current dynamics. Curr Biol 24:104–108. https://doi.org/10.1016/j.cub.2013.11.031

Barth JMI, Gubili C, Matschiner M, Tørresen OK, Watanabe S, Egger B, Han YS, Feunteun E, Sommaruga R, Jehle R et al (2020) Stable species boundaries despite ten million years of hybridization in tropical eels. Nat Commun 2020(11):1–13. https://doi.org/10.1038/s41467-020-15099-x

Belpaire C, Pujolar JM, Geeraerts C, Maes G (2016) Contaminants in eels and their role in the collapse of the eel stocks. In: Arai T (ed) Biology and ecology of anguillid eels. CRC Press, pp 225–250

Belpaire C, Hodson P, Pierron F, Freese M (2019) Impact of chemical pollution on Atlantic eels: facts, research needs, and implications for management. Curr Opin Environ Sci Health 11:26–36. https://doi.org/10.1016/j.coesh.2019.06.008

Bernard P (1994) Les zones humides—Rapport de l'instance d'évaluation. Documentation Française, Paris, p 396

Besson ML, Trancart T, Acou A, Charrier F, Mazel V, Legault A, Feunteun E (2016) Disrupted downstream migration behaviour of European silver eels (*Anguilla anguilla*, L.) in an obstructed river. Environ Biol Fish. https://doi.org/10.1007/s10641-016-0522-9. In press
Bevacqua D, Melia P, Gatto M, De Leo GA (2015) A global viability assessment of the European eel. Glob Change Biol https://doi.org/10.1111/gcb.12972
Biseau A, Diaz E (eds) (2021) Workshop of the future of eel advice (WKFEA). ICES Sci Rep 3(13):67
Boëtius J, Harding EF (1985) List of Atlantic and Mediterranean Anguilla leptocephali: Danish material up to 1966. Dana 4:163–249
Bonhommeau S, Chassot E, Planque B, Rivot E, Knap AH, Le Pape O (2008) Impact of climate change on eel populations of the Northern Hemisphere. Mar Ecol Prog Ser 373:71–80
Bonhommeau S, Blanke B, Tréguier A-M, Grima N, Rivot E, Vermard Y, Greiner, Le pape O (2009) How fast can the European eel (*Anguilla anguilla*) larvae cross the Atlantic ocean? Fish Oceangr 18(6):371–385
Bonhommeau S, Castonguay M, Rivot E, Sabatié R, Le Pape O (2010) The duration of migration of Atlantic Anguilla larvae. Fish Fish 11:289–306. https://doi.org/10.1111/j.1467-2979.2010.00362
Bornarel V, Lambert P, Briand C, Antunes C, Belpaire C, Ciccotti E, Diaz E, Diserud O, Doherty D, Domingos I, Evans D, de Graaf M, O'Leary C, Pedersen M, Poole R, Walker A, Wickström H, Beaulaton L, Drouineau H (2018) Modelling the recruitment of European eel (*Anguilla anguilla*) throughout its European range. ICES J Mar Sci 75:541–552
Bourgeaux J, Teichert N, Gillier JM, Danet V, Feunteun E, Acou A et al (2022) Modelling past migrations to determine efficient management rules favouring silver eel escapement from a large regulated floodplain lake. J Nat Conserv 67:126192
Bourillon B, Acou A, Trancart T, Belpaire C, Covaci A, Bustamante P, Faliex E, Amilhat E, Malarvannan G, Virag L, Aarestrup K, Bervoets L, Boisneau C, Boulenger C, Gargan P, Becerra-Jurado G, Lobón-Cerviá, Maes, G. E, Pedersen MI et al (2020) Assessment of the quality of European silver eels and tentative approach to trace the origin of contaminants—A European overview. Sci Total Environ 743:140675. https://doi.org/10.1016/j.scitotenv.2020.140675
Bourillon B, Feunteun E, Acou A, Trancart T, Teichert N, Belpaire C, Aarestrup K, Walker A, Righton D (2022) Anthropogenic contaminants shape the fitness of the endangered European eel: a machine learning approach. Fishes 7(5):274
Brämick U, Fladung E, Simon J (2016) Stocking is essential to meet the silver eel escapement target in a river system with currently low natural recruitment. ICES J Mar Sci 73(1):91–100
Broecker WS (1979) The great ocean conveyor. Oceanography 4:79–89
Bru N, Prouzet P, Lejeune M (2009) Daily and seasonal estimates of the recruitment and biomass of glass-eels runs (*Anguilla anguilla*) and exploitation in the Adour open estuary (Southwestern France). Aquat Living Resour 22(4):509–524
Buslé J (1994) L'anguille Européenne *Anguilla anguilla*, un poisson sensible aux stress environnementaux et vulnérable à diverses atteintes pathogènes. Bull Fr Pêche Piscic 335:237–260
Byl S, Schouls S (1990) Quelques préjugés d'Aristote en ichtyologie et leur survivance chez certains de ses successeurs jusqu'au XVIIIe siècle. Revue belge de plilologie et d'histoire 68(2):305–314
Castonguay M, Hodson PV, Couillard CM, Eckersley MJ, Dutil JD, Verreault G (1994) Why is recruitment of the American eel, Anguilla rostrata, declining in the St Lawrence river and Gulf? Can J Fish Aquat Sci 51:479–488. https://doi.org/10.1139/f94-050
Chang Y-LK, Miyazawa Y, Miller MJ, Tsukamoto K (2018) Potential impact of ocean circulation on the declining Japanese eel catches. Sci Rep Nat Res 8:5496
Chang Y-LK, Feunteun E, Miyazawa Y, Tsukamoto K (2020) New clues on the Atlantic eels spawning behavior and area: the Mid-Atlantic Ridge hypothesis. Sci Rep Nat Res 10:15981
Chang YL, Feunteun E, Miyazawa Y (2022) Reply to: no new evidence for an Atlantic eels spawning area outside the Sargasso Sea. Sci Rep 12 (1):1–4

Ciccotti E, Morello EB (2023) European eel in the Mediterranean Sea—outcomes of the GFCM research programme. Studies and review. No 103 (General fisheries commission for the Mediterranean), Rome, FAO. https://doi.org/10.4060/cc7252en

Clark JR, Cole M, Lindeque PK, Fileman E, Blackford J, Lewis C, Galloway TS (2016) Marine microplastic debris: a targeted plan for understanding and quantifying interactions with marine life. Front Ecol Environ. https://doi.org/10.1002/fee.1297

Claveroso M, Hermoso V (2015) Historical data to plan the recovery of the European eel. J Appl Ecol 52:960–968

Dekker W, Beaulaton L (2016) Faire mieux que la nature ? The history of Eel restocking in Europe. Environ Hist 22:255–300

Delecluse P (2014) Chapter 2. The ocean and the climate system. In: Monaco A, Prouzet P (eds) Ocean and the earth system—Oceanography and Marine biology series—seas and oceans set—ISTE—WILEY Eds, pp 55–102

Desaunay Y, Guérault D (1997) Seasonal and long-term changes in biometrics of eel larvae: a possile relationship between recruitment variation and North Atlantic ecosystem productivity. J Fish Biol 51:317–339

Desprez M, Crivelli AJ, Lebel I, Massez G, Gimenez O (2013) Demographic assessment of a stocking experiment in European Eels. Ecol Freshw Fish. https://doi.org/10.1111/eff.12035

Drouineau H, Durif C, Castonguay M, Mateo M, Rochard E, Verreault G, Yokouchi K, Lambert P (2018) Freshwater eels: a symbol of the effects of global change. Fish Fish. https://doi.org/10.1111/faf.12300

Drouineau H, Lambert P, Acou A, Beaulaton L, Boulenger C (2022) Estimations des possibilités de captures d'anguilles de moins de 12 cm pour la saison 2022–2023. Rapport INRAe, OFB, MNHN (2023) Avis du Comité scientifique, p 26

Ducrotoy J-P (2018) Les milieux estuariens et littoraux—une approche scientifique pour les préserver et les exploiter durablement. Editeur Lavoisier—collection environment, p 466

Durif CM, Stockhausen HH, Skiftesvik AB, Cresci A, Nyqvist D, Browman HI (2021) A unifying hypothesis for the spawning migrations of temperate anguillid eels. Fish Fish 23:358–375

Elie P, Feunteun E, Rigaud C (1990) The inshore brackish water domain of the French Atlantic coast: ecological functions for the exploited species—impact of physical development. Bull Ecol 21:33–38

Erdman W, Kmita H, Kosicki JZ, Kaczmarek L (2021) How the magnetic field influences life on earth—an integrated approach to geomagnetobiology. Origins Life Evol Biospheres 51:231–257

Feunteun E (2002) Restoration and management of the European eel : an impossible bargain? Ecol Eng 18: 575–591

Feunteun E, Robinet T (2013) Freshwater eels and people in France. In: Tsukamoto K, Kuroki M (eds) Eels and humans. Springer Tokyo, p 177

Feunteun E, Prouzet P (2020) Forty years of decline and 10 years of management plan: are European Eels (*Anguilla anguilla*) recovering? In: Ceccaldi HJ, Hénocque Y, Komatsu T, Prouzet P, Sautour B, Yoshida J (eds) Evolution of marine coastal ecosystems under the pressure of global changes. Proceedings of coast Bordeaux symposium. Springer Nature Switzerland, pp 269–295

Feunteun E, Rigaud C, Elie P, Lefeuvre JC (1992) Le marais doux endigué de Bourgneuf Machecoul (Pays de Loire). Premiers éléments de connaissance de la faune piscicole. Relations ichtyofaune-habitat et problèmes majeurs de gestion. Rev Sci Eau 5:509–528

Feunteun E, Rigaud C, Elie P, Lefeuvre JC (1999) Les peuplements piscicoles des marais littoraux de l'ouest de la France: un patrimoine à gérer ? Bull Fr Pêche et Piscic 352:63–79

Feunteun E, Laffaille P, Robinet T, Briand C, Baisez A, Olivier JM, Acou A (2003) A review of upstream migration and movements in inland waters by Anguillid eels: towards a general theory. In: Aida K, Tsukamoto K, Yamauchi K (eds) Eel biology, chapter 14, pp 191–213

Feunteun E, Miller MJ, Carpentier A, Aoyama J, Dupuy C, Kuroki M, Pagano M, Réveillac E, Sellos D, Watanabe S, Tsukamoto K, Otake T (2015) Stable isotopic composition of anguilliform leptocephali and other food web components from west of the Mascarene Plateau. Progr Oceanogr 137:69–83

Feunteun E, Serranito B, Le Peru Y (2022) Etude ADRAF: analyse des données de 10 années de repeuplement Anguille en France—2011/2021, p 49

Freese M, Sühring R, Pohlmann JD, Wolschke H, Magath V, Ebinghau R, Hanel R (2016) A question of origin: dioxin-like PCBs and their relevance in stock management of European eels. Ecotoxicology 25:41–55. https://doi.org/10.1007/s10646-015-1565-y

Gandolfi-Hornyold A (1931) Un étrange poisson: l'Anguille—ses mœurs et son intérêt économique. La Terre et La Vie. Revue d'Histoire Naturelle 1(11):1931, 657–674

Gagnaire PA, Minegishi Y, Zenboudji S, Valade P, Aoyama J, Berrebi P (2011) Within-population structure highlighted by differential introgression across semipermeable barriers to gene flow in Anguilla marmorata. Evolution 65:3413–3427. https://doi.org/10.1111/j.1558-5646.2011.01404.x

Ghiotti S, Molle F (2008) Vers de grands barrages en Méditerranée? Maghreb- Machrek, vol 196, p 16

Grassi B (1896) The reproduction and metamorphosis of the common eel (*Anguilla vulgaris*). Proc R Soc Lond 50:260–271

Higuchi T, Watanabe S, Manabe R, Kaku T, Okamura A, Yamada Y, Miller MJ, Tsukamoto K (2018) Tracking *Anguilla japonica* silver eels along the West Marina Ridge using pop-up archival transmitting tags. Zool Stud 57:24. https://doi.org/10.6620/ZS.2018.57-24

Holmqvist N, Stenroth P, Berglund O, Nyström P, Olsson K, Jellyman D, McIntosh AR, Larsson P (2006) Low levels of persistent organic pollutants (POPs) in New Zealand eels reflect isolation from atmospheric sources. Environ Pollut 141:532–538. https://doi.org/10.1016/j.envpol.2005.08.052

IPBES (2019) Global assessment report on biodiversity and ecosystem services of the Intergovernmental science-policy platform on biodiversity and ecosystem services. In: Brondizio ES, Settele J, Díaz S, Ngo HT (eds) IPBES secretariat. Bonn, Germany, p 1148. https://doi.org/10.5281/zenodo.3831673

Jacobsen MW, Pujolar JM, Pedersen L, Hansen MM (2018) Single nucleotide polymorphism markers for assessing potential maternal population structure in European eel (*Anguilla anguilla*). Conserv Genet Resour 10:907–909. https://doi.org/10.1007/s12686-017-0917-8

Jespersen P (1942) Indo-Pacific Leptocephalids of the genus Anguilla-systematic and biological studies. Dana Rep 22:1–128

Kaifu K, Yokouchi K, Miller MJ, Washitani I (2021) Management of glass eel fisheries is not a sufficient measure to recover a local Japanese eel population. Mar Policy 134:104806

Kangur A, Kangur P, Kangur K (2002) The stock and yield of the European eel, *Anguilla anguilla* (L.), in large lakes of Estonia. Proc Estonian Acad Sci Biol Ecol 51(1):45–61

Kim H, Kimura S, Shinoda A, Kitagawa T, Sasai Y, Sasaki H (2007) Effect of El Nino on migration and larval transport of the Japanese eel (*Anguilla japonica*). ICES J Mar Sci 64:1387–1395

Kishida, Kanto (2013) Reconstruction on the catch of glass eel in Japan. Bull Jpn Soc Fish Oceanogr (Suisan Kaiyo Kenkyu) 77(3):164–166

Kleckner RC, McCleave JD (1985) Spatial and temporal distribution of American eel larvae in relation to North Atlantic ocean current system. Dana 4:67–92

Knights B (2003) A review of the possible impacts of long-term oceanic and climate changes and fishing mortality on recruitment of anguillid eels of the Northern Hemisphere. Sci Total Environ 310:237–244

Koster WM, Aarestrup K, Birnie-Gauvin K, Church B, Dawson D, Lyon J, O'Connor J, Righton D, Rose D, Westerberg H, Stuart I (2021) First tracking of the oceanic spawning migrations of Australasian short-finned eels (*Anguilla australis*). Sci Rep 11:22976. https://doi.org/10.1038/s41598-021-02325-9

Kuroki M, Miller MJ, Tsukamoto K (2014) Diversity of early life-history trait as in freshwater eels and the evolution of their oceanic migrations. Can J Zool 92:749–770

Labedan M, Sagnes P (2021) GT2 Estimation des mortalités aux obstacles—Délivrables 2.1.1 & 2.2.3.—Quelles données collecter et comment estimer les mortalités d'anguilles dévalantes au niveau des aménagements hydroélectriques. Rapport SUDOANG, Programme Interreg Sudoe, p 66

Laffaille P, Baisez A, Rigaud C, Feunteun E (2004) Habitat preferences of different European eel size classes in a reclaimed marsh: a contribution to species and ecosystem conservation. Wetlands 24(3):642–651

Lasne E, Acou A, Vila-Gispert A, Laffaille P (2008) European eel distribution and body condition in a river floodplain: effect of longitudinal and lateral connectivity. Ecol Freshw Fish Ecol Freshw Fish 17(4):567–576. https://doi.org/10.1111/j.1600-0633.2008.00307.x

Le Clerc J (1930) L'anguille dans le bassin de La Loire. Bull Fr De Piscic 19:146–152

Le Luherne E, Réveillac E, Ponsero A, Sturbois A, Ballu S, Perdriau M, Le Pape O (2016) Fish community responses to green tides in shallow estuarine and coastal areas. Estuar Coast Shelf Sci 175:79–92

Le Pape O, Holley J, Guérault D, Désaunay Y (2003) Quality of coastal and estuarine essential fish habitats: estimations based on the size of juvenile common sole (*Solea solea* L.) Estuar. Coast Shelf Sci 58:793–803

Lecomte-Finiger R (1992) Growth history and age at recruitment of European glass eels (*Anguilla anguilla*) as revealed by otolith microstructure. Mar Biol 114:205–210

Lefeuvre J-C, Fustec E, Barnaud G (2000) Chapitre 1. De l'élimination à la reconquête des zones humides. In: Fustec E, Lefeuvre J-C (eds) Fonctions et valeurs des zone humides. Dunod, Paris, pp 1–16

Lin Y-S, Poh Y-P, Tzeng C-S (2001) A phylogeny of freshwater eels inferred from mitochondrial genes. Mol Phylogenet Evol 20(2):252–261

Lin YL, Lozys L, Shiao J-C, Iizuka Y, Tzeng W-N (2007) Growth differences between naturally recruited and stocked European eel *Anguilla anguilla* from different habitats in Lithuania. J Fish Biol 71:1773–1787

Machut LS, Limburg KE, Schmidt RE, Dittman D (2007) Anthropogenic impacts on American Eel demographics in Hudson river tributaries, New York. Trans Am Fish Soc 136:1699–1713

Maes G, Volckaert FAM (2007) Challenges for genetic research in European eel management. ICES J Mar Sci 64:1463–1471. https://doi.org/10.1093/icesjms/fsm108

Martin J, Daverat F, Pécheyran C, Als TD, Feunteun E, Réveillac E (2010) An otolith microchemistry study of possible relationships between the origins of leptocephali of European eels in the Sargasso Sea and the continental destinations and relative migration success of glass eels. Ecol Freshw Fish. https://doi.org/10.1111/j.1600-0633.2010.004444

Mateo M, Lambert P, Tétard S, Drouineau H (2017) Impacts that cause the highest direct mortality of individuals do not necessarily have the greatest influence on temperate eel escapement. Fish Res 193:51–59. https://doi.org/10.1016/j.fishres.2017.03.024

Mateo M, Drouineau H, Pella H, Beaulaton L, Elsa A, Bardonnet A, Domingos I, Fernandez-Delgado C, De Miguel Rubio R, Herrera M, Korta M, Zamora Lluis, Diaz E, Briand C (2021) Atlas of European Eele distribution (*Anguilla anguilla*) in Portugal, Spain and France. (1.0.1). Zenodo. https://doi.org/10.5281/zenodo.6384022

Matondo BN, Benitez J-Ph, Dierckx A, Renardy S, Rollin X, Colson D, Baltus L, Romain VRM, Ovidio M (2021) What are the best upland river characteristics for glass eel restocking practice? Sci Total Environ 784:147042

Miller MJ, Tsukamoto K (2004) An introduction to leptocephali biology and identification. Ocean Research institute—the university of Tokyo report, p 110

Miller MJ et al (2015a) A century of research on the larval distributions of the Atlantic eels: a re-examination of the data. Biol Rev 90:1035–1064. https://doi.org/10.1111/brv.12144

Miller MJ, Feunteun E, Aoyama J, Watanabe S, Kuroki M, Lecomte-Finiger R, Minegishi Y, Robinet T, Réveillac E, Gagnaire PA, Berrebi P, Tsukamoto K, Otake T (2015b) Biodiversity and distribution of leptocephali west of the Mascarene Plateau in the southwestern Indian Ocean. Prog Oceanogr. https://doi.org/10.1016/j.pocean.2015.05.026

Miller MJ, Feunteun E, Tsukamoto K (2015c) Did a "perfect storm" of oceanic changes and continental anthropogenic impacts cause northern hemisphere anguillid recruitment reductions? ICES J Marine Sci. https://doi.org/10.1093/icesjms/fsv063

Miller MJM, Hanel R, Feunteun E, Tsukamoto K (2020) The food source of Sargasso Seau leptocephali. Mar Biol 167:57. https://doi.org/10.1007/s00227-020-3662-6

Minegeshi Y, Aoyama J, Inoue JG, Miya M, Nishida M, Tsukamoto K (2005) Molecular phylogeny and evolution of the freshwater eels genus *Anguilla* based on the whole mitochondrial genome sequencies. Mol Phylogenet Evol 34(1):134–146. https://doi.org/10.1016/j.ympev.2004.09.003

Minto C, Myers RA, Blanchard W (2008) Survival variability and population density in fish populations. Nature 452. https://doi.org/10.1038/nature06605

Mochioka N, Iwamizu M (1996) Diet of anguilloid larvae: leptocephali feed selectively on larvacean houses and fecal pellets. Mar Biol 125:447–452. https://doi.org/10.1007/BF00353257

Oliver IW, Macgregor K, Godfrey JD, Harris L, Duguid A (2015) Lipid increases in European eel (*Anguilla anguilla*) in Scotland 1986–2008: an assessment of physical parameters and the influence of organic pollutants. Environ Sci Pollut Res 22:7519–7528. https://doi.org/10.1007/s11356-015-4116-4

Ondulami M, Vallet G (2012) Modelling and mathematical analysis of the glass-eel migration in the Adour river estuary. Math Model Nat Phenom 7(3):168–185

Otake T, Nogami K, Maruyama K (1993) Dissolved and particulate organic matter as possible food sources for eel leptocephali. Mar Ecol Prog Ser 92:27–34

Pedersen MI, Rasmussen GH (2015) Yield per recruit from stocking two different sizes of eel (*Anguilla anguilla*) in the brackish Roskilde Fjord. ICES J Mar Sci. https://doi.org/10.1093/icesjms/fsv167

Perez M (2021) Menace sur les zones humides note CNRS/INSU du 2/02/2021. Menace sur les zones humides ! | CNRS Terre and Univers

Poncet F (1984) Les zones humides du littoral breton. Caractères et évolution. Thèse de l'Université de Bretagne Occidentale. UBO, Brest, p 349

Prouzet P, Bouvet J-C, Bru N, Duquesne E, Antunes J-C, Demasceno-Oliveira A, Boussouar A, De Casamajor M-N, Sanchez F, Lissardy M (2008) Chapitre 7—Indicateurs de recrutement estuarien. In L'anguille Européenne—Indicateurs d'abondance et de Colonisation. Editions Quae. Collection Savoir-faire, pp 223–274

Prouzet P, Odunlami M, Duquesne E, Boussouar A (2009) Analysis and visualization of the glass-eel behaviour (*Anguilla anguilla*) in the Adour estuary and estimate of its upstream migration speed. Aquat Living Resour 22(4):525–534

Prouzet P, Lecomte M, Herfaut J, Mas L, Porcher N (2023) Sustainable development and responsible exploitation. As an example, the management and exploitation of diadromous species in the context of small-scale fisheries. La mer vol 61. Ed Société Franco-japonaise d'Océanographie Tokyo, pp 275–302

Pujolar JM, Maes G (2016) Evolutionary genomics of North Atlantic Eels: current status and perspectives. In: Arai T (ed) Biology and ecology of anguillid eels, CRC Press, pp 36–51

Pujolar JM, Jacobsen MW, Als TD, Frydenberg J, Munch K, Jónsson B, Jian JB, Cheng L, Maes GE, Bernatchez L, Hansen MM (2014) Genome-wide single-generation signatures of local selection in the panmictic European eel. Mol Ecol 23:2514–2528. https://doi.org/10.1111/mec.12753

RAMSAR report (2018) Perspectives mondiales des zones humides—L'état mondial des zones humides et de leurs services à l'humanité 2018. Gland Suisse: Secrétariat de la Convention RAMSAR, p 85

Réveillac E, Feunteun E, Berrebi P, Gagnaire PA, Lecomte-Finiger R, Bosc P, Robinet T (2008) Anguilla marmorata larval migration plasticity as revealed by otolith microstructural analysis. Can J Fish Aquat Sci 65(10):2127–2137

Riemann L, Alfredsson H, Hansen MM, Als TD, Nielsen TG, Munk P, Aarestrup K, Maes GE, Sparholt H, Petersen MI, Bachler M, Castonguay M (2010) Qualitative assessment of the diet of European eel larvae in the Sargasso Sea resolved by DNA barcoding. Biol Lett 6:819–822

Righton D, Aarestrup K, Jellyman D, Sébert P, Den Thillart V, Tsukamoto K (2012) The Anguilla spp. Migration problem: 40 million years of evolution and two millenia of speculation. J Fish Biol 81:365–386

Righton D, Westerberg H, Feunteun E, Økland F, Gargan P, Amilhat E, Metcalfe J, Lobon-Cervia J, Sjöberg N, Simon J, Acou A, Vedor M, Walker A, Trancart T, Brämick U, Aarestrup K (2016) Empirical observations of the spawning migration of European eels: the long and dangerous road to the Sargasso Sea. Sci Adv 2:e1501694

Righton D, Piper A, Aarestrup K, Amilhat E, Belpaire C, Casselman J, Castonguay M, Díaz E, Dörner H, Faliex E, Feunteun E, Fukuda N, Hanel R, Hanzen C, Jellyman D, Kaifu K, McCarthy K, Miller M, Pratt T, Sasal P, Schabetsberger R, Shiraishi H, Simon G, Sjoberg N, Steele K, Tsukamoto K, Walker A, Westerberg H, Yokouchi K, Gollock M (2021) Important questions to progress science and sustainable management of anguillid eels. Fish Fish. https://doi.org/10.1111/faf.12549

Robinet T, Feunteun E (2002) Sublethal effects of exposure to chemical compounds: a cause for the decline in Atlantic eels? Ecotoxicology 11:265–277

Scaccuto A (2022) La reproduction des anguilles dans les textes zoologiques de l'Antiquité. RursuSpicae. Transmission, réception et réécriture des textes de l'Antiquité au Moyen Âge. https://doi.org/10.4000/rursuspicae.2474

Schabetsberger R, Miller MJ, Dall'Olmo G, Kaiser R, Økland F, Watanabe S, Aarestrup K, Tsukamoto K (2016) Hydrographic features of anguillid spawning areas: Potential signposts for migrating eels. Mar Ecol Prog Ser 554:141–155. https://doi.org/10.3354/meps11824

Schmidt J (1922) The breeding places of the eel. Philos Trans R Soc B 385(211):279–316

Schmidt J (1923) Breeding places and migrations of the eel. Nature 111:51–54. https://doi.org/10.1038/111051a0

Scotese CR, Wright N (2018) PALEOMAP paleodigital elevation models (PaleoDEMS) for the Phanerozoic. Paleomap Proj

Shiraishi H, Crook V (2015) Eel market dynamics, *Anguilla* production, trade and consumption in East Asia. Traffic report, Tokyo, p 45

Simon J, Dörner H (2014) Survival and growth of European eels stocked as glass and farm-sourced eels in five lakes in the first years after stocking. Ecol Freshw Fish 23:40–48

Smith (DG) 1989 order Anguilliformes. FAO species identification guide for fishery purposes, the living marine resources of the western central Pacific, p 3

Susperregui N (coord) (2007) Caractérisation de la pêcherie professionnelle de civelles sur l'Isle—Bassin versant Gironde-Garonne-Dordogne. Rapport final Indicang, p 54

Teichert N, Tétard S, Trancart T, De Oliveira E, Acou A, Carpentier A, Bourillon B, Feunteun E (2020a) Towards transferability in fish migration models: a generic operational tool for predicting silver eel migration in rivers. Sci Total Environ 739:140069. https://doi.org/10.1016/j.scitotenv.2020.140069

Teichert N, Tétard S, Trancart T, Feunteun E, Acou A, De Oliveira E (2020b) Resolving the trade-off between silver eel escapement and hydropower generation with simple decision rules for turbine shutdown. J Environ Manage 261:110212. https://doi.org/10.1016/j.jenvman.2020.110212

Teichert N, Bourillon B, Suzuki K, Acou A, Carpentier A, Kuroki M, Righton D, Trancart T, Virag L, Walker A, Feunteun E (2023) Biogeographical snapshot of life-history traits of European silver eels: insights from otolith microchemistry. Aquat Sci 85:39. https://doi.org/10.1007/s00027-023-00940-4

Tesch FW (2003) The eel. Blackwell science, Oxford. https://doi.org/10.1002/9780470995389

Trancart T, Tudorache C, van den Thillart GEEJM, Acou A, Carpentier A, Boinet C, Gouchet G, Feunteun E (2014) The effect of thermal shock during diel vertical migration on the energy required for oceanic migration of the European silver eel. J Exp Mar Biol Ecol 463:168–172. https://doi.org/10.1016/j.jembe.2014.12.004

Trancart T, Feunteun E, Danet V, Carpentier A, Mazel V, Charrier F, Druet M, Acou A (2018a) Migration behaviour and escapement of European silver eels from a large lake and wetland system subject to water level management (Grand-Lieu Lake, France): new insights from regulated acoustic telemetry data. Ecol Freshw Fish 27(2):570–579. https://doi.org/10.1111/eff.12371

Trancart T, Têtard S, Acou A, Schaeffer F, Feunteun E, De Oliveira E (2018b) Silver eel downstream migration in the River Rhine, route choice, and its impacts on escapement: a 6-year telemetry study in a highly anthropized system. Ecol Eng 123:202–211. https://doi.org/10.1016/j.ecoleng.2018.09.002

Trancart T, Carpentier A, Acou A, Danet V, Elliott SAM, Feunteun E (2019) Behaviour of endangered European eels in proximity to a dam during downstream migration: novel insights using high accuracy 3D acoustic telemetry. Ecol Freshw Fish. https://doi.org/10.1111/eff.12512

Trancart T, Carpentier A, Acou A, Charrier F, Mazel V, Danet V, Feunteun E (2020) When safe dams kill: analyzing combination of impacts of overflow dams on the migration of silver eels. Ecol Eng 145. https://doi.org/10.1016/j.ecoleng.2020.105741

Trancart T, Teichert N, Lamoureux J, Gharnit E, Acou A, Oliveira D, Roy R, Feunteun E (2022) Possible strong effect of tidal power plant on European silver eels migration. Estuar Coast Shelf Sci 278(3):108–116. https://doi.org/10.1016/j.ecss.2022.108116

Tsukamoto K, Kuroki M, Watanabe S (2020) Common names for all species and subspecies of the genus Anguilla. Environ Biol Fish 103:985–991

Van Ginneken VJT, Maes GE (2005) The European eel (*Anguilla anguilla*, Linnaeus), its lifecycle, evolution and reproduction: a literature review. Rev Fish Biol Fisheries 15:367–398

Vøllestad LA (1992) Geographic variation in age and length at metamorphosis of maturing European eel: environmental effects and phenotypic plasticity. J Anim Ecol 61:41–48. https://doi.org/10.2307/5507

WGEEL country report (2023) ICES scientific reports 05(98):358

WGEEL report (2021) Joint EIFAAC/ICES/GFCM working group on eels (WGEEL) 3(85):213

WGEEL report (2023) Joint EIFAAC/ICES/GFCM working group on eels (WGEEL) 5(98):172

Wielgoss S, Gilabert A, Meyer A, Wirth T (2014) Introgressive hybridization and latitudinal admixture clines in North Atlantic eels. BMC Evol Biol 14:61. https://doi.org/10.1186/1471-2148-14-61

Wirth T, Bernatchez L (2003) Decline of North Atlantic eels: a fatal synergy? Proc Roy Soc B: Biol Sci 270:681–688. https://doi.org/10.1098/rspb.2002.2301

WKSTOCKEEL (2016) ICES report of the workshop on eel stocking. ICES CM 2016/SSGEPD vol 21, p 77

Wright RM, Piper AT, Aarestrup K, Azevedo JMN, Cowan G, Don A, Gollock M, Ramallo SR, Velterop R, Walker A, Westerberg H, Righton D (2022) First direct evidence of adult European eels migrating to their breeding place in the Sargasso Sea. Sci Rep 12:15362

Innovations in Coastal Management from Unconventional Origins: In the Pursuit of Exploring and Managing Coastal Zones, Unconventional Approaches and Methodologies Can Act as Catalysts for Fostering Research and Innovation

Patrick J. P. Vincent

1 Introduction

The coastal environment is "infinitely" complex, subject to the multiple interactions between nature (seasonal cycles, weather, water movements, etc.) and human activities (housing, fisheries, aquaculture and agriculture, tourism, industries, etc.) attributable to the two billion people living within 100 km from the coastlines, with half in low-elevation coastal zones (LECZ[1]) particularly exposed to changes in sea levels.

One can only acknowledge that human occupation and largely uncontrolled expansion of agricultural, industrial or urbanised areas, come with the degradation and destruction of coastal environments worldwide. The mobilisation and efforts of all stakeholders, experts from academic institutions, public authorities, private businesses, and citizens from local communities are essential to understand, mitigate, and hopefully reverse this trend.

By some estimates, 90% of European citizens think climate change is the biggest threat of the twenty-first century (European Commission, Directorate-General for

[1] low-elevation coastal zones (LECZ) are defined as "land with an elevation of up to 10 m and in hydrological connection to the sea" (McGranahan, Balk and Anderson 2007).

P. J. P. Vincent (✉)
Angers, France
e-mail: patrickjpvincent.pro@gmail.com

A.-C. Bennis et al. (eds.), *Constraints and Adaptations to Global Change at the Land-Sea Interface: For a Shared Ecological and Energy Transition*,
https://doi.org/10.1007/978-3-031-90050-1_10

Environment 2020), with particular concern for coastal areas. This global awareness favours an “exposure” to issues facing coastal environments of specialists from unrelated disciplines.

Whether labelled “serendipity”, “cross-disciplinarity”, or “curiosity-driven”, contributions from experts in one discipline to other disciplines are not rare, sometime with a profound impact like Herschel (a musician and astronomer) in physics, with the discovery of infrared in 1800 (Minkina 2021).

Although hundreds of institutions and thousands of scientists are specialised in Coastal sciences in a broad sense, the involvement of experts from unrelated disciplines may be highly valuable as will be illustrated by three examples in biology, physics, and environmental studies. Specific attention is given to why these experts decided to look into a question relevant to coastal sciences, although outside of their original academic domain.

2 A Research Unit Specialised in Machine Learning and AI Opens New Ways Towards Commercial Squid Aquaculture

2.1 The Physics and Biology Unit of Pr. Jonathan Miller at OIST, Japan

The Okinawa Institute of Science and Technology Graduate University is a new graduate school and research institute recently founded by the Japanese Government in 2011, with a remarkable campus on the subtropical island of Okinawa, Japan (Fig. 1). Headed by Prof. Karin Markides, it is designed as truly multinational and multidisciplinary, with more than one thousand employees from more than 50 nationalities, and 90 research units operating in a wide range of research domains (OIST Graduate University 2024).

Prof. Jonathan Miller heads the “Physics and Biology Unit” of OIST (Fig. 2). Co-discoverer with Aron K lug in 1985 of the “zinc finger” protein motif (Miller et al. 1985), his research interests include, among others, condensed matter physics, machine learning and AI via reverse engineering of animal intelligence (Miller 2019a, b), environmental perception, camouflage (Nakajima et al. 2022), and social interaction in cephalopods.

For researchers working on animal behaviour in a controlled laboratory environment, the yearlong availability of animals is a limiting factor if they cannot be bred in captivity. This was the case for Miller in 2015 when he started his research on cephalopod camouflaging and relied on fishermen’s catches in Okinawa waters, which are highly variable with seasons and weather conditions, not to mention the general concern about overfishing of cephalopods. Fortuitously in July 2016, OIST inaugurated its Marine Science Station (OMSS) in Seragaki Port, approximately seven kilometres northeast of the main campus (OIST MSS 2024), with core facilities

Fig. 1 Okinawa Institute of Science and Technology (OIST)

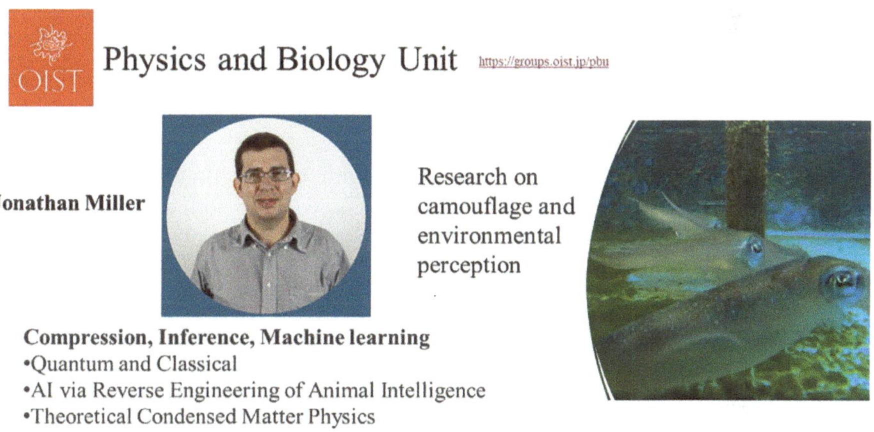

Fig. 2 Prof. Jonathan Miller and OIST physics and biology unit

and resources including indoor and outdoor water tanks, a closed circulation system and direct sea water pumping (since 2020). This new facility provided the opportunity to reduce the dependency of Miller's unit on wild catches of squids, allowing two of its researchers, Dr Zdenek Lajbner and Dr Ryuta Nakajima (visiting), to explore the development of a squid breeding method (Fig. 3), with substantial support and contribution from researchers and technicians from the Unit and from the OMSS.

The first step was the selection of an appropriate species from wild squids living in Okinawan waters that would be interesting scientifically, but also potentially suitable for future breeding in the laboratory.

Fig. 3 Dr Zdenek Lajbner and Dr Ryuta Nakajima from Miller's unit, and OIST Marine science station in Seragaki Port, Onna Village, Okinawa

2.2 Selection of a Squid Species Suitable for Breeding and Behavioural Research in the Laboratory

Okinawa coastal waters host a great diversity of cephalopod families and species with new ones regularly discovered (Sanchez et al. 2019), (Reid et al. 2023). Several species are highly valued economically in the Japanese market and support an active Okinawan fishing industry. It was natural for Miller and his team to collaborate with local fishermen to select potential candidates for breeding with the following desirable features: interesting complex behaviour, advanced camouflaging abilities, no threat of overfishing, relatively small size for holding in indoor tanks and previous works about breeding in captivity. One squid species met these conditions, *Sepiotheutis* sp.2 known as Aori-ika or Shiro-ika in Japanese, belonging to the oval squid species complex *Sepioteuthis lessoniana.*[2] It is fished commercially in Okinawa and sold fresh as a premium product in Japan, with a fluctuating population in decline due to overfishing like most commercialised squid species. Miller's researchers opted then for *S. Lessoniana* as an animal model and started studying reproductive cycle and breeding protocol in captivity.

2.3 "Non-marine Scientists" Experiment with Squid Breeding

Benefiting from the Marine Science Station resources and combining field observations, literature review and experimentation, Zabner, Nakajima and colleagues passed through successive stages leading to a viable squid breeding method (Fig. 4).

[2] The oval squid complex S. lessoniana is divided into three genetically distinct members (Imai and Misuzu 2012).

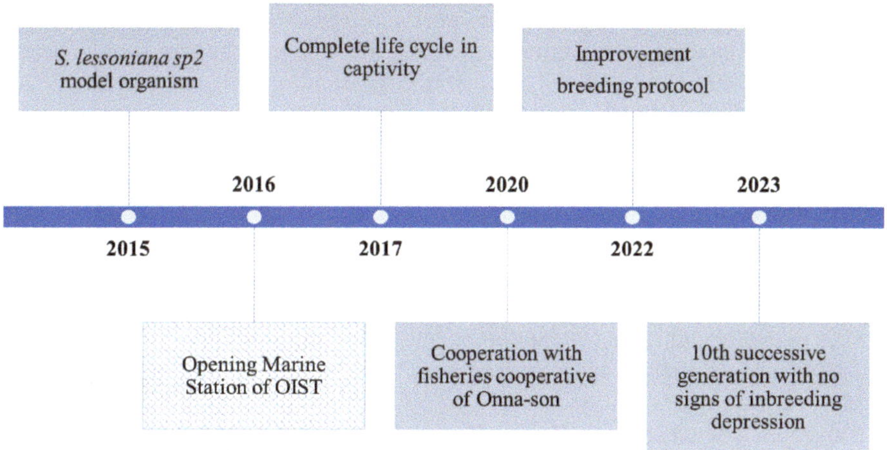

Fig. 4 Key steps in the breeding of squid *S. lessoniana* in captivity

In 2017 they collected strings of eggs from natural spawning grounds and could close one reproductive cycle in captivity (Lajbner et al. 2018); in 2019 five generations were successfully bred (Miller 2019a, b); in 2021 OIST and the local Fishery cooperative of Onna-son opened a pilot squid farm; in 2022, after successive improvements in breeding protocol and tank design and despite complications from typhoons and the Covid-19 pandemic, ten successive generations were raised with no signs of inbreeding depression (Lajbner et al. 2022). Recently in 2024, the peracetic acid treatment of egg infestation by a newly described parasitic copepod was reported (Zoral et al. 2024a, b; Lajbner et al. 2024a, b).

While development continues, the project's initial objective of a year-round supply of squids to Miller's unit for research on cephalopods behaviour can be considered achieved.

As is customary in research, OIST opens this development to the scientific community and provides reared squids to research institutions in Japan and abroad.[3]

2.4 Perspectives for the Development of Squid Aquaculture

The development of the *S. lessoniana* breeding method in captivity at OIST aimed at addressing specific research needs from Miller's Physics and Biology Unit. Once the project was completed, Lajbner, Nakajima and their colleagues transferred their know-how and experience "internally" to the OIST Animal Resources Section responsible for providing laboratory animals to the Institute (OIST ARS 2024).

[3] All OIST animal facilities and practices are fully accredited by AAALAC International since 2014.

Besides, and somewhat unexpectedly, the successful breeding of ten successive generations of oval squids, optimising juvenile diet and maintaining aquatic health situation (Lajbner et al. 2024a, b) attracted media attention in Japan (Juichiro 2022) and internationally (The Guardian 2022; Waycott 2023) and the interest of commercial ventures and fishermen cooperatives. Squid breeding is not new, but overfishing is worsening and any progress towards a sustainable aquaculture of squids could reduce pressure on the natural population. Commercial squid farms would be very different from the OIST research-oriented breeding with the known problems of waste management and the development of sustainable protein-rich diet formulas for predatory animals.

This potential commercial exploitation and technology transfer is in the hands of the University's patenting and licensing team "OIST Innovation" (OIST Innovation 2024).

Time will tell whether answering the research needs of Miller's Physics and Biology Unit stimulated the development of a sustainable squid farming economy and a reduction of fishing pressure on wild populations.

3 How the Sight of Concrete Tetrapods on the Japanese Coastline, Motivated a Physicist to Produce Energy

3.1 The OIST Quantum Wave Microscopy Unit

Prof. Tsumoru Shintake heads the Quantum Wave Microscopy Research Unit of the OIST Graduate University in Okinawa, Japan, developing and applying novel electron microscopy techniques (Shintake 2020). Shintake has a Ph.D. in Engineering and before joining OIST in 2012, contributed to experimental programs and infrastructure building in high-energy physics. In particular, he led the design, construction, commissioning and operation of the SACLA X-Ray Free Electron Laser located at the Spring-8 facility in Hyogo Prefecture (Japan), an endeavour for which he received the Gersch Budker Prize from the European Physical Society (Shintake 2014). Shintake's background is in "non-marine sciences" where "waves" belong to the quantic domain (Fig. 5).

Moving to OIST, Okinawa, Japan in November 2011, Shintake was appalled by the sight of concrete "tetrapods"[4] along kilometres of coastline. Omnipresent on Japanese coasts and Okinawa in particular, tetrapod-shaped wave breakers protect local populations, infrastructure, and natural habitats from erosion or extreme ocean conditions during storms and typhoons.

Acknowledging their usefulness did not deter Shintake from thinking about possible improvements or alternatives, and as a physicist became interested in the

[4] Tetrapods wave breakers are reinforced concrete structures, with a tetrahedral shape. Designed to dissipate the force of incoming waves eroding coastal areas. They were invented by P. Danel and P. Anglès d'Auriac and patented in 1950 by NEYRPIC SA, Grenoble, France (Patent. FR1017546A).

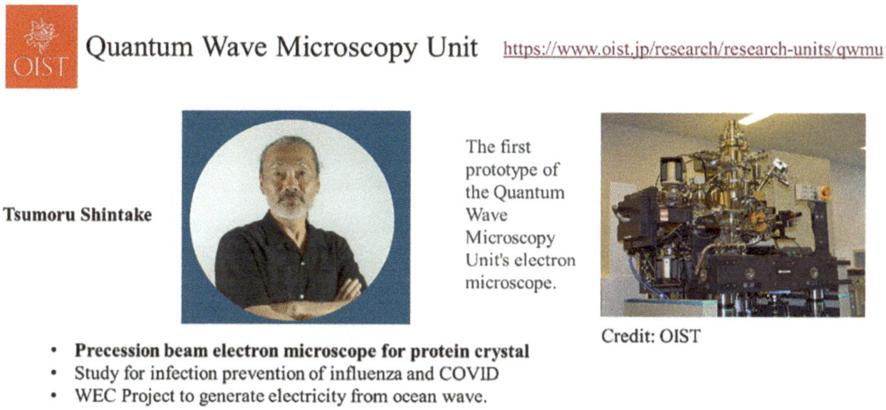

Fig. 5 Prof. Tsumoru Shintake and OIST quantum wave microscopy unit

considerable energy dissipated day and night by waves breaking on the shore. Could he contribute to converting this "infinite" energy into electric power, and add a function to these concrete "passive" structures (a particularly relevant reflection given the Fukushima Daichi nuclear disaster five months earlier on 11th March 2011)?

3.2 *From Intellectual Curiosity into a New Engineering Project*

Shintake knows about energy and is familiar with the engineering of complex research facilities, but converting coastal wave kinetic energy into electrical power was a new interest for him, and outside of his research unit core activity. However, the University agreed in 2014 to support a small, curiosity-driven side project on alternative electricity production technologies using breaking wave energy: the "Wave Energy Converter Project (WEC)".

Shintake started small, with mathematical modelling and conceptual drawings (Shintake 2016), testing makeshift prototypes, standing on the beach in a wetsuit, facing the waves with volunteers holding at arm-strength different propeller designs and collecting data (Shintake et al. 2017) (Fig. 6).

The development phase benefited from the Onna-son Fishing Association collaboration for towing and testing different designs of generator propellers after testing in a water tunnel (Halder et al. 2020), while industrial partners in Okinawa and mainland Japan built the working prototypes to measure performance and assess durability under field conditions (OIST EDU 2019).

An important milestone was reached with the signature in 2018 of an MOU[5] between OIST, the Ministry of Energy and Environment (MEE) of the Maldives

[5] MOU: Memorandum Of Understanding.

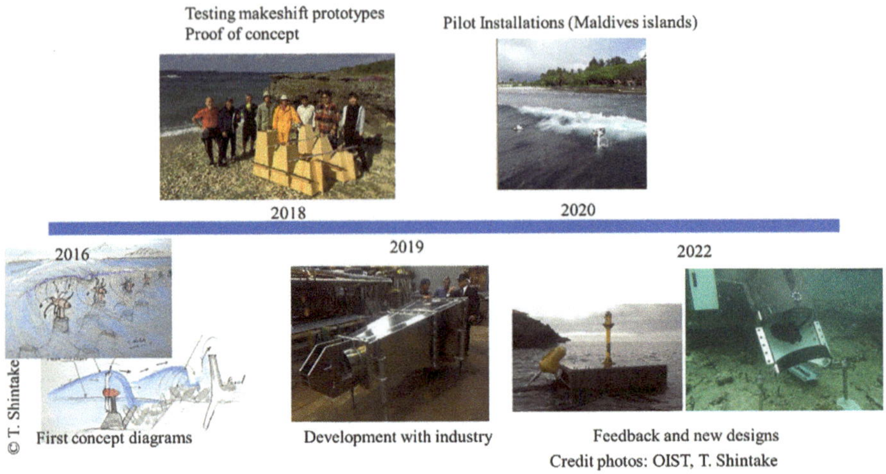

Fig. 6 Key milestones in Shintake's wave energy converter (WEC) project

and Kokyo Tatemono Co., Ltd, a Japanese real estate company in Tokyo, to demonstrate the WEC technology's viability under real conditions. The Maldives islands were considered a very appropriate environment. Quoting Tsumoru Shintake: "*The Maldives is an ideal place to test our Wave Energy Converters for three reasons: Firstly, the Maldives needs new ways to generate electricity... Secondly, the Maldives has a vested interest in finding renewable energy sources and reducing carbon emissions... Thirdly, the location of the Maldives makes it a suitable place to harness wave energy. The Maldives receives a constant stream of waves which have propagated across the Indian Ocean from the South Pole. Unlike Okinawa, the Maldives is not a hurricane or typhoon region, so the WEC units are at low risk of damage from extreme weather conditions.*" (Prothereo 2024).

Under the terms of the Agreement, a WEC pilot plant was installed in shallow waters on the coastline of the Holiday Inn Resort Kandooma island, Maldives (Takebe et al. 2018; Shintake et al. 2018). After one season, results were encouraging enough to but the Covid-19 pandemic derailed the plan. The program in Maldives was put on hold and has not resumed yet in 2024.

3.3 *Perspectives*

Shintake points out that typhoons hit Okinawa Island exposing the WEC installations to extreme winds, rains, and meters high breaking waves. Structures are damaged and energy conversion efficiency drops dramatically under extremely turbulent conditions. He is testing new designs, sturdier, and with higher energy conversion factors during stormy conditions. One prototype following this design, the "Ducted-WEC" was immersed in 2022 (Fig. 6) and generated a power output six times higher than using a bare turbine (OIST EDU 2019). The Fukui Prefecture, on the Sea of Japan, is interested in a new type of submerged tidal turbine, and the Okinawa Prefecture is considering the connection of a WEC pilot unit to the power grid.

Shintake and his team have accumulated more than 20,000 h of WEC operation and measurements, continuing. Numerous challenges remain, specific to the WEC technology such as facing storm explosive energy or levelling the highly fluctuating output of electrical power before injection into a power grid, or more generally encountered by structures at sea like biofouling, corrosion or access for maintenance.

Many projects aim at harnessing wave energy, either offshore or in shallow water and breakwaters. It is unclear which wave energy conversion technology will become economically viable outside of niche markets (Guo and Ringwood 2021) but one serious option is under development in Okinawa, Japan.

4 How a Research Centre in Molecular Biology and a Manufacturer of Customised Trucks Teamed up to Improve the Scientific Knowledge of Coastal Environments

4.1 From Molecules to Ecosystems

The European Molecular Biology Laboratory (Fig. 7), "*With 29 member states, the European Molecular Biology Laboratory (EMBL) has more than 110 independent research groups and service teams covering the spectrum of molecular biology at six sites in Barcelona, Grenoble, Hamburg, Heidelberg, EMBL-EBI Hinxton, and Rome.*" (EMBL 2024).

Fig. 7 EMBL's director general Edith heard and the main campus

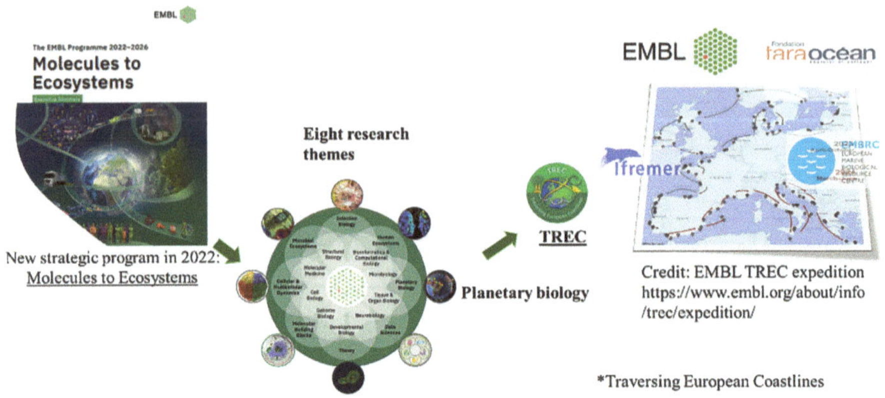

Fig. 8 The EMBL strategic program 2022–2026 and the TREC expedition with *Tara* ocean foundation

In 2022, the EMBL, headed by Dr Edith Heard, Director General appointed in 2019, launched a new five-year strategic programme, "Molecules to Ecosystems", with the ambition to "*... advance our understanding of ecosystems at the molecular level and study life in its context*" (EMBL PRGM 2022).

Among the eight main research themes, the "Planetary biology" theme aims to "*understand, from the molecular to the population level, how microbes, plants, and animals respond to each other and their environment*" (EMBL PRGM-PB 2024). Its flagship project "TREC" (EMBL TREC 2024) focuses on the environment at the interface between land and sea in Europewith the collaboration of the Tara Foundation (Fondation Tara Ocean 2024) and the European Marine Biological Resource Centre (EMBRC 2024) (Fig. 8). It could have been other environments.

4.2 The TREC Expedition: Sophisticated Organisation and Methods for AN Ambitious Scientific Endeavour

The TREC expedition is presented in detail on the EMBL's and the Tara Ocean Foundation's websites. A succinct description is given here as contextual information. The TREC expedition is planned from April 2023 to July 2024. Researchers will investigate 68,000 km of European coastline from Tallinn in Estonia to Thessaloniki in Greece. They will massively sample the coastal ecosystem in 120 locations, on land and in the sea along land-sea transects, in soil, sediment and shallow water at micro and macro scales. These samples will be analysed "fresh" in situ, for oceanographic, physical, chemical and biological measurements, preserved and shipped to partners' laboratories across Europe for more observations.

Large scientific expeditions always raise complex financial, human, organisational and technical challenges. The sheer scale and breadth of TREC sampling and analytical campaign make it stand apart, with its strategy of on-site observation of fresh samples using sensitive state-of-the-art equipment and technologies originally designed for the confinement of standard laboratories.

4.3 Reversal of the Service Platform Concept

To meet the scientific requirements of the TREC expedition, Rainer Pepperkok, Director of Scientific Core Facilities and Services and his colleagues had to "think out of the box". The EMBL is renowned internationally as a centre of excellence in molecular biology and for designing and operating groundbreaking research instrumentation and infrastructure for research in life sciences. With this background of expertise, Peppernock and his Team came up with the bold proposal "reversing the service platform concept": the sample would no longer go to the analytical platform, but the platform would go to the sample. This is already routinely done for simple analytical equipment but would be a new paradigm for highly sensitive instrumentation normally found in laboratory rooms with anti-vibration tables, sophisticated air conditioning and stabilised power supplies. An expensive and delicate analytical platform moving between dozens of marine stations, some of them isolated, along thousands of kilometres of European coastal roads is an invaluable tool for researchers, but an unusual challenge for engineers and Operations managers.

It became the task of Yannick Schwab (Team Leader and Head of the Electron Microscopy Core Facility) and Nikolaus Tobias Leisch (Operational Manager of Mobile Services) (Fig. 9) to turn the concept into reality.

To "put microscopes on wheels" means finding an industrial partner to customise and manufacture a sophisticated mobile unit able to travel on roads across Europe. After international consultation, EMBL contracted in July 2022 with the company Toutenkamion Group (Toutenkamion Group 2024), only six months after the Strategic program "Molecules to Ecosystem" was officially launched. The

Fig. 9 Advanced mobile laboratory (AML) project team

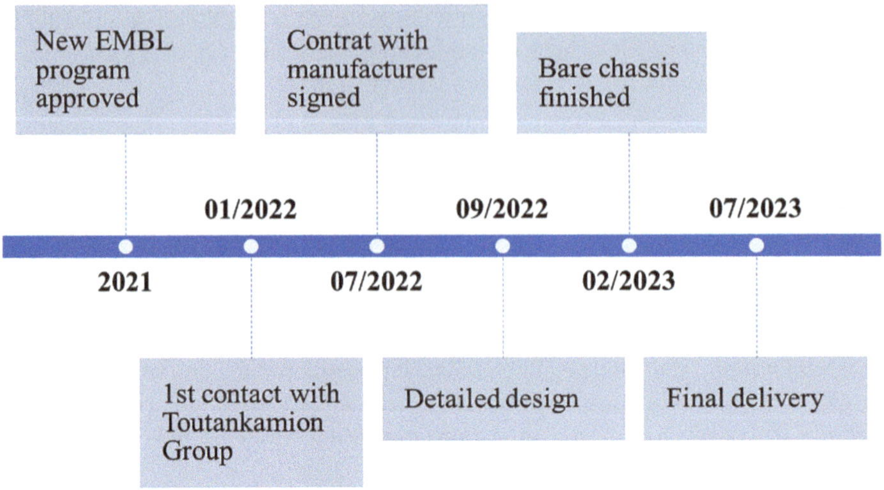

Fig. 10 Main milestones for construction of the EMBL mobile unit

mobile unit was delivered, fully equipped and ready to operate, in July 2023 (Toutenkamion-EMBL 2023) (Fig. 10). It is an impressively short time frame.

The mobile unit or "Advanced Mobile Laboratory" (AML), is a semi-trailer truck with expandable and retractable sides. The "expanded configuration" offers 45 m^2 of state-of-the-art laboratory space. It accommodates high-end research equipment such as a high-pressure freezer, plunge freezer, confocal microscope, and image-enabled cell sorter, with all ancillary functions and enough working space (Toutenkamion 2023). The AML opens new opportunities for scientists who can now perform on-site analysis of fresh samples (cryo-EM, cryo-ET, single-cell omics, Live high-resolution imaging etc.) in addition to the preservation and shipment to remote laboratories that was the only option during conventional expeditions.

In the "road configuration", sides are retracted to the size of a standard European trailer with instruments, equipment and fit-outs docked and tightly fixed (Fig. 11).

The AML's maiden trip in August 2023 was to Kristineberg Marine Station in Sweden (Ghosh (EMBL) 2023) and the expedition's final stop is Athens, Greece in the summer of 2024.

AML is not moving in isolation and is seconded by a fleet of vehicles specifically designed for field sample collection and advanced processing (Fig. 12) and moving closer to the sampling points, parallel with the *Tara* schooner skirting along the coastline in shallow waters the "EMBL Mobile service" (EMBL MS 2024).

The concept of a "reverse service platform" might seem obvious but turning it into an operational research facility with this level of equipment sophistication is not. It required expertise, experience and confidence to believe it was realistic and achievable.

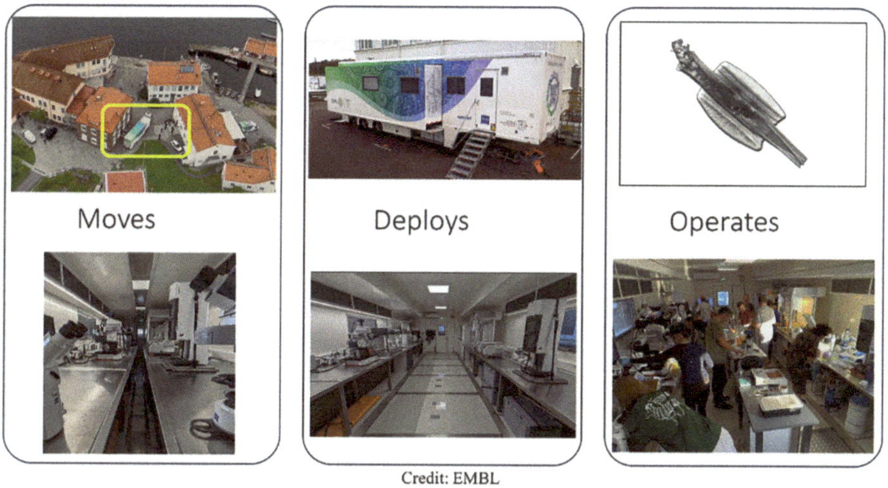

Fig. 11 The "expandable" EMBL advanced mobile laboratory (AML)

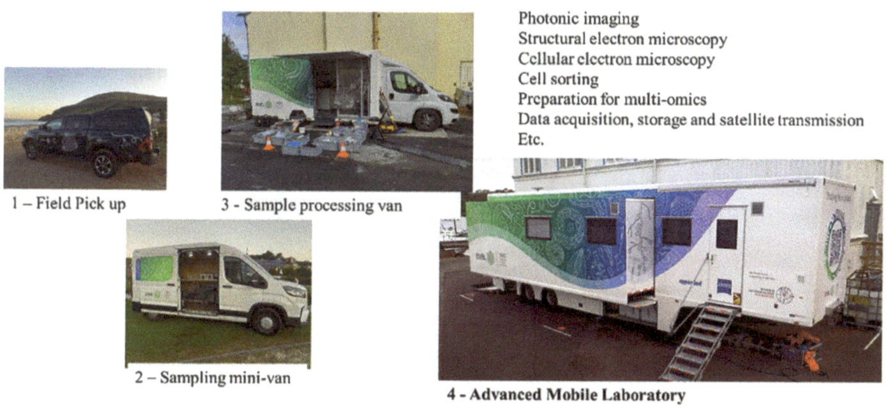

Fig. 12 EMBL mobile services

The development by EMBL of the Advanced Mobile Laboratory Service illustrates perfectly how a "non-marine science" institution can contribute to a better knowledge and understanding of coastal environments and their dynamics.

AML's concept and design are versatile and can be tailored to situations and projects where on-site analysis of "fresh" samples would add significant scientific value. Some come to mind like "regional inventory of biodiversity and natural resources", "environmental impact of natural or anthropogenic events", or "evaluation of alternative cultural practices on degraded agricultural soil" to name a few.

Will EMBL emulate other "AML-like" facilities in other scientific domains for field investigations, filling the gap between natural ecosystems and a controlled laboratory environment? Time will tell, but it is already a reality for "coastal sciences".

5 Conclusion

Three examples of innovations and developments in the field of "coastal sciences" (in aquaculture, renewable energy and environmental studies) have in common of be ing initiated by experts from disciplines unrelated to this field: mathematics and AI, quantum physics, and management of large laboratory facilities for research in life sciences.

These experts origin and motivations are equally diverse: a researcher at OIST in Japan working on animal behaviour and facing an irregular supply of animals, another researcher, also at OIST, "frustrated by the sight of concrete breakwater structures", or a state of the art European research institution, the EMBL, willing to "*explore ecosystems at a molecular level*".

Coastal areas are highly complex systems under environmental anthropogenic stress and demographic pressure, making their sustainable "management" extremely difficult

One question raised by the three examples is whether they suggest the existence of an untapped resource for innovation, with its original approach or if they are only anecdotal and marginal.

Acknowledgements I am very grateful to Prof. Jonathan Miller, Dr Zdeněk Lajbner, Dr Ryuta Nakajima at OIST PBU Unit, Prof. Tsumoru Shintake at OIST QWM unit and Dr Yannick Schwab at EMBL for their support, cooperation and constructive comments.

References

EMBL MS (2024) EMBL mobile service. https://www.embl.org/about/info/trec/mobile-labs/

EMBL PRGM (2022) Programme 2022–2026: molecules to ecosystems—executive summary. https://www.embl.org/documents/wp-content/uploads/2022/08/embl-programme-2022-2026-executive-summary.pdf

EMBL PRGM-PB (2024) Planetary biology. https://www.embl.org/about/programme/research-plans/planetary-biology/

EMBL TREC (2024) TREC. https://www.embl.org/about/info/trec/expedition

EMBL (2024) About. https://www.embl.org/about/

EMBRC (2024). About. https://www.embrc.eu/about

European Commission, Directorate-General for Environment (2020) Attitudes of European citizens towards the environment: report. https://data.europa.eu/doi/10.2779/902489

Fondation Tara Ocean (2024) About us. https://fondationtaraocean.org/en/who-we-are/

Ghosh S—EMBL (2023) Taking science on the road. https://www.embl.org/news/embletc/issue-101/taking-science-on-the-road/

Guo B, Ringwood J (2021) A review of wave energy technology from a research and commercial perspective. IET Renew Power Gener 15(14):3065–3090

Halder P, Takebe H, Pawitan K, Fujita J, Misumi S, Shintake T (2020) Turbine characteristics of wave energy conversion device for extraction power using breaking waves. Energies 13(4). https://www.mdpi.com/1996-1073/13/4/966

Imai H, Misuzu A (2012) Genetic diversity and genetic heterogeneity of bigfin reef squid *Sepioteuthis Lessoniana* species complex in Northwestern Pacific Ocean. In: Analysis of genetic variation in animals. InTech. https://doi.org/10.5772/35024

Juichiro I (2022) New and inventive squid farming method could save declining population. Japan2earth, vol 24

Lajbner Z, Gutnick T, Nishibayashi T, Ziadi F, Asada K, Iglesias T, Miller J, Kuba M (2018) Captive breeding of the oval squid (Aori-ika; S*epioteuthis sp.*). St Petersburg, Cepalopod international advisory council conference 2018 in book of abstracts, p 105

Lajbner Z, Nakajima R, Iglesias TL, Gutnick T, Masucci GD, Asada K, Nishi-bayashi T, Kim C, Miller J, Kuba MJ (2022) Culturing of oval squids (*Sepioteuthis sp.*). Cephalopod international advisory council conference. Sesimbra, Portugal, p 202

Lajbner Z, Zoral M, Zifcakova L, Emelianenko V, Arakaki N, Wang CC, Nakajima R, Miller J (2024) Tracing pathogens in cephalopod culture. World Fisheries Congress, Seattle

Lajbner Z, Nakajima R, Masucci GD, Iglesias TL, Asada K, Ueda N, Nishibayashi T, Ziadi-Künzli F, Kuba MJ, Miller J, Gutnick T (2024) Squid in the lab: methods for rearing and multigeneration breeding of an emerging cephalopod model species. (In Prep). Appl Sci

McGranahan G, Balk D, Anderson B (2007) The rising tide: assessing the risks of climate change and human settlements in low elevation coastal zones. Environ Urbanization 19(1):17–37

Miller J, McLachlan A, Klug A (1985) Repetitive zinc-binding domains in the protein transcription factor IIIA from *Xenopus* oocytes. EMBO J 4(6):1609–1614

Miller J (2019a) Psychophysics of cephalopod camouflage: what is the i/o response function of a cuttlefish? 10th KEK joint colloquium. Tsukuba

Miller J (2019b) Physics and biology unit annual report. https://groups.oist.jp/pbu/fy2019-annual-report

Minkina W (2021) How infrared radiation was discovered—range of this discovery and detailed, unknown information. Appl Sci 21(10):21. https://doi.org/10.3390/app11219824

Nakajima R, Lajbner Z, Kuba M, Gutnick T, Iglesias TL, Asada K, Nishibayashi T, Miller J (2022) Squid adjust their body color according to substrate. Sci Rep 12(1). https://doi.org/10.1038/s41598-022-09209-6

OIST ARS (2024) Animal resources section. https://groups.oist.jp/ars

OIST EDU (2019) Professor shintake's new ducted wave energy turbine (video). https://www.facebook.com/oistedu/videos/473071633261963/. Accessed 31 Aug 2024

OIST Graduate University (2024) About. https://www.oist.jp/. Accessed 31 Aug 2024

OIST Innovation (2024) Innovation. https://groups.oist.jp/innovation. Accessed 31 Aug 2024

OIST MSS (2024) Marine science station. https://groups.oist.jp/omsss/okinawa-marine-science-station

Prothereo S, OIST (2024) OIST to harness energy from ocean waves in the Maldives. https://www.oist.jp/news-center/press-releases/oist-harness-energy-ocean-waves-maldives

Reid A, Sato N, Jolly J, Strugnell J (2023) Two new pygmy squids, *Idiosepius kijimuna n. sp.* and *Kodama jujutsu n. gen., n. sp.* (Cephalopoda: Idiosepiidae) from the Ryukyu Islands, Japan. Mar Biol 170. https://doi.org/10.1007/s00227-023-04305-1

Sanchez G, Jolly J, Reid A, Sugimoto C, Azama C, Marlétaz F, Simakov O, Rokhsar DS (2019) New bobtail squid (Sepiolidae: Sepiolinae) from the Ryukyu islands revealed by molecular and morphological analysis. Commun Biol. https://doi.org/10.1038/s42003-019-0661-6

Shintake T, Shirasawa K, Takebe H (2017) Technical R&D on a surf zone WEC. 12th European wave and tidal energy conference (EWTEC2017). Cork, Ireland

Shintake T, Shirasawa K, Fujita J, Misumi S, Nagahama T, Shindou T, Taggart H, Takebe H (2018) Wave energy experiment in the Maldives. 4th Asian wave and tidal energy conference (AWTEC2018). Taipei, Taiwan. https://tethys-engineering.pnnl.gov/sites/default/files/publications/AWTEC2018-414.pdf

Shintake T (2014) Gersch Budker prize lecture. 5th international particle accelerator conference (IPAC2014). Dresden, Germany. https://doi.org/10.18429/JACoW-IPAC2014-THPPA02

Shintake T (2016) Harnessing the power of breaking waves. Asian wave and tidal energy conference (AWTEC2016). Singapore. https://doi.org/10.3850/978-981-11-0782-5_229

Shintake T (2020) OIST QWM unit. https://groups.oist.jp/qwmu/oist-wave-energy-project. Accessed 31 Aug 2024

Takebe H, Shirasawa K, Fujita J, Misumi S, Shintake T (2018) Development of OIST wave energy converter monitoring system for Maldives Island experiment. J Energy Power Eng 12:375–384. https://doi.org/10.17265/1934-8975/2018.08.001

The Guardian (2022) Japan makes squid farming breakthrough as wild catches plummet. https://www.theguardian.com/environment/2022/nov/04/japan-makes-squid-farming-breakthrough-as-wild-catches-plummet. Accessed 31 Aug 2024

Toutenkamion Group (2024) Home. https://www.toutenkamion-group.com/en/

Toutenkamion (2023) An advanced mobile laboratory to explore the diversity of European coasts. [https://www.toutenkamion-group.com/en/news/news-reader/mobile-lab-reseach-technologies.html. Accessed 31 Aug 2024

Toutenkamion-EMBL (2023) Press release. https://www.toutenkamion-group.com/files/tkg/product/unit/environment/22-00285/20230802-PressRelease-ToutenkamionGroup-EMBL.pdf. Accessed 31 Aug 2024

Waycott B (2023) Japan's squid farming breakthrough. World fishing and aquaculture. https://www.worldfishing.net/aquaculture/japans-squid-farming-breakthrough/1487378.article. Accessed 31 Aug 2024

Zoral MA, Lajbner Z, Nakamura S, Zifcakova L, Miller J (2024) Co-infection of *Liburna oophaga sp. nov.* and *Ikanecator primus* on cuttlefish (*Sepia pharaonis*) eggs and the effectiveness of a treatment. (in prep). BMC veterinary research

Zoral M, Lajbner Z, Zifcakova L, Miller J (2024) Peracetic acid treatment of squid eggs infected with parasitic copepod (*Ikanecator primus gen. et sp. nov.*). Sci Rep 14. https://doi.org/10.1038/s41598-024-65290-z

Development of Marine Renewable Energies and Impact on Socio-ecosystems

Variation of Fish Communities on Two Artificial Structures Along the French Atlantic Coast

C. Besson and M. Jorge

1 Introduction

The loss of key biological habitats is a rapidly expanding phenomenon affecting all ecosystems due to the increasing impact of human activities. The marine environment, in particular, is significantly impacted by the rise in anthropogenic pressures, with breeding, nursery, and feeding zones being the most affected (Worm et al. 2006). Moreover, the Atlantic coastline, like many coastal areas worldwide, has suffered and continues to suffer from the intensive exploitation of fishery resources (Worm et al. 2006).

To mitigate the effects of these degradations, stakeholders in the marine environment may choose to install structures to artificially recreate key habitats (Elliott et al. 2007; Seaman 2007). These artificial structures are also known as artificial reefs (AR) and are defined by Lacroix (1999) as:

> Deliberately submerged structures designed to create, protect, or restore a rich and diversified ecosystem. These structures can induce attraction, concentration, protection responses in animals, and, in some cases, an increase in the biomass of certain species.

Artificial structures have been implemented since the Middle Ages in Japan, but their introduction in France is more recent, with the first installations in the Gulf of Lion in 1985. It was only from the 2000s that the installation of these structures for biodiversity and resource conservation, as well as sustaining the economy associated with fishery production, gained momentum (Cépralmar 2015). Increasingly integrated into coastal zone management (Pioch 2008), these structures, depending on their design goals, can contribute to biological production, juvenile protection, and ecosystem revitalization (Leitão et al. 2007; Santos et al. 2011).

C. Besson (✉) · M. Jorge
Atlantique Landes Récifs, Vieux-Boucau-Les-Bains, France
e-mail: Celiabesson73@gmail.com

A.-C. Bennis et al. (eds.), *Constraints and Adaptations to Global Change at the Land-Sea Interface: For a Shared Ecological and Energy Transition*,
https://doi.org/10.1007/978-3-031-90050-1_11

It is important to note that AR alone cannot address overfishing issues and the lack of proper management. However, if well-targeted and sized, AR can contribute to local improvement (Charbonnel 2002). Furthermore, AR can be tools facilitating collaboration and synergy among marine stakeholders, serving as vectors for public awareness.

Following fishermen's observations of a decline in fishery resources on the Landes coast in the 1990s, the Atlantique Landes Récifs (ALR) association was established in 1996, bringing together various sea users. In 1999, the first artificial structure was immersed off the coast of Capbreton. In total, three different structures were installed, with the first ones being culverts, but they are currently present in only one of the three concessions. Subsequently, two other modules, taller in height, were installed in 2010 and 2015. Prefectural Order defining concessions and their uses, prohibits all types of fishing, trawling, or any type of diving not conducted by the site manager for scientific purposes. The assemblies of ichthyofauna and colonization patterns are monitored through Underwater Visual Census (UVC) conducted by the association's volunteer divers and various studies between 1999 and 2024. Biodiversity monitoring through citizen science is increasingly utilized, allowing citizen inclusion in science, environmental awareness, and cost-effective data acquisition covering a broader spatial and temporal range (Hill and Wilkinson 2004; Gillett et al. 2012).

Since the initial monitoring implemented in 1999 following the installation of the first reefs, ichthyofauna monitoring protocols have evolved significantly to adapt to citizen science constraints and biological constraints.

While constraints or requirements for biological monitoring exist, those related to AR design are just as important. It is recognized that the design of the AR structure can exert a significant influence on the ichthyofauna population dynamics (Moura et al. 2007; Lemoine et al. 2019). The more an artificial habitat resembles a natural habitat, the greater the possibility of reproduction on AR and the persistence of these communities in the ecosystem (Cannizzo et al. 2020). Thus, the architecture, materials, and geometry of artificial reefs require careful consideration to best meet ecological requirements (Lemoine et al. 2019). If the objective defined during AR installation includes increasing biodiversity, it involves designing the structures to create habitats sought by a maximum number of species, regardless of their life cycle stages. This may involve complicating AR to increase carrying capacity and biological functioning (Lemoine et al. 2019).

The two structures installed on the French Atlantic coastline have assumed different geometries and complexities, raising the question of whether there is a difference in ichthyofauna assemblage on these two forms of artificial reefs. The first hypothesis suggests, based on visual observations by divers, that there would be a difference in population dynamics between the two structures. The burial of some modules and a different arrangement of reefs within the concession imply the second hypothesis that there would be a difference between concessions.

2 Materials and Methods

2.1 Survey Sites

2.1.1 Artificial Structures

The first artificial reef structure takes the form of a pyramid module measuring 2.60 m in height with a base of 4.60 m diagonal, known as "Typi" (Fig. 1). This module, submerged in 2010 at depths ranging from 22 to 25 m depending on tides, is composed of a total concrete volume of 4,872 m^3 with a weight of 1.3 t. Manufactured from XF1-type concrete, this material is inert in the marine environment, posing no ecotoxicological risk to the surroundings.

The second structure consists of an assembly of three layers of concrete, measuring 2.50 m in height by 2.77 m in width, named "Babel" (Fig. 1). All "Babel" modules were submerged in 2015 at depths ranging from 22 to 25 m depending on tides. Similar to the "Typi" reef, the "Babel" module is made of XFl-type concrete.

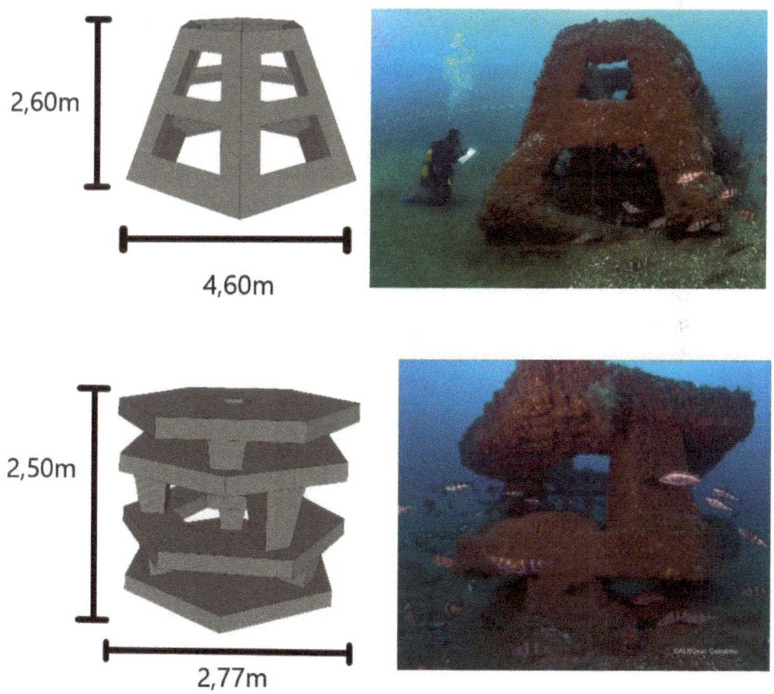

Fig. 1 Schematic representations and photographs of ARs: Typi (above) and Babel (below)

2.1.2 Sites

The artificial reefs are situated in three locations, referred to as "concessions", along the French Atlantic coast (Fig. 2): near Capbreton (Cap), close to Soustons/Vieux Boucau (Vb), and in the vicinity of Messanges/Azur/Moliets-et-Maâ (Mm).

These sites feature flat, sandy bottoms, with depths ranging from approximately 20 to 25 m depending on tides and are situated approximately 1 nautical miles from the coast.

The strategic placement of artificial reefs accounts for the benefits of ongoing freshwater flow, typically seen in areas such as canals or estuaries (for Cap and Mm). These locations are integral to the life cycle of many marine fish species. Some species complete their life cycle entirely within the estuary, while others use it as a migratory pathway from their sea spawning areas to their main feeding grounds in freshwater, or vice versa (Potter et al. 2015). The placement of artificial reefs in Vieux Boucau is specifically influenced by the presence of a small port with rocky features, accentuating its role in fostering the development and protection of juvenile marine organisms (Pastor 2008).

Each concession comprises three types of artificial reefs: clusters of pipes, a Typi module, and a Babel module, covering an area of approximately 16 hectares. Each concession has a particular set up of structure (Fig. 2). Following the submersion of the pipe clusters, only the Typi and Babel reefs are subjected to study.

2.2 Data Collection

The ichthyofauna sampling method relies on the Under Visual Census (UVC) technique, as developed by Charbonnel (1997). This approach conducts counts in strata, specifically tailored to the behavior of mobile fauna in relation to Artificial Reefs.

The protocol begins with the careful observation, for one minute, of three elusive and hard-to-approach species (*Mola mola, Diplodus cervinus, Dicentrarchus labrax*) upon arrival at the AR. Subsequently, divers undertake a circular route at 3 m from the reef in the opposite direction (Fig. 3), aiming to survey species living in the vicinity of the reefs on the substrate.

A second circular route is conducted at 1 m from the reef to count species living immediately near the AR (Fig. 3). Following this, a thorough exploration of the cavities and internal faces of the reef is performed during a complete rotation around and/or within the reef to count species associated with the reef and benthic species. A sinuous movement above the reef can also be executed.

The final step involves counting all pelagic and schooling species 2 m above the structure. The counting is exhaustive for groups of 20 individuals or fewer. For larger groups, individuals are counted in "patches" of 10 to 20 individuals (Labrosse et al. 2001), and species are identified to the species level if possible.

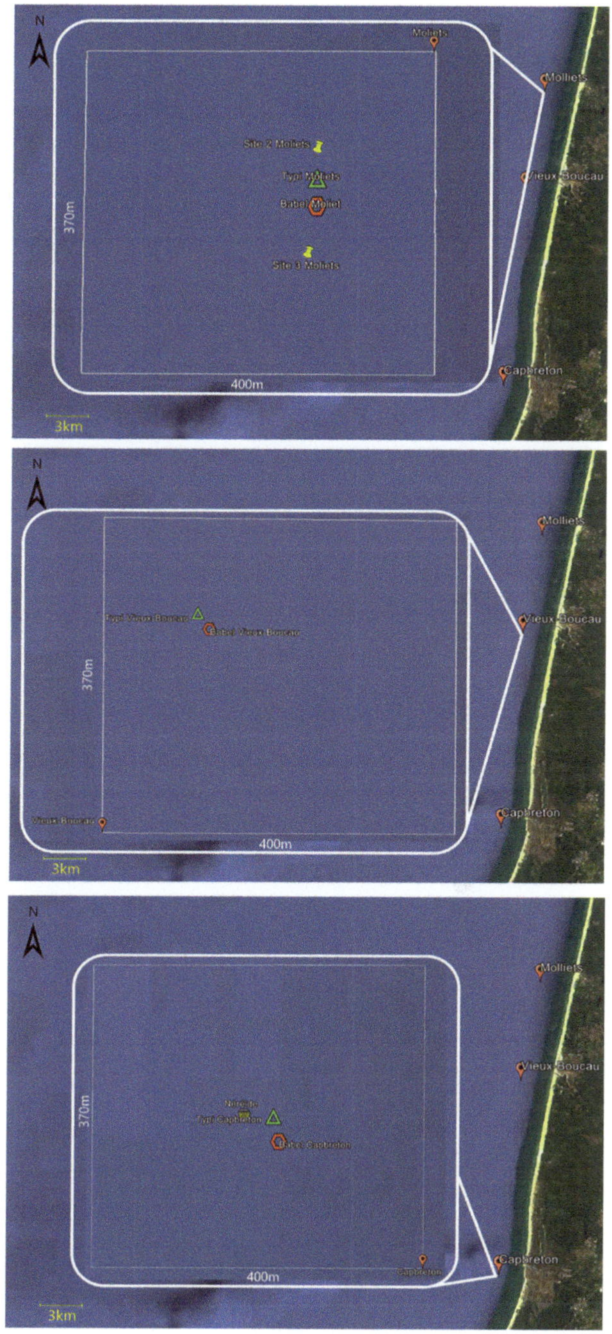

Fig. 2 Location of the ARs in Moliets-et-Maa concession (above), Vieux Boucau concession (middle), Capbreton concession (below). The green figure corresponds to the 'Typi' AR, and the red figure corresponds to the 'Babel' AR

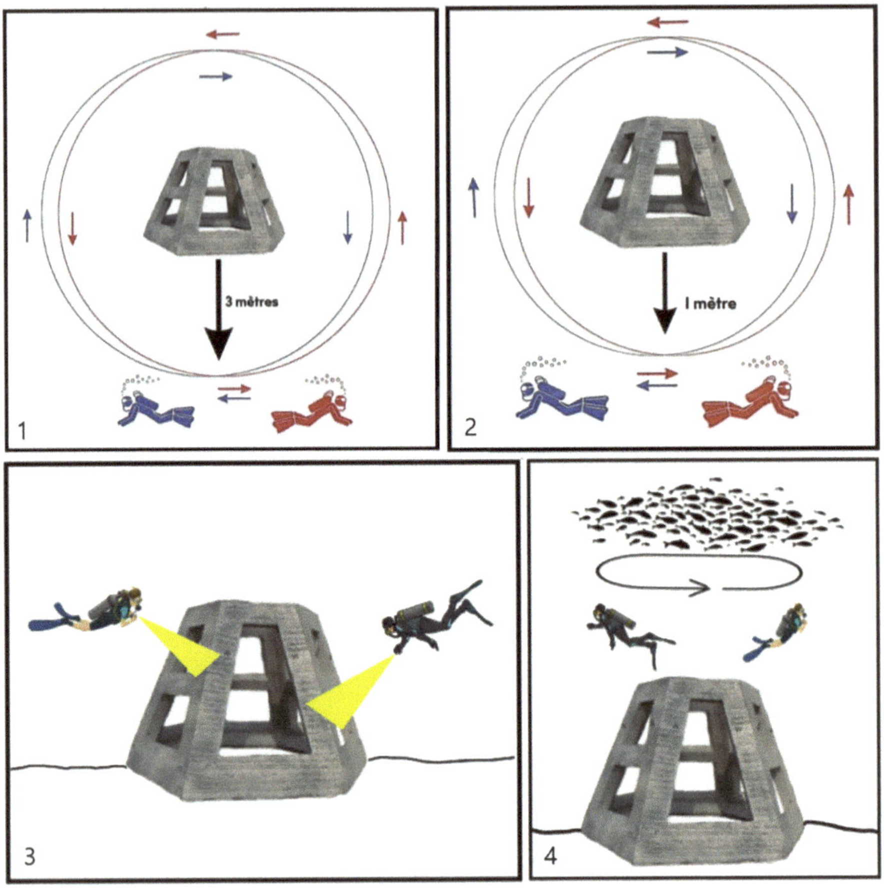

Fig. 3 Schematic representation of UVC protocol across the three AR sites

The data were collected from 2018 to 2022, during the spring, summer and early autumn season due to meteorological conditions. Samling were made at all the locations and all the structure types during the period.

2.3 *Statistical Analyses*

Abundance and species richness data are examined based on factors "year", "location", and "structure type". Multivariate tests are conducted on abundance data transformed by the square root function of the "vegan" package (Oksanen et al. 2015) to mitigate the influence of abundant species (Smith et al. 2017). Subsequently, a Bray–Curtis dissimilarity matrix is applied to both abundance and species richness data.

The datasets undergo PERMANOVA analysis, which is a permutation-based technique that is insensitive to assumptions of normal data distribution, homogeneity of variances, and is suitable for asymmetric designs (Anderson and Walsh 2013, Anderson 2017). This method measures dissimilarity between sample groups using multivariate analyses on a distance matrix.

P-values for each term in the model are obtained through 9999 permutations and Bray–Curtis distances. Pairwise tests are conducted for model terms where significant differences are revealed. The artificial reef structure factor is considered nested within the concession factor. The null hypothesis is stated as: "There is no significant difference at the 0.05 significance level between the studied groups." Assurance of our data meeting the assumption of multivariate homogeneity of dispersion for PERMANOVA is confirmed through a PERMDISP test on the fish abundance table.

For a graphical approach to visualizing abundance differences on significant terms, a non-metric multidimensional scaling (nMDS) is performed (Anderson 2017) on abundance data transformed into a Bray–Curtis dissimilarity matrix. Following this, a SIMPER (Similarity Percentage) analysis is conducted to calculate the contribution of each species (as a percentage) to the dissimilarity between groups identified (Piló et al. 2018) as significantly different by PERMANOVA. This analysis is calculated from the Bray–Curtis dissimilarity matrix, with 9999 permutations and a significance threshold of 0.05.

3 Results

A total of 65 valid surveys have been conducted between 2018 and 2022, recording 39 fish species. On average, across the three concessions and two structures, 884 individuals were counted, with an average species richness of 8. To ascertain whether structural and locational parameters influence ichthyological communities, PERMANOVA analyses are conducted on both species' richness and abundance data.

Species richness appears to differ depending on the structure, reaching significance at a threshold of 0.05 (Table 1). Location also seems to have a significant effect on species assemblages. Capbreton concession exhibits a specific richness of 10.03 ± 4.5, Moliets-et-Maâ concession of 7.89 ± 2.49, and Vieux Boucau concession of 6.89 ± 2.21. The substantial variation observed across years in the graphs is confirmed by PERMANOVA. However, there appears to be no interaction effect on species richness between structure type and temporality.

Pairwise analysis on species richness highlights a significant difference between Capbreton and Vieux Boucau location ($p = 0.01$). When the abundance of each species is considered, the structural effect is no longer significant. There still exists a location effect, confirmed by Pairwise analysis. (Cap-Vb $p = 0.0031$, Cap-Mm $p = 0.005$). Temporality remains a strongly influential factor on individual assemblages.

Table 1 3-factors PERMANOVA looking at the effect of location (fixed, 3 levels) and structure (niched on location, 2 levels) on fish communities and the time factor (random)

	Species richness					Community abundance				
	df	Sum of Sqs	R2	F	p	df	Sum of Sqs	R2	F	p
Structure	1	0.12886	0.044	4.842	0.02*	1	0.304	0.019	1.451	0.159
Location	2	0.309	0.106	5.086	0.003**	2	1.746	0.109	4.319	0.0001***
Structure: location	4	0.313	0.108	2.942	0.09*	4	2.130	0.133	2.545	0.124
Structure: years	4	0.188	0.064	1.768	0.09*	4	0.729	0.045	0.871	0.670
Location: years	8	0.641	0.221	2.642	0.011*	8	1.942	0.121	1.201	0.126

* $p < 0.1$, ** $p < 0.01$, *** $p < 0.001$

The representation in the form of NMDS (Fig. 4) confirms the absence of difference in abundance between the two structures. The Babel structure seems to have more variation in individual assemblages than the Typi structure.

The NMDS representation (Fig. 5) confirms a difference in individual assemblages based on location. Moliets-et-Maâ and Vieux Boucau concessions seem to have similar assemblages, with Vieux Boucau concession displaying notable dispersion.

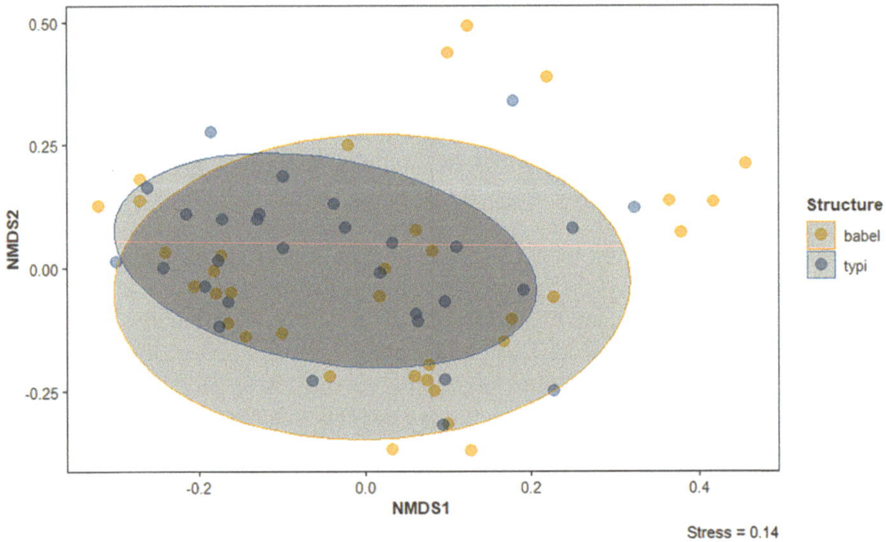

Fig. 4 Non-metric multidimensional analysis (NMDS) of the fish community at the specific level. Based on a Bray–Curtis dissimilarity matrix of fish abundance classified by structure type. The ellipses represent the 75% confidence intervals

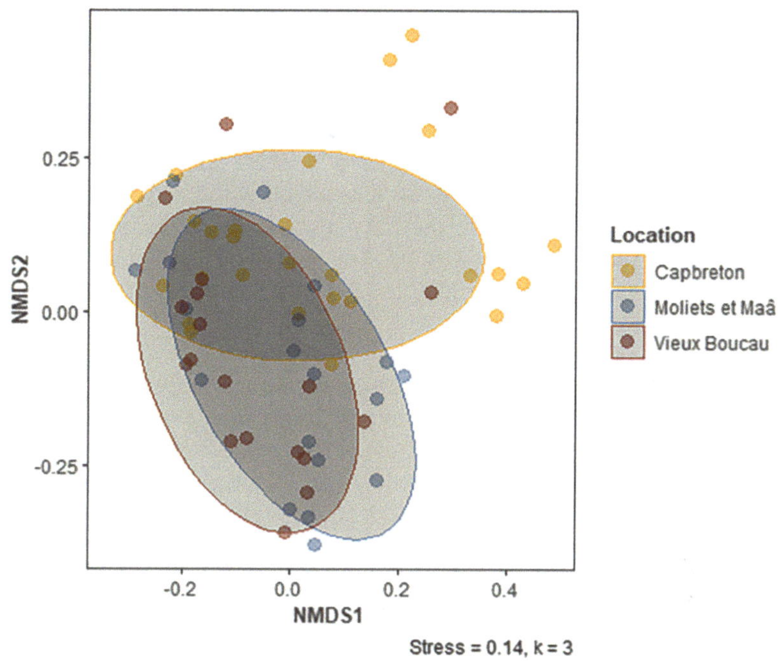

Fig. 5 Non-metric multidimensional analysis (NMDS) of the fish community at the specific level. Based on a Bray–Curtis dissimilarity matrix of fish abundance classified by location. The ellipses represent the 75% confidence intervals

Analysis in the form of SIMPER reveals, based on the abundance table, which species have the most influence on dissimilarities. Capbreton and Moliets-et-Maâ location seem to have the greatest number of structuring species. This may indicate a greater difference between these two concessions, presumably related to the significant distance between these two locations. The species *Trisopterus luscus* discriminates the Capbreton—Vieux Boucau group ($p = 0.0003$) and the Moliets-et-Maâ—Vieux Boucau group ($p = 0.0055$) with a total of abundance: Mm = 10,845, Cap = 4246, Vb = 2022 across years, thus significantly contributing to the differentiation of the Vieux Boucau concession. The species *Ctenolabrus rupestris* plays a similar role for the Moliets-et-Maâ concession (Cap – Mm group $p = 0.0016$, Vb – Mm group $p = 0.0058$) with a total of abundance: Mm = 14, Cap = 0, Vb = 2 across years.

4 Discussion

It seems that there is no marked structural effect. The structural differentiation between the two types of artificial reefs apparently does not influence the abundance table but appears to only affect the number of species present on each structure (Table 1). However, this result should be interpreted with caution, as the specific richness on the Babel structure is 7.57 ± 2.95, and the specific richness on the Typi structure is 9.6 ± 4.09. Thus, the specific richness on the Typi structure is highly variable. Numerous studies confirm that the structural aspect is a significant factor influencing the diversity and abundance of ichthyofauna assemblages (Charbonnel 2002; Moura et al. 2007; Lemoine et al. 2019). This influence can be explained by the short distance (from 22 to 32 m depending on the concession) separating the two types of structures in each concession. Divers' observations have reported movements of fish shoals from one structure to another, confirmed by a photo identification study. Indeed, Jorge's study (2023) demonstrated that an individual of the species *Conger conger* was observed on both types of reefs on different dates. The same was true for the species *Seriola dumerili*. Thus, a similar study on settled populations could be an additional indicator of the presence or absence of the structural effect.

The lack of a natural zone similar to the surroundings of the structures could be the cause of the settling of species on both structures, where they would find sufficient resources for their development (Logan and Lowe 2018). The installation of artificial reefs on sandy bottoms may imply changes in the ichthyological community composition originally present. The so-called characteristic species of each concession, highlighted by SIMPER analysis, are carnivorous species mainly feeding on larvae and juveniles, such as *Trisopterus luscus* (Linnaeus and Salvius 1758). Artificial reefs thus seem to play a role as a feeding ground for these species. The majority of encountered species are those observables in coastal areas. These species can exert strong predation on artificial habitats. The overrepresentation of these predatory species may be due to the attraction effects of Artificial Reefs (ARs). The attraction function involves several causes associated with the presence of ARs described by Sakai (1975): primary causes such as sounds of animals on ARs, sounds caused by currents, hydrodynamic instability, vision, and secondary causes such as the presence of food, shelter, shade, inducing stability and habitability (resting place). Thus, fish are attracted to artificial reefs for primary causes and remain there for secondary causes. Therefore, a reef that follows an attraction pattern would have a majority of subadult and adult individuals (Pastor 2008).

Another explanation could be the lack of a significant structural difference between ARs; indeed, both structures have a similar size and volumes. Additional studies will be conducted to determine a complexity index for each structure. The Capbreton concession has the most dissimilarity compared to other concessions. Therefore, it seems to have different species assemblages and specific richness than Vieux Boucau and Moliets et Maâ. This can be explained by several factors.

The first factor is the immediate presence of the Gouf de Capbreton; nutrient upwellings induced by its presence favor primary productivity and the concentration of pelagic species (Sanchez and Santurtun 2013). This complex biogeographic system is an ideal nursery, feeding, and breeding area for many species (Sanchez and Santurtun 2013). The proximity to a natural rocky area (i.e., the slopes of the Gouf de Capbreton) would represent a reservoir and a preferred exchange zone for species for artificial structures installed on the Capbreton concession. The second source of dissimilarity could be related to significant human activity around these reefs. Despite a formal prohibition of any human activity on the concessions, the Capbreton concession is strongly affected by the presence of fishing nets at each winter outing on the structures. The removal of these nets leads to a partial destruction of fixed fauna and flora, which can destabilize trophic chains. The Moliets-et-Maâ concession is the only concession where the AR made of pipes has not totally silted up; this can influence ichthyological assemblages since it appears that this structure attracts a large number of individuals (Personal communication ALR 2023).

Telemetric studies on inter-reef connectivities and on connectivities between ARs and natural rocky areas are increasingly used to understand ecological patterns. It emerges that fish populations can be considered residents of a reef complex and do not circulate between different locations (Pastor 2008; Abecasis et al. 2009; Koeck 2012; Logan and Lowe 2018). Therefore, assuming that there would be no exchange between the different concessions, the lack of difference in communities between the two structures would confirm that these structures can be considered as similar artificial habitats.

5 Conclusion

In conclusion, this study emphasizes the critical importance of carefully selecting and placing artificial reefs, considering both their design intricacies and the broader ecological context. Based on the hypotheses examined—namely, the potential differences in population dynamics between the two types of structures and variations between concessions—and the subsequent analyses conducted, the observed similarity between the two types of structures suggests the potential for comparable fish populations, underscoring the significance of optimal design considerations. Notably, the geographic placement of these reefs emerges as a crucial factor shaping fish community compositions. Interestingly, proximity to natural rocky substrates appears to influence community dynamics, leading to distinct compositions compared to reefs situated in areas lacking such features, as supported by previous research (Koeck et al. 2014; Davis and Smith 2017).

Therefore, it is imperative for overseeing authorities responsible for artificial reef deployments to meticulously consider the desired placement location and the ecological characteristics of the surrounding environment. This involves evaluating various design options and accounting for the proximity of structure installations to enhance ecological connectivity. Additionally, further research on benthic communities of fauna and flora attached to these structures is essential to corroborate these findings and provide a comprehensive understanding of artificial reef ecosystem dynamics on the french atlantic coast.

References

Abecasis D, Bentes L, Erzini K (2009) Home range, residency and movements of Diplodus sargus and Diplodus vulgaris in a coastal lagoon: connectivity between nursery and adult habitats. Estuar Coast Shelf Sci 85(4):525–529. https://doi.org/10.1016/j.ecss.2009.09.001

Anderson MJ, Walsh DCI (2013) PERMANOVA, ANOSIM, and the mantel test in the face of heterogeneous dispersions: what null hypothesis are you testing? Ecol Monogr 83(4):557–574. https://doi.org/10.1890/12-2010.1

Anderson MJ (2017) Permutational multivariate analysis of variance. In Balakrishnan N, Colton T, Everitt B, Piegorsch W, Ruggeri F, Teugels JL (eds) Wiley StatsRef: statistics reference online, (1st edn). Wiley, pp 1–15. https://doi.org/10.1002/9781118445112.stat07841

Cannizzo ZJ, Lang SQ, Benitez-Nelson B, Griffen BD (2020) An artificial habitat increases the reproductive fitness of a range-shifting species within a newly colonized ecosystem. Sci Rep 10(1):554. https://doi.org/10.1038/s41598-019-56228-x

Cépralmar, Région Languedoc-Roussillon (2015) Guide pratique d'aide à l'élaboration, l'exploitation et la gestion des récifs artificiels en Languedoc-Roussillon, p 236

Charbonnel E (2002) Effects of increased habitat complexity on fish assemblages associated with large artificial reef units (French Mediterranean coast). ICES J Mar Sci 59:S208–S213. https://doi.org/10.1006/jmsc.2002.1263

Charbonnel E (1997) Finfish population assessment techniques on artificials reefs: a review in the European Union. Eur Artif Reef Res 261–277

Davis TR, Smith SDA (2017) Proximity effects of natural and artificial reef walls on fish assemblages. Reg Stud Mar Sci, 9:17–23. https://doi.org/10.1016/j.rsma.2016.10.007

Elliott M, Burdon D, Hemingway KL, Apitz SE (2007) Estuarine, coastal and marine ecosystem restoration: confusing management and science—a revision of concepts. Estuar Coast Shelf Sci 74(3):349–366. https://doi.org/10.1016/j.ecss.2007.05.034

Gillett DJ, Pondella DJ, Freiwald J, Schiff KC, Caselle JE, Shuman C, Weisberg SB (2012) Comparing volunteer and professionally collected monitoring data from the rocky subtidal reefs of Southern California, USA. Environ Monit Assess 184(5):3239–3257. https://doi.org/10.1007/s10661-011-2185-5

Hill J, Wilkinson C (2004) Methods for ecological monitoring of coral reefs: version 1. Aust Inst Mar Sci

Jorge M (2023) Suivi scientifique des récifs artificiels Typi et Babel. Concessions de Capbreton—Soustons/Vieux-Boucau—Messanges/Azur/Moliets. Année—2022. Atlantique Landes récifs, p 86

Koeck B (2012) Rôle des récifs artificiels dans l'écologie des poissons: application aux récifs artificiels de Leucate—Le Barcarès (Golfe du Lion, Mer Méditerranée)

Koeck B, Tessier A, Brind'Amour A, Pastor J, Bijaoui B, Dalias N, Astruch P, Saragoni G, Lenfant P (2014) Functional differences between fish communities on artificial and natural reefs: A case study along the French Catalan coast. Aquat Biol 20(3):219–234. https://doi.org/10.3354/ab00561

Labrosse P, Kulbicki M, Ferraris J (2001) Comptage visuel de poissons en plongée: conditions d'utilisation et de mise en œuvre. Secrétariat général de la Communauté du Pacifique

Lacroix D (1999) Rapport de synthèse sur le 7ème congrès international sur les récifs artificiels et les habitats marins. [Rapport technique]. IFREMER, pp 7–11/10

Leitão F, Santos MN, Monteiro CC (2007) Contribution of artificial reefs to the diet of the white sea bream (Diplodus sargus). ICES J Mar Sci 64(3):473–478. https://doi.org/10.1093/icesjms/fsm027

Lemoine HR, Paxton AB, Anisfeld SC, Rosemond RC, Peterson CH (2019) Selecting the optimal artificial reefs to achieve fish habitat enhancement goals. Biol Cons 238:108200. https://doi.org/10.1016/j.biocon.2019.108200

Linnaeus CV, Salvius L (1758) Caroli Linnaei...Systema naturae per regna tria naturae :secundum classes, ordines, genera, species, cum characteribus, differentiis, synonymis, locis. Impensis Direct. Laurentii Salvii. https://doi.org/10.5962/bhl.title.542

Logan R, Lowe C (2018) Residency and inter-reef connectivity of three gamefishes between natural reefs and a large mitigation artificial reef. Mar Ecol Prog Ser 593:111–126. https://doi.org/10.3354/meps12527

Moura A, Boaventura D, Cúrdia J, Carvalho S, da Fonseca LC, Leitão FM, Santos MN, Monteiro CC (2007) Effect of depth and reef structure on early macrobenthic communities of the Algarve artificial reefs (southern Portugal). Hydrobiologia 580(1):173–180. https://doi.org/10.1007/s10750-006-0455-y

Oksanen J, Blanchet FG, Kindt R (2015) Vegan: commity ecology package. R package version 2.3–0

Pastor J (2008) Rôle des enrochements côtiers artificiels dans la connectivité des populations, cas du sar commun (Diplodus sargus, Linné, 1758) en Méditerranée nord-occidentale. Université de Perpignan

Potter IC, Tweedley JR, Elliott M, Whitfield AK (2015) The ways in which fish use estuaries: A refinement and expansion of the guild approach. Fish and Fisheries, 16(2):230–239. https://doi.org/10.1111/faf.12050

Piló D, Barbosa AB, Teodósio MA, Encarnação J, Leitão F, Range P, Krug LA, Cruz J, Chícharo L (2018) Are submarine groundwater discharges affecting the structure and physiological status of rocky intertidal communities? Mar Environ Res 136:158–173. https://doi.org/10.1016/j.marenvres.2018.02.013

Pioch S (2008) Les habitats artificiels: Élément de stratégie pour une gestion intégrée des zones côtières ? Essai de méthodologie d'aménagement en récifs artificiels adaptés à la pêche artisanale côtière, p 280

Sakai L (1975) The swarming mechanism of a school of fish in an artificial reef. SHEEHY, pp 254–257

.Sanchez F, Santurtun M (2013) SYNThèse et Analyse des données eXistantes sur un écosystème profond transfrontalier: Le gouf de Capbreton—« SYNTAX » [Archimer Ifremer]. https://archimer.ifremer.fr/doc/00137/24787/

Santos LN, A-Berthou EG, Agostinho AA, Latini JOD (2011) Fish colonization of artificial reefs in a large Neotropical reservoir: material type and successional changes. Ecol Appl 21(1):12

Seaman W (2007) Artificial habitats and the restoration of degraded marine ecosystems and fisheries. Hydrobiologia 580(1):143–155. https://doi.org/10.1007/s10750-006-0457-9

Smith JA, Cornwell WK, Lowry MB, Suthers IM (2017) Modelling the distribution of fish around an artificial reef. Mar Freshw Res 68(10):1955. https://doi.org/10.1071/MF16019

Worm B, Barbier EB, Beaumont N, Duffy JE, Folke C, Halpern BS, Jackson JBC, Lotze HK, Micheli F, Palumbi SR, Sala E, Selkoe KA, Stachowicz JJ, Watson R (2006) Impacts of biodiversity loss on ocean ecosystem services. Science 314(5800):787–790. https://doi.org/10.1126/science.1132294

Methods for Modeling Marine Food Webs and Studying the Cumulative Impact of Climate Change and Offshore Wind Turbines on Ecosystem Functioning

N. Niquil, A. Raoux, J.-C. Dauvin, and Q. Nogues

1 Introduction

Currently, the functioning of terrestrial and marine ecosystems is threatened by the intensification of anthropogenic activities (Costanza et al. 1997). Coastal ecosystems are particularly vulnerable due to the convergence of various demographic, economic, and ecological pressures (Costanza et al. 1997). They face numerous and increasing local disturbances, such as overfishing, pollution, mineral resource extraction, invasive species, habitat degradation, and the development of marine renewable energy. Additionally, they are impacted by global climate change effects, including shifts in temperature, water salinity, sea level rise, and ocean acidification (Worm et al. 2006; Eastwood et al. 2007; Dauvin 2019a, b).

It appears that the functioning of ecosystems can no longer be separated from the dynamics of anthropogenic activities (Vitousek et al. 1997). The changes that ecosystems undergo can be attributed to various local and global disturbances, as well as their interactions. Marine ecosystems, due to their high biodiversity and the diversity of ecological roles (known as functional diversity), form complex networks. Therefore, predicting their evolution under the constraints of upcoming changes presents a challenge. The simultaneous consideration of climate change and the emergence

N. Niquil (✉)
Unité Biologie Des Organismes Et Écosystèmes Aquatiques (BOREA), MNHN, CNRS, IRD, Sorbonne Université, Université de Caen Normandie, Université Des Antilles, Caen, France
e-mail: nathalie.niquil@unicaen.fr

A. Raoux · J.-C. Dauvin
UNICAEN, Laboratoire Morphodynamique Continentale Et Côtière, CNRS UMR 6143 M2C, Normandie Université, Caen, France

Q. Nogues
France Energie Marines, Le Havre, France

A.-C. Bennis et al. (eds.), *Constraints and Adaptations to Global Change at the Land-Sea Interface: For a Shared Ecological and Energy Transition*,
https://doi.org/10.1007/978-3-031-90050-1_12

of marine energy presents the following paradox: while developing marine energy provides an alternative to fossil fuels and reduces the carbon footprint associated with energy production, it also introduces new pressures at the local scale. For example, the construction and operation of energy production facilities can lead to noise (CNRS 2021), sediment modifications, and the creation of new substrates that allow unusual species to settle in the area (Degraer et al. 2020). In recent years, various studies have aimed to develop scenarios regarding the impacts on marine ecosystems, but much remains to be done to better characterize possible futures for these ecosystems in the face of both local and global pressures. Furthermore, the growing interest in assessing ecological status based on ecosystem functioning and the need to address complex environmental issues have stimulated the application of tools for overall system assessment. Understanding community changes and ecosystems in relation to food web can provide a more holistic view of how communities and their functioning might shift in response to disturbances. In fact, trophic relationships between species offer a comprehensive perspective of the system since communities primarily organize around their feeding relationships. This notion of the flow of matter and energy between different species within a community is one of the most important concepts in ecological interactions. Thus, a better understanding of these transfers between ecosystem components is a primary challenge for ecologists. To this end, food-web models describe interactions between different biological compartments and are based on quantifying the flows among all living organisms in the ecosystem. These models enable the application of numerical methods to characterize the emergent properties of ecosystems, a process known as Ecological Network Analysis (ENA). ENA is defined as "a systems-oriented methodology to analyze within-system interactions used to identify holistic properties that are otherwise not evident from direct observations" (Fath et al. 2007). The rapid advancement in the application of Ecological Network Analysis (ENA), particularly in coastal ecology, has led to new insights into food web characteristics and their potential for local stability. ENA indices, which highlight the holistic aspects of food webs, are emerging as valuable tools for evaluating ecosystem stability and may have significant implications for management practices (Heymans et al. 2007; Saint-Béat et al. 2015). By providing a comprehensive and integrative approach to ecosystem functioning, food web models, and specifically ENA indices, can serve as indicators of ecosystem health, aiding in the assessment of marine ecosystems' conditions (e.g., Chkili et al. 2024).

In this study, we model the potential impacts associated with the construction of an offshore wind farm in the Bay of Seine, as well as the effects of climate change on the ecosystem. These methods are based on a mean square meter approach (LIM-MCMC) or on 2D approaches (Ecospace). We propose using these methods to investigate the cumulative impacts of OWF and climate change on species distribution, and to evaluate the usefulness of Ecological Network Analysis (ENA) indices in assessing the health of the ecosystem.

2 ENA, a Set of Tools for Studying the Complexity of Trophic Networks

A major challenge in coastal management is finding straightforward methods to assess the health of complex ecosystems, specifically their ecological quality status. The development of appropriate indicators and assessment tools has primarily occurred in response to the implementation of the Water Framework Directive (WFD), the Marine Strategy Framework Directive (MSFD), and the OSPAR Convention (Niquil et al. 2014). By combining the structural and functional attributes of the system with modeling results that integrate biological and physical environments, we can demonstrate the utility and complementarity of methods for assessing ecosystem health (Noguès 2021). Nevertheless, the choice of relevant indicators depends on the ecological questions raised, as well as the biological components and habitats considered, which can range from a single level (individual or population) to multiple levels (community- or ecosystem-based indicators) within the ecosystem (Niquil et al. 2014). The combination of ecological attributes and tools should be used to enhance our understanding and evaluation of marine ecosystems for improved management and conservation in the future (Niquil et al. 2014). In this context, ENA indices appear to be a set of potential indicators for describing the health status of ecosystems, as well as valuable decision-support tools for managers, informing them about the functional state, organization, and structure of ecosystems (Niquil et al. 2014; Safi et al. 2019). These indices allow for the characterization of ecosystem structure and functioning at different spatial scales from global (Heymans et al. 2007), regional (de la Vega et al. 2018) to local scales (Noguès et al. 2021).

They could therefore be used in the implementation of management programs on both small and large scales. As a result, eight ENA indices were proposed to OSPAR and the Ministry of the Environment to describe the health status of ecosystems from the perspective of ecosystem functioning (Safi et al. 2019).

In this context, Safi et al. (2019) presented several case studies that demonstrated the potential benefits of applying ENA indices. These include their relevance in the construction of OWF (Raoux et al. 2017) and the potential for ENA to be used as a supplement to environmental impact assessments to improve understanding of the local effects of large national infrastructure projects such as the construction of a new harbor 'Port 2000' in the Seine estuary (Tecchio et al. 2015, 2016). They also explored the impacts of continuous nutrient additions on the structural and functional dynamics of a marine planktonic ecosystem in outdoor mesocosm experiments (Luong et al. 2014). One of the primary challenges for some ENA indices is their sensitivity to multiple, often unrelated, pressures (fishing pressure and climate change). Safi et al. (2019) emphasized the need for additional research. Specifically, the effects of various pressures on ENA indices should be examined (i) individually and (ii) in combination to assess their cumulative effects.

3 An Ecosystem Approach to Assess the Impact of Offshore Wind Farms: The Case Study of Courseulles-Sur-Mer

Since 2014 (Raoux 2017; Noguès 2021; and research from the TROPHIK project, ANR FEM/ITE), a study has been conducted in the Bay of Seine aimed at developing methods to model the fate of the food web under the combined pressures of climate change and the planned installation of an offshore wind farm (Noguès et al. 2022, 2023a; b). To achieve this, the food web in the area of the Courseulles-sur-Mer wind farm was modeled prior to its construction, as well as in a larger area covering the Bay of Seine and the region extending from the bay to the northern boundary of the French EEZ. We then worked on developing methods to simulate the area in 2050 and in 2100, accounting for the northward shift of species while combining these scenarios. This enabled us to characterize how the impacts on ecosystem functioning, linked to these pressures, accumulate.

3.1 *Simulation of the Wind Farm Effect*

In a wind farm scenario specifically designed to slightly reduce the impact on fishing activities, the primary expected effect during its operation is the creation of an artificial reef effect. The reef effect associated with wind farms refers to the phenomenon where the foundations of wind turbines attract various marine species (Degraer et al. 2020; Hutchinson et al. 2020). These structures can serve as refuges and habitats for different marine organisms, potentially leading to an increase in biodiversity in the surrounding area (Fig. 1). This effect is considered as one of the most important effects on the marine environment generated by the construction of offshore wind farms (Peterson and Malm 2006; Degraer et al. 2020).

For this simulation, we relied on observations from the North Sea, particularly in the Southern Part of the North Sea along the Belgian waters, where wind farms have been operational for over twenty years. These observations were cross-referenced with expert opinions on local species in the English Channel. We then employed various mathematical models to estimate overall fluxes within the trophic web.

The scenarios related to the artificial reef effect indicate a significant shift in the food web at the Courseulles-sur-Mer site. The hypothetical substantial arrival of mussels on the turbine masts and other hard substrates leads to a food web that is more dependent on detritus consumption. Increased filtration of phytoplankton, zooplankton, and detritus in the water column supports this shift. By incorporating different substrates and species communities at the site, we increase the diversity of fluxes within the food web. This enhances the network's resilience to disturbances, as redundancy in flux provides stability.

However, these findings should be considered with caution. In fact, we have only addressed the direct effect on the food web, without considering other factors such as noise or additional environmental effects. Moreover, recent observations of the

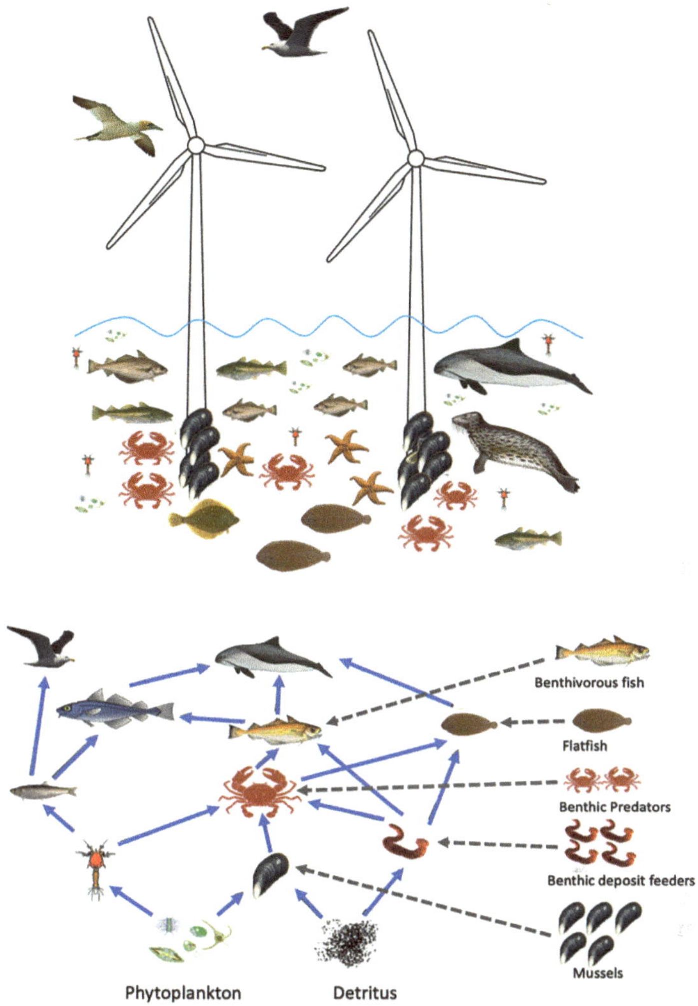

Fig. 1 Modified from Raoux (2017) and Noguès (2021). At the top is a schematic representation of the trophic compartments considered in trophic food web models and their attraction to wind turbines due to the artificial reef effect. At the bottom is a representation of the trophic food web, illustrating the interactions between these trophic compartments, with modifications to these interactions related to the artificial reef effect shown on the right

mussel population may show a lower colonization than what had been modelled based on Belgian observations.

3.2 *Study of the Cumulative Effects on Ecosystem Functioning and the Services Provided to Humans*

The next step of our project was to obtain a spatial view of the cumulative impacts at the larger scale (the operation of the wind farm and the displacement of species due to climate change, considering factors such as average temperature and salinity). Various functional properties of the trophic web have been compared across these scenarios. For recycling, the focus is on whether wind turbines increase recycling, whether climate change has a similar effect, and whether the combined impact of both pressures is additive, synergistic (greater than the sum of individual effects), or dampened (less than the sum of the isolated effects) (Fig. 2).

The modeling tools used in this study allow for the creation of a detailed map of the Bay of Seine, where each spatial cell represents a quantified food web. This enables the mapping of recycling or flow redundancy across different scenarios (current state, the impact of wind turbines, climate change, or a combination of both) (Nogués et al. 2023a, b). The results are highly complex. Depending on the region and ecological indicators, a wide range of outcomes can be observed. For example, the average omnivory of the trophic web shows a synergistic effect of both pressures in areas to the east and west of the wind farm, while a dampened effect is observed in the north. This research is still in its early stages, but it highlights the complexity of creating future scenarios.

Based on the flows within the food web, three ecosystem indicators have been defined: the production of species targeted by fishing, the production of top predators (considered emblematic species of high cultural value), and the potential resilience

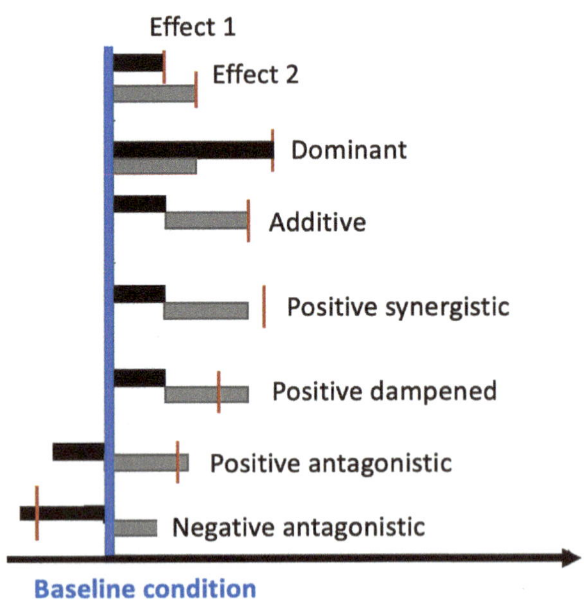

Fig. 2 How to characterize the cumulative effect of two influences on emergent properties of the trophic food web: the current state serves as the reference, climate change related to species displacement represents effect 1, and the presence of wind turbines during the operational phase represents effect 2

of the ecosystem, based on the redundancy of flows (Noguès 2021). The Seine Bay ecosystem's capacity to produce biomass for fished species seems to be negatively impacted by climate change. Given the economic importance of fishing in Normandy, these changes could affect the entire social and economic system of the bay. However, we are facing challenges in modeling the species entering the English Channel, which might lead fishermen to adapt their practices by targeting new species. The offshore wind farm at Courseulles-sur-Mer seems to enhance ecosystem services and reduce the local impacts of climate change, offering stakeholders a way to mitigate its effects.

This ecosystem-based approach offers an integrated view of environmental services, relying on the system's overall functioning. However, it simplifies the complex links within the ecosystem, which requires further investigation, as demonstrated by the Sensitroph project (funded by Fondation de France, N. Niquil and J. Clavel co-PI). The aim of this project is to characterize the interactions between human actors and nature, and the perception of the marine ecosystem by actors along the Bay of Seine coastline. The evolution of the Bay of Seine socio-ecosystem involves not only the complexity of interactions between non-human organisms but also the relationships between these species and human actors. These actors, depending on their sensitivities, will have different perspectives on adapting their activities to climate change in this marine environment.

4 Conclusions and Thoughts on the Outlook

Ecosystems are increasingly being shaped by the effects of climate change. This situation can be observed in the marine world, with gradual changes in the distribution of species, abrupt changes and, in particular, drastic degradation of sensitive ecosystems such as coral reefs (SROCC 2019). There is an urgent need to act, first and foremost by limiting global greenhouse gas emissions. To achieve this, it is essential to change our energy consumption and move towards a gradual reduction in consumption to obtain a large energetic sobriety. While we are waiting for these changes to take place, we need to make a sudden break by abruptly limiting emissions from fossil fuels. To achieve this, we need to develop the production of renewable energy, but we also need to think about the distribution of areas devoted to these renewable energies, taking into account the local effects on biodiversity associated with their development. This approach must therefore form part of a spatial planning approach, in which ecosystems of interest, because of their good ecological status, are protected, and taking into account other human activities. It seems important that the establishment of marine protected areas should be part of an approach that involves local stakeholders.

In this statement, the National Commission for Public Debate (CNDP) had organized a public debate for six months from November 2023 to April 2026 on maritime planning to debate the future of the sea and the littoral, the marine biodiversity and the offshore wind farms. With several hundred participants, it appeared necessary to move future wind farms offshore and to take into account other activities at sea and the impact of climate change on biodiversity. A map of sites, suitable for wind power has been produced, suggesting several sites off the coast of Normandy. This debate underlined the importance of preserving the biodiversity and ecosystems of the 3 nautical mile coastal band and the need to better regulate continental pollutant inputs.

Protecting an ecosystem while at the same time protecting traditional human activities, including fisheries and granulate extraction, is possible to a certain extent, and fully enclosing other ecosystems must be done concomitantly. In this way, the resistance of ecosystems to climate change will be strengthened. However, in spite of cumulative pressures, the marine ecosystems of the English Channel, and in particular those of the Bay of Seine, exhibit from low to high resistance to stressors, and high resilience which refers here to the ability of a habitat to recuperate its initial structure after a certain time following a disturbance (Dauvin et al. 2021; Dauvin, this volume).

In certain areas, however, where the ecosystem may have suffered a great deal of degradation, it is conceivable that the development of offshore wind power could also lead to an increase in local functional diversity. This idea is similar to that of ecosystem gardening, close to the Japanese concept of Satoumi. Satoumi is a word that combines 'Sato', the village, and 'Umi', the sea, and is at the heart of the Japanese vision of managing the commons. According to Toyoshima et al. (2022) 'Satoumi is a sea whose productivity and conservation of biological diversity are the result of human intervention in harmony with the natural ecosystem'. To take this comparison to its logical conclusion, however, the activity of gardening the marine ecosystem would have to go beyond the simple presence of a new substrate and take on a broader social role, involving many local players, from schoolchildren to fish farmers. However, the spatial organization of the wind farms, which are set at a distance from the coast for landscape reasons, makes them difficult to access.

Elinor Ostrom's works on the governance of common goods (Antona and Bousquet 2017), by proposing an in situ approach involving stakeholders, is in line with the Satoumi Research Center's experiments. These have led to a strong involvement of local people in the restoration of eelgrass ecosystems in the Kobe Sea in Japan, with extraordinary success. The idea is to test how these experiences could lead to the organization of a collective of field modelers, discussing with local stakeholders the possibilities of adapting their activities to changes in ecosystems under local and global pressure. Such an approach should be based on a greater on-site presence of researchers to deepen the dialogue and the search for solutions to changes, while integrating local knowledge into our prospective simulation models. In this way, for the English Channel, the Scientific Interest Group ECUME, which is a multidisciplinary consortium to assess the cumulative effects of human activities (Mensens et al., this volume)., is able to favor the co-construction of the long-term future of this coastal sea which is among the more anthropized in the World Ocean (Halpern et al. 2008). Such activities, to be developed, could benefit from the Japanese experience and the creation of a Satoumi research center "à la Française" may be proposed.

Acknowledgements Nathalie Niquil warmly thanks Osamu Matsuda, Takehiro Tanaka and Hiroaki Sugino, who welcomed her from June 8 to 14, 2024 in Okayama, Hinase and Yamaguchi. They enabled her to discover Satoumi's activities in the field and in contact with local players. She would also like to thank Patrick Prouzet and Teruhisa Komatsu for their help in organizing this field trip.

References

Antona M, Bousquet F (2017) Une troisième voie entre l'État et le marché: échanges avec Elinor Ostrom. Éditions Quae, p 148

Chkili O et al (2024) Typology of planktonic food webs and associated emerging properties as indicators of the ecological status of a permanently disturbed Gulf of Gabès. J Mar Syst 242:103948

CNRS (2021) Impacts acoustiques des projets éoliens en mer sur la faune marine. Compartiments des mammifères marins, des poissons et des invertébrés. Synthèse de l'expertise scientifique collective—Novembre 2021. Direction de la communication CNRS, p 110. https://www.cnrs.fr/sites/default/files/page/2022-09/Rapport_complet_impacts_acoustiques_VF.pdf

Costanza R, D'Arge R, de Groot R, Farber S, Grasso M, Hannon B, Limburg K, Naeem S, Oneill RV, Paruelo J, Raskin RG, Sutton P, van den Belt M (1997) The value of the world's ecosystem services and natural capital. Nature 387:253–260

Dauvin JC, Raoux A, Pezy JP, Baux N, Niquil N (2021) The Bay of Seine: a resilient socio ecosystem under cumulative pressures. In: Ceccaldi HJ, Hénoque Y, Komatsu T, Prouzet, P, Sautour B, Yoshida J (eds) Evolution of Marine coastal ecosystems under the pressure of global changes. Springer Nature Switzerland, pp 95–109

Dauvin JC (2019) English channel: La Manche. In: Sheppard C (ed) World of the seas: an environmental evaluation, 2nd edn, vol I. Europe, The Americas and West Africa. Elsevier Academic Press, pp 153–188

Dauvin JC (2019) Chapter 6: the English channel: La Manche. World seas: an environmental evaluation, 2nd edn, vol 1. Elsevier Academic Press, pp 153–188

Dauvin JC (this volume) Resilience of marine benthic communities in an area highly impacted by human activities: the case of the English channel. Springer, Proceedings Coast Caen 2023

de la Vega C, Schückel U, Horn S, Kröncke I, Asmus R, Asmus H (2018) How to include ecological network analysis results in management? A case study of three tidal basins of the Wadden Sea, south-eastern North Sea. Ocean Coast Manag 163:401–416

Degraer S, Carey D, Coolen JP, Hutchison Z, Kerckhof F, Rumes B, Vanaverbeke J (2020) Offshore wind farm artificial reefs affect ecosystem structure and functioning, a synthesis. Oceanography 33:48–57

Eastwood PD, Mills CM, Aldridge JN, Houghton CA, Rogers SI (2007) Human activities in UK offshore waters: an assessment of direct, physical pressure on the seabed. ICES J Mar Sci 64:453–463

Fath BD, Scharler UM, Ulanowicz RE, Hannon B (2007) Ecological network analysis: network construction. Ecol Model 208:49–55

Halpern BS, Walbridge S, Selkoe KA, Kappel CV, Micheli F, D'Agrosa C et al (2008) A global map of human impact on marine ecosystems. Science 319:948–952

Heymans JJ, Guénette S, Christensen V (2007) Evaluating network analysis indicators of ecosystem status in the Gulf of Alaska. Ecosystems 10:488–502

Hutchison ZL, La France Bartley M, Degraer S, English P, Khan A, Livermore J, Rumes B, King JW (2020) Offshore wind energy and benthic habitat changes: lessons from Block Island wind farm. Oceanography 33:58–69

Luong AD, De Laender F, Olsen Y, Vadstein O, Dewulf J, Janssen CR (2014) Inferring time-variable effects of nutrient enrichment on marine ecosystems using inverse modelling and ecological network analysis. Sci Total Environ 493:708–718

McLean M, Mouillot D, Maureaud AA, Hattab T, MacNeil MA, Goberville E, Lindegren M, Engelhard G, Pinsky M, Auber A (2021) Disentangling tropicalization and deborealization in marine ecosystems under climate change. Curr Biol 31:4817–4823

Mensens C, Paporé L, Dauvin JC (submitted) The scientific interest group ECUME: a multidisciplinary consortium to assess cumulated impacts of human activities in the eastern part of the English channel. Springer, Proceedings Coast Caen 2023

Niquil N, Le Loc'h F, Tecchio S, Chaalali A, Vouriot P, Mialet B, Fizzala X, Féral JP, Lamare S, Dauvin JC, Safi G (2014) Ongoing research on ecosystem health indicators for food webs in the MSFD context. Trans-channel forum proceedings science and governance of the channel Marine ecosystem Caen, France, pp 14–15

Nogués Q (2021) Etude du cumul d'impact autour du parc éolien offshore de la Baie de Seine: quelles conséquences pour le fonctionnement trophique de l'écosystème, sa résilience et les services écosystémiques ? Université de Caen Normandie, Thèse de doctorat, p 340

Nogués Q, Raoux A, Araignous E, Hattab T, Leroy B, Ben Rais Lasram F, Le Loc'h F, Dauvin JC, Niquil N (2021) Cumulative effects of marine renewable energy and climate change on ecosystem properties: sensitivity of ecological network analysis. Ecol Indic 121:107128

Nogués Q, Araignous E, Bourdaud P, Halouani G, Raoux A, Foucher E, Le Loc'h F, Loew-Turbout F, Ben Rais Lasram F, Dauvin JC, Niquil N (2022) Spatialized ecological network analysis for ecosystem-based management: effects of climate change, marine renewable energy and fishing on ecosystem functioning in the Bay of Seine. J Mar Sci 79:1098–112

Nogués Q, Bourdaud P, Araignous E, Halouani G, Ben Rais Lasram F, Dauvin JC, Le Loc'h F, Niquil N (2023a) An ecosystem-wide approach for assessing the spatialized cumulative effects of local and global changes on coastal ecosystem functioning. ICES J Mar Sci 80:1129–1142

Nogués Q, Baulaz Y, Clavel J, Araignous E, Bourdaud P, Ben Rais Lasram F, Dauvin JC, Girardin V, Halouani G, Le Loc'h F, Loew-Turbout F, Raoux A, Niquil N (2023b) The usefulness of food web models in the ecosystem services framework: quantifying, mapping, and linking services supply. Ecosyst Serv 63:101550

Petersen JK, Malm T (2006) Offshore windmill farms: threats to or possibilities for the marine environment. Ambio 35:75–80

Raoux A (2017) Approche écosystémique des Energies Marines Renouvelables: étude des effets sur le réseau trophique de la construction du parc éolien au large de Courseulles-sur-mer et du cumul d'impacts. Université de Caen Normandie, Thèse de doctorat, p 278

Raoux A, Tecchio S, Pezy JP, Degraer S, Wilhelmsson D, Cachera M, Ernande B, Lassalle G, Leguen C, Grangeré K, Le loch F, Dauvin JC, Niquil N (2017) Benthic and fish aggregation inside an offshore wind farm: which effects on the trophic web functioning?. Ecol Indic 72:33–46

Safi G, Giebels D, Arroyo NL, Heymans JJ, Preciado I, Raoux A, Schückel U, Tecchio S, de Jonge VN, Niquil N (2019) Vitamine ENA: a framework for the development of ecosystem-based indicators for decision makers. Ocean Coast Manage 174:116–130

Saint-Béat B, Baird D, Asmus H, Asmus R, Bacher C, Pacella SR, Johnson GA, David V, Vézina AF, Niquil N (2015) Trophic networks: how do theories link ecosystem structure and functioning to stability properties? A review. Ecol Ind 52:458–471

SROCC (2019) IPCC special report on the ocean and cryosphere in a changing climate. Cambridge University Press, Cambridge, p 755. https://doi.org/10.1017/9781009157964

Tecchio S, Tous Rius A, Dauvin JC, Lobry J, Lassalle G, Morin J, Bacq N, Cachera M, Chaalali A, Villanueva MC, Niquil N (2015) The mosaic of habitats of the Seine estuary: insights from food-web modelling and network analysis. Ecol Model 312:91–101

Tecchio S, Chaalali A, Raoux A, Tous Rius A, Lequesne J, Girardin V, Lassalle G, Cachera M, Riou P, Lobry J, Dauvin JC, Niquil N (2016) Evaluating ecosystem-level anthropogenic impacts in a stressed transitional environment: the case of the Seine estuary. Ecol Ind 61:833–845

Toyoshima J, Fujii I, Maekawa M, Tsunoda T, Kamada N, Hidaka H et al (2022) Assessing effectiveness of Satoumi activities in Japanese coastal areas from ecological and socioeconomic perspectives. Ocean Coast Manag 230:106354

Vitousek PM, Mooney HA, Lubchenco J, Melillo JM (1997) Human domination of Earth's ecosystems. Science 277:484–499

Worm B, Barbier EB, Beaumont N, Duffy JE, Folke C, Halpern BS, Jackson JBC, Lotze HK, Micheli F, Palumbi SR, Sala E, Selkoe KA, Stachowicz JJ, Watson R (2006) Impacts of biodiversity loss on ocean ecosystem services. Science 314:787–790

Field Data Acquisition Strategy for an Ecosystem Approach to Studying the Reef and Reserved Effects of Offshore Wind Farms: The Fécamp Measuring Mast

J.-Ph. Pezy, A. Raoux, M. Charbonnelle, F. Schlicklin, M. Deloor, P. Claquin, R. Gallon, I. Poirier, A.-M. Rusig, I. Mussio, P. Balay, C. Dancie, F. Leroy, and L. Furgerot

1 Introduction

Wind power energy constitutes an important strategy for a low carbon energy strategy. Wind is the second largest source of renewable electricity production, after hydropower, and has recently undergone important developments to reduce carbon emissions. Actually, this new energy is concentrated on offshore wind power, which seems to be a way of the future. The development of wind power and, more generally, renewable energy meets two objectives included in the multi-year energy plan for 2019–2028: accelerating the energy transition to achieve carbon neutrality by 2050 and projecting new electricity needs. If growth of wind power continues, the

J.-Ph. Pezy (✉) · A. Raoux · M. Charbonnelle · F. Schlicklin · L. Furgerot
UNICAEN, Laboratoire Morphodynamique Continentale et Côtière, CNRS UMR 6143 M2C, Normandie Université, Caen, France
e-mail: jean-philippe.pezy@unicaen.fr

M. Deloor · P. Claquin · A.-M. Rusig · I. Mussio
Unité Biologie des Organismes et Écosystèmes Aquatiques (BOREA), MNHN, CNRS, IRD, Sorbonne Université, Université de Caen Normandie, Université des Antilles, Caen, France

R. Gallon · I. Poirier
Conservatoire National des Arts et Métiers/INTECHMER–Laboratoire Universitaire des Sciences Appliquées de Cherbourg LUSAC, UNICAEN, Cherbourg, France

P. Balay · C. Dancie
Cellule de Suivi du Littoral Normand (CSLN), Le Havre, France

F. Leroy
France Energie Marines, Le Havre, France

A.-C. Bennis et al. (eds.), *Constraints and Adaptations to Global Change at the Land-Sea Interface: For a Shared Ecological and Energy Transition*,
https://doi.org/10.1007/978-3-031-90050-1_13

construction of new onshore installations is hampered by space constraints (standards, regulations) and acceptability of the population (visual and noise pollution). In France, development in this sector today is strongly focused on the installation of offshore wind turbines, with seven projects undertaken to exploit the thousands of kilometers of coastline. An offshore wind farm is more complex and more expensive to implement, but has the advantage of less physical space constraints and greater generating capacity. The two main types of offshore wind turbines are planned: (i) wind turbines moored on the seabed a few kilometers from the coast (especially the North Sea and the English Channel); (ii) floating wind turbines placed on platforms, further offshore, which can benefit from stronger and more constant winds (Atlantic, Mediterranean). Seven concessions off the French maritime coast are currently under development in Dunkirk (Pas-de-Calais), Fécamp, Le Tréport (Seine-Maritime), Courseulles-sur-Mer (Calvados), Saint-Brieuc (Côtes d'Armor), Saint-Nazaire (Loire-Atlantique) and Noirmoutier (Vendée). Their commissioning is planned between 2022 and 2027. For the moment, the most advanced project is located 12 km from the coast of Saint-Nazaire. The installation of 80 wind turbines, with a power of 6 MW each, is operational from 2022. They produce the equivalent of 20% of the electricity consumption of the department. The second most advanced project is in Normandy, 13 km from Fécamp and is scheduled to finish construction in the first quarter of 2024.

During the operational phase, the introduction of scour protection associated with the mast structure leads to the creation of new habitats and shelters that are immediately colonized by several marine species (Wilhelmsson et al. 2006, 2010; Wilhelmsson and Malm 2008). This effect, generally known as the "reef effect", is considered to be one of the most important effects on the marine environment generated by the construction of offshore wind farms (Petersen and Malm 2006). The effect has been described on epibenthic and benthic communities. Furthermore, it has been also reported for fish species (cod *Gadus morhua*, pout *Trisopterus luscus,* horse mackerel *Trachurus trachurus*), which have higher biomasses inside the park than outside the park (Wilhelmsson et al. 2006; Wilhelmsson and Malm 2008; Maar et al. 2009; Reubens et al. 2011, 2013; Leonhard et al. 2011; Lindeboom et al. 2011; Bergström et al. 2013; Degraer et al. 2014).

Observations at offshore wind farms in the North Sea and Baltic Sea reveal that epibenthic communities found on wind turbine foundations are dominated by mussels (*Mytilus edulis*), amphipods (*Jassa* sp.), and barnacles (Leonhard et al. 2013; Wilhelmsson et al. 2006; Wilhelmsson and Malm 2008; Krone et al. 2013a, b; Coates et al. 2014; De Mesel et al. 2015). They also show that foundations around the mast provide a particularly suitable substrate suitable for mold development. In fact, a wind turbine mast can house between 1 and 2 t of mussels in the intertidal zone (Shields and Payne 2014). For example, in wind farms in the North Sea and Baltic Sea, mussels represent between 75 and 80% of the total epifaunal biomass of monopiles (Wilhelmsson et al. 2006, Krone 2012). It appears that the establishment of a wind farm can double the total biomass of filter feeders in the installation area (Maar et al. 2009). This massive colonization of mussels can be also the cause of various local changes in the ecosystem such as a depletion of plankton (Joschko et al.

2008; Maar et al. 2009; De Mesel et al. 2015) and /or an increase in organic matter in the water column and the seabed due to the production of feces and pseudo-feces, or by the decomposition of dead mussels dropped on the seabed which can lead to organic enrichment of the seabed. According to Norling and Kautsky (2008), this blue mussel expansion could be responsible for a shift from the food chain dominated by the primary producers and the primary consumers such as grazers towards that dominated by a more detritus feeding community. If mussels dominate the intertidal zone, the subtidal zone is dominated by amphipods, hydroids and anemones (Maar et al. 2009; Degraer et al. 2014). Finally, near the sea bottom substrate, benthic invertebrate predators dominate such as the starfish *Asterias rubens*, the swimming crab *Liocarcinus holstus* and the crab *Cancer pagurus* (Wilhelmsson et al. 2006; Krone et al. 2017).

Studies have also highlighted the aggregation of fish in the immediate vicinity of wind turbine foundations such as gobies (*Gobiusculus flavescens* and *Gobius niger*), cod (*Gadus morhua*), pouting (*Trisopterus luscus*) and horse mackerel (*Trachurus trachurus*) (Reubens et al. 2011; Bergström et al. 2013). For example, Reubens et al. (2011) observed a school of pouting (*T. luscus*) in the immediate vicinity of a turbine of a Belgian wind farm in the North Sea composed of 22,000 individuals with a total biomass of 2.7 t. Furthermore, Couperus et al. (2010) reported in the grey literature that average fish density per volume of water was found to be on average a factor 37 times higher in the direct vicinity of the monopile than in the open water around it.

It appears that fish are attracted to the new sources of food and shelter that offshore wind farms provide. Analysis of pouting stomach contents sampled at the Thornton Bank wind farm, North Sea, revealed a dietary preference for prey species that inhibit on the surface of the turbine support structure below sea level (Reubens et al. 2011). Additionally, tagging and acoustic telemetry experiments revealed that cod used the wind farm for foraging and refuge (Winter et al. 2010; Reubens et al. 2013).

Finally, the reef effect is likely to attract top predators such as marine mammals, which could affect entire trophic network (Lindeboom et al. 2011; Henkel et al. 2014). For example, at the Horn Rev site, the number of porpoises observed within the wind farm (up to one individual per km^2) was higher than the average observed in the North Sea (0.1 individual per km^2, Tougaard et al. 2009). Their presence is probably closely associated with fish resources, but this could not be clearly established based on the available data.

The aim of this paper is to present the methodology deployed at the Fécamp research platform, the first French Offshore Research Platform dedicated to Offshore Wind Energy, in order to improve our understanding of the interaction between the offshore wind farm and the biological compartments of marine ecosystem. The methodology consists of deploying an ecosystem approach through in situ observations and isotopic measurements of different biological compartments in order to characterize the structure and function of trophic webs according to a gradient of distance from the structure. This strategy will make it possible to better understand the reef effect generated by the OWF foundations and to answer to the following scientific questions:

- What effects does the Fécamp research platform have on the ecosystem?
- Do the masts have a reef effect on the ecosystem and thus do the fish feed on this new structure?
- Do the masts have a reserve effect on the fish compartments through biomass accumulation?
- What is the link on the mast between sessile species (biogenic role) and vagile species?

2 Material and Methods

In support of the offshore wind farm development, the company Eoliennes Offshore des Hautes Falaises (EOHF) obtained a temporary authorization of public maritime domain occupancy on 01/08/2014 for the installation and operation of a measuring mast located off Fécamp coast (order no. 2014213–0011). Thus, the installation of this measuring mast was carried out in February 2015. The main purpose of the mast was to specify the wind potential at the site and to benefit from environmental measures before and during the construction of the offshore wind farm.

France Energies Marines (FEM) is an institute for energy transition dedicated to offshore renewable energies. As part of the implementation of an ambitious program of technological innovation platforms aimed at studying the interactions between offshore wind power and the marine environment, the institute approached EOHF in order to take back the mast of Fécamp. The transfer has been effective since March 21, 2023. FEM will deploy a lot of different equipment (sensors, probes) to implement the technological innovation program for five years.

2.1 Study Site

2.1.1 Environment

The measuring mast is located in a sedimentary habitat corresponding to pebbles, with an average percentage of pebbles greater than 80% (grain size up to 20 mm). The eastern sleeve is a semi-enclosed sea where hydrodynamics are influenced by tides, winds and waves. Tidal currents dominate in the English Channel (Salomon and Breton 1991).

2.1.2 Measuring Mast

Installed in February 2015, 13 km off the coast of Fécamp, the measuring mast allows us to further deepen the technical and environmental data collected since 2007. Located at 40 m above sea level, it rests on a reinforced concrete structure (similar to the gravity foundation of a wind farm) with a diameter of 23 m and a total weight of 1800 t, placed on the seabed. This structure was initially built to obtain precise knowledge of atmospheric and oceanographic conditions.

2.2 *Field Data Acquisition for Ecosystem Approach*

In 2024–2025, biological compartments such as benthic invertebrates, suprabenthos, zooplankton, phytoplankton and demersal fishes will be sampled in two separate seasons, late summer (September/October 2024) land ate winter (March/April 2025). The different compartments will be sampled in the same week, except for demersal fishes, which will be sampled 15 days later for safety reasons around the mast. The sampling plan will be carried out at 12 stations, according to two transect (Fig. 1). The first transect corresponds to the position of the six stations along the direction of the main current from Stn. 6 to Stn. 3 through Stns. 5, 4, 1 and 2. The second transect corresponds to the position of six stations perpendicular to the main current from Stn. 9 to Stn. 12 through Stns. 8, 7, 10 and 11.

2.2.1 Sediment Analyses

A sediment sample will be analysed for particle size distribution. The first sediment is sieved through a 50 μm mesh obtained by wet sieving and rinsed with fresh water to remove salts. The sieved sediment fraction (50 μm) is dried at 60 °C and sieved using 33 sieve columns (< 50; 50; 63; 80; 100; 125; 160; 200; 250; 315; 400; 500; 630; 800; 1000; 1250; 1; 600; 2000; 2500; 3150; 4000; 5000; 6300; 8000; 10,000; 12,500; 16,000; 20,000; 25,000; 31,500; 40,000; 50,000; 63,000 μm). The sieve choice followed the modified Wentworth's classification to determine the sediment type of each station using a Folk diagram (Folk 1954, 1966; Folk and Ward 1957). For each station, the sediments are characterized by three main sedimentary fractions: gravel (>2 mm); sand (2 mm–63 μm) and silts-clays (<63 μm). Samples for organic matter (OM) analysis are dried for three days at 60 °C and then calcined at 500 °C for 4 h.

2.2.2 Field Observations of the Marine Environment

Vertical profiles of marine environments are measured in situ by lowering several sensors from the boat. Photosynthetically active radiation is recorded at a frequency of 0.5 Hz with an RBR solo3 PAR recorder connected to an underwater quantum sensor (marine Li-COR LI-192). Water temperature, pH, conductivity, salinity, dissolved oxygen and depth are measured using YSI EXO1 multiparameter probe at a frequency of four hertz.

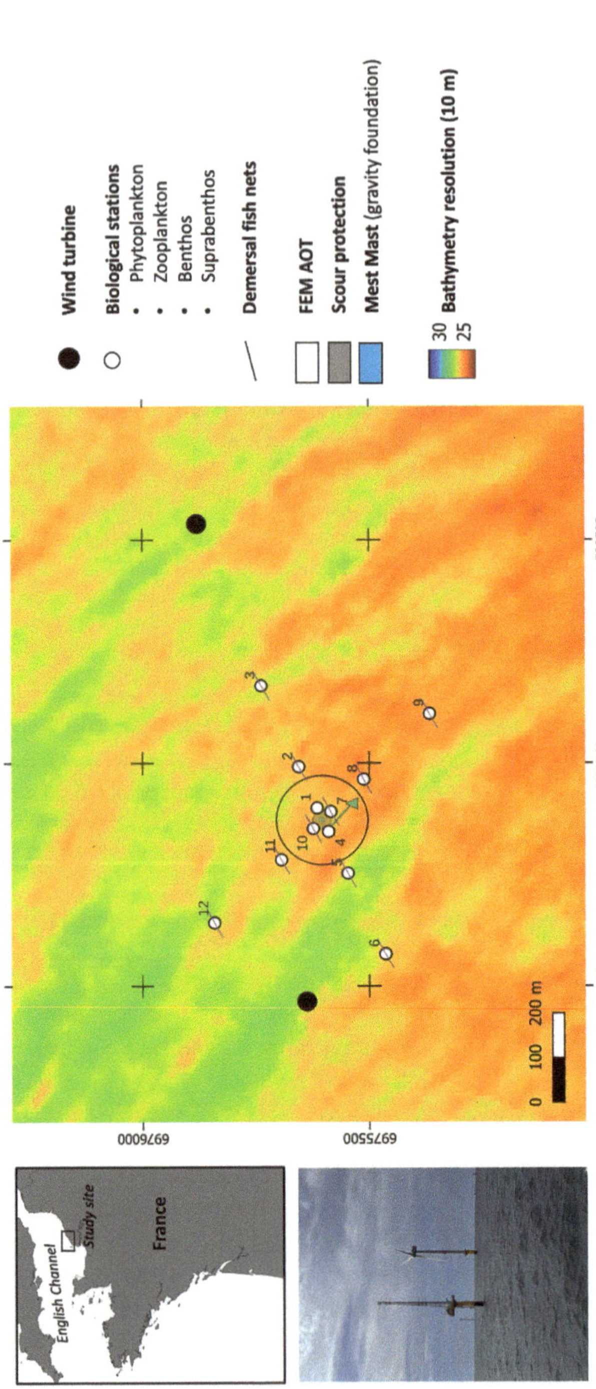

Fig. 1 Map showing 12 sampling stations for the phytoplankton, zooplankton, benthos, suprabenthos and the demersal fish along the transect in direction of main current passing Stns. 6, 5, 4, 1, 2 and 3, and the transect in direction perpendicular to the main current passing Stns. 9, 8, 7, 10, 11 and 12 by the mast

2.2.3 Benthic Macrofauna

Soft-Bottom Macrofauna

For the 12 stations, three replicates are collected at each station with a 0.1 m^2 Van Veen grab for benthic analysis. Sieving is performed on board using a circular 1 mm mesh, and the retained materiel is preserved in 10% formalin. The samples are sorted and, if possible, identified at the species taxonomic level. Taxonomic richness, abundance and biomass for each sample are recorded. To determine the biomass, each taxon is dried at 60 °C for three days to obtain dry weight and then calcined at 500 °C for 5 h to obtain weight after calcination. Biomass of each species (endofauna, vagile and sessile epifauna) is determined and expressed in g of Ash Free Dry Weight (AFDW) per m^2.

Flora and Macrofauna of Hard Substrate

The flora is sampled in 4 zones on the measuring mast (Fig. 2). three stations are located in the subtidal zone at −4, −14 and −25 m (Fig. 2). At each station, a total of 5 samples of 0.1 m^2 are taken. Samples are scraped of organisms constituting the flora and fauna with a spatula, which were sucked up with a suction device and placed in a bag with a mesh size of 0.5 mm. On 9 October 2023, the first observation dive was carried out to estimate the amount of biofouling according to this protocol. During this dive, the intertidal zone was observed to be covered with barnacles and macroalgae (Fig. 2a). The −4 m zone was 100% covered with some red algae, barnacles (a significant portion dead), hydroids, bryozoans and sponges (Fig. 2b). In the −25 m zone, algae are absent, and the faunal composition appears similar to the −4 m zone (Fig. 3d). Numerous fish species were observed near the structure, some of them wrasses and conger eels, and at −25 m depth, very large schools of pouting (*Trisopterus luscus*) and conger eels found in the scour protection (Fig. 2).

2.2.4 Phytoplankton

At each of the 12 stations studied, a probe profile and water sample are carried out. The water will be sampled at a depth of 2 m below the sea surface using a Niskin bottle, and the collected water is then directly darkened for the various conditioning and analyzed directly in the cabin of the vessel.

Photosynthetic Parameters

Photosynthetic parameters of phytoplankton is determined using a LabSTAF "Single Turn-Over" fluorimeter (Chelsea Technologies, UK). The LabSTAF allows measurements at different wavelengths. Samples are analyzed after a 5-min incubation in the

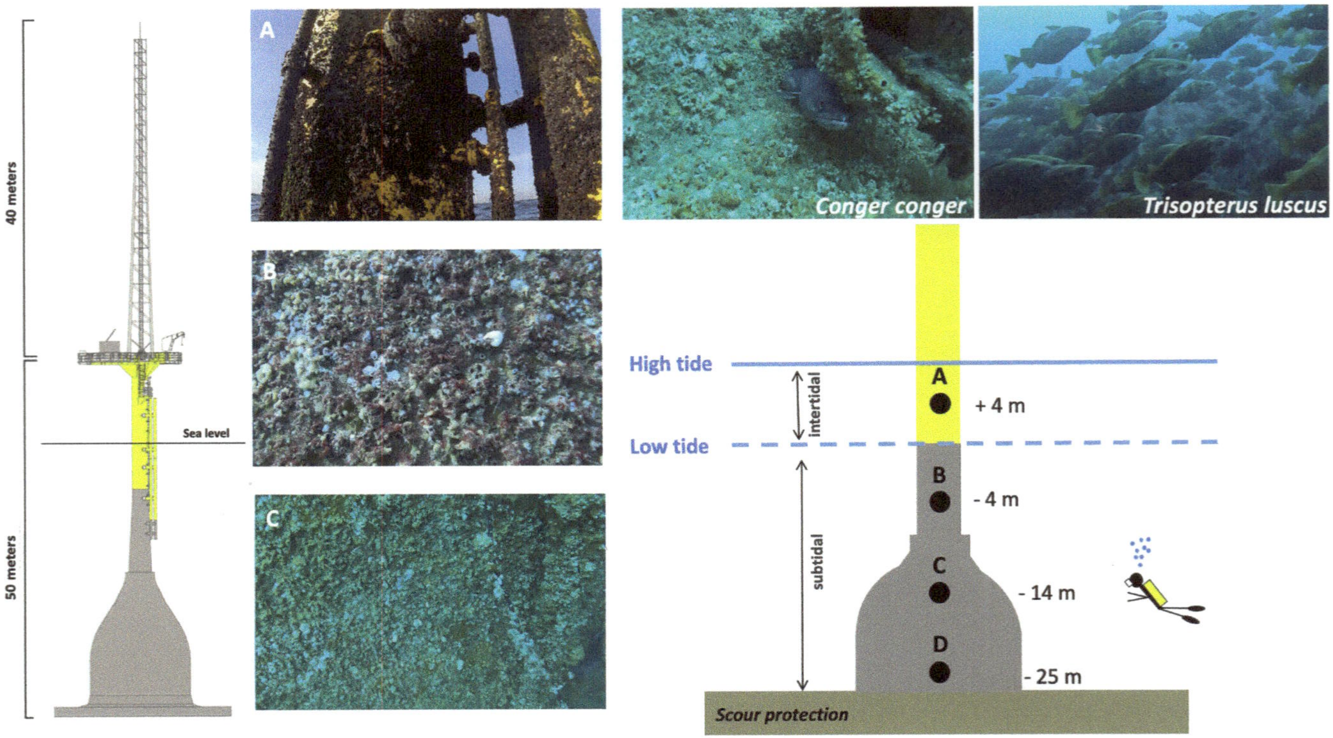

Fig. 2 Observation and Sampling on hard substrate of Fécamp mast in four zones (**a**: intertidal + 4 m, **b**: subtidal 4 m, **c**: subtidal −14 m, **d**: subtidal −25 m) and photographs of the two main fish species, *Conger conger* and *Trisopterus luscus*)

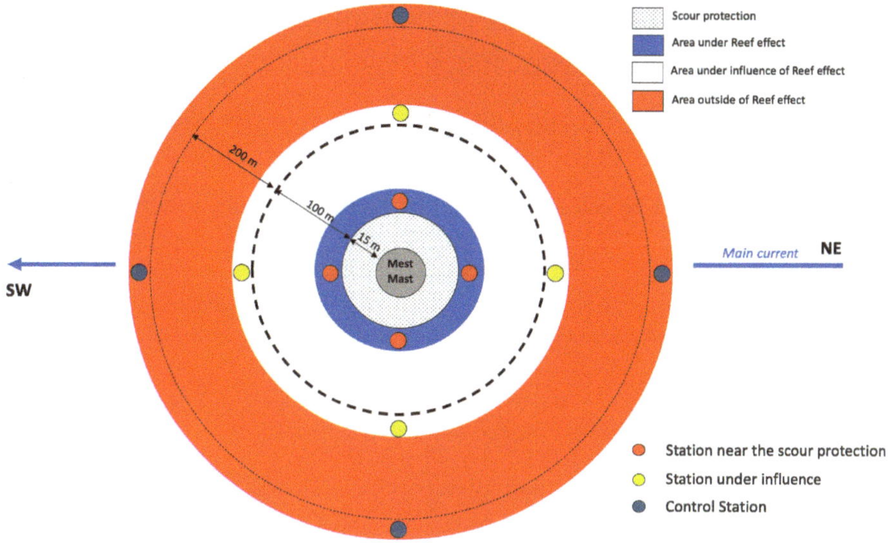

Fig. 3 Location of the three areas to study the reef effects on the structure and functioning of the trophic web around the measuring mast

dark to oxidize the quinone A (QA) pool. A single renewal (ST) saturation phase with a 100 μs saturation flash is delivered to measure minimum and maximum fluorescence as described in Boatman et al. (2019). Samples are exposed to 12 increasing PAR intensities (0–1500 μmol photon s^{-1} m^{-2}) at 30-s intervals to obtain a light response curve (RLC). The maximum quantum efficiency of PSII (Fv/Fm) is calculated using the following equation from Genty et al. (1989):

$$\frac{F_v}{F_m} = \frac{(F_m - F_0)}{F_m} \tag{1}$$

The effective quantum efficienc of PSII $\left(F(q')/F(m')\right)$ is measured at each light step according to Genty et al. (1989):

$$\frac{F_{q'}}{F_{m'}} = \frac{\left(F_{m'} - F'\right)}{F_{m'}} \tag{2}$$

where Fm′ is the maximum fluorescence in light and F′ is the steady state fluorescence in light.

The method of Webb et al. (1974), modified by Boatman et al. (2019), is used to estimate α and Ek of the RLC using RUNSTAF software (Chelsea Technologies, UK), where α is the initial slope of th RLC and Ek is the light saturation index:

$$\frac{F_{q'}}{F_{m'}} = \left(\alpha \times E_k \times \left[1 - e^{-\frac{E}{E_k}}\right] - \beta \times E_{k_\beta} \times \left[1 - e^{\frac{E-E_k}{E_{k\beta}}}\right]\right) \times E^{-1} \tag{3}$$

The relative electron transport rate (rETR, relative unit) is calculated for each irradiance (E) which corresponds to the following calculation:

$$rETR(E) = \frac{F_{q'}}{F_{m'}} \times E \quad (4)$$

The rETRMAX can be calculated according to the following equation:

$$rETR_{MAX} = \alpha_{webb} \times Ek_{webb} \quad (5)$$

The JVPIIMAX (mmol electrons m^{-3} h^{-1}) will be then calculated according to the Eq. (6).

$$JVPII_{\max} = a_{LHII} \times rETR \times 3.6 \times 10^{-3} \quad (6)$$

with a_{LHII} the absorption coefficient of PSII in m^{-1} (see Boatman et al. (2019) for more details).

Maximum productivity (ETRmax) and maximum production (JVPIImax) data will be transformed into carbon by applying a value of Φe.c. of 8 mol of electrons per mole of carbon.

Chlorophyll a

Phytoplankton biomass is estimated as chlorophyll *a* (Chl *a*) per unit volume. A 250 mL triplicate sample is filtered through a glass fiber filter (Whatman, GF/F, 47 mm, 0.7 μm) and the filter is immediately stored at −20 °C until analysis in the following days. 10 mL of acetone (90%) is added to extract the chlorophyll pigments and the sample is then incubated at 4 °C for 12 h. After two centrifugations for 5 min at 1700 g, the concentration of Chl *a* in the extracts is measured using a Trilogy fluorimeter (Turner Designs, Sunnyvale, USA) according to the method of Strickland and Parsons (1968) adapted in the SNO reference protocol SOMLIT.

2.2.5 Zooplankton

Zooplankton is sampled using a WP2 net (200 μm) with a flow meter at its centre to measure the volume of filtered water. Zooplankton corresponds to pelagic organisms present in both seasons of winter and summer of the water column. At each station, one diagonal line is operated for sampling at a speed of approximately 1 m s^{-1}. All sampled organisms are washed and sieved on a 100 μm mesh. Then, the organisms are sorted and identified to the species level. Biomass of each group per sample is determined based on g of Ash Free Dry Weight (AFDW) per m^2.

2.2.6 Suprabenthos

The suprabenthos is defined as organisms living in the water layer immediately above the seabed, which perform daily vertical migrations and/or seasonal movements at varying distances from the bottom (Brunel et al. 1978; Pezy 2017). In this study, we consider the suprabenthos *sensus stricto:* peracarids (amphipods, cumaceans, mysids, tanaids, isopods, pycnogonids, leptostracea) and decapods (Dauvin and Vallet 2006).

Sampling is carried out at 12 stations. At each station, samples are taken at night and during the day to study diurnal vertical migration. The suprabenthos is sampled using a suprabenthic sledge designed to filter the water column at a height of 0.1 to 0.40 m above the sea bottom. The suprabenthic sledge is equipped with a WP2 zooplankton net with a mesh size of 500 μm, including a flow meter at its centre to measure the volume of filtered water. Towing time (the period during which the sledge bottom is in contact with the seabed) is 5 min at a sledge speed of approximately 1.0 knots.

All sampled organisms are washed, fixed with 10% neutralized formaldehyde, and then transferred to a 70% ethanol solution. All specimens are sorted, identified and counted under a dissecting microscope to the species level. Species richness corresponds to the total number of taxa found in the four nets, and abundance and biomass values (AFDW: Ash Free Dry Weight) are standardized to a mean volume of 100 m^3 per haul or surface-area (m^2) corresponding to sledge.

2.2.7 Demersal Fishes and Cephalopods

Abundance and biomass data for the demersal fish are collected by sampling with a 50 m trammel net (3.3 m high at the time of fishing). A total of 10 nets are deployed around the mast at the center of the 10 stations during one day (24 h). At each sampling station, megainvertebrates (large-size invertebrates), cephalopods and fish are identified, measured, weighed in a laboratory. Analysis of fish stomach contents is the only way to provide accurate information on prey species composition. This analysis allows prey determination at the species level. The biomass of demersal fish individuals is determined based on g of fresh weight (FW).

2.3 Ecosystem Approach: Food Web Structure by Isotopic Analyses

2.3.1 Phytoplankton

Samples of phytoplankton are carried out for stable isotope analysis. A water sample is filtered through a specially conditioned fiberglass filter (Whatman, GF/F, 47 mm,

0.7 μm) for a maximum of 30 min with a maximum volume of 3 l and immediately stored in the laboratory at −20 °C. Samples are dried in an oven at 50 °C for 48 h and then placed in a desiccator for analysis (National SNO SOMLIT protocol).

2.3.2 Zooplankton, Benthos and Demersal Fishes

All samples are kept frozen until processed in the laboratory. Preparation of tissues for stable isotopic analyses varies by taxonomic group. For example, we chose to collect muscle because muscle shows less change in a nitrogen isotopic ratio over time than other organs such as digestive gland or heart tissue (Pinnegar and Polunin 1999). We sample abductor muscle in bivalves, mantle tissue in cephalopods, a mantle tissue from chelipeds in crabs, arms in ophiuroids, lantern of Aristotle in sea urchins, white dorsal muscle in fish.

All samples collected are oven dried (60 °C for 48 h) and then each dried sample is ground to a homogeneous powder using a mixer mill. Approximately 1 mg of powder is weighed into small tin cups. Isotopic analysis is performed with an elemental analyser EA 3000 (EuroVector) coupled with a mass spectrometer (Pinnegar and Polunin, 1999) at the Plateau d'Analyse d'Isotopie de Normandie (PLAtin, SFR ICORE) at the University of Caen Normandie, France.

Results are expressed in standard δ notation based on international standards [N2 for $\delta^{15}N$ (Peterson and Fry 1987)] according to the following the equation:

$$\delta \times (\text{‰}) = \left[\left(R_{\text{sample}}/R_{\text{standard}}\right) - 1\right] \times 1000$$

where $X = {}^{15}N$, $R_{sample} = {}^{15}N/{}^{14}N$ in sample and $R_{standard} = {}^{15}N/{}^{14}N$ in international standards. The trophic level of each species is also calculated according to the equation of Minagawa and Wada (1984)

3 Results and Discussion

3.1 *Biofouling Communities*

The measuring mast of the Fécamp offshore wind farm is the first French offshore structure in the Channel. It is a tool for meteorological-oceanic measurements and environmental monitoring. Underwater observations, photos and videos, of the submerged structure of the mast were carried out in mid-April 2021 to assess the use, functionality and potential influence of the mast on the marine environment. Identification of fixed or mobile species of flora and fauna was carried out by analysing underwater photography taken in situ between April 19 and 21, 2021. Photographs were taken with homogeneous lighting and perpendicular to the mast to reduce deformation. The reference used was a quadrat with a surface area of 1 m^2, graduated every

10 cm, and two replicates were made for each depth. Biofouling communities were studied at three depth: −4, −12 and −30 m. In the subtidal zone, biofouling covered 100% of the steel and concrete structures. A total of 24 taxa were recorded in 2021 by photography rather than a direct observation of a diver. These 24 taxa correspond to 5 taxa in the flora and 19 taxa in the fauna (18 sessile and one motile) (Table 1).

The position of the mast implies exposure to two types of currents: the average residual current and tidal current. The residual current is directed towards the north and constitute the end of the circulation generally from the Atlantic to the North Sea. The order of magnitude of the residual current in the Dover Strait is 5 to 10 cm s^{-1}. Tidal currents, the dominant phenomenon of the English Channel, reverse between flood and ebb tides with a period of 12 h 25 min. During the rising tide, the mast is subjected to a current in the WSW–ENE–direction, while during the ebb tide it is subjected to a current in ENE–WSW direction. The two directional surfaces of the structure are alternately subjected to the same tidal current force. The tidal current is dominant compared to the residual current. Since tidal currents alternate between WSW-ENE and ENE-WSW every 6 h, it is hypothesized that the two directional surfaces of the structure are exposed to equivalent tidal currents and that tidal currents do not play an important role in species distribution on the two surfaces. In agreement with in situ observations, the distribution of species on the mast does not seem to be linked to tidal currents. Therefore, the effect of tidal currents on the distribution of species between the two directional surfaces of the structure is probably limited.

In comparison, Schlicklin et al. (2022) studied the colonization on a surface of a prototype vertical axis tidal turbine, of which height is 15 m above the bottom, by Hydroquest located off Paimpol Bréhat coast, France. This tidal turbine was deployed for two years and was uninstalled at the end of September 2021. The deployment of the infrastructure provided the opportunity to create a vertical habitat in a highly hydrodynamic environment. The samples were taken by scraping after the removal of the structure. The protocol used was 5 scrapings of 0.1 m^2 at three different depths and positions on the tidal turbine: the top, the middle and the bottom. The first observations showed that only three surfaces were colonized: the top (− 22 m) and the underside (− 33 m) of the tidal turbine, which was not covered with antifouling paint, and the small part of the middle of the structure (−29 m). No flora was present on the tidal turbine because the structure was deeply submerged and the algae did not receive enough light for their development. However, the fauna is diverse with a total of 118 taxa. The sessile organisms (barnacles, scallops, bryozoans) provide biogenic habitat for mobile organisms which consisted of mainly peracarid and decapod crustaceans.

Currently, little research has been done on the structures deployed at sea in France. Indeed, the installation of offshore wind farms remains a recent activity. At present, there is no regulatory monitoring of biofouling associated with these structures (Raoux 2017; Pezy 2017). The few existing studies consist of photographing a quadrat to investigate the recovery of fauna and flora. However, when comparing the two approaches from the qualitative and quantitative viewpoints carried out at the level of the Fécamp measuring mast (qualitative approach by photography) and at the level of the Paimpol tidal turbine (quantitative approach by scraping), we find very

Table 1 Flora and fauna presence (in black) in 2021 along three different depths (−4 m; −12 m; −30 m) along the masuring mast

	Family	Taxa	-4 m	-12 m	-30 m
Flora	Chlorophyta	*Ulva* sp	■		
	Rhodophyta	Gigartinales	■	■	
		Plocamium cartilagineum	■		
	Fucophyceae	*Dictyota* sp	■		
		Sargassum muticum	■		
Sessile fauna	Spongiaire	*Thymosia guernei*		■	
		Sycon ciliatum			■
		Spongosorites calcicole	■	■	
	Cnidaires	*Abietinaria abietina*			■
		Nemertesia antennina		■	
		Actinothoe sphyrodeta	■	■	■
		Alcyonium digitatum			■
		Tubularia indivisa	■	■	■
		Sagartia elegans	■		■
		Sertularia cupressina			■
		Ectopleura larynx	■	■	■
		Hydrallmania falcata		■	
	Bryozoaires	*Schizomavella sarniensis*		■	■
		Schizoporella unicornis	■	■	■
	Tuniciers	*Botryllus* sp			■
		Didemnum fulgens	■	■	■
	Arthropode	*Chthamalus stellatus*	■	■	■
	Mollusque	*Mytilus edulis*	■		
Motile fauna	Mollusque	*Calliostoma zizyphinum*			■

large differences in the observation of the communities established in these structures. In order to deeply understand the role of these new structures in the marine ecosystems, it is needed to obtain quantitative data from quadrat samples of structure surface organisms scraped by divers. Indeed, peracarids are a mobile fauna that is underestimated or unobserved in quadrat photos, whereas these organisms play an important functional role in the structure and function of this reef-related ecosystem (Pezy 2017).

3.2 From an Ecosystemic Approach to Trophic Web Modeling

To understand the structure and the function of the measuring mast in an ecosystem, it is important to collect precise data. These data allow to build trophic models using different methodology approaches, i.e. Ecopath with Ecosim approach (Polovina 1984; Christensen and Walters 2004; Christensen et al. 2008) or Linear Inverse Model Monte Carlo Markov Chain). These approaches, in which all biotic components of the system are considered at the same time, provide measures of the structure and functioning properties through the calculation of Ecological Network Analysis (ENA) indices (Ulanowicz 1986).

Actually, in France, the only OWF site where industry has adopted an ecosystemic approach is at Dieppe-Le Tréport. Indeed, as part of the regulatory monitoring of the DLT OWF construction and operation, the Scientific Interest Group (GIS) "Eolien en mer" plans to carry out a study of the complete trophic web and its evolution in relation to the OWF construction. The goal of this study is to promote a sampling strategy to collect data on the different ecosystem compartments of the future DLT OWF site and to adopt an ecosystem approach so that trophic network analysis can be applied to other sites of OWFs. To this end, it is essential to collect new data on biological compartments (phytoplankton, zooplankton, suprabenthos, meiofauna, benthos, demersal fishes, marine mammals and birds). Two cruises (winter: February 2022 and summer: September 2022) allowed us to estimate the contribution of each zoological group in the two main benthic habitats (coarse sand and medium clean sand) distributed at the site. In addition, isotopic analyses of various biological compartments, ranging from phytoplankton to birds, and stomach contents of demersal fish were also conducted. These studies will complete the EIA and characterize the future trophic web in and around the future OWF.

The second step is to integrate all these data into various trophic models. The first modelling exercise is a zero-dimensional approach of a complete trophic web at the future site of DLT OWF. For this, the modelling tool Ecopath with Ecosim will be used. The ecosystem model is employed to characterize the ecosystem and its complexity, to take into account interspecific interactions, and to allow the calculation of Ecological Network Analysis (ENA) indices (Ulanowicz 1986). The ecopath is built from local data sampled during the two cruises and integrates the results of the stomach contents of the local demersal fish.

To go further, we also construct a Linear Inverse Model Monte Carlo Markov Chain to quantify the uncertainty associated with the estimations of energy flow and ENA. In addition, isotopic analyses result from the biological compartments are used to more accurately quantify confidence intervals for energy flow.

Acknowledgements This study is carried out in the DRACCAR project supported the Normandy Region. The authors thank Teruhisa Komatsu for his help in review and editing of the manuscript.

References

Bergström L, Sundqvist F, Bergström U (2013) Effects of an offshore wind farm on temporal and spatial patterns in the demersal fish community. Mar Ecol Prog Ser 485:199–210

Boatman TG, Geider RJ, Oxborough K (2019) Improving the accuracy of single turnover active Fluorometry (STAF) for the estimation of phytoplankton primary productivity (PhytoPP). Front Mar Sci 6

Brunel P, Besner M, Messier D, Poirier L, Granger D, Weinstein M (1978) Le traîneau Macer-Giroq: appareil amélioré pour l'échantillonnage quantitative étagé de la petite faune nageuse au voisinage du fond. Int Rev Hydrobiol 63:815–829

Christensen V, Walters CJ, Pauly D, Forrest R (2008) Ecopath with Ecosim version 6 user guide. Lensfest ocean futures project, p 235

Christensen V, Walters CJ (2004) Ecopath with Ecosim: methods, capabilities and limitations. Ecol Model 172:109–139

Coates D, Deschutter Y, Vincx M, Vanaverbeke J (2014) Enrichment and shifts in macrobenthic assemblages in an offshore wind farm area in the Belgian part of the North Sea. Mar Environ Res 95:1–12

Couperus B, Winter E, van Keeken O, van Kooten T, Tribuhl S, Burggraaf D (2010) Use of high-resolution sonar for near turbine fish observations (didson). WeSea 2007–002 (Report No. C138/10). Report by IMARES-Wageningen UR

Dauvin JC, Vallet C (2006) The near-bottom layer as an ecological boundary in marine ecosystems: diversity, taxonomic composition and community definitions. Hydrobiologia 555:49–58

De Mesel I, Kerckhof F, Norro A, Rumes B, Degraer S (2015) Succession and seasonal dynamics of the epifauna community on offshore wind farm foundations and their role as stepping stones for non-indigenous species. Hydrobiologia 756:37–50

Degraer S, Brabant R, Rumes B (2014) Environmental impacts of offshore windfarms in the Belgian part of the North Sea. Learning from the past to optimize future monitoring programmes. Royal Belgian institute of natural sciences, management unit of the North Sea Mathematical models, Marine ecosystem management section, p 242

Folk RL (1954) The distinction between grain size and mineral composition sedimentary rock nomenclature. J Geol 62:344–359

Folk RL (1966) A review of grain-size parameters. Sedimentary 6:73–93

Folk RL, Ward WC (1957) Brazos river bar: a study of significant of grain size parameters. J Sediment Petrolol 27:3–26

Genty B, Briantais JM, Baker NR (1989) The relationship between the quantum yield of photosynthetic electron transport and quenching of chlorophyll fluorescence. Biochim Biophys Acta Gen Sub 990:87–92

Henkel SK, Suryan RM, Lagerquist BA (2014) Marine renewable energy and environmental interactions: baseline assessments of seabirds, Marine Mammals, sea turtles and benthic communities on the Oregon shelf. In: Shields MA, Payne A (2014) Marine renewable energy technology and environmental interactions. Springer Sciences, p 176

Joschko TJ, Buck BH, Gutow L, Schröder A (2008) Colonization of an artificial hard substrate by Mytilus edulis in the German Bight. Mar Biol Res 4:350–360

Krone R (2012) Offshore Wind-power Reef Effects and Reef Fauna Roles. Ph.D. thesis, University Bremen, Germany , p. 211

Krone R, Gutow L, Brey T, Dannheim J, Schröder A (2013a) Mobile demersal megafauna at artificial structures in the German Bight-Likely effects of offshore wind farm development. Estuar Coast Shelf Sci 125:1–9

Krone R, Gutow L, Joschko T, Schröder A (2013b) Epifauna dynamics at an offshore foundation-implications of future wind power farming in the North Sea. Mar Environ Res 85:1–12

Krone R, Dederer G, Kanstinger P, Kramer P, Scheinder C, Schmalenbach I (2017) Mobile demersal megafauna at common offshore wind turbine foundations in the German Bight (North Sea) two years after deployment—increased production rate of *Cancer pagurus*. Mar Environ Res 123:53–61

Leonhard S, Stenberg C, Støttrup J (2011) Effect of the horns rev 1 offshore wind farm on fish communities follow-up seven years after construction. DTU Aqua report, p 99

Leonhard SB, Stenberg C, Stottrup J, van Deurs M, Christensen A, Pedersen J (2013) Fish benefits from offshore wind farm development. In: Danish offshore wind—key environmental issues—a follow-up. Danish energy agency, pp 31–45

Lindeboom HJ, Kouwenhoven HJ, Bergman MJN, Bouma S, Brasseur S, Daan R, Fijn RC, de Haan D, Dirksen S, van Hal R, HilleRisLambers R, terHofstede R, Krijgsveld LKL, Scheidat M (2011) Short term ecological effects of an offshore wind farm in the Dutch coastal zone; a compilation. Environ Res Lett 6:1–13

Maar M, Bolding K, Petersen JK, Hansen JLS, Timmermann K (2009) Local effects of blue mussels around turbine foundations in an ecosystem model of Nysted off-shore wind farm, Denmark. J Sea Res 62:159–174

Minagawa M, Wada E (1984) Step wise enrichment of 15N along food-chains: further evidence and the relation between d15N and animal age. Geochimica 48:1135–1140

Norling P, Kautsky N (2008) Patches of the mussel *Mytilus* sp. are islands of high biodiversity in subtidal sediment habitats in the Baltic Sea. Aquat Biol 4:75–87

Petersen JK, Malm T (2006) Offshore windmill farms: threats to or possibilities for the marine environment. Ambio 35:75–80

Peterson BJ, Fry B (1987) Stable isotopes in ecosystem studies. Annu Rev Ecol Syst 18:293–320

Pezy JP (2017) Approche écosystémique d'un futur parc éolien en Manche orientale: exemple du site de Dieppe-Le Tréport. Ph.D. thesis (in French) University of Caen Normandie, p 336

Pinnegar JK, Polunin NVC (1999) Differential fractionation of d13C and d15N among fish tissues: implications for the study of trophic interactions. Funct Ecol 13:225–231

Polovina JJ (1984) Model of a coral reef ecosystem. The ECOPATH model and its application to French Frigate Shoals. Coral Reefs 3:1–11

Raoux A (2017) Approche écosystémique des énergies marines renouvelables: étude des effets sur le réseau trophique de la construction du parc éolien au large de Courseulles-sur-mer et du cumul d'impacts. Ph.D. thesis University of Caen Normandie, p 292

Reubens JT, Degraer S, Vincx M (2011) Aggregation and feeding behaviour of pouting (*Trisopterus luscus*) at wind turbines in the Belgian part of the North Sea. Fish Res 108:223–227

Reubens JT, Braeckman U, Vanaverbeke J, Van Colen C, Degraer S, Vincx M (2013) Aggregation at windmill artificial reefs: CPUE of Atlantic cod (*Gadus morhua*) and pouting (*Trisopterus luscus*) at different habitats in the Belgian part of the North Sea. Fish Res 139:28–34

Salomon JC, Breton M (1991) Numerical study of the dispersive capacity of the bay of Brest, France towards dissolved substances. Environmental hydraulocs, Lee & Cheung, eds, Balkerma Rotterdam: pp. 459–464

Schlicklin F, Raoux A, Gallon R, Pezy JP (2022) Biodiversité et transition énergétique: la surprenante biodiversité qui a colonisé l'hydrolienne de Paimpol-Bréhat. Paralia 66:639–649

Shields MA, Payne AIL (2014) Marine renewable energy technology and environmental interactions, Humanity and the Sea. Edited by Springer Sciences, p 176

Strickland JDH, Parsons TR (1968) A Practical Handbook of Seawater Analysis. Bulletin of Fisheries Research Board of Canada, 167:1–311

Strickland JDH, Parsons TR (1972) A practical handbook of seawater analysis, 2nd edn

Tougaard J, Carstensen J, Teilmann J, Skov H, Rasmussen P (2009) Pile driving zone of responsiveness extends beyond 20 km for harbor porpoises (*Phocoena phocoena*). J Acoust Soc Am 126:11–14

Ulanowicz RE (1986) Growth and development: ecosystems phenomenology. Springer, New York

Webb WL, Newton M, Starr D (1974) Carbon dioxide exchange of *Alnus Rubra*. Oecologia 17:281–291

Wilhelmsson D, Malm T (2008) Fouling assemblages on offshore wind power plants and adjacent substrata. Estuar Coast Shelf Sci 79:459–466

Wilhelmsson D, Malm T, Öhman MC (2006) The influence of offshore wind power on demersal fish. ICES J Mar Sci 63:775–784

Wilhelmsson D, Malm T, Thompons R, Tchou J, Sarantakos G, McCormick N, Luitjens S, Gullstrom M, Patterson Edwards J, Amir O, Dubi A (2010) Greening blue energy: identifying and managing the biodiversity risks and opportunies of offshore renewable energy Edited by Dan Wilhelmsson, p 104

Winter HV, Aarts G, Van Keeken OA (2010) Residence time and behaviour of sole and cod in the offshore wind farm Egmond aan zee (OWEZ). Edited by IMARES, Wageningen, p 50

Co-Existence Between Offshore Wind Farm and Fisheries: Experiences in Japan

Hiroyuki Nakahara

1 Introduction

The author had been worked at Research Institute for Ocean Economics (RIOE), currently Promotion and Research Institute for Ocean Economics (PRIOE), more than 45 years until 2020, and at that institute, the author involved in developing Proposals on Co-Existence Between Offshore Wind and Fisheries publicized in 2013 for fixed-type turbines and its revised edition in 2015 along with proposals for the floating-type turbines. These are the very first of its kind proposals, so that they made great influence on both sides of wind farm developers and fisheries related communities as well as on public sector. For example, the concept of co-existence between wind farm developers and fisheries were adopted in the Second Basic Plan on Ocean Policy, a cabinet decision, based upon the Basic Act on Ocean Policy, in 2013. In 2018, The Act on Promoting the Utilization of Sea Areas for the Development of Marine Renewable Energy Power Generation Facilities, shortly expressed by Offshore Wind Farm Promotion Act, OWFPA, was enacted, and joint government ministries council was organized with Ministry of Economics, Trade and Industries, (METI), and Ministry of Land, Infrastructures and Transportation, (MLIT), with five members from each side. The author was a member of this Council until December 2022, and, in present, the author remains as a member of Local Councils organized by OWFPA in Aomori, Yamagata and Nagasaki Prefecture. Also, the author has been a member of Council on Offshore Wind Farm and Environmental Impact Assessment of Ministry of Environment, (MOE), until March, 2023. Discussions in these councils included consensus building and conflict resolution between developers and fishermen on offshore wind farm projects.

H. Nakahara (✉)
Part-time Lecturer, Yokohama National University, Yokohama, Japan
e-mail: nakahara0304@gmail.com

A.-C. Bennis et al. (eds.), *Constraints and Adaptations to Global Change at the Land-Sea Interface: For a Shared Ecological and Energy Transition*,
https://doi.org/10.1007/978-3-031-90050-1_14

2 Type of Offshore Wind Turbine and Anticipated Impact

There are two types of the turbine; bottom-fixed and floating type as shown in Fig. 1.

Regarding Bottom Fixed Type, there will be three basic designs of the base structure, namely Gravity Type, Mono-Pile Type and Jacket type. Gravity Type may fit for solid sea bottom which means not fit for pile driving. Mono-Pile Type is usually used in the case of shallow and sandy sea bottom, like North Sea, which is simple in design and easy to be installed. Jacket Type may be adopted in the case of either sandy or rocky sea bottom or mixed characteristics of sea bottom. Regarding Floating Type, there will be also a couple of designs, including Barge Type, TLP, Tension-Leg Platform Type, Semi-Sub Type and SPAR Type. These floating type designs are based upon and are modified from offshore oil development technologies. They will be adopted in the case in rather deeper sea areas, more than 50 m, and concurrently not near coastal areas but rather some distant offshore areas. Wind farm developers may choose which type of base structure is preferable upon the water depth, sea bottom characteristics and also difficulties of installation works.

Structural characteristics of offshore turbine, particularly those of base structure, may have different impact on environment and on fisheries both in positive and negative manner. Focusing on the positive side, in the case of bottom-fixed type turbine which will be deployed in the shallow water, particularly in the case of Mono-Pile type, usually needs scour protection bed which has positive impact because it plays a kind of role of artificial reef, working as adhesion substrate attracting seaweeds and bottom fish. In the case of floating type turbine which will be deployed in rather deep water more than 30–50 m to around 100 m or more, it may also have function of floating artificial reef, attracting fish living close to surface and migratory species.

In addition, wind farm is consisted with many numbers of turbines with two or three rows like grid shape layout. Then, such layout of wind farm is to be examined in the perspective of environment and fisheries, including whether or not coastal current will be affected, whether or not sea-bottom sand will be drifted, whether or not fishing operation will be possible inside the grid shape or other shape of wind turbine layout in the ocean space. Most of wind farm developers in Japan tend to

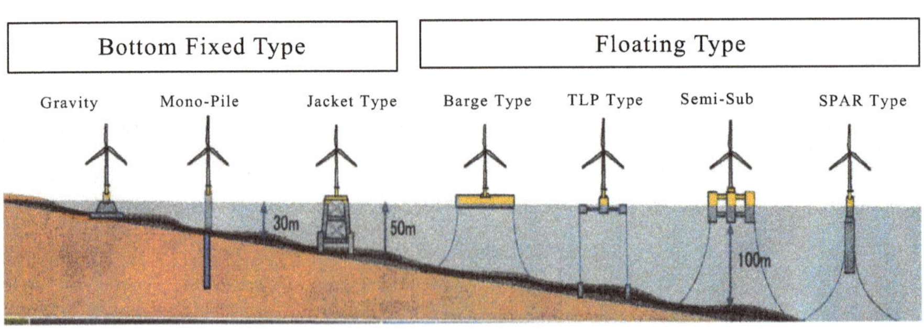

Fig. 1 Type of offshore wind turbine (*Source* METI and NEDO)

exclude fishing operation inside the wind farm sea area. However, there may be some developers who accept fishing operation inside wind farm sea area in the safe manner of fishing vessel maneuvering and appropriate use of fishing gear in abroad.

3 Positive and Negative Impact of Offshore Wind Turbines

As written above, there may be positive impact and negative impact of offshore wind turbine on marine ecosystem and fisheries. As shown in Fig. 2, basement structure has plus impact; habitat gain and resource preservation, but excluding fishing operation means minus impact; loss of fishing ground, too.

Upon the perspective on timeline of Offshore Wind Farm from construction phase to operation phase, fishermen may have some kind of concern on noise, vibration, sediment suspension and plume, possibility of current change, sand drift and others. Moreover, they are anxious whether or not they will be able to do their fishing operation inside of the offshore wind farm sea area, these concerns are shown in Fig. 3.

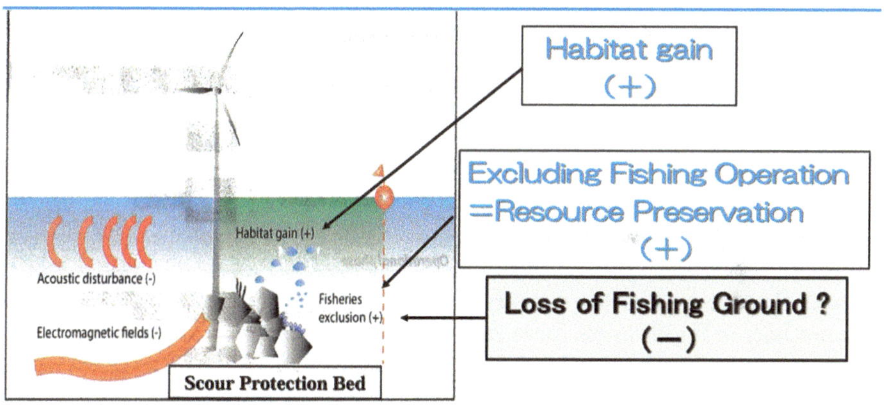

Fig. 2 Overview of main pressures from offshore wind turbine. *Note* 1: + indicates aggregation/increase,—indicates avoidance/decrease (*Source* Bergström et al., Effects of offshore wind farms on marine wildlife—a generalized impact assessment. 2014)—https://iopscience.iop.org/article/https://doi.org/10.1088/1748-9326/9/3/034012/pdf—*Note* 2; Two light blue comments are added by Nakahara)

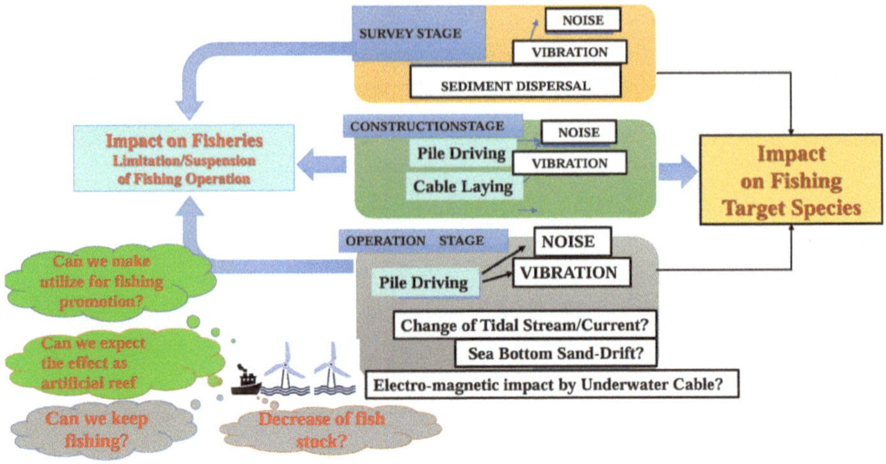

Fig. 3 Anticipated impact of offshore wind farm on Marine ecosystem and fisheries. (*Source* RIOE's proposal. Translation and lower left balloon illustration and comments by Nakahara)

4 Present Situation of Offshore Wind Farm Development in Japan

4.1 Present Situation in Japan

The very first offshore wind turbines were built in Port of Setana, located in western Hokkaido, with two 600 kw small jacket type turbines in 2004, and they are still working as of April 1st, 2025, but they are destined to be dismantled in 2026 because they exceed their life span of twenty years and also the size of 600kw is gettig obsolete. However, it was also the very first experience of co-existence with fisheries as shown later in Fig. 9. In the same year, 10 MW turbine was installed in the port of Sakata, Yamagata Prefecture, located in western side of Tohoku area of Honshu main island. In 2010, total capacity of 14 MW wind turbines were constructed in the coast of Port of Kashima, Ibaraki prefecture, located eastern Honshu. In 2012, Tokyo Electric Power Company, hereinafter TEPCO, installed gravity type bottom fixed wind turbine off Choshi City, and J-Power, another electric company, also installed the same type of wind turbine off Kita-Kyushu, located western part of Japan. A very first floating spar type offshore wind turbine was set up off Fukue-shima island in Goto Islands, Nagasaki Prefecture, located in the East Chine Sea. This floating turbine was moved to nearby off Kaba-shima later. Following these projects, METI initiated large scale wind farm demonstration project with three floating type turbines of 2 MW, 5 MW and 7 MW, far off the coast of Fukushima, about 20 km from the coast line, for a countermeasure for nuclear power plant disaster caused by the tsunami of Great Eastern Japan Disaster in March 11, 2011. These turbines were decommissioned in 2016 and 2017. The recent developments are found in Port of

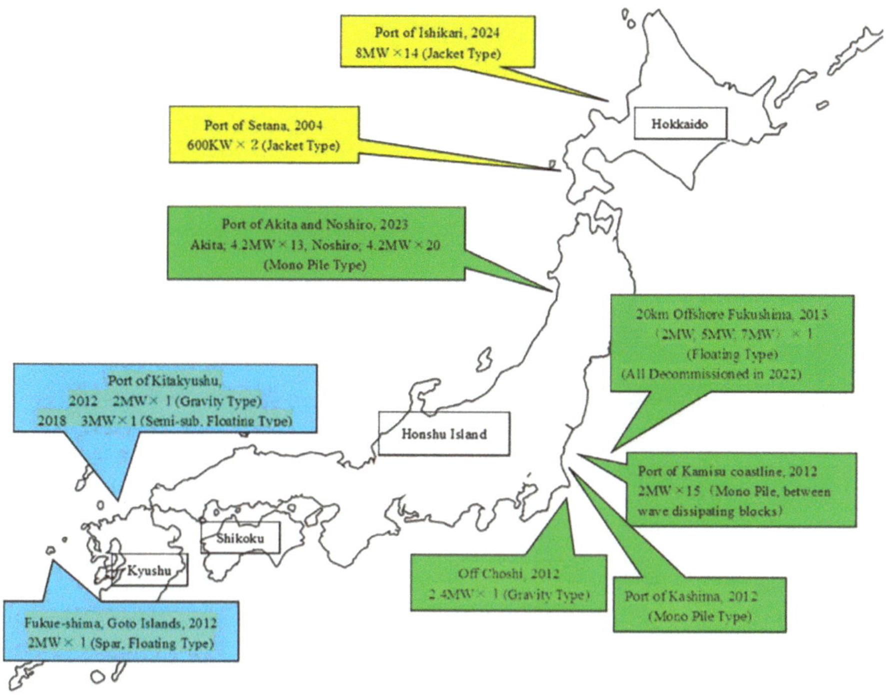

Fig. 4 Location map of offshore wind farm in Japan. (*Note* Colored ones are for demonstration, the others are in commercial operation). (From Nakahara, May 10, 2024)

Akita and Noshiro, Akita prefecture, with thirty-three mono-pile type turbines started its commercial operation in the end of 2022 to early 2023, and another big wind farm project had started also in the Port of Ishikari, Hokkaido, with fourteen jacket type turbines. All these wind farm developments are illustrated in Fig. 4.

4.2 *National Goals of Offshore Wind Energy*

In October 2020, Prime Minister Suga declared that Japan aims to reduce greenhouse gas emissions to net-zero by 2050, that is, to realize a carbon–neutral, decarbonized society. And at the Leaders' Summit on Climate in April 2021, Mr. Suga also announced that Japan aims to reduce its GHG emissions by 46 percent in 2030 from its 2013 levels. The cabinet decided 6th Energy Strategic Plan in October, 2021, and it declares to raise the percentage of renewable energy to 36–38% by 2030 with solar 14–16%, wind power 5% of that total. In March, 2023, an advisory council to the prime minister, the GX Executive Council have compiled the Basic Policy for the Realization of GX. GX stands for Green Transformation. The government says Japan will implement bold upfront investment support of 20 trillion yen in order to realize

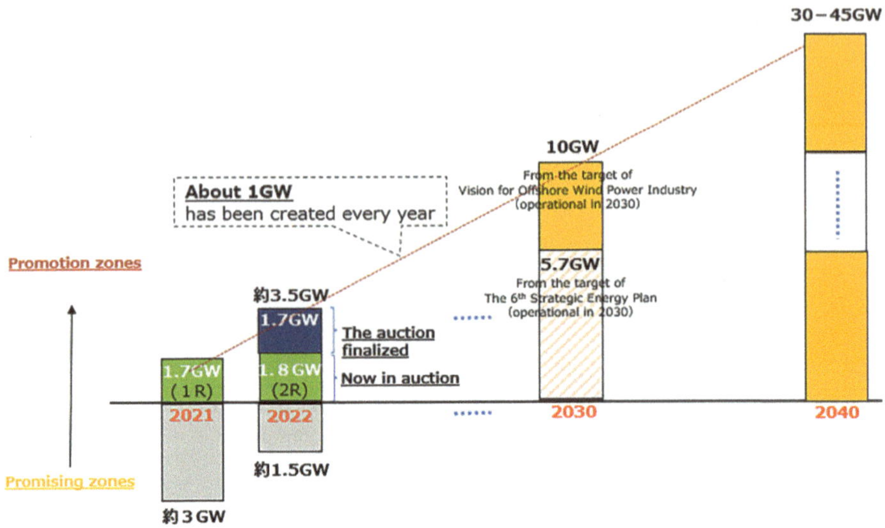

Fig. 5 National Goals on renewable energy development. (*Source* Presentation by Takahiro Ishii, Director, Wind Energy Policy Office, Agency for Natural Resources and Energy, ANRE, METI, March 2023)

GX investment of over 150 trillion yen (≒$1.1 trillion) in the public and private sectors, being flexible and agile, taking into account technology and will review the progress of each project and continuously upgrade the Basic Policy. Japan's Vision for Offshore Wind Power Industry was publicized in December, 2020, presenting the Introduction Target; 10 GW by 2030, 30–45 GW by 2040, and cost reduction target; 8–9 yen/kwh by 2030–2035. The numbers of each target need more than thousands of turbines around Japan and it is apparently very much challenging (Fig. 5).

4.3 Legal Framework of Offshore Wind Promotion Act

In April 2019, Offshore Wind Promotion Act (OWPA) was enacted and 8 Promotion Zones, 5 Promising Zones, 11 Preparation Zones were publicized. Also, to achieve above mentioned target capacity, Draft Amendment of OWPA was authorized by the cabinet on March 12, 2024, which aims to expand the act to EEZ sea areas, and the bill was eancted in June, 2025. The reason why the government needs to make amendment the Act is that current Act says Promotion Zones are to be set up only in the Territorial Sea areas, but it is insufficient to realize the target scale. So, there is a necessity for the government to set up Promising Zones in 200-mile EEZ. Japan's EEZ is about 4.47 million Km^2 and it is the sixth largest in the world and has great wind energy potential, and the government will promote large-scale floating offshore wind power projects in EEZ. Currently, the Green Innovation Fund is accelerating technology development and is conducting large-scale demonstrations.

4.4 Auction of Promotion Zones

Round 1 auction was finalized in December, 2021, which covers 1.7 GW in offshore Akita, northern part of Sea of Japan side of Honshu main island. Round 2 auction, around 1.8 GW, was concluded in July, 2023. Current status as of March 2023 is shown in Fig. 6.

5 Experiences in Japan

5.1 Proposal on Fisheries-Co-Existence Offshore Wind Farm by RIOE, 2013–2015

RIOE, currently PRIOE, Promotion and Research Institute for Ocean Economics, March, 2012, issued the Proposal on Co-Existence between Offshore Wind Farm and Fisheries (Interim Report) and in May, 2013, and Proposal on Co-Existence between Offshore Wind Farm and Fisheries (menu on fixed type Turbine) and its second edition I June, 2015. The second edition includes revised menu for fixed type and for floating type. Concept of "Fisheries-friendly wind farm" had been accepted by and penetrated in the central ministries and agencies, local government, wind farm developers, fisheries and other related communities, and also this concept was adopted into the Second Basic Plan on Ocean Policy, 2013, by cabinet decision (Fig. 7, see above).

Menu of Co-existence between Offshore Wind Farm and Fisheries is consisted by the following items, and developers, local fishermen and local governments are encouraged to choose adequate and optimum combination of these menu upon local conditions. Detail of menu is as follows:

Offshore site

a. Provide real-time oceanic data;Temperature, Salinity, Wave Hight, etc.

   → MUST! For Effective Navigation Pattern of Fishing Operation. (Save fuel).

b. Make use of foundation as artificial reef

   → Scour protection bed of mono-pile structure is a kind of artificial reef.

c. Attach offshore aquaculture facilities in the wind farm

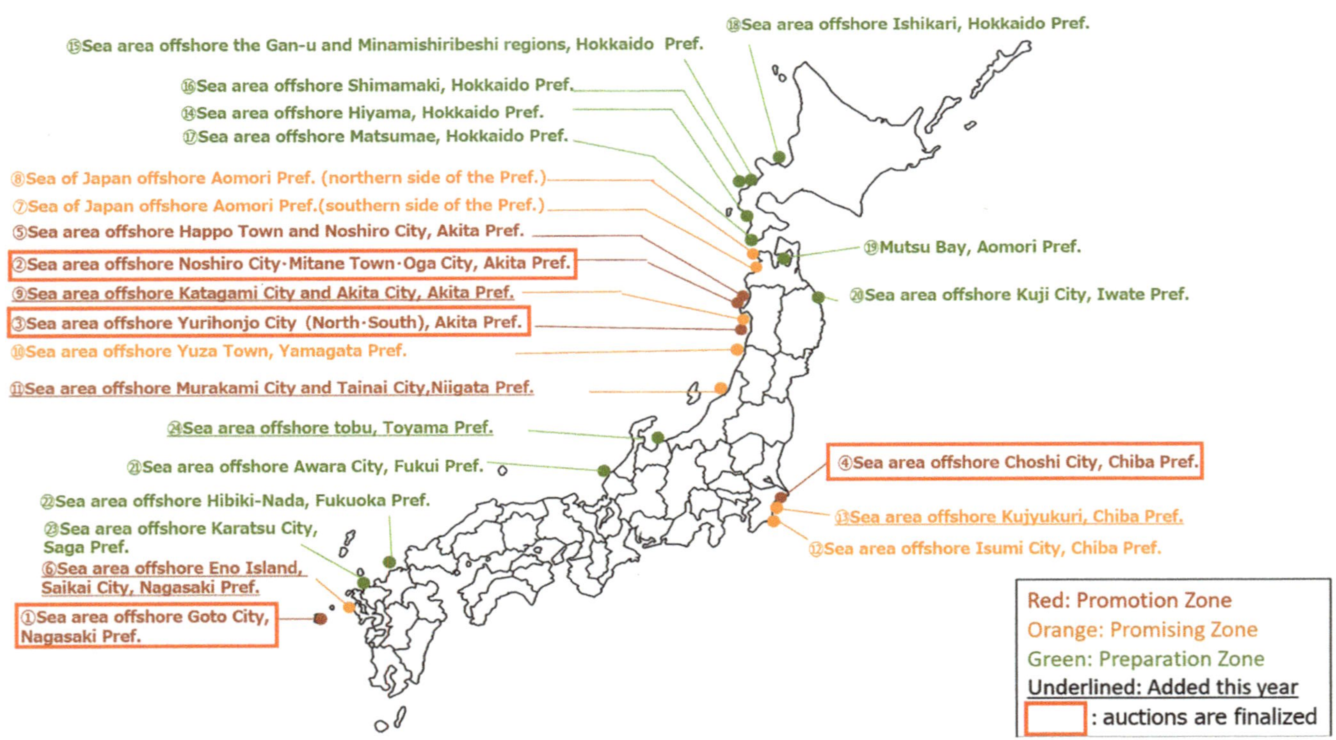

Fig. 6 Current status of offshore wind farm development as of March, 2023. (*Source* Wind Energy Policy Office, Agency for Natural Resources and Energy, ANRE, METI)

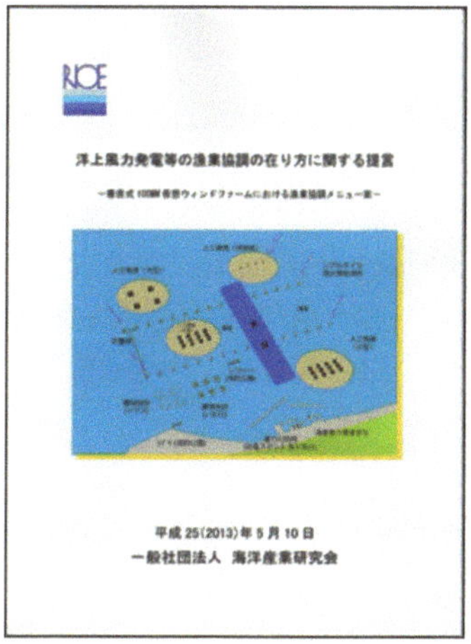

Cover page, RIOE's Proposal May, 2013 (Japanese only)

Cover page, RIOE's Proposal Second Edition, June 2015(Japanese only)

Fig. 7 Cover page of proposals by RIOE on Fisheries-friendly Offshore Wind Farm

→ Set fish cages among turbines in the area of OWF (High level ocean engineering is required).

d. Recreational use; Boat Tour to the wind farm, Offshore Fishing Park,

→ Diving Spot, Offshore Observation Platform (combine with Fishing Park).

e. Use the fishing boats based upon the cooperation with fishermen for patrol, inspection and maintenance work

On Land

a. Provide electricity produced by the wind farm to the fishing port facilities
b. Joint Venture or Fishermen's participation to the wind farm business when possible (Fig. 8).

Basic idea of this proposal is "From Old Scheme of Compensation by Money to New Co-existence Execution". The conflict between fisheries and wind farm development has been setteled by monetary compensation scheme for many years. Monetary compensation means abandon of the fishing ground. But the proposal was made with non monetary scheme with co-location of both of fisheries and wind farm development with following perspectives instead: (i) Win–Win Policy for both sides; (ii) Contribution to local community vitalization and (iii) Transparent consensus building procedure.

Required attitude for both sides are described as shown below.

For Developers: (i) Learn and get knowledge on local fisheries, Make consensus building with respect as existing user of the sea; (ii) Make plan on co-existence system

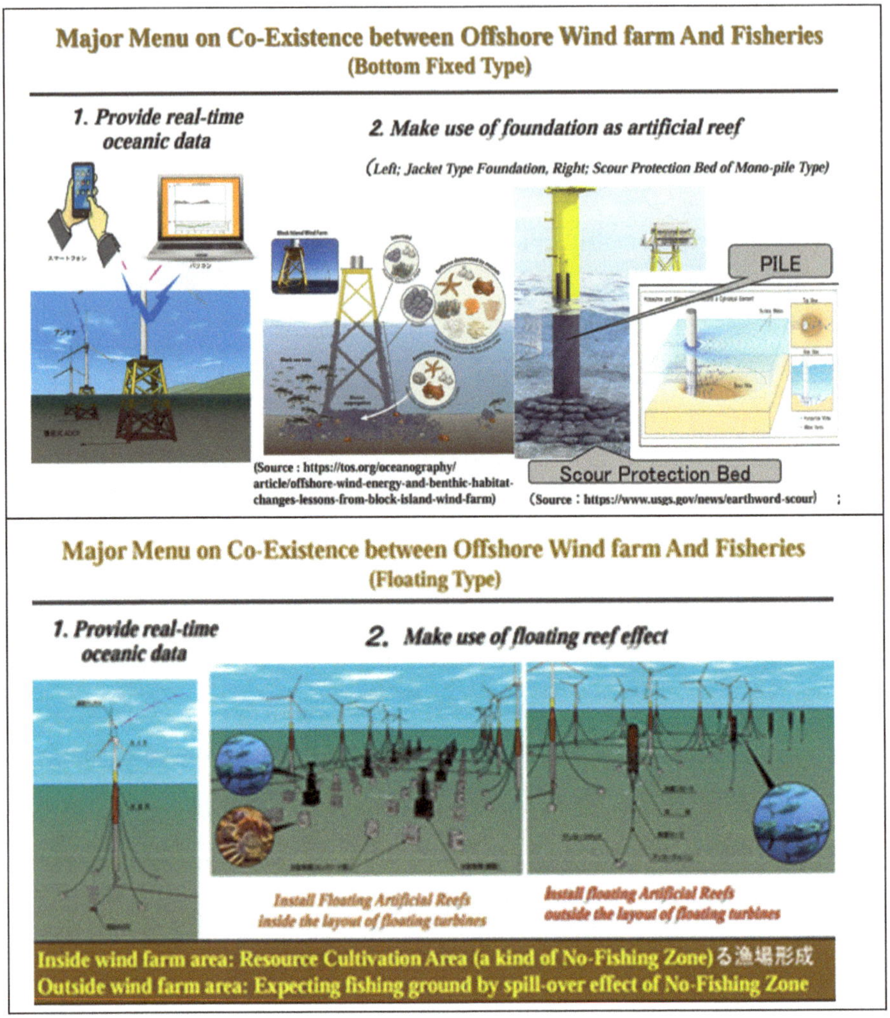

Fig. 8 Image illustrations of co-existence between OWF and fisheries. (*Source* RIOE's proposal, 2013 and 2015)

positively and make contribution to regional community development, including fisheries promotion.

For Fishery communities: (i) Recognize the importance of renewable energy development and collaborate with developers and related policies upon multiple and comprehensive use of the sea area, and (ii) Try and make challenge to make use of Offshore Wind Farm positively for sustainable fisheries and development of local fisheries communities.

Proposed menu will be categorized into three patterns as follows.

a. Direct Contribution to Fishing Activities (Contribution to Production Volume and Value): Improvement of fishing pattern, saving fuel/ resource cultivation; formation of fishing ground; Add cultivation facilities.
b. Secondary Contribution to Fishing Activities; Contribution on Upgrading Fishermen's Income, e.g. contract with developers to hire fishing vessels as patrol boat and join the project managment side when possible and in proper timing.
c. Contribution to Infrastructure Improvement: Provide electricity to the fishing port facilities, e-boat in future.

This categorization could be the index to the decision making of allocation of the Fisheries Promotion Fund which will be set up by means of compensation.

5.2 *Real Examples of Co-Existence Between Offshore Wind and Fisheries*

5.2.1 Setana Port, Hokkaido in 2002

This example was executed in Setana Port, Hokkaido, in 2002, as shown in Fig. 9. As written in down right, originally a fisherman was doing cultivation to raise abalone and sea urchin at the construction site in the Port of Setana area. Then he was invited to be a member of the local examination committee on wind turbine installation. Consequently the committee decided to introduce co-existence idea into the design of the turbine. A kelp cultivating rope was set between two turbines to provide bait for abalone and sea urchin, and then, the fisherman became no need to buy the bait for the cage.

5.2.2 Off Fukushima Floating-Type National Demonstration Project, 2013–2020

On March 11, 2011, East Japan Great Disaster had occurred, along with big Tsunami, resulting serious accident of Fukushima Nuclear Power Plant located along the coastline. To cope with such oceanic disaster, the government carried out large-scale demonstration project, installing several floating wind turbines 20km off Fukushima coast in the depth of about 130m. The location was so close to the outer edge of territorial waters, about 22km, and it was a challenging program being executed not by private sector but by the government itself, METI (Fig. 10).

Local fishermen enjoyed receiving real time oceanic data, 24 hours in 365 days, including temperature, current direction and others, then they were able to imagine where the fishing ground would be formed upon their own long-time experience. Without such real time data before, they were obliged to guess potential

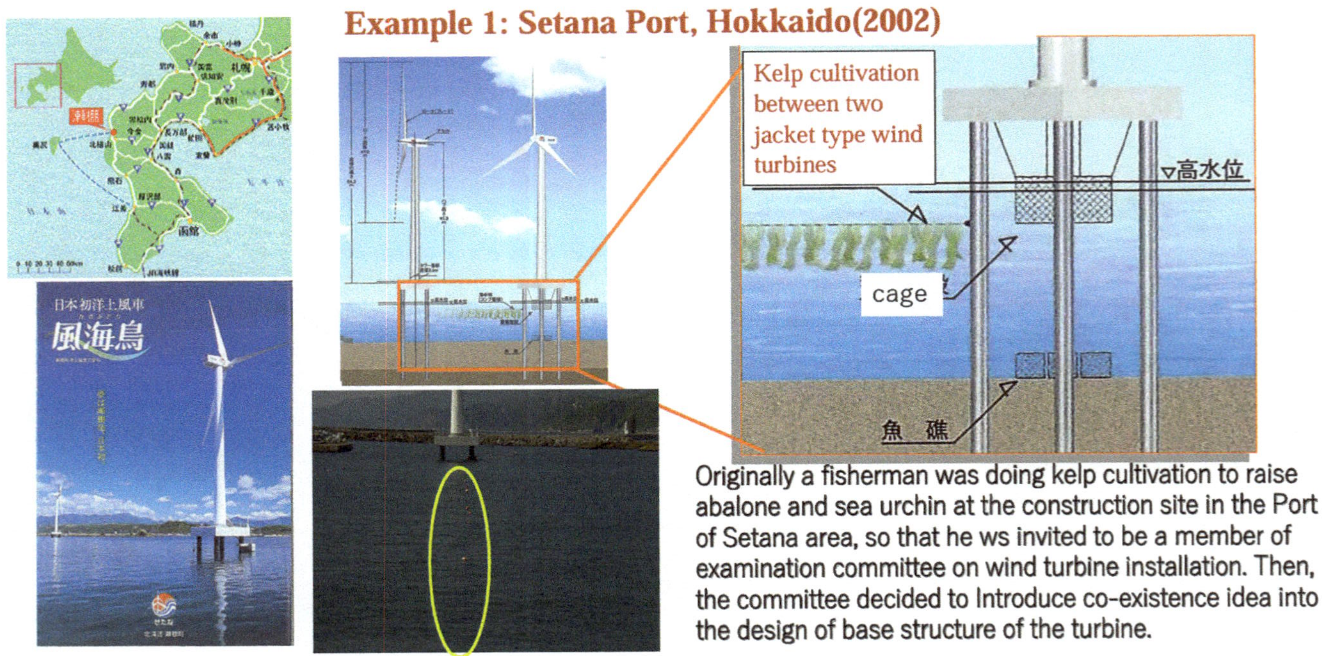

Fig. 9 Example of co-existence between offshore wind turbine and fisheries in Setana Port, Hokkaido, 2002. (*Source* RIOE's proposal)

Fig. 10 Example 2: Off Fukushima (2013–2020)—co-existence between offshore wind turbine and fisheries, providing oceanic data to the local fishermen's cooperatives, off Fukushima. (*Source* Pamphlet of the off Fukushima floating wind Turbine demonstration project)

fishing ground depending on the information provided by the research vessel operated by local government' fisheries research center only once or even several times in a month.

6 Offshore Wind Promotion Act and Fisheries

6.1 Procedure of Consensus Building

As described earlier in this article, The Offshore Wind Promotion Act entered into force in April, 2019. The act includes the procedure for the consensus building between developers and fishermen as shown in Fig. 11.

6.2 Importance of the Guideline

Guideline on Designation of the Promotion Zone for Offshore Renewable Energy Power Generating Facilities (herein after Guideline) was issued in June, 2019, by METI and MLIT, and the important point of this Guideline is as follows. Firstly, the implementation of the power generation project is not expected to interfere with fishing operation. Secondly, Promising Zone will not be designated when the implementation of the power generation project is expected to interfere with the fishing industry. Thirdly, when the implementation of the power generation project is expected to interfere with the fishing industry, the official Council will not be organized. The selected developer becomes one of the regular members of the Council until the end of project, so that PDCA cycle discussion would be held throughout

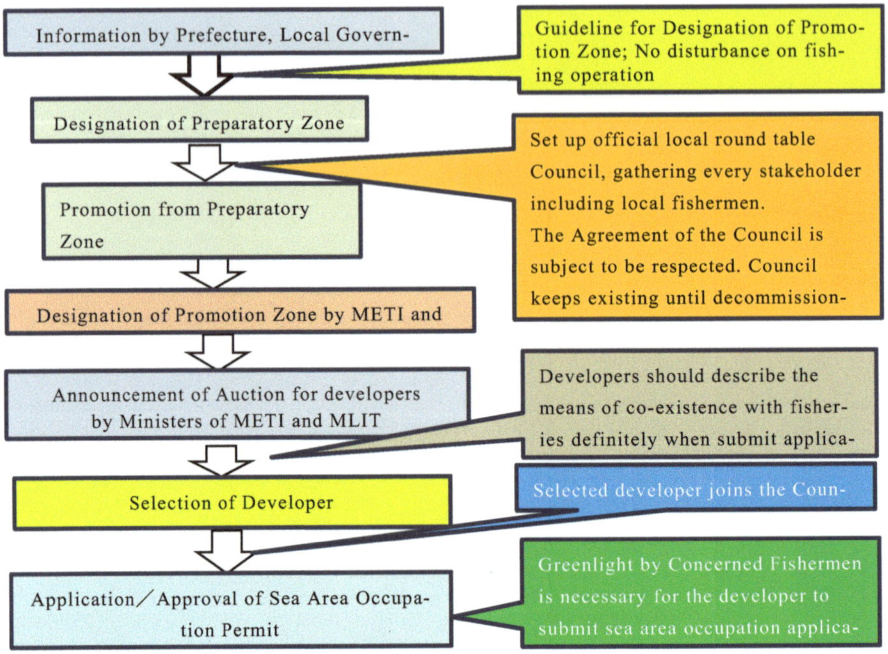

Fig. 11 Procedures of consensus building between wind farm developers and fishermen by the offshore wind promotion act. (*0*; Nakahara's presentation at the 19th French–Japanese symposium of oceanography, Coast Caen 2023, University of Caen, France, October 27, 2023)

the period of the project. It is very much significant for the conflict avoidance and consensus building between developers and fishermen.

Present status of project development around Japan in the level of Preparatory, Promising and Promotion Zones based upon the Act are illustrated in Fig. 12.

6.3 Contents of Agreement of the Council

There is a kind of standardized Table of Contents of the Agreement of the local Council as follows:

(a) General Idea
(b) Co-Existence with Fisheries and Local Communities, and Survey on the Effect on Fisheries
(c) Consideration on Site Decision
(d) Consideration on Construction
(e) Consideration on Implementation of Power Generation
(f) Consideration on Environmental Impact
(g) Others.

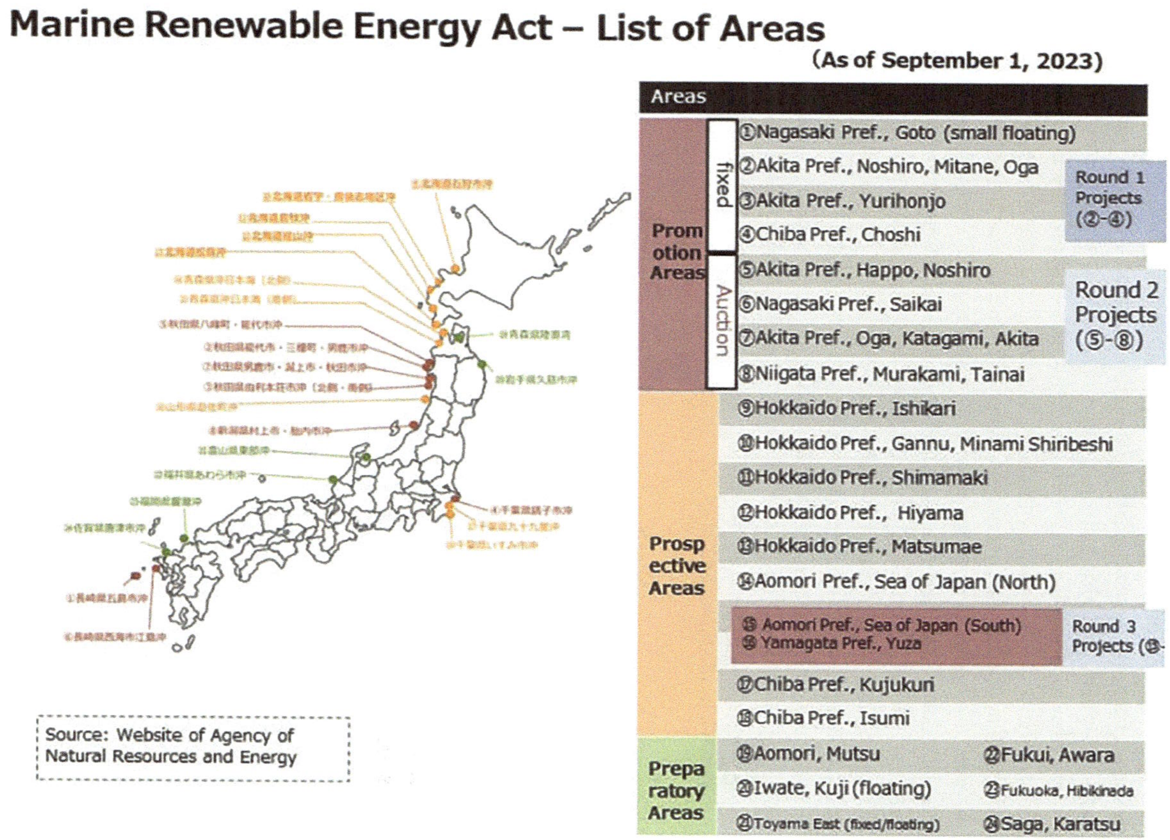

Fig. 12 Present status of project development based upon the offshore wind promotion act. (*Source* Watson Farley & Williams, Japanese offshore wind—status and recent developments, 12 March 2024), (https://www.wfw.com/articles/japanese-offshore-wind-developments/)

Among these contents, the most important point is (b) which mainly includes the establishment of Fisheries Promotion Fund upon developers' contribution. But we have to be careful because developers may tend to think that they can solve all their problems financially by contributing to the Fund, and thus they become rather reluctant to make efforts on their fishery's cooperative activities. As in the same manner, fishermen should be careful not to become complacent with the money they receive from the Fund, and also not to become reluctant to make effort to utilize offshore wind farm for local fisheries development. The first priority is to consider what kind of fisheries coordination measures are needed, rather than to consider whether a fund is needed at all. It is necessary to take care that contributions are not made in a sense of monetary compensation. While this approach to organize Fund is reasonable, there are some arguments that it may give the developers misconception that they can easily coexist with the fishing industry with paying money. In any way, we have to pay attention that, when the Fund once established, to make sure to keep transparency, fairness and impartiality. For example, some of the key points are who runs the Fund, how to make decision for allocation of the Fund for what purpose, how to make sure to objective audit in transparent manner. Also, advisory board may be needed for right management of the Fund. Current situation on the discussion for organization of Fund is shown in Table 1.

Table 1 Description on the fisheries promotion fund appeared in the agreements of the local councils

Area	Contents of the fund
Goto/Nagasaki (Nov. 25, 2019, June 29, 2022)	Establish the fund for fisheries promotion by floating wind turbine developer with city of Goto Fund be allocated mainly to subside of fishing boat insurance fee
Noshiro, Mitane, Oga/Akita (March 30, 2020)	0.5% of the income of electricity selling, 20 years
Yuri-Honjo/Akita (March 30, 2020)	Ditto
Choshi/Chiba (June 5 and Oct 22, 2020)	Establish the Fisheries promotion fund, and later, request of 1.18 billion yen
Saikai/Nagasaki (May 31, 2022)	Generation capacity × 250 × 30 Years
Murakami and Tainai/Niigata (June 20, 2022)	Ditto
Oga, Katagami/Akita (Aug 2, 2022)	Ditto

(*Source* Nakahara's presentation, 19th French–Japanese Symposium of oceanography, Coast Caen 2023, University of Caen, France, October 27, 2023)

Fig. 13 Offshore wind and fisheries; a win–win relationship is essential for the energy transition. (*Source* https://windeurope.org/newsroom/news/Jan.22, 2020)

7 Conclusion

To the end of the discussion, the author would like to touch upon a notable phrase found in a slide shown below. It says "Can fisheries and offshore wind farms co-exist?" "The answer to this question, of course, is yes. But requires specific conditions" (Fig. 13).

References

Bergström L, Kautsky L, Malm T, Rosenberg R, Wahlberg M, Capetillo NÅ, Wilhelmsson D (2014) Effects of offshore wind farms on marine wildlife—a generalized impact assessment. Environmental Research Letters, IOP Publishing Environmental Research Letters™ (ERL). https://doi.org/10.1088/1748-9326/9/3/034012

JWPA, Japan wind power association, an industrial organization of wind farm developers in Japan. Outline of JWPA is available at https://jwpa.jp/

Offshore wind and fisheries; a win-win relationship is essential for the energy transition, https://windeurope.org/newsroom/news/ Jan 22, 2020 (Note: This page doesn't exist anymore. Viewed on May 10, 2024, and re-checked on August 10, 2024).

Offshore Wind Promotion Act, formal name of the act is act on promoting the utilization of sea areas for the development of Marine renewable energy power generation facilities. JPN/ENG parallel translation of the Act is available at https://www.japaneselawtranslation.go.jp/ja/laws/view/3580

Promotion and Research Institute for Ocean Economics (2013) Proposals by RIOE on fisheries-friendly offshore wind farm, interim report, May 2013. Available at https://www.rioe.or.jp/cms/wp-content/uploads/2024/07/0510teigen.pdf and full version is available at the following URL.

Either one is in Japanese only. https://www.rioe.or.jp/cms/wp-content/uploads/2024/07/2015teigen.pdf

Co-Existence of Offshore Wind and Other Users of the Sea: 20 years of Learning What to Do (and not Do!)

L. Iborra, L. Leonard, S. Nielsen, A. Schultinga, E. Wilce, V. Türk, and C. Weller

1 Introduction

Urgently meeting global targets to address the interconnected global crises of biodiversity loss and climate change will require a strategic and integrated approach. Climate change is one of the main drivers of biodiversity loss and is steadily intensifying. In this context, the rapid deployment of offshore renewable energy (ORE) offers the potential to tackle both challenges—provided that global actors work together to unlock and enable a biodiversity-positive energy transition (Watson et al. 2024). This means ensuring that ORE deployment is timely, implemented in the right places and delivered in the right way.

As a global energy company, RWE has been active in the offshore wind sector in Europe for almost 25 years, developing, building and operating offshore wind farms (OWFs) in partnership with a wide range of stakeholders and civil society actors (local elected representatives, associations, NGOs, public authorities, scientists, etc.). Dialogue and consultation have always been at the core of the company's project development. Committed to ensuring the respect and coexistence of all

L. Iborra (✉) · C. Weller
RWE Renouvelables France, Offshore Wind, Clichy, France
e-mail: laura.iborra@rwe.com

L. Leonard · E. Wilce
RWE Renewables Management UK Ltd, Windmill Hill Business Park, Swindon, Wiltshire, UK

S. Nielsen
RWE Offshore Wind GmbH, Hamburg, Germany

A. Schultinga
RWE Offshore Wind Netherlands BV, Geertruidenberg, Netherlands

V. Türk
RWE Offshore Wind GmbH, Essen, Germany

A.-C. Bennis et al. (eds.), *Constraints and Adaptations to Global Change at the Land-Sea Interface: For a Shared Ecological and Energy Transition*,
https://doi.org/10.1007/978-3-031-90050-1_15

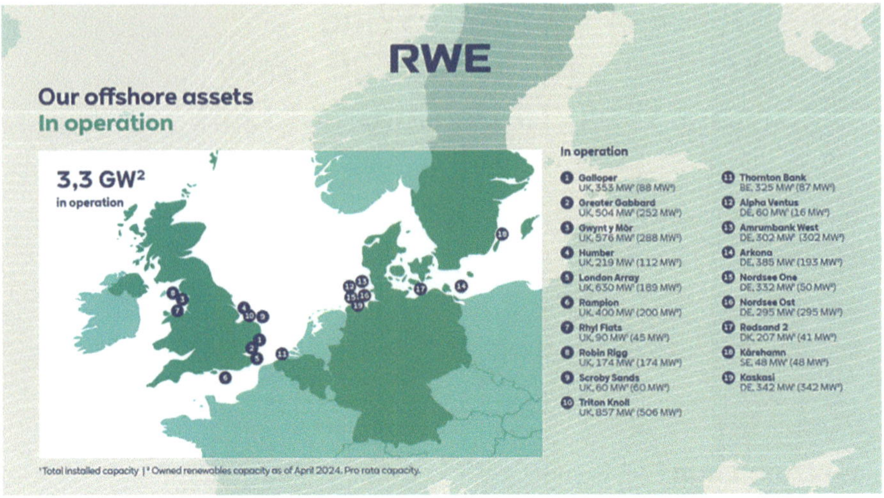

Fig. 1 Map of RWE's offshore wind assets currently in operation across Europe

maritime activities, RWE works closely with all users of the sea, including fishermen, leisure boaters, port and commercial traffic operators, environmental associations, scientists, economic players, etc. As the world's second-largest producer of offshore wind energy, RWE is involved across the entire value chain, with recognised expertise ranging from design and development to construction and operation. The experience gained over the past 20 years has led to the commissioning of 19 OWFs. RWE currently owns and operates 3.3 GW of OWFs across Europe (Fig. 1 and Table 1).

The company is therefore ideally positioned to drive forward the energy transition and to play a leading role in the sector at international level, with an objective of reaching grid capacity of 8 GW by 2030. To define this ambition and the means to achieve it, RWE is guided by two main principles: a carbon neutrality goal set for 2040—with reduction targets validated by the SBTN to align with the 1.5 °C trajectory—and a net positive impact objective on biodiversity, applicable to all new assets by 2030.

Over time, and through the development of RWE's offshore wind projects, a number of lessons have been learned regarding the co-existence of offshore wind with other users of the sea. It has become clear that a collaborative and participatory approach is essential to the successful planning and implementation of OWFs. One key question is how to reach acceptable compromises that reconcile strategic interests with local needs, including environmental concerns? The aims of this article are (i) to share RWE's experience from recent years; (ii) to provide concrete examples of how this has been applied across the RWE portfolio and; (iii) to identify the challenges that lie ahead.

Table 1 Characteristics of offshore wind farms developed or operated by RWE, as referenced in this paper.

Name	Type	Country	Location	Type	Capacity (in MW)	Surface (in km^2)	Number of turbines	Maximum depth (in m)	Commissioning
Tetraspar	Demonstrator	Norway	Stavanger	Floating	3.6	–	1	200	2021
Rampion 1	Commercial	UK	English channel	Fixed	400	72	116	30	2018
Rampion 2	Commercial	UK	English channel	Fixed	1200	196	90	60	Expected for 2029/2030
Gwynt y Mor	Commercial	UK	North Wales	Fixed	750	124	160	34	2018
Kaskasi	Commercial	Germany	German North sea	Fixed	342	18	38	25	2022
Karehamn	Commercial	Sweden	Swedish part of the Baltic Sea	Fixed	48	14	16	20	2013
New York Bight	Commercial	US	4 miles from New York Harbor	Fixed	1300	Not defined at this time	Not defined at this time	Not defined at this time	Expected for the end 2030
Sofia	Commercial	UK	Dogger Bank, North Sea	Fixed	1400	593	100	33	Expected for 2026
Canopy	Commercial	US	California	Floating	1600	Not defined at this time	Not defined at this time	723	Expected for the mid-2030s

2 Early Involvement

Understanding how different actors—including developers, policy makers, different stakeholder groups and the public—perceive offshore wind projects is key to determining not only the pace of development, but also the long-term success of the sector, regardless of the technical progress already achieved (Wüstenhagen et al. 2007).

The reasons behind local-level opposition to offshore wind projects have been widely discussed in the literature (Firestone and Kempton 2007; Wolsink 2010; Haggett 2011; Waldo 2012), often associated with the so-called "NIMBY syndrome" (Not In My Backyard). This concept describes local resistance to development projects due to perceived nuisances, with communities preferring such projects to be relocated elsewhere (Petrova 2013). However, these demands from coastal communities often reflect a lack of meaningful concertation and stem from perceived threat to the environment and/or to the broader interests of the local population, rather than purely selfish motives as implied by the NIMBY concept (Wolsink 2010; Devine-Wright 2005; Kempton et al. 2005). In addition, feelings of inequity and injustice appear to be the driving forces behind territorial claims (Wolsink 2007).

RWE's philosophy is therefore to initiate community consultation from the earliest stages of project design. This phase of listening, involvement and dialogue with the local community and users of the sea is crucial in order to develop projects that are concerted and exemplary. This can be achieved through various approaches, such as representing the industry at a strategic level, involving stakeholders in the design and development of projects, and sharing knowledge.

For example, in September 2023, RWE invited more than 100 stakeholders from seven different countries, with diverse backgrounds and interests, to Stavanger (Norway) to visit the operational floating wind demonstrator TetraSpar, installed since 2021. The study tour provided an opportunity to present the challenges and prospects of this innovative technology, but also to hear and understand the concerns raised by the different stakeholders.

Community acceptance has long been a major barrier to the development of onshore wind projects (Wiersma and Devine-Wright 2014), and offshore wind projects are not exempt from resistance within coastal communities. However, it has been shown that projects which communicate their benefits clearly, minimise their impacts, and engage local stakeholders throughout the planning and development phases are more likely to be accepted (Gross 2007). There is therefore an opportunity to draw on international best practice in terms of community acceptance and stakeholder engagement.

Stakeholder and local community engagement is at the heart of the development of the Rampion 1 and 2 OFWs, located in the English Channel on the south coast of the United Kingdom. Indeed, as part of the project development process, five Project Liaison Groups (PLGs) were established to ensure that representatives of various groups and organisations could work collaboratively with the project development teams to address emerging issues. The topics discussed cover a wide range of activities and interests—including professional fishing, business and tourism, the local community, the environment, and other sea users—as documented in the stakeholder meeting minutes available on the Rampion Offshore Website (https://www.rampionoffshore.com/news/library/meeting-minutes/). Consultation with fishermen has led to the implementation of a range of measures and procedures to minimise disruption to fishing activities during the various phases of the project. Dedicated sea user engagement events have been organised to share information and discuss the wind farm construction programme. A code of practice has also been developed and implemented to limit interference during project activities and cooperation agreements have been signed with fishers where appropriate. Based on these experiences, and supported by RWE's internal task force of fisheries specialists in various markets, RWE is currently developing marine user engagement guidelines to provide exemplary engagement practices across all projects in its portfolio.

3 Responsibility to Go Beyond Minimum Requirements

The early stages of project development represent an excellent opportunity to support the sustainable development of local communities. This has been put into practice at Gwynt y Mor OWF in North Wales (UK), where a Community Benefit Fund was launched to support the long-term sustainable development of local communities. A fund of £19 million was committed over 25 years, starting in 2015. Members of the community, including a fisherman, are also part of the Community Benefit Fund panel, ensuring that their views are taken into account and their needs thoughtfully considered. To date, a range of community projects have been funded, including improvements to port facilities, the provision of specialist vessel services, and the purchase of personal protective equipment for local sea rescue volunteers.

Another example of extensive engagement with local communities is the fishermen reskilling programme in Poland, launched by RWE at the end of 2022. The main objective of this skills development programme is to train existing fishermen to become part of the active workforce in the offshore wind sector. This free local service is designed, among other things, to provide the community of Ustka with tangible proof of RWE's commitment to local sustainable development. During the four weeks of training, the fishermen were able to acquire practical technical knowledge in the following areas: working at height, electrical and mechanical systems, installation and operation of wind turbines, first aid, and working at sea in accordance with the highest health and safety standards. In addition to hands-on technical training, an intensive English language course was included, as knowledge of English is a prerequisite for working in an international environment. All participants have successfully completed the training, and an evaluation of the programme by the participants will be carried out in 2024.

Another example is RWE's continued commitment to innovation. Indeed, in October 2023, RWE launched its fourth innovation competition in 3 years. The theme of this latest competition is the enhancement of co-use and biodiversity within floating OWFs. RWE invited start-ups, large companies, research institutes, individuals and organisations to propose innovative solutions to ensure that floating offshore wind projects make a positive contribution to preserving marine biodiversity and/or benefiting local stakeholders in the areas where the windfarms are located. More than 70 applications were received, and the winners of the competition will benefit from personalised support from RWE experts to further the viability of their ideas. They will also have the opportunity to work closely with RWE's international teams currently developing floating wind projects in Europe, Asia and the Americas. In addition, the winners will receive a research grant to support the implementation of their initiatives.

4 Focus on Impact Avoidance and Reduction

Human pressures leading to biodiversity loss are evident in more than three-quarters of marine areas under national jurisdiction and in almost two-thirds of all ocean systems (Halpern et al. 2015). Only 13.2% (~55 million km^2) of the global ocean can now be considered largely free from human impacts (Jones et al. 2018). In response, the application of the mitigation hierarchy—to avoid, reduce, restore, and, as a last resort, compensate for impacts—lies at the core of many biodiversity risk management policies (CSBI 2015). This framework is particularly relevant for the marine renewable energy sector, which aims to deliver both climate mitigation and biodiversity co-benefits. The hierarchy positions avoidance as the first and most important step in project development. Avoiding impacts on priority habitats and species is best achieved by integrating this objective early in the project design phase. However, the large-scale expansion of marine renewable energy needed to drive the energy transition and to adress climate change is likely to make it increasingly difficult to avoid all negative impacts on marine habitats and species. This highlights the importance of continuing to apply the mitigation hierarchy rigorously, iteratively and sequentially, with the objective of achieving—at the very least—no-net loss. Where impacts cannot be avoided, they should be reduced as far as possible. Finally, if residual impacts cannot be avoided or sufficiently minimised, compensation may be considered. Existing and emerging national mitigation policies are also increasingly recognising the potential of marine biodiversity offsets to address residual impacts, although their application has so far been limited (Niner et al. 2017; Shumway et al. 2018).

To focus on avoidance and mitigation measures, RWE has adopted a series of complementary approaches. First, open and transparent dialogue with stakeholders is essential. Identifying environmental issues and assessing the sensitivity of the various environmental receptors to offshore wind activities is a critical first step. This requires the involvement of local expertise—local associations, NGOs, scientists, academics and sea users—who possess precise and reliable knowledge of local ecosystems. Second, innovation and alternative methods are central to the mitigation process. The Kaskasi wind farm provides a useful illustration. During its the construction, an innovative pile installation—vibro pile driving—was tested as an alternative to conventional impact piling, with the goal of reducing underwater noise emissions, particularly for harbour porpoise [*Phocoena phocoena*, (Linnaeus 1758)] populations, which are highly sensitive to such disturbances. The objectives were multiple: to reduce the noise-related impact on a sensitive marine mammal species; to trial an innovative installation technique; and to determine whether Germany's regulatory underwater noise tresholds could be met without the use of additional noise mitigation systems (such as bubble curtains). Third, adaptive rather than prescriptive approaches should be favoured. The Rampion OWF offers an interesting example. Although fish stocks of black seabream and herring—species targeted by mitigation measures—are considered stable, they remain under surveillance. Notably, black seabream is a designation feature for the Kingsmere Marine Protected Area (MPA), which is located near the wind farm site. Any change to this feature could affect the MPA's conservation status. As a result, seasonal restrictions have been implemented, particularly for pile-driving, to avoid the key breeding periods of these two species.

5 Learn, Feedback and Improve

In recent years, policies requiring project developers to ensure no net loss of biodiversity have become increasingly common worldwide, supported by the concept of the mitigation hierarchy, as discussed previously (Niner et al. 2017, Maron et al. 2016, CSBI 2015, Tallis et al. 2015 and references therein). However, despite this approach, the environment continues to deteriorate, and an increasing number of governments are proposing that developers not only reduce their environmental impacts but also actively contribute to the improvement of nature. There is a growing momentum globally to extend the concept of no net loss to that of net gain (NG), primarily in terrestrial habitats, but increasingly in the marine environment as well (Hooper et al. 2021).

Over time, and throughout the development of RWE's OWFs, many lessons have been learned—unsuccessful trials, last-minute changes, things that worked, and things that did not. RWE's philosophy is to learn from these experiences, share knowledge, and continuously improve. The same mindset applies to the much-needed shift towards restoration and improvement as a systematic approach—not just as compensation—to achieving net gain. The UK is leading the way in this area: recent Environmental Impact Assessments (EIAs), including RWE's EIA for Rampion 2 presented a few months ago, have attempted to quantify net gain in the offshore marine environment.

Several projects in the RWE portfolio focus on restoring and enhancing the marine environment. One example is the pilot-scale Multi-Use Offshore Platform (MUOP) project near the Karehamn OWF in the Baltic Sea. This project explores the feasibility of integrating aquaculture systems—growing blue mussels and seaweed—into turbines foundations, with potential benefits for carbon storage and eutrophication reduction. It also considers co-use dynamics and social acceptance of the wind farm (more information: https://lnu.se/en/research/research-projects/multi-use-offshore-platform/). Another example is the offshore wind project in the New York Bight, where RWE is working with local communities to restore oyster habitats. The initiative will release millions of oyster larvae into the Huntington-Northport water body through the establishment of spawning sanctuaries. Finally, as part of the Sofia wind farm, RWE funded hatchery equipment ans supported the creation of an open-access marine education and awareness centre in Whitby (UK). The initiative aims to strenthen collaboration between fishermen and scientists—fishermen being directly involved in the collection of berried females, which are brought to the hatchery to support larval rearing efforts—to study lobster life cycle, raise public awareness of marine conservation, and potentially enhance local lobster populations through conservation-focused breeding.

6 Conclusion

Looking ahead, several challenges are anticipated that will require increased consideration, experimentation and learning. To meet international net-zero emissions targets by 2050, it is estimated that 2,000 GW of offshore wind capacity will need to be installed worldwide, compared to just 35 GWin 2020 (IRENA 2021). Achieving this would involve installing approximately 5,000 new wind turbines per year, occupying more than 500,000 km^2 of ocean by 2050 (Putuhena et al. 2023).

The need to scale up rapidly within constrained maritime space will require access to areas where opposition to offshore wind development has historically been strong, as well as to new frontier environments. The expansion of floating offshore wind, in particular, implies development in deeper and more remote locations. For instance, the floating wind development concession off the coast of California, awarded to RWE in late 2022, is located 45 km from shore, at a depth of approximately 700 m.

Accessing less-explored areas of the marine environment brings significant challenges in terms of understanding ecosystems structure and functioning. The deployment of teams and equipment to study these areas will remain a logistical and operational challenge, especially under changing climatic conditions.

The offshore wind sector must now evolve from a focus on mitigation impacts to one of integrating nature—and ultimately, delivering nature-positive outcomes. This transition calls for clear management and conservation objectives at the ecosystem scale. However, these objectives remain poorly defined in certain marine compartments, and long-term datasets will be needed to identify relevant trends. In addition, robust monitoring strategies and appropriate environmental indicators will be essential to evaluate the effectiveness of such initiatives.

In summary, ensuring the coexistence of offshore wind with other users of the sea requires an integrated approach, that reconciles multiple, and sometimes competing and divergent, stakeholders interests with the broader imperative of preserving, restoring and enhancing marine ecosystems health. The measurable outcomes of restoration and nature-positive initiatives are unlikely to be fully relaised at the scale of individual projects. Instead, progress will depend on ecosystem-scale efforts developed through collaborative, multi-stakeholders approaches.

References

Cross-Sector Biodiversity Initiative (CSBI) (2015) A cross-sector guide for implementing the mitigation hierarchy. Biodiversity Consultancy on behalf of IPIECA, ICMM, and the equator principles association, Cambridge, UK

Devine-Wright P (2005) Beyond NIMBYism: towards an integrated framework for understanding public perceptions of wind energy. Wind Energy: Int J Prog Appl Wind Power Convers Technol 8(2):125–139

Firestone J, Kempton W (2007) Public opinion about large offshore wind power: underlying factors. Energ Policy 35(3):1584–1598. https://doi.org/10.1016/j.enpol.2006.04.010

Gross C (2007) Community perspectives of wind energy in Australia: the application of a justice and community fairness framework to increase social acceptance. Energ Policy 35(5):2727–2736. https://doi.org/10.1016/j.enpol.2006.12.013

Haggett C (2011) Understanding public responses to offshore wind power. Energy Policy 39(2):503–510

Halpern BS, Frazier M, Potapenko J, Casey KS, Koenig K, Longo C, Lowndes JS, Rockwood RC, Selig ER, Selkoe KA, Walbridge S (2015) Spatial and temporal changes in cumulative human impacts on the world's ocean. Nat Commun 6(1):1–7

Hooper T, Austen M, Lannin A (2021) Developing policy and practice for marine net gain. J Environ Manage 277:111387

IRENA (2021) World energy transitions outlook: 1.5 °C Pathway, international renewable energy agency, Abu Dhabi

Jones KR, Klein CJ, Halpern BS, Venter O, Grantham H, Kuempel CD, Shumway N, Friedlander AM, Possingham HP, Watson JEM (2018) The location and protection status of Earth's diminishing marine wilderness. Curr Biol 28(15):2506–2512

Kempton W, Firestone J, Lilley J, Rouleau T, Whitaker P (2005) The offshore wind power debate: views from cape cod. Coast Manage 33:119–149

Maron M, Ives CD, Kujala H, Bull JW, Maseyk FJ, Bekessy S, Evans MC (2016) Taming a wicked problem: resolving controversies in biodiversity offsetting. Bioscience 66(6):489–498

Niner HJ, Milligan B, Jones PJ, Styan CA (2017) A global snapshot of marine biodiversity offsetting policy. Mar Policy 81:368–374

Petrova MA (2013) NIMBYism revisited: public acceptance of wind energy in the United States. Wiley Interdisc Rev: Clim Change 4(6):575–601

Putuhena H, White D, Gourvenec S, Sturt F (2023) Finding space for offshore wind to support net zero: a methodology to assess spatial constraints and future scenarios, illustrated by a UK case study. Renew Sustain Energy Rev 182:113358

Shumway N, Watson JE, Saunders MI, Maron M (2018) The risks and opportunities of translating terrestrial biodiversity offsets to the marine realm. Bioscience 68(2):125–133

Tallis H, Kennedy CM, Ruckelshaus M, Goldstein J, Kiesecker JM (2015) Mitigation for one et all: an integrated framework for mitigation of development impacts on biodiversity and ecosystem services. Environ Impact Assess Rev 55:21–34

Waldo Å (2012) Offshore wind power in Sweden—a qualitative analysis of attitudes with particular focus on opponents. Energy Policy 41:692–702

Watson SC, Somerfield PJ, Lemasson AJ, Knights AM, Edwards-Jones A, Nunes J, Pascoe C, McNeill CL, Schratzberger M, Thompson MSA, Couce E, Szostek CL, Baxter H, Beaumont NJ

(2024) The global impact of offshore wind farms on ecosystem services. Ocean Coast Manag 249:107023
Wiersma B, Devine-Wright P (2014) Public engagement with offshore renewable energy: a critical review. Wiley Interdisc Rev: Clim Change 5(4):493–507
Wolsink M (2007) Planning of renewables schemes: deliberative and fair decision-making on landscape issues instead of reproachful accusations of non-cooperation. Energy Policy 35(5):2692–2704
Wolsink M (2010) Near-shore wind power—protected seascapes, environmentalists' attitudes, and the technocratic planning perspective. Land Use Policy 27(2):195–203
Wüstenhagen R, Wolsink M, Bürer MJ (2007) Social acceptance of renewable energy innovation: an introduction to the concept. Energy Policy 35(5):2683–2691

Contributing to Scientific Knowledge in the Context of Industrial Project: The GIS Eolien en Mer

E. Praca, R. Lafite, L. Martinez, F. G. Schmitt, and S. Lebourg

1 Introduction

The need to move towards a zero-carbon footprint for energy production by 2050 and the growing need for energy have led various countries with coastlines to develop offshore wind farms (OWF). Northern Europe and Asia are the regions of the world where this development is strong (Soares-Ramos et al. 2020; Xu et al. 2020). In France, after a period of strategy definition and testing at Veulettes-sur-mer (Nadai and Labussière 2014), the government has launched a schedule to install OWF on the four mainland seaboards: the East Channel-North Sea seaboard (MEMN), the North Atlantic-West English Channel seaboard (NAMO); the South Atlantic seaboard (SA); and the Mediterranean seaboard (MED) (Fig. 1).

An initial call for tenders launched by the French government took place on the MEMN coast in 2011. During the second call for tenders in 2014, Eoliennes en Mer Dieppe Le Tréport (EMDT) won the tender on the area offshore of Dieppe and Le Tréport, located in the eastern part of the English Channel (Fig. 2). EMDT is a

E. Praca (✉)
Eoliennes en Mer Dieppe Le Tréport (EMDT), Dieppe, France
e-mail: emilie.praca@eoliennes-mer.fr

R. Lafite
Université Rouen Normandie, Université Caen Normandie, CNRS, M2C UMR 6143, Rouen, France

L. Martinez
Cohabys, La Rochelle, France

F. G. Schmitt
CNRS, Université de Lille, Université Littoral Côte d'Opale, UMR 8187, Laboratoire d'Océanologie et de Géosciences (LOG), Wimereux, France

S. Lebourg
Arques La Bataille, France

A.-C. Bennis et al. (eds.), *Constraints and Adaptations to Global Change at the Land-Sea Interface: For a Shared Ecological and Energy Transition*,
https://doi.org/10.1007/978-3-031-90050-1_16

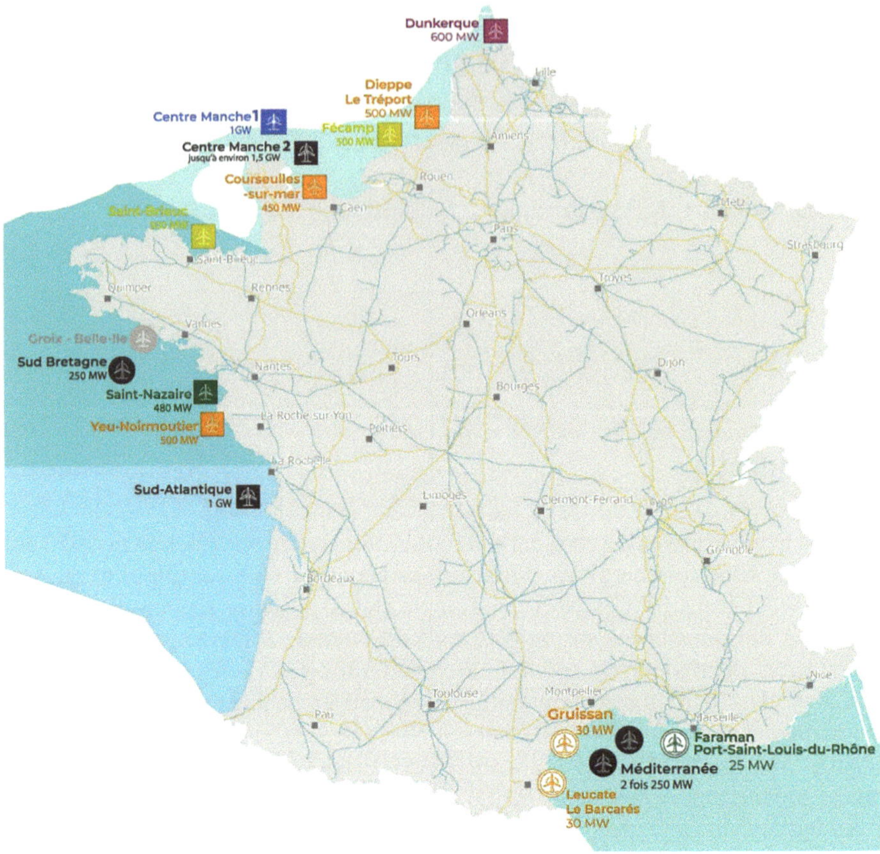

Fig. 1 Offshore wind farm projects along the French coasts (www.eoliennesenmer.fr; last access 2024/08/28). The four (4) French mainland seaboards are indicated by the different light blue areas

consortium of Ocean Winds (a joint venture between Engie & Energias de Portugal Renováveis [EDPr]), Sumitomo Corporation and Caisse des Dépôts. The provisional timetable is as follow:

- 2014 to 2024: development with detailed design, consultation, environmental impact study and permits;
- 2024 to 2026: construction;
- 2026 to 2051: operation and maintenance;
- from 2051: decommissioning or renewing of the concession.

The eastern English Channel is a shallow epicontinental sea, with a maximum depth of 70 m, located between France and the United Kingdom, and characterized by a mesotidal to megatidal tidal regime (Dauvin 2019). It has all the characteristics required for the installation of offshore wind farms, both in terms of its bedrock and

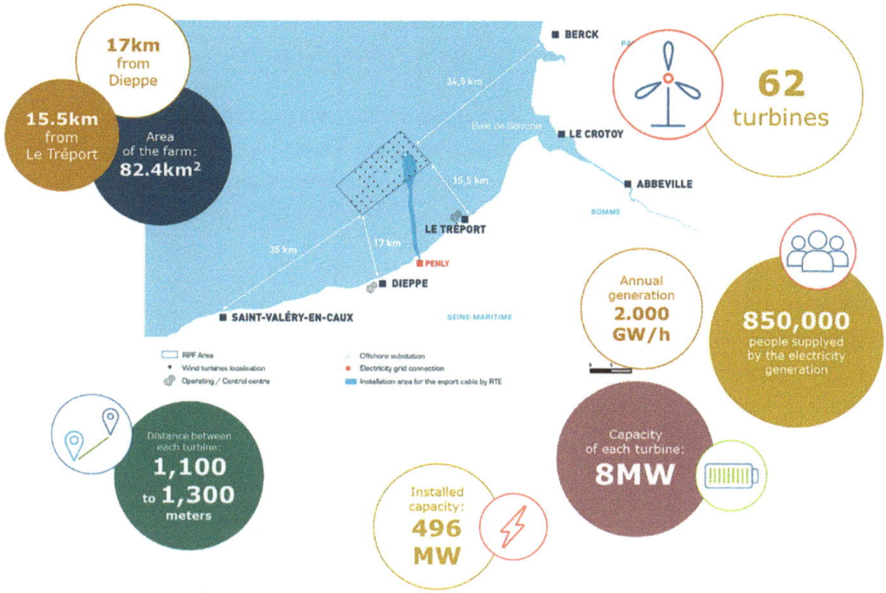

Fig. 2 Key figures for the Dieppe Le Tréport OWF (https://dieppe-le-treport.eoliennes-mer.fr/le-projet/le-projet-en-bref/, last access 2024/08/28)

its quaternary sedimentary cover (Eamer et al. 2014; Larsonneur et al. 1982) and in terms of wind power (Archer and Jacobson 2005). However, the English Channel is one of the busiest seas in the world, with a very high level of cumulative impacts (Halpern et al. 2008; Dauvin 2019) leading to conflicts of use. The impact on fisheries remains a major issue in the English Channel and North Sea (Stelzenmüller et al. 2022).

2 Material and Method

2.1 Creation of the GIS Eolien en Mer

In this context, the idea of creating a Scientific Interest Group (Groupement d'Intérêt Scientifique in French, GIS) emerged during the public debate on the Dieppe Le Tréport OWF in 2015, so that the consortium could participate in improving knowledge of the marine environment in the region of the farm. In France, a GIS is a research network governed by a scientific partnership agreement focusing on a specific subject. It has no legal personality. It is a public–private partnership (Desprez and Lafite 2014) or a partnership between public sub-entities (Roybin et al. 2001; Arrouays et al. 2022). Many GISs exist in different fields, because this type of structure is flexible and only requires an agreement between partners. GIS enables a Research & Development collaboration or synergy to be made visible and to operate within a conventional framework for a fixed period.

The commitment to create the GIS Eolien en Mer was included in the prefectoral decree authorising construction of the Dieppe Le Tréport OWF in 2019. The GIS Eolien en Mer has been launched in February 2020, with an initial private budget of eight millions of euros over the life of the OWF. For the GIS Eolien en Mer, the initial agreement is for 6 years and renewable.

Twenty-six partners are associated with the GIS and have signed the agreement: six universities (Caen Normandie, La Rochelle, Lille, Littoral Côte d'Opale, Picardie Jules Verne and Rouen Normandie), five public bodies (Cerema, Dieppe Maritime [urban community around Dieppe], Ifremer, Office Français de la Biodiversité [OFB] and Parc National Marin des Estuaires Picards et de la Mer d'Opale [EPMO]), eleven associations (Centre d'Hébergement et d'Etude de la Nature et de l'Environnement [CHENE], Comités Régionaux des Pêches Maritimes et des Elevages Marins [CRPMEM] de Normandie, Estran, France Nature Environnement [FNE] Normandie, Groupe D'Etude des Milieux Estuariens et Littoraux [GEMEL], Groupe Ornithologique et Naturaliste [GON], Groupe Ornithologique Normand [GONm], Groupe Mammalogique Normand [GMN], Groupement Régional des Associations de Protection de l'Environnement [GRAPE] en Normandie, Picardie Nature and Syndicat des Energies Renouvelables [SER]) and four private companies (ADERA-Cohabys, Créocéan, France Energies Marines [FEM] and EMDT). Other entities have not signed the agreement but are present as observers (Centre National de la Recherche Scientifique [CNRS], Réseaux de Transports d'Electricité [RTE], Union Nationale des Industries de Carrières et Matériaux de construction [UNICEM], CRPMEM des Hauts de France, Electricité de France [EDF] renouvelables, City Council of Dieppe, Région Normandie and state services).

Additional partnerships and responses to calls for projects may lead to additional funding and collaborations. This structure remains flexible and highly suited to the operational development of partnerships.

2.2 Objectives

The GIS Eolien en Mer has set itself three objectives as part of its partnership agreement:

(a) To support the implementation and monitoring of the effectiveness of the avoidance, reduction and compensation measures of the OWF by evaluating the regulatory protocols and reports;
(b) To improve knowledge of the marine environment in the eastern English Channel through fundamental studies and Research and Development (R&D) on environmental and socio-economic subjects;

(c) To promote the results of these studies in the form of popularization, scientific publications and presentations at conferences.

These objectives contribute to scientific advances and innovation, not only by disseminating the results of research, but also by encouraging the search for concrete and operational results needed to meet the challenges facing English Channel users, particularly fishermen represented by their fisheries committees and mobilized around fisheries resources and fishing grounds.

2.3 *Organization*

The GIS is governed by two bodies, the Steering Committee and the Scientific Council, and a coordinator (Fig. 3).

The Steering Committee is made up of a representative from each partner of the agreement. It decides on the operations to be carried out, the corresponding budget and organizational changes, and elects a chairman and vice-chairman. The Scientific Council, whose scientific experts are appointed by the Steering Committee, is an advisory body. It makes proposals on the direction of scientific issues and guarantees the relevance and quality of scientific work. It elects a chairman and a vice-chairman from among its members. The coordinator plays a key role in assisting the two chairmen and running the GIS on a day-to-day basis. He or she participates in the development project, shares data and results, maintains a bibliographic watch and takes part in conferences and other events. The coordinator is a key point of contact with external bodies, government departments and local authorities.

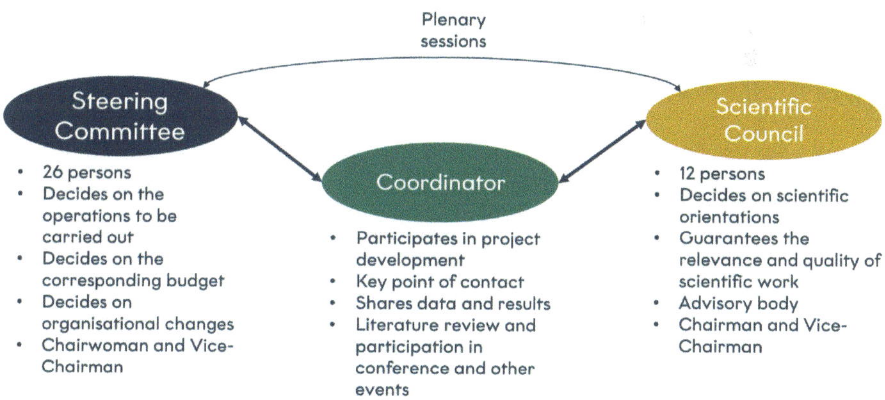

Fig. 3 Organization and tasks of the GIS Eolien en Mer committees (www.gis-eolienenmer.fr, last access 2024/08/28)

2.4 Study Selection Strategy

In addition to this regulatory monitoring, the Scientific Council and the Steering Committee wanted to launch targeted calls for tenders to provide additional scientific data in areas where knowledge gaps had been identified. It also drew on the summary of the ECUME national working group which is working on cumulated impact of OWF (Brignon et al. 2022). Once the topics had been identified or submitted to the Scientific Council, workshops were held with national experts to define the key objectives, innovative aspects, avoid duplication with other projects and determine a meaningful scale. The launch of a call for tenders is preferred, but if few researchers can be mobilized on a given subject, a targeted launch package is organized. Numerous studies are currently scheduled on OWF in France and scientific experts in certain fields are finding it difficult to respond to requests. In the end, these studies must be approved by the Scientific Council and the Steering Committee.

3 Studies and Activities of the GIS Eolien en Mer

3.1 Support for Regulatory Environmental Monitoring of the Dieppe Le Tréport OWF

The assessment of the environmental status of the Dieppe Le Tréport OWF began in 2014 with the initial impact assessment. It covered all the biological compartments in the area and enabled the impacts of the construction, operation and dismantling of the OWF to be assessed, and the Avoid–Reduce–Compensate measures and monitoring of these impacts to be defined. In particular, an ecosystem study was used to model the structure and functioning of the ecosystem and food webs of the future farm (Pezy 2017; Pezy et al. 2020).

As early as 2020, the GIS members reviewed and issued advices on the protocols proposed for the implementation of the environmental baseline monitoring, which took place between mid-2022 and the end of 2023. Then, in 2023, the GIS assessed the pre-construction environmental baseline reports and the implementation of specific measures planned for the construction phase scheduled to start in 2024.

3.2 Specific Studies of the GIS Eolien en Mer

Gulls monitoring. An initial study has been set up from 2021 to monitor several gull colonies on the Caux cliffs (between Dieppe and Le Havre). These colonies show a significant decline in numbers, -44% between 2000 and 2010 (Debout 2013), and it is therefore important to monitor these colonies in relation to the installation of the OWF. To this end, an annual census is carried out to count breeding pairs and the number of young fledged. In addition, ringing has been carried out since 2023. This monitoring will last a minimum of eight years in order to study the long-term evolution of the colonies and the movements of individuals between colonies.

Environmental DNA R&D studies. To complement traditional environmental monitoring with more effective molecular tools, the SIG wanted to characterize biodiversity using environmental DNA in two ecological compartments: seals and their diet, and zooplankton communities. The molecular approach should provide more comprehensive information to enhance traditional monitoring methods. By studying seal faecal samples, new sequencing technologies can be used to obtain an inventory of the prey species present in a predator's diet (Sousa et al. 2019). The robustness of these approaches and their routine application for marine top predators are still imperfect, but their application has already demonstrated a definite added value, particularly when coupled with the traditional hard remains approach (Jeanniard-du-Dot et al. 2017; McInnes et al. 2017). As far as zooplankton monitoring is concerned, genetic approaches based on environmental DNA studies have demonstrated their ability to accurately detect the presence of a wide range of species in the marine environment (Boussarie et al. 2018; Fediajevaite et al. 2021), complementing or even surpassing many traditional inventory techniques. These analyses with environmental DNA will be continued during the various phases of the OWF environmental monitoring in order to test and see the contribution of these techniques over the long term.

Bats transit at sea. The chiroptera compartment has not been studied as part of OWF impact studies, but acoustic detections of common pipistrelle (*Pipistrellus pipistrellus* [Schreber 1774]) and Nathusius' pipistrelle (*Pipistrellus nathusii* [Keyserling and Blasio 1839]) were carried out during the initial offshore impact study of Dieppe Le Tréport OWF. The Conseil National de la Protection de la Nature (CNPN) therefore recommended that the GIS carries out a study on the transit of bats at sea. Initially, this study included a genetic component between France and England, which could not be carried out due to a lack of capacity in the scientific community to respond to the call for tenders. However, it was possible to set up an acoustic monitoring component, which compares bat detections on land and at sea. The data acquired will form part of the national program led by the OFB *Migratlane* (2023–2027) and will enable a local analysis around the Dieppe Le Tréport OWF area as well as a large-scale analysis along the Atlantic, English Channel and North Sea coasts.

Marine megafauna. The GIS also wanted to extend and develop the local studies carried out as part of the monitoring of marine megafauna (birds, marine mammals, turtles, large pelagic fish) by associating them with national projects. The results of the megafauna monitoring (airborne monitoring) have been incorporated into the FEM *OWFSOMM* project (2021–2023), which applies to the four French seaboards, makes it possible to establish inter-site intercalibration and improves the efficiency of several sensors in the detection, identification and characterization of marine megafauna using deep learning tools. The GIS will also be participating in a multi-sensor megafauna observatory project on a mast in Fécamp OWF, the FEM *MMER-MAID* project (2023–2027), which will incorporate a regional scale for the Eastern English Channel and southern North Sea.

Food webs. Beyond the simple impact on each species, the GIS wanted to consider the complete food webs by analyzing the marine food chains (Pezy et al. 2020; Wang et al. 2024) in order to continue the initial work (Pezy 2017). A study of food webs: biomass, flows and seasonal variations (March 2022–March 2024) supplemented by higher trophic levels, through a study of the relationships between pelagic prey and top predators (September 2023–September 2024) has therefore been set up. The results will feed a trophic model such as Ecopath (Pezy et al. 2020). This study is intended to be repeated once the OWF is in operation and to have a broader approach at the scale of the Eastern English Channel and southern North Sea. This study is also taking part in a cumulative impact approach between OWF in the FEM *NESTORE* project (2023–2026).

Socio-economic aspects. The main opposition to OWF projects in France comes from professional maritime fishing, with the exclusion of some of the fishing zones and a possible impact on the resources in connection with the installation of the farms (Soma and Haggett 2015). It was therefore important for the GIS to look at the socio-economic aspects of the OWF. After several working meetings with the fisheries committees (CRPMEM Normandie and Hauts-de-France) and researchers in the field, the GIS decided to fund the Red-Fish study (Capturing the REDuction of FISHing grounds: OWF' development in the context of fishing grounds' contraction (past, present and future)). This study will target OWF areas in the broader context of the various existing constraints on professional fishing, by mapping them both retrospectively and prospectively.

3.3 Communication

In order to fulfill its last objective, the GIS uses various means of communication to disseminate the results of these studies and activities.

The main communication channel is its bilingual French–English website (www.gis-eolienenmer.fr, last accessed 2024/08/28), which publishes all the latest news, meeting minutes, articles and study reports. This information is shared on the dedicated LinkedIn page (https://www.linkedin.com/vitrine/gis-eolien-en-mer/, last accessed 2024/08/28). The GIS's activities and studies are also presented at conferences and round tables at national and international events. A webinar presenting the GIS studies for scientists was held in June 2023 and a conference for the general public is planned for the end of 2024. The GIS also intends to publish these studies in peer-reviewed journals so that the results can be disseminated to the scientific community.

4 Discussion and Conclusions

During the pre-construction phase of the Dieppe Le Tréport OWF, the GIS Eolien en Mer was able to put in place a series of fundamental and R&D studies to go broader the regulatory environmental monitoring of the OWF. These studies are both local and part of larger-scale projects via partnerships and are planned to be repeated over the long term.

Integration into multi seaboard projects or projects on the English Channel-North Sea scale is possible, as such projects are funded by various bodies (FEM, Observatoire de l'Eolien, Fonds Biodiversité de l'Eolien en Mer, Agence Nationale de la Recherche). It also works in partnership with the GIS ECUME (reference in this issue). The Scientific Council of the Comité Maritime de Façade Manche Est-Mer du Nord should also help to better define the studies of the GIS in a more global and integrated approach. However, this large number of players and calls for tenders on the subject of offshore wind energy must not lead to a dispersion of scientific experts' resources and energies.

During the Dieppe Le Tréport OWF construction and operation phases, studies on gull populations, the use of environmental DNA analysis to monitor seal diets and identify zooplankton, and the study of food webs via a doctoral thesis will be extended. New subjects should also be developed, such as socio-economic issues and the spatio-temporal monitoring of electrosensitive and/or commercial species using acoustic telemetry.

The creation of the GIS Eolien en Mer to accompany the Dieppe Le Tréport OWF has proved to be a useful way of improving knowledge on the marine environment in the Eastern English Channel through fundamental and R&D studies.

With the participation of twenty-six partners from different backgrounds (universities, public research centers, associations and private companies), the GIS is an innovative and effective way of public/private cooperation. It enables partners to exchange information on the many projects being developed around offshore wind energy. While the industrial sector often confines itself to "grey literature", this public/private collaboration and the ambition to promote the results through peer-review publications will be a good way of disseminating the results of studies to the scientific community.

References

Archer CL, Jacobson MZ (2005) Evaluation of global wind power. J Geophys Res 110:D12110

Arrouays D, Stengel P, Feix I, Lesaffre B, Morard V, Bardy M, Bispo A, Laroche B, Caquet T, Juille F (2022) Le GIS Sol, sa genèse et son évolution au cours des vingt dernières années. Etude Gestion Sols, Assoc Française Pour L'etude Sols 29:365–379

Boussarie G, Bakker J, Wangensteen OS, Mariani S, Bonnin L, Juhel JB, Kiszka JJ, Kulbicki M, Manel S, Robbins WD, Vigliola L, Mouillot D (2018) Environmental DNA illuminates the dark diversity of sharks. Sci Adv 4(5):eaap9661

Brignon JM, Lejart M, Nexer M, Michel S, Quentric A, Thiébaud L (2022) A risk-based method to prioritize cumulative impacts assessment on marine biodiversity and research policy for offshore wind farms in France. Environ Sci Policy 128:264–276

Buyse J, Hostens K, Degraer S, De Backer A (2022) Offshore wind farms affect the spatial distribution pattern of plaice *Pleuronectes platessa* at both the turbine and wind farm scale. ICES J Mar Sci 79:1777–1786

Dauvin J-C (2019) The English channel: La Manche, 2nd edn. In: Sheppard C (ed) World seas: an environmental evaluation, volume I: Europe, the Americas and West Africa. Chapter 6. Academic Press, United Kingdom, pp 153–188

Debout G (2013) Oiseaux marins nicheurs de Normandie : bilan d'un demi-siècle de recensements. Le Cormoran 19(78):67–78

Degraer S, Carey D, Coolen J, Hutchison Z, Kerckhof F, Rumes B, Vanaverbeke J (2020) Offshore wind farm artificial reefs affect ecosystem structure and functioning: a synthesis. Oceanography 33:48–57

Desprez M, Lafite R (eds) (2014) Monitoring the impacts of marine aggregate extraction: knowledge synthesis 2012, GIS SIEGMA, PRPH, p 43

Eamer JBR, Shaw J, King EL, MacKillop K (2014) The inner shelf geology of Atlantic Canada compared with the North Sea and Atlantic United States: insights for Atlantic Canadian offshore wind energy. Cont Shelf Res 213:104297

Eoliennes en Mer Dieppe Le Tréport Homepage. https://dieppe-le-treport.eoliennes-mer.fr/le-projet/le-projet-en-bref/. Accessed 28 Aug 2024

Eoliennes en Mer en France Homepage. https://eoliennesenmer.fr. Accessed 28 Aug 2024

Fediajevaite J, Priestley V, Arnold R, Savolainen V (2021) Meta-analysis shows that environmental DNA outperforms traditional surveys, but warrants better reporting standards. Ecol Evol 11:4803–4815

GIS Eolien en Mer Homepage. https://gis-eolienenmer.fr/en/. Accessed 28 Aug 2024

GIS Eolien en Mer LinkedIn page. https://www.linkedin.com/vitrine/gis-eolien-en-mer/. Accessed 28 Aug 2024

Halpern BS, Walbridge S, Selkoe KA, Kappel CV, Micheli F (2008) A global map of human impact on Marine ecosystems. Science 319:948–952

Jeanniard-du-Dot T, Thomas AC, Cherel Y, Trites AW, Guinet C (2017) Combining hard-part and DNA analyses of scats with biologging and stable isotopes can reveal different diet compositions and feeding strategies within a fur seal population. Mar Ecol Prog Ser 584:1–16

Larsonneur C, Bouysse P, Auffret JP (1982) The superficial sediments of the English channel and its western approaches. Sedimentology 29:851–864

McInnes JC, Jarman SN, Lea MA, Raymond B, Deagle BE, Phillips RA, Catry P, Stanworth A, Weimerskirch H, Kusch A, Gras M, Cherel Y, Maschette D, Alderman R (2017) DNA metabarcoding as a marine conservation and management tool: a circumpolar examination of fishery discards in the diet of threatened albatrosses. Front Mar Sci 4:277

Nadai A, Labussière O (2014) Recomposer la mer pour devenir offshore: le projet éolien de Veulettes-sur-Mer. Nat Sci Soc 22:204–218

Pezy JP (2017) Approche écosystémique d'un futur parc éolien en Manche orientale: exemple du site de Dieppe-Le Tréport. Université de Caen Normandie, Thèse de doctorat, p 324

Pezy JP, Raoux A, Dauvin JC (2020) The environmental impact from an offshore windfarm: challenge and evaluation methodology based on an ecosystem approach. Ecol Ind 114:106302

Raoux A, Dambacher JM, Pezy JP, Mazé C, Dauvin JC, Niquil N (2020) Assessing cumulative socio-ecological impacts of offshore wind farm development in the Bay of Seine (English channel). Mar Policy 89:11–20

Roybin D, Fleury P, Béranger C, Curtenaz D (2001) Conduite de recherches plurisdisciplinaires en partenariat et apprentissages collectifs. Le cas du GIS Alpes du Nord. Nat Sci Soc 9(3):16–28

Soares-Ramos EPP, de Oliveira-Assis L, Sarrias-Mena R, Fernandez-Ramirez LM (2020) Current status and future trends of offshore wind power in Europe. Energy 202:117787

Soma K, Haggett C (2015) Enhancing social acceptance in marine governance in Europe. Ocean Coast Manag 117:61–69

Sousa LL, de Silva SM, Xavier R (2019) DNA metabarcoding in diet studies: unveiling ecological aspects in aquatic and terrestrial ecosystems. Environ DNA 1(3):199–214

Stelzenmüller V, Letschert J, Gimpel A, Kraan C, Probst WN, Degraer S, Döring R (2022) From plate to plug: the impact of offshore renewables on European fisheries and the role of marine spatial planning. Renew Sustain Energy Rev 158:112108

Wang L, Wang B, Cen W, Xu R, Huang Y, Zhang X, Han Y, Zhang Y (2024) Ecological impacts of the expansion of offshore wind farms on trophic level species of marine food chain. J Environ Sci 139:226–244

Xu W, Liu Y, Wu W, Dong Y, Lu W, Liu Y, Zhao B, Li H, Yang R (2020) Proliferation of offshore wind farms in the North Sea and surrounding waters revealed by satellite image time series. Renew Sustain Energy Rev 133:110167

The Scientific Interest Group ECUME: A Multidisciplinary Consortium Challenging to Assess Cumulated Impacts of Human Activities in the Eastern Part of the English Channel

C. Mensens, L. Paporé, and J.-C. Dauvin

1 Introduction

The English Channel is among the most anthropized marine ecosystems in the world (Halpern et al. 2008; Dauvin 2019). Its coastline is characterised by a multitude of industrial and port activities (marine aggregates extractions, dredging and dumping of port sediments, maritime traffic, burying of submarine cables) and a high level of fishing activity (Dauvin 2019), to which can now be added the development of marine renewable energies (Raoux et al. 2017, 2021; Pezy et al. 2020). These activities at sea face major challenges in terms of sustainable development, social acceptability and the acquisition of knowledge and access to existing data (Degraer et al. 2020; Hutchnison et al. 2020; Chen et al. 2023; Farias Pardo et al. 2023). Following the transposition of the EU Marine Strategy Framework Directive into French law and the reform of impact studies, which include the study of cumulative effects, the development of research into the cumulative impacts of offshore activities has become a key area of research and development (Noguès et al. 2021, 2023).

The Scientific Interest Group GIS ECUME aims to respond to this challenge with a cross-sectoral approach, bringing together the main maritime activities in the eastern part of the Channel, and the implementation of a research programme to study

C. Mensens (✉) · J.-C. Dauvin
Laboratoire Morphodynamique Continentale et Côtière, UMR CNRS 6143 M2C,
24 Rue de Tilleuls, F-14000 Caen, France
e-mail: christoph.mensens@unicaen.fr

L. Paporé
Directrice Granulats Marins, Heidelberg Materials France, TOUR ALTO – 24ème étage,
4 Place des Saisons, CS 90143, F-92400 Courbevoie, France

A.-C. Bennis et al. (eds.), *Constraints and Adaptations to Global Change at the Land-Sea Interface: For a Shared Ecological and Energy Transition*,
https://doi.org/10.1007/978-3-031-90050-1_17

their cumulative impacts. With this approach, GIS ECUME guarantees the acquisition and sharing of scientific information by pooling the monitoring data from its partners and carrying out scientific studies to acquire new knowledge on the cumulative impacts of human activities at sea. The research projects facilitated by GIS ECUME aim to address five topics identified as crucial with regard to the impact of human activities at sea (see 2.2. for details): hydro-sedimentary dynamics, underwater acoustics, the artificial reef effect, social acceptability and the development of a methodology for assessing cumulative impacts, designed to meet the needs of industrial project developers to construct the chapter on cumulative effects required in environmental impact studies. Through these research activities, the GIS ECUME aims to further the acquisition of knowledge based on innovation and research and to favour the adaptation of practices to reduce their impact. The dissemination of the acquired knowledge by GIS ECUME contributes to a better understanding with the aim to reduce the impacts of anthropogenic activities on the coastal English Channel ecosystem.

2 GIS ECUME Key Information

2.1 Origins and Governance of GIS ECUME

The Groupement d'Intérêt Scientifique Effets Cumulés en Mer (GIS ECUME) was created on November the 17th of 2020, as a continuation of the GIS SIEGMA (Suivi de l'Impact des Extractions de Granulats Marins: Monitoring the Impact of Extractions of Marine Aggregations), which focused on the impact of aggregate extractions on the marine environment and fisheries (Desprez 2012). Between 2003 and 2014, the GIS SIEGMA had allowed to not only acquire the knowledge about the impact of marine aggregate extractions (Desprez 2000; Le Bot et al. 2010; Desprez and Lafite 2014), but also to establish a genuine consultation between the various users of the sea (fishermen, aggregates producers, scientists, public authorities). Subsequent to GIS SIEGMA, representatives of the main activities along the French coast of the English Channel recognized the need for a systemic approach to the cumulative effects of activities at sea.

The 10 member organizations of GIS ECUME unite the main maritime activities in the eastern Channel and higher education and research establishments in Normandy:

- Université de Caen Normandie
- Université de Rouen Normandie
- Université du Havre Normandie
- Union Nationale des Industries de Carrières et Matériaux de Construction (UNICEM: national union of marine aggregate extraction industries) (marine aggregate extraction)

- Comité Régional des Pêches Maritimes et des Élevages Marins de Normandie (CRPMEM: Normandie regional committee of marine fisheries and marine aquaculture) (fisheries)
- Réseau de Transport d'Électricité (RTE: Electricity Transport Network) (submarine cables)
- Ports de Normandie (ports)
- HAROPA PORT with the ports of Rouen and Le Havre
- EODC (Éoliennes Offshore du Calvados, joined in 2023) (offshore wind energy of Calvados)
- EOHF (Éoliennes Offshore des Hautes-Falaises, joined in 2023) (offshore wind energy of Hautes-Falaise).

The organisation of the GIS ECUME comprises the Steering Committee, the Scientific Council, the coordinator and the General Assembly (Fig. 1). The Steering Committee defines the research priorities and deliberates on all matters relating to the activities of the GIS ECUME. It consists of representatives of the member organisations of the GIS ECUME. The Scientific Council is responsible for conceiving research programmes in coherence with the research priorities defined by the Steering Committee, in particular by responding to regional, national and European calls for tenders. The Scientific Council is made up of 17 scientific researchers, appointed in a personal capacity and drawn from multi-disciplinary teams from throughout Normandy as well as from neighbouring regions. The General Assembly (GA) of the GIS ECUME is organised once a year to present the results of the actions carried out over the past period and to discuss future activities. The GA brings together industrial partners, scientific researchers, as well as representatives of regional authorities, environmental associations and other organisations and institutions with an interest in cumulative effects. The GA thus corresponds to a platform for exchanging and discussing information on all aspects of the cumulative effects of maritime activities with the various stakeholders.

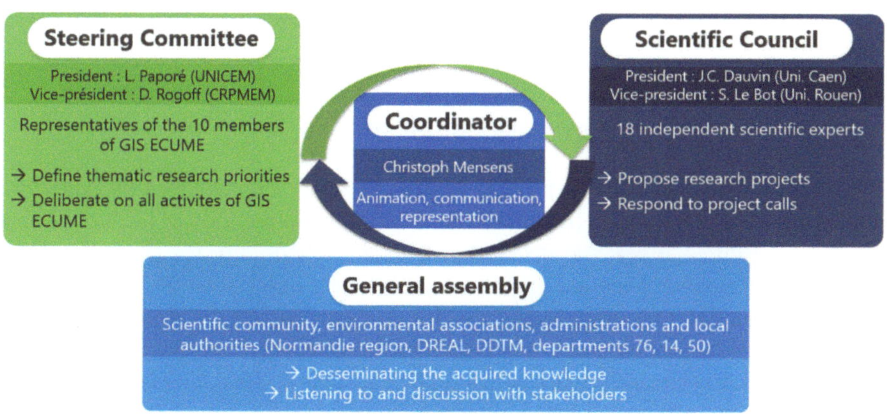

Fig. 1 Organization and governance of GIS ECUME

The research projects of GIS ECUME are financed through the participation in regional, national and European research tenders. The creation of the GIS ECUME and its functioning for the period 2020–2022 was enabled by funding from the European Maritime and Fisheries Fund obtained by the University of Caen with the support of the Regional Fisheries Committee of Normandy and co-financing from UNICEM Normandie, as well as funds from the industrial founding members of GIS ECUME (RTE, Grand Ports Maritime du Havre, Grand Port Maritime de Rouen, Ports de Normandie, UNICEM Normandie). EODC and EOHF contributed to this funding upon their accession to GIS ECUME in 2023. In order to meet the operational needs of the GIS ECUME for the period 2023–2026, its industrial members have approved the provision of additional annual funding for the position of the GIS's coordinator as well as for co-funding the GIS's communication activities, the thesis on the methodology for cumulative impact assessment (see 2.2.1), and the societal analysis of human marine activities (see 2.2.3).

2.2 GIS ECUME Research Topics and Projects

The central objective of the GIS ECUME is to set up an inter-regional approach, initially on the scale of the eastern Channel basin, in order to understand the cumulative effects of maritime activities:

- Extraction of marine aggregates
- Immersion of sediments from harbour dredging
- Fishing activities
- Offshore wind parks
- Submarine cables and offshore electrical substations.

The research programme of GIS ECUME on the cumulative effects of activities at sea comprises five main scientific topics which are summarized below and represented in Fig. 2.

2.2.1 Methodology for Assessment of Cumulative Effects

Following the transposition of the EU Marine Strategy Framework Directive into French law and the reform of impact studies dealing with the study of the cumulative effects of activities at sea, the development of a methodology for assessing these cumulative effects has been identified as a crucial research priority by all industrial members of the GIS ECUME. To address this research priority, the doctoral thesis of Simon Police ("Methodological and operational approaches to assessing the cumulative effects of human activities at sea"), which begins on 1st March 2024, is co-hosted by the University of Caen (M2C Laboratory, supervisors: Laurent Dezileau, Jean-Claude Dauvin, Sophie Le Bot) as well as by the company "Les Graves de l'Estuaire" (LGE, supervisor: Laetitia Paporé). The thesis is co-funded by LGE and the ANRT

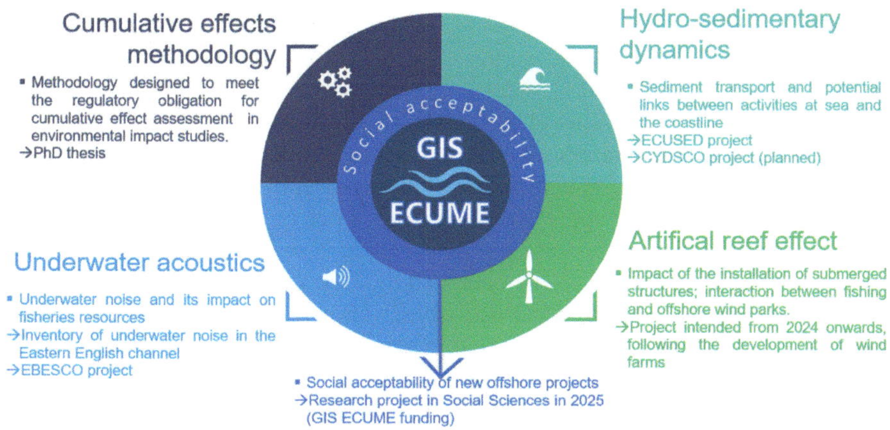

Fig. 2 Research topics of GIS ECUME

(Association Nationale de la Recherche et de la Technologie, CIFRE funding from 2024 to 2026). The thesis' main objective is the development of a methodology for assessing cumulative effects at sea, applicable to projects subjected to regulatory environmental impact assessment. The scientific objectives are (i) a bibliographical analysis on cumulative effects methodologies, (ii) an operational translation by creating a methodology adapted to the industrial activities represented in the GIS ECUME, (iii) the application of the methodology to existing projects in the Bay of Seine and along the Atlantic coast.

2.2.2 Hydro-Sedimentary Dynamics

This research topic comprises the cumulated impacts of anthropogenic activities on sediment transit and the potential interaction between activities and the coastline, which concern most of the activities carried out in the Channel (wind power, fishing, ports, aggregates).

On this topic, the ECUSED project ("Effets CUmulés sur le SEDiment") was conducted by the research teams of CNAM/Intechmer Cherbourg and the M2C laboratory of the University of Caen Normandy (lead scientists: Anne Murat and Pierre Weill). Funded by the Normandy region (2021–2023), ECUSED assessed the impact of marine aggregate extraction activities (GMH concession, Granulats Marins Havrais), and the dumping of sediments dredged by the port of Rouen for the maintenance of shipping channels (MACHU site). The ECUSED project applied a new geochemical methodology (Baux et al. 2019, 2022) allowing to characterise the origin of the various sedimentary fractions and their transport. The methodology used is based on elemental analysis by X-ray fluorescence spectrometry (major, minor and trace elements) of granulometric fractions separated by sieving (21 fractions from <

20 μm to 4 mm). Around twenty chemical elements (from Al to U) can be quantitatively measured in the various fractions of coastal deposits, and informative ratios can be calculated (Sr/Ca, Rb/K, etc.), 25 to 30 geochemical tracers are therefore available to explore the origins of the particles and their transport dynamics. The identification of tracer elements allowed to assess the impact and zone of influence of the aggregate extraction and sediment dumping sites on the seafloor sediments, although no cumulative effects between the two types of activities were detected. The successfully tested methodology will be applied to study sedimentary dynamics at nine sites in mainland France characterized by the presence of different maritime activities (CDYSCO project, "Chemical composition, a key to understanding the origins and DYnamics of natural and anthropized COastal Sediments", currently in planning).

2.2.3 Social Acceptability

The social acceptability of human activities at sea will be assessed by the GIS ECUME in a one-year research project from September 2024 to August 2025, by proposing a societal analysis of maritime industries and their cumulative impacts, accompanied by workshops in participatory sciences. The societal analysis is conducted by the CERREV lab (CEntre de Recherche Risques et Vulnérabilités) of the University of Caen Normandy, the work in participatory sciences is conducted in a collaborative innovation space open to the general public.

The project is funded by the University of Caen Normandy through the recruitment of a research engineer financed by funds made available by the industrial partners of the GIS ECUME, with co-funding for the actions in participatory sciences solicited with the *'Fonds d'Intervention Maritime' and the 'Fondation de France'.*

The societal analysis will focus on the stakeholders involved in human activities at sea (harbour dredging and disposal of dredged sediments, aggregate extractions, fisheries, offshore wind farms, submarine cables). Through interviews with stakeholders (NGOs, the private sector, industry, scientific, technical and political institutions), the expected deliverable is a research report, with a societal analysis summarising the state of controversies surrounding human activities at sea and the views of the various stakeholders. This societal analysis will serve as a baseline for an approach in participatory sciences, which includes the organisation of three "Living Labs", a type of workshop stimulating collaboration between heterogeneous stakeholders mobilized by CERREV and its local partners (the general public, local residents, businesses, local authorities). The aim of these workshops is to develop a shared vision of human activities at sea and to co-construct consensual solutions (technical solutions, usage scenarios in a changing marine environment, modes of governance, identification of research needs). The expected deliverables are a participatory research report as well as a symposium to share the conclusions of the project (societal analysis + solutions developed during the workshops), in order to promote exchanges between the scientific teams, government departments and the various stakeholders involved in human activities at sea.

2.2.4 Underwater Acoustics

Underwater noise has become an emerging issue in the presence of multiple activities in a confined area, as it is typically the case in the eastern Channel (Gigot et al. 2023, Cones et al. 2024). The impact of noise on fishery resources is a priority for the GIS ECUME, given the importance of the fishing industry for the Normandy region. A first inventory of underwater noise in the eastern Channel was conducted in 2022 by the LOMC laboratory of the University of Le Havre Normandy (Simon Bernard, Gilbert Faye). By means of a bibliographical compilation of acoustic monitoring data available in existing databases (OSPAR, SHOM), this inventory allowed the identification and mapping of the main sources of continuous and impulsive noise recorded from 2015 to 2019. The EBESCO project ("Effets du Bruit sous-marin sur les ESpèces à Intérêt Commercial», 2024–2026) is conducted by research teams of the Université du Havre Normandie (LOMC laboratory, lead scientists Fernan Léon and Simon Bernard) and Ifremer (lead scientists: Laurent Dubroca and Julien Normand), in a partnership with TBM Environnement and with funding from the European Maritime, Fisheries and Aquaculture Fund and the Normandy Region. The project's two main objectives are mapping the cumulative underwater noise in the eastern Channel and analysing its impact on species of commercial interest for fisheries. The development of a sound propagation model will allow to map the superposition of underwater sound in the eastern Channel and the identification of risk areas for species of commercial interest (by cross-referencing acoustic and species distribution maps). The impact of underwater noise on species of commercial interest will then be assessed by measuring the growth rates of scallops and through in situ testing of the behaviour of whelks.

3 Perspectives

Created for a first period from the end of 2020 to the end of 2026, the GIS ECUME now embarks on the second stage of that period (2024–2026) with extended scientific as well as geographic ambitions. The activities of GIS ECUME will align on the French maritime façade of the eastern Channel and southern North Sea (from Bay of Mont Saint-Michel to the Belgian border), with the aim of being a major stakeholder with regard to an ecosystem approach within the French national strategy for the planification of this maritime space. This coming period will also allow to consider international collaborations with scientific partners from neighbour countries in the southern part of the North Sea (Belgium, the Netherlands, Germany).

The scientific ambitions of GIS ECUME over the coming three-year-period include both the implementation of the existing projects on the already identified research priorities (see 2.2.), as well as the identification of potential new scientific priorities for the sustainable development of human activities at sea on the maritime façade of the eastern Channel and North Sea. With regard to the increasing development of human activities, the challenge for future research in the Channel will be to facilitate a sustainable development of these activities as well as their coexistence in a shared maritime space while safeguarding the marine biodiversity and environmental heritage of the Channel ecosystem.

References

Baux N, Murat A, Poizot E et al (2022) An innovative geostatistical sediment trend analysis using geochemical data to highlight sediment sources and transport. Comput Geosci 26:263–278. https://doi.org/10.1007/s10596-021-10123-5

Baux N, Murat A, Faivre Q et al (2019) Sediment dynamic equilibrium, a key for assessing a coastal anthropogenic disturbance using geochemical tracers: application to the eastern part of the Bay of Seine. Cont Shelf Res 175:87–98. https://doi.org/10.1016/j.csr.2019.02.002

Chen J, Mao B, Wu Y, Zhang D, Wei Y, Yu A, Peng L (2023) Green development strategy of offshore wind farm in China guided by life cycle assessment. Resour Conserv Recycl 188:106652

Cones F, Jézéquel Y, Jarriel S, Aoki N, Brewer H, Collins J, Chauvaud L, Aran T, Mooney T (2024) Offshore windfarm construction elevates metabolic rate and increases predation vulnerability of a key marine invertebrate. Environ Pollut 360:124709

Dauvin J-C (2019) The English Channel: la manche. In: World seas—an environmental evaluation. Academic Press, pp 153–188

Degraer S, Carey DA, Coolen JW, Hutchison ZL, Kerckhof F, Rumes B, Vanaverbeke J (2020) Offshore wind farm artificial reefs affect ecosystem structure and functioning: a synthesis. Special issue on understanding the effects of offshore wind energy development on fisheries. Oceanography 33:48–57

Desprez M (2000) Physical and biological impact of marine aggregate extraction along the French coast of the Eastern English Channel: short- and long-term post-dredging restoration. ICES J Mar Sci 57:1428–1438. https://doi.org/10.1006/jmsc.2000.0926

Desprez M, Lafite R (2014) Monitoring the impacts of marine aggregate extraction: knowledge synthesis 2012 (GIS SIEGMA). Presses universitaires de Rouen et du Havre

Desprez M (2012) Synthèse bibliographique—L'impact des extractions de granulats marins sur les écosystèmes marins et la biodiversité. Union Nationale des Producteurs de Granulats

Farias Pardo JC, Aune M, Harman C, Walday M, Figenschau Skjellum S (2023) A synthesis review of nature positive approaches and coexistence in the offshore wind industry. ICES J Mar Sci. https://doi.org/10.1093/icesjms/fsad191

Gigot M, Olivier F, Cervello G, Tremblay R, Mathias D, Meziane T, Chauvaud L, Bonnel J (2023) Pile driving and drilling underwater sounds impact the metamorphosis dynamics of *Pecten maximus* (L., 1758) larvae. Mar Poll Bull 191:114969

Halpern BS, Walbridge S, Selkoe KA et al (2008) A global map of human impact on marine ecosystems. Science (80–) 319:948–952. https://doi.org/10.1126/science.1149345

Hutchison ZL, LaFrance BM, Degraer S, English P, Khan A, Livermore J, Rumes B, King JW (2020) Offshore wind energy and benthic habitat changes: lessons from Block Island wind farm. Oceanography 33:58–69

Le Bot S, Lafite R, Fournier M et al (2010) Morphological and sedimentary impacts and recovery on a mixed sandy to pebbly seabed exposed to marine aggregate extraction (Eastern English Channel, France). Estuar Coast Shelf Sci 89:221–233. https://doi.org/10.1016/j.ecss.2010.06.012

Noguès Q, Raoux A, Araignous E, Chaalali A, Hattab T, Leroy B, Ben Rais Lasram F, David V, Le Loc'h F, Dauvin J-C., Niquil N (2021) Cumulative effects of marine renewable energy and climate change on ecosystem properties: sensitivity of ecological network analysis. Ecol Ind 121:107128

Noguès Q, Bourdaud P, Araignous E, Halouani G, Ben Rais Lasram F, Dauvin J-C., Le Loc'h F, Niquil N (2023) An ecosystem approach to spatialized cumulative effect assessment of local and global changes on coastal ecosystem functioning. ICES J Mar Sci 80:1129–1142

Pezy J-P, Raoux A, Dauvin J-C (2020) The environmental impact from an offshore windfarm: challenge and evaluation methodology based on an ecosystem approach. Ecol Indic 114:106302. https://doi.org/10.1016/j.ecolind.2020.106302

Raoux A, Tecchio S, Pezy J-P et al (2017) Benthic and fish aggregation inside an offshore wind farm: which effects on the trophic web functioning? Ecol Indic 72:33–46. https://doi.org/10.1016/j.ecolind.2016.07.037

Raoux A, Pezy JP, Robin I, Bennis AC, Dauvin J-C (2021) Multidisciplinary and multi-scale assessment of marine renewable energy structure in tidal system. J Energ Power Techn 3. https://doi.org/10.21926/jept.2101012

The Satoumi Concept and the Management of Commons: An Integrated Approach, a Link Between Land and Sea, Between Nature and Culture

SDG14 as a Local Integrated Management Tool

Y. Henocque

1 Introduction

SDG14—Life below water—focuses on conserving and sustainably using the oceans, seas and marine resources, including coastal zones at the land-sea interface, for sustainable development. It draws attention to human interactions with the ocean, seas, and marine resources including coastal zones, hence cities and territories, at the land-sea interface. Oceans, seas, and coastal areas are subject to pollution, overexploitation, and climate change impacts such as warming, coastal erosion, sea-level rise, ocean acidification, deoxygenation, and changes in marine and coastal biodiversity. These change drivers, direct or indirect, have substantially damaged ecosystems across the globe, and in some cases led to irreversible losses, compromising the services they provide. The careful management of human activity practices is essential to guarantee the planet's future—and our cities', many of them being coastal (Fig. 1).

Y. Henocque (✉)
Aix en Provence, France
e-mail: henocquey@yahoo.com

A.-C. Bennis et al. (eds.), *Constraints and Adaptations to Global Change at the Land-Sea Interface: For a Shared Ecological and Energy Transition*,
https://doi.org/10.1007/978-3-031-90050-1_18

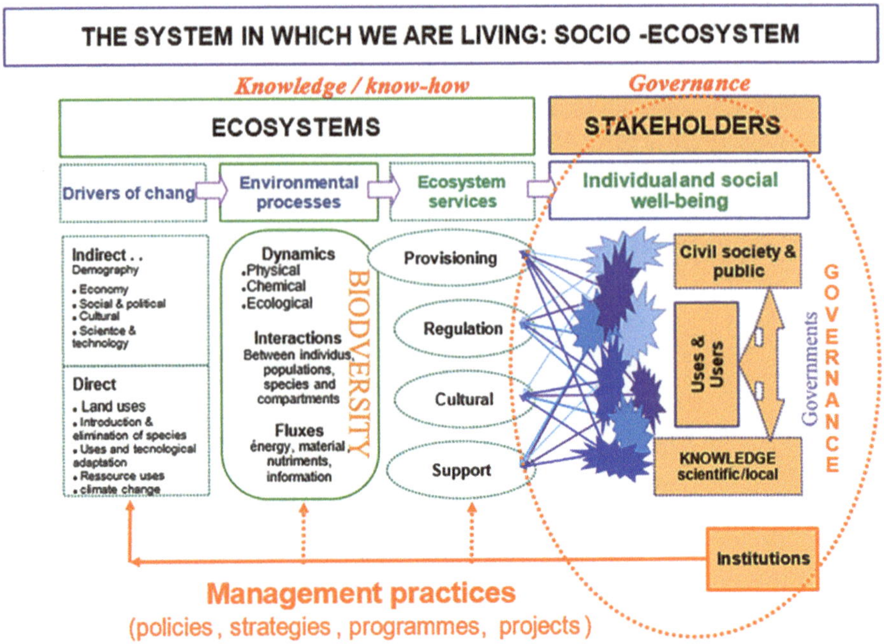

Fig. 1 An inter-connected socio-ecosystem

Local and regional governments have a key role when approaching marine life particularly with regard to the impact of urban growth and of solid and liquid waste discharges into the health of coastal areas and the oceans. After all, the pollution that reaches the oceans today could mean that, by 2050, there will be more plastic than fish in marine waters in 2050. The source of so much garbage is known: even if they are not coastal, our cities and territories—in particular through the disposal of waste—deteriorate water resources such as rivers, whose final destination is the ocean. As the government level closest to the population, LRGs need to transform the way their population uses and consumes natural resources, a theme that is also addressed in SDG 12—Responsible consumption and production.

SDG14 and its targets cover most of the anthropogenic pressures in the coastal and marine environment (in particular, targets 14.1: marine pollution; 14.3: Ocean acidification; and 14.4: overfishing), and the necessary management measures to mitigate those pressures and ensure sustainable development including in the Small Island Developing States (SIDS) and the Least Developed Countries (LDC) coastal communities which are particularly dependent on the oceans and are thus particularly impacted by the negative socioeconomic impacts (targets 14.2: Management, protection and restoration; 14.5: Conservation; 14.6: Fishing regulation; 14.7: specific to SIDS and LDCs; 14.a: Scientific knowledge; 14.b: Small-scale fishing; and 14.c: International law).

In such multidimensional context, this paper looks at the way Local and Regional Governments (LRGs) have adopted policies, plans, incentives and strategies to protect our global common which is the ocean. LRGs are more particularly responsible for a series of actions that cover at least five of the above-mentioned SDG 14 targets: 14.1 (marine pollution), 14.2 (management, protection and restoration), 14.5 (conservation), 14.7 (SIDS and LDCs), and 14.b (small-scale fishing). Since the effects of human activities and urbanization transcend city boundaries in many senses, a special attention will be paid to policies that apply territorial approaches to the conservation of ecosystems from regional governments, intermediary cities and small towns. Likewise, urban-ocean linkages are further explored especially considering the direct and indirect water flows coming from urban centres that eventually end up in the ocean.

The ocean space in general and SDG14 in particular have a cross-cutting role in the 2030 Agenda, Progress towards SDG14 is therefore critical because:

- Ecosystems underpin and maintain all life on planet Earth: SDG14 + 15 (Life and land);
- Oceans directly and indirectly contribute to societal wellbeing: SDG11 (sustainable cities), SDG3 (Good health and wellbeing), SDG1 (No poverty), SDG2 (Zero hunger), SDG16 (Peace, justice and strong institutions);
- The blue economy is growing at great speed: SDG7 (affordable and clean energy), SDG9 (industry innovation and infrastructure).

2 Goals and Targets Are a Regional/National Affair

International ocean governance involves many multilateral organisations and non-governmental organisations (NGOs), often working in partnership across marine issues or on specific ones. Some of the key organisations involved include:

- Global and regional intergovernmental organisations such as UN regional commissions and the Intergovernmental Oceanographic Commission of the United Nations Educational, Scientific and Cultural Organization (IOC-UNESCO), UN agencies such as the Food and Agriculture Organization of the United Nations (FAO) and UNESCO, and regional partnerships such as the Arctic Council, the Pacific Islands Forum and others,
- Regional treaties and conventions, such as the Barcelona Convention and the Union for the Mediterranean coming up with their own strategies including the use of SDGs,
- NGOs such as Worldwide Wild Fund for Nature (WWF), international and regional finance institutions such as the World Bank and funding bodies such as the Global Environment Facility,
- National governments and stakeholders, including government departments, communities, business and marine users,

- Monitoring collective platforms or forums, including Global Fish Watch, the United Nations Regular Process for Global Reporting and Assessment of the State of the Marine Environment, including Socio-Economic Aspects (the UN Regular Process), Intergovernmental Science-Policy Platform on Biodiversity and Ecosystem Services (IPBES) and the International Council for Exploration of the Seas (ICES). The UN Statistical Division also plays a role in data collection and analysis.

To ensure the sound planning and delivery of SDG 14, these actors must work together across levels and geographies. A key way of ensuring this is through the High-Level Political Forum (HLPF) on Sustainable Development, a voluntary, state-led body that provides a platform for partnerships, including through the participation of major groups and other relevant stakeholders. The HLPF tracks SDG implementation and achievement through voluntary national reviews (VNRs).

Besides, it is worth mentioning the ‘communities of Ocean Action’ launched in June 2017 (UN Ocean Conference, New York) gathering more than 1600 voluntary commitments (late 2020) in advancing SDG14 (https://www.youtube.com/watchPv=H9Lg1d0a4B8). Interestingly, their in-depth analysis (UN-DESA 2017) includes a discussion on the linkages between these Voluntary Commitments and the Call for Action adopted by the Ocean Conference participants: besides highlighting the integrated nature of SDGs and targets, and thus the need for involvement of all stakeholders including Local and Regional Governments (LRGs), the Call for Action brings additional focus to the impacts of climate change, indeed one of the strongest current concerns of coastal LRGs, and the connection between SDG14 and the Paris Agreement (COP 21, 2015), a topic that is not as strongly reflected in the SDG14 targets.

3 Some Lessons from the COVID Pandemic

Local governments have innovated, repurposed, and applied lessons from past disasters; however, their ability to undertake these efforts varies significantly depending on the human resources at hand and their fiscal capacity (Dzigbede et al. 2020). The COVID pandemic brings to the fore the complex interconnections between human and social-ecological system health and coastal governance in an age of climate and global change (Perillo et al. 2021).

3.1 A Largely Impacted Local Economic Activity: Tourism

COVID-19 has abruptly impacted all dimensions of society, including coastal activities and their management. Fear of infection, and travel and meeting restrictions have dramatically reduced the use of coastal areas, especially beaches for tourism, resulting in a local economy severely impacted, and in some places, almost completely flat. The time pressure and the unpredictable spread of the COVID-19 pandemic require that local governments analyse the possibility of users' contagion and then take the appropriate measures to reduce/avoid risks.

Box 1: Coast Risk Mapping in Taranto Province (Italy) (Armenio et al. 2021)

The coastal morphology influences the risk connected to COVID-19: flat and sandy beaches are easily accessible by users and therefore attract many more people. On the contrary, in rocky and steeply sloping areas, there will be fewer people. In the first case, it is essential to define measures to access control and to keep the social distance. In the second case, the conditions of access to the coast may be facilitated. A good example can be found in the province of Taranto (Italy), in both municipalities of Maruggio and Torricella. The entire coast was classified according to 3 COVID-19 risk levels (low, medium, high-risk). According to this classification, 7 different zones could be distinguished: two level 3 (high-risk) zones characterized by flat and sandy beaches, where it is fundamental maintaining social distancing at 'pinch points', including access points to beaches and car parks; two other level 1 (low-risk) zones featured by rocky coast, where measures can be envisaged to attract people to reduce the pressure on beaches; and three level 2 (medium-risk) are small beaches where groups of people can be expected, requiring safety distance surveillance and keeping the beach and surrounding areas clean from litter that may be contaminated (e.g. discarded face masks).

3.2 *Small-Scale Fisheries*

Efforts by national and local governments, social organizations, and small-scale fishers themselves to bolster local good networks have also been observed (Bennett et al. 2020), with many reporting increases in local sales initiatives around the globe, such as direct sale schemes, internet selling, door-to-door deliveries, and community supported fisheries. Governments, development organizations, NGOs, the small-scale fisheries sector and scientists need to come together to strengthen local aquatic food systems and push for a long-term shift in distribution channels like shortening supply chains.

3.3 *Benefits to the Environment*

On the other hand, the overall reduction of human activities had had noticeable positive impacts on coastal and marine environment. However, such observed positive environmental impacts will probably be temporary with the quick resuming of human activities. Nevertheless, these quickly acquired positive impacts clearly demonstrate all the benefits that could be got from efficient rules, regulations, and governance regimes (Chuan et al. 2021).

For the first time in modern times, many coastlines have developed features of marine protected areas. COVID-19 with its social, economic, health and environmental consequences, gives an opportunity to reshape tourism and coastal fisheries management plans in the frame of Integrated Coastal Zone Management (ICZM) including cultural awareness of environmental quality.

Box 2: Ecuador: Case Studies on Visual Perception of Beaches and Nearshore Aquatic Wildlife (Ormaza-Gonzalez et al. 2021)
Through questionnaires and surveys about two very popular beaches in continental Ecuador and two port sites in the Galapagos Islands, there was almost a unanimous view that beach zones (sand and coastal water) have notably improved during confinement, at least from a visual point of view. The beaches have less garbage in general and plastic in particular. Since there is a direct relationship between fish demand and tourism, and as tourism activities stopped, the local fishery markets from the selected sites were partially closed, resulting in less artisanal fishing impact. Different species in the three selected locations, including small pelagic fish and marine mammals (dolphins, orcas, whales, sea lions) as well as sharks and turtles have been reported in the survey. One of the reasons for this is the decrease in noise during pandemic restrictions which, like in the case of the Venice channels as mentioned before, is certainly a major factor in the re-apparition of aquatic species.

4 Implementation is Local

Coastal zones have historically been an attractive place for human settlements, mainly through growing urbanization providing huge economic benefits, but at considerable costs to the environment. Many of the environmental issues that face urbanized coastal areas today are either global (e.g. climate change, sea-level rise) or shared across them at local scale (e.g. pollution, erosion, eutrophication, habitat degradation and loss). But protecting oceans and coasts and managing human activities impacting them, is not just the responsibility of coastal towns and cities since any urban/industrial activity along the watershed can affect the coast and the oceans.

Therefore, the roles of local and regional governments, depending on their geographic and jurisdictional context, cover a large array of activities towards the SDG14 targets which will more specifically concern 5 of them, which are: *14.1* (marine pollution), *14.2* (management and restoration), *14.5* (protection), *14.7* (Small Iland Developing States and Least Developed Countries), and *14.b* (small-scale fishing).

4.1 Addressing Coastal and Marine Pollution (SDG Target 14.1)

It concerns all aspects of pollution, in particular pollution from land-based sources including nutrients, wastewaters and plastics. Addressing coastal and marine pollution also means promoting waste prevention and minimization, adopting the 3Rs -Reduce, reuse, and recycle-, in particular improving mechanisms for environmentally-sound waste management, disposal and recycling, and developing alternatives such as reusable water, and reduction of use of plastics.

Ocean sustainability directly links to sustainable water management (SDG 6). Preventing marine pollution contribute to improving water quality and vice-versa. Oceans and seas are major sources of water in the hydrological cycle and therefore require integrated water management that addresses the multiplicity of stakeholders along the water basin.

About 65% of all megacities worldwide are located in coastal areas, which implies a direct relation between ocean sustainability and sustainable cities and communities (SDG 11). Coastal cities and urbanized areas are key hotspots of runoff of all nature including plastic (macro and micro) along rivers and sewage networks, not only on the coast bur in the middle and deep ocean as well. Hence, many local initiatives are taking place, most often through local NGOs initiatives supported by local governments.

Box 3: Establishing Plastic Value Chains in Vilanculos (Mozambique) (https://page.impacttrack.org/3r-ecopoint-network)
Vilanculos is one of the many places along the Indian Ocean in Mozambique, which has been suffering from marine litter for decades. The purpose of the initiative is to reduce the amount of plastic leakage into the ocean. The initiative established eco-points in the city, where plastic waste material is collected at cost by waste pickers, local residents and organizations. The material is sorted, processed, and sent to recyclers within and outside Mozambique. The various partnerships and advocacy actions carried out has resulted in an increasing collection of plastic waste. The current waste production is approximately 19 tons/day, 2.7% being plastics as a result of the initiative. In total, the initiative allows to collect 6 tons of waste material that is been recycled. Consequently, the marine ecosystem has drastically improved due to the initiative's dynamic, including regular beach clean-up exercises by young people and waste pickers.

4.2 Adaptation to Climate Change Coastal Impacts (SDG Target 14.2)

Nowadays, Local and Regional Governments (LRGs) are particularly concerned by the development and implementation of effective adaptation measures as a response to sea-level rise, flooding, and erosion, including the management and restoration of natural defences like seagrasses, mangroves, coral reefs, tidal marshes, and dunes. More and more, LRGs consider risks as contributing to cumulative impacts and as such to be included in their overall coastal management plans. Strengthening the resilience of ocean and coastal ecosystems by reducing pollution (*14.1*), restoring their health (*14.2*), tackling ocean acidification (*14.3*), managing fish stocks sustainably (*14.4, 14.6*) and protecting coastal and marine areas and biodiversity (*14.5*) helps strengthen the overall resilience and adaptive capacity of coastal systems to climate change (*13.1*).

More particularly, as recalled in the last IPCC Report (IPCC 2022), "responses to ongoing sea level rise and land subsidence in low-lying coastal cities and settlements and small islands include protection, accommodation, advance and planned relocation. These responses are more effective when combined/or sequenced, planned well ahead, aligned with sociocultural values and development priorities, and underpinned by inclusive community engagement process", hence largely involving LRGs.

Box 4: Shorelands and Environmental Assistance Program in Washington State (USA) (Shorelands and environmental assistance program 2021)

In Washington State (USA), there are 56 jurisdictions with marine waters. Local Shoreline Master Programs (SMPs) implement the state's shoreline management Act. Since 2004, the Washington State Department of Ecology has been working with local governments on comprehensive updates of their SMPs, recognizing that SMPs are only one potential tool of planning for sea level rise considerations within critical areas ordinances whilst trying to put more coherence between hazard mitigation plans, building and zoning codes, local flood ordinances, stormwater management plans, etc. As an example, the Port and City of Bellingham County are engaged in the long-term process of redeveloping about 96 ha. of the Bellingham waterfront over the course of 50 years. In 2018, a new Park beach's design accounts for 73cm of sea level rise. Other jurisdictions have recognized that future impacts from sea level rise to private septic and public wastewater systems are relatively unknown, and that vulnerability assessments should be carried out to better understand the threat to these systems.

Among the responses to mitigate climate change, are conserving and enhancing sinks and reservoirs of greenhouse gases, with a strong synergistic links to targets 14.2 and 14.5 when considering the overlooked carbon sink enormous potential of coastal ecosystems and the need to protect, conserve, restore, and enhance this potential. Here as well exist many initiatives around the world, especially in tropical areas regarding seaweeds and mangroves replanting.

Box 5: Sustainable Financing for Restoration of Mangroves, Maharashtra State (India) (Government of Maharashtra 2015)

The Western Indian State of Maharashtra, with a coastline of 720 km, has 30,000 ha of mangrove, a third of which is the metropolitan city of Mumbai and its suburbs. The provincial government set up the Mangrove Cell in January 2012, which succeeded in increasing the mangrove cover of Maharashtra by 63% (2017). To ensure sustainability of its mangrove protection efforts, the Mangrove Cell created the 'Mangrove and Marine Biodiversity Conservation Foundation' of Maharashtra. With almost a third of the mangrove forests in the state falling within one of the most populated cities in the world, Mumbai, a specialised unit was formed to check and prevent incidences of encroachment and mangrove destruction within and around the city.

4.3 Advancing Area-Based Management and Protection Tools (SDG Targets 14.2 and 14.5)

Risk management is linked to effective and appropriate area-based management and protection tools like community-based marine managed and/or protected areas, embedded or not in cross-sectoral approaches, including Integrated coastal zone management (ICZM) and Maritime spatial planning (MSP), based on best available science and stakeholder engagement.

Healthy and productive oceans and coasts help reduce vulnerability to climate hazards whilst it benefits small-scale fishers, improve tourism revenue and increase potential for blue carbon markets. ICZM and protection of coastal ecosystems reinforces the achievement of various SDG11 targets, including the safeguarding of coastal natural heritage (e.g. coastal wetlands). Therefore, strong local governments and urban institutions as well as coherent policymaking and governance across administrative and jurisdictional boundaries are essential for formulating integrated solutions and effective implementation.

Box 6: The Mediterranean ICZM Protocol and Local Government's Achievements

The Integrated Coastal Zone Management protocol (2011) of the Barcelona Convention (UNEP Regional Seas Programme) sets governance objectives for countries bordering the Mediterranean Sea. This protocol emphasizes collaborative approaches to acknowledge the role of local people and authorities in coastal management. Till now, in the Mediterranean basin, local governments tend to focus their activities on the landward side of the land-sea interface like in the two following wetland case studies: Southern Camargue in France, and the Gediz Delta in Turkey (Ernoul and Wardell-Johnson, 2013).

The Camargue is one of the largest wetlands in the Mediterranean basin and is of international importance for waterbird breeding, staging and wintering. The Natural Regional Park of the Camargue encompasses more than 80,000 ha within the central delta, with about 15,000 ha of national natural reserve of restricted use. Long-term cultural values of this delta are reflected in the socioeconomic activities, which include horticulture, agriculture, livestock production, hunting, fishing, tourism and salt production. The system of governance is made of a tri-party management between a local mixed syndicate (composed of national and local governmental representatives, private businesses, and civil society) and two local NGOs. The Camargue was previously characterized by large and influential private landowners; however, progressively more wetland has been allocated to public ownership.

The Gediz Delta, adjacent to Izmir, is also among the most important wetlands of the Mediterranean region. The delta extends over 40,000 ha and includes a range of natural habitats. As in Camargue, the delta is a Ramsar site and an 'Important Bird Area'. Significant socioeconomic activities include salt production, fishing and agricultural production. The daily management of the reserve is undertaken by a governmental union formed by local municipalities.

Hence, local governance approaches and existing local sociocultural values are of the utmost importance for the success of any integrated coastal zone management (ICZM) framework like the Mediterranean ICZM protocol.

4.4 The Crafting of Sustainable Blue Economies (SDG Target 14.7)

It concerns the promotion and strengthening of sustainable activities such as fisheries, tourism, aquaculture, marine transportation, renewable energies, marine biotechnology, and seawater desalination, as means to achieve the economic, social and environmental dimensions of sustainable development, including Small Island Developing States (SIDS) and Least Developed Countries (LDCs).

While designating parts of marine and coastal areas for protection might constrain options for growth and jobs in some cases, they may help generate jobs and growth opportunities in others. Tackling marine pollution through improving waste management and increasing recycling (like plastic as seen before) can enable a shift to circular economy hence contributing to what many countries and organizations call nowadays 'Blue economy', as a policy tool or means to drive economic growth and create jobs on the coast and at sea (SDG 8.1; 8.5). Focused on revitalizing the economy, blue economy encompasses seaside tourism, fisheries, sand and mineral dredging, aquaculture, construction, transportation, shipbuilding, communication cables, renewable energies, etc., all of them of direct interest to LRGs. Blue economy development in the frame of Maritime Spatial Planning (MSP) is part of a complex web of coastal and marine governance including ICZM, MPAs, OECMs (Other Environmental Conservation Measures), and others as providing planning across several scales of government.

Sustainable blue economies will have to cope with Subnational and Local Biodiversity Strategies and Action Plans (LBSAPs), which are increasingly being developed at state/provincial/territorial, local and cities levels (https://www.cbd.int/nbsap/related-info/sbsap/).

Box 7: The North Ayrshire Local Biodiversity Plan and its Partnership in Scotland (North Ayrshire Council 2019)
The North Ayrshire Biodiversity Partnership (NAPB) intends to bring together a variety of stakeholders including government and statutory agencies, local authorities, farmers and landowners, voluntary conservation organizations, businesses, educators and local communities. Under the North Ayrshire

Council, the Partnership partners prepared and endorsed the current North Ayrshire Local Biodiversity Action Plan (NALBAP) 2019–2031. Its vision includes the sustainable management of marine and coastal ecosystems as one of the Six Big Steps for Nature set out in 'Scotland's biodiversity: a route map to 2020'. While the establishment of a network of MPAs has been an important step in recognising the conservation value of marine habitats, other tasks include, (1) the engagement with organisations involved in coastal and marine management in the region; (2) undertaking research to monitor the health and biodiversity of the Clyde and the south Arran Marine Protected Area; (3) developing effective control measures for the dune conservation. This is the 5th Local Biodiversity Action Plan to include North Ayrshire, and the second to exclusively cover North Ayrshire. The establishment of a Local Environmental Records Centre (LERC) covering North Ayrshire is an essential resource underpinning all aspects of NALBAP 2019–2031 for prioritising LBAP work, monitoring its impact, and assessing whether priorities need to change over time.

4.5 Access for Small-Scale Artisanal Fishers to Marine Resources and Markets (SDG Target 14.b)

It concerns the form of local governance (fishing rights), including strengthening of capacity building and technical assistance provided to small-scale and artisanal fishers to enable and enhance their access to marine resources and markets, including actions such as auction infrastructure facilities, traceability, certification and ecolabelling as well as access to market-based instruments for fishing communities.

Small-scale fisheries support the livelihoods of many coastal communities around the world. However, these fisheries face growing threats such as overfishing, competition with industrial fleets, water pollution, destruction of fish habitats, and an increasing human population and demand for land and food in coastal areas (FAO 2016). Since decades, the co-management of small-scale fisheries developed as a response to these threats and challenges, promoting the joint management of the fisheries' resources by direct users, local governments and other actors (D'Armengol et al. 2018).

Box 8: Regulating Small-Scale Fishing in Pandeglang, Banten Province (Indonesia) (Kusumajanti et al. 2020), and in Hainan (China) (Zhao and Jia 2020)

Pandeglang Regency is about 110 km from Jakarta. It is renowned for its many beach destinations. In 2016, the number of fishermen in Pandeglang Regency was registered as 12,500 people, supported by 14 units of fish auction places. Pandeglang Regency Fishery Office has the duty to carry out the tasks of local government in the field of fishery based on the principle of autonomy. The Fishery Office has been carrying out various efforts for the enforcement of ministerial regulation (2015) on the use of fishing gear as follows: (1) socialization on the impact of the use of a prohibited trawl on the sustainability of fish availability in the area; (2) establishing an exchange post of fishing equipment; (3) together with the marine police, conduct raids and arrests of fishing boats still using illegal fishing gear. In the year 2017, the Fisheries Office made plans for the provision of assistance intended to increase the amount of catches to meet market needs, providing fishing gear and ship workshops, outboard motors for boats, and fish processing equipment. Among the varied coastal regions of China, the world's largest fishing nation, the island of Hainan (China's southernmost province) has the largest seas area with 13 coastal cities (counties), a population of 414,000 fishermen, and an estimated 25,000 fishing vessels of which 80% are small-scale fishing vessels (less than 12m long). In light of local conditions (overexploitation), some local governments have implemented more stringent protection measures for fishery resources in addition to national regulations.

5 Conclusion: SDGs, a Common Language for Sharing

From the above, the SDGs should not be viewed as a rigid accounting framework but as a co-learning process to look at how sustainable development as a concept, can actually translate into local impact. SDGs including SDG14 may be seen as a communication tool with colleagues and citizens, drawing on the international status and provenance of the SDGs to highlight ambitions for transforming cities, cross-referencing strategies in development and identifying gaps, combating silo-thinking, encouraging partnerships, and using them for developing local targets and demonstrate progress toward overall responsibility.

Yet in order to make the monitoring and evaluation of SDG14 (like many other SDGs) more accurate, a much sharper focus is needed on enhancing the roles and capacities of local and subnational governments by providing more support and resources to fulfil their critical responsibilities. In the domain of coast, seas, and ocean, this support (technical and financial) may come from countries themselves but also from well-established regional frameworks like the UNEP Regional Seas

Programme or the Global Environment Facility (GEF) Large Marine Ecosystem (LMEs) (GEF LME: LEARN 2017), in collaboration with the private sector and citizens concerned.

In June 2020, the Global Taskforce of LRGs (www.global-taskforce.org) made a statement on Planetary urbanization and life below water towards an integrated localization of global agendas, with the following words: "*We acknowledged the pertinence of our local and regional governments, especially intermediary cities and small towns, to a sustainable development model, more particularly in regard to SDG14, stressing the presence of 'urban-ocean linkages*". Some 7 years before (2013), the city of Santa Monica (California, USA) had passed a Sustainability Rights Ordinance which recognizes that 'natural communities and ecosystems possess fundamental and inalienable rights to exist and flourish' (https://panorama.solutions/fr/building-block/legal-recognition-natures-inherent-rights-through-earth-law). The goal towards healthy and sustainable coasts, seas, and ocean is a Human and Biodiversity right where Man 'lives in harmony with Nature'.

References

Armenio E, Mossa M, Petrillo AF (2021) Coastal vulnerability analysis to support strategies for tackling COVID-19 infection. Ocean Coast Manag 211:105731. https://doi.org/10.1016/j.ocecoaman.2021.105731

Bennett NJ, Finkbeiner EM, Ban NC et al (2020) The COVID-19 pandemic, small-scale fisheries and coastal fishing communities. Coast Manag 48(4):336–347. https://doi.org/10.1080/08920753.2020.1766937

Chuan OM, Ghazali A, Amin RM et al (2021) Positive and negative effects of COVID-19 pandemic on aquatic environment: a review. Sains Malaysiana 50(4):1187–1198. https://doi.org/10.17576/jsm-2021-5004-28

D'Armengol L, Castillo MP, Ruiz-Mallen I, Corbera E (2018) A systematic review of co-managed small-scale fisheries: social diversity and adaptive management improve outcomes. Glob Environ Chang 52(2018):212–225

Dzigbede KD, Gehl SB, Willonghby K (2020) Disaster resiliency of US local governments: insights to strengthen local response and recovery from the COVID-19 pandemic. Public Adm Rev. https://doi.org/10.1111/puar.13249

Ernoul L, Wardell-Johnson A (2013) Governance in integrated coastal zone management: a social network analysis of cross-scale collaboration. Environ Conserv. https://doi.org/10.1017/S0376892913000106

FAO (2016) Technical and socio-economic characteristics of small-scale coastal fishing communities, and opportunities for poverty alleviation and empowerment. FAO, Rome, Italy

GEF LME: LEARN (2017) The large marine ecosystem approach. An engine for achieving SDG14. Paris, France

Government of Maharashtra (2015) Establishment of the Mangrove and Marine biodiversity conservation foundation of Maharashtra. Government of Maharashtra, revenue & forest department. Government resolution No S-30/2015/CR219/F-3. Matralaija, Mumbai 400032, 23rd September 2015

IPCC (2022) Climate change 2022. Impacts, adaptation and vulnerability. Summary for policymakers. IPCC WGII sixth assessment report

Kusumajanti, Widiastuti NE, Kamaluddin A (2020) Strategies and role of local government in improving the competitiveness of traditional fishermen in Pandeglang, Banten. http://ejournal.upnvj.ac.id/index.php/JEP/index

North Ayrshire Council (2019) North ayrshire local biodiversity action plan 2019–2031. Housing strategy and corporate sustainability service. http://www.north-ayrshire.gov.uk

Ormaza-Gonzalez FI, Castro-Rodas D, Statham PJ (2021) COVID-19 impacts on beaches and coastal water pollution at selected sites in Ecuador, and management proposals post-pandemic. Front Mar Sci. https://doi.org/10.3389/fmars.2021.669374

Perillo GME, Botero CM et al (2021) Integrated coastal zone management in the context of COVID-19. Ocean Coast Manag 210:105687. https://doi.org/10.1016/j.ocecoaman.2021.105687

Shorelands and environmental assistance program (2021) Lessons learned from local governments incorporating sea level rise in shoreline master programs. Washington State Department of Ecology. Olympia, Washington. Publication 21 June 2014

UN-DESA (2017) In-depth analysis of ocean conference voluntary commitments to support and monitor their implementation. 14 life below water. Division for sustainable development of economic and social affairs, United Nations. Advance Copy

Zhao X, Jia P (2020) Towards sustainable small-scale fisheries in China: a case study of Hainan. Mar Policy 121:103935. https://doi.org/10.1016/j.marpol.2020.103935

Concept of Satoumi and Its Related Activities in Coastal Areas of Japan

O. Matsuda

1 Basic Idea and Concept of *Satoumi*

Basic idea of *Satoumi* was first proposed by the late Professor Tetsuo Yanagi in 1998 (Yanagi 2007). Widely known definition of *Satoumi* is "a coastal area where biological productivity and biodiversity have increased through human interaction". However, this straightforward definition of *Satoumi* is sometimes too strict to include wide variety of *Satoumi* related activities. Since strict definition of *Satoumi* often works as exclusive function of the diversified *Satoumi* creation activities, more inclusive idea of *Satoumi* has been widely spread (Berque and Matsuda 2013). As a result, *Satoumi* is now generally recognized in broader sense such as "bountiful coastal area shaped by the coexistence between human and nature". Activities to realize bountiful *Satoumi* has gradually expanded due to both change of policy and active participation of local people in Japan (Yanagi 2013).

Satoumi is, in another context, a kind of Japanese ecosystem that use to be found in coastal areas such as small bays throughout Japan in days gone by, but regulations and taboos in their communities were rarely formalized and were liable to lose much of their effectiveness as a result of rapid environmental changes due to drastic economic development. Water quality, once highly polluted by domestic and industrial waste, has improved to some extent due to painstaking efforts to reduce water pollution by legal and other regulations, but damaged ecosystems and living resources have not yet fully recovered.

The successful introduction of *Satoumi* through human interaction is expected to improve the impoverished ecosystem and to provide greater biological diversity as habitats, greater biological production and fishing grounds. The concept of *Satoumi* is gradually being recognized not only in its homeland of Japan but also at international

O. Matsuda (✉)
International EMECS Center-DRI East Bldg, Kobe, Hyogo, Japan
e-mail: matsuda036@go3.enjoy.ne.jp

A.-C. Bennis et al. (eds.), *Constraints and Adaptations to Global Change at the Land-Sea Interface: For a Shared Ecological and Energy Transition*,
https://doi.org/10.1007/978-3-031-90050-1_19

meeting held in both western and Asian countries. The initial concept of *Satoumi* was introduced for the first-time outside Japan at EMECS7, which was held in Caen, France in 2006. In this meeting Tetsuo Yanagi, Matsuda and other scientists made presentations on the concept and case studies of *Satoumi*.

These presentations attracted attention and was highly evaluated in the review session as a "Symbiosis among human communities and coastal/marine areas—a more rational vision of co-existence". After that many international meetings and workshops etc. on *Satoumi* has been held in many places.

As to factors affecting the type of *Satoumi* activities, there are two basic factors which characterize type of *Satoumi* activity. One is the place where *Satoumi* activity is made and the other is the main group of people who conducts and participate in the activity. According to the combination of these two factors, *Satoumi* activities can be classified into 7 types from the above two major viewpoints. One is the place where *Satoumi* activity is made, for example, fishing village, urban area, river basin etc. and the other is the main group of people who participate in the activity such as fisher, urban citizen and residents of a river basin, etc.

According to the Ministry of the Environment, Japan, *Satoumi* creation activities have been tentatively classified into following seven categories (Table 1). However, actual *Satoumi* creation activity in each *Satoumi* site has more or less compound nature from the viewpoint of classification of *Satoumi* because above mentioned classification indicates a kind of "purified" prototype.

The Seto Inland Sea has possessed the characteristics of *Satoumi* that is closely connected to human life since ancient times (Matsuda 2012a). By the Jomon Era, not only were its ecosystem services already being utilized in a primitive way through

Table 1 Classification of *Satoumi* creation activity

Type of activity	Contents of activity
1. River basin-integrated activities	Activities that consider the entire area from the forest to the ocean as an integrated whole
2. Mitigation activities	Activities to restore environments that has become lost as a result of urban development, etc.
3. Fishing village activities	Activities for which fishing villages are the implementing entities and that are implemented in the course of fishing activities
4. Environmentally "sacred" activities	Activities to create ocean regions by establishing marine reserves and "no fishing" seasons
5. Experiential activities	Hands-on activities that are conducted near the urban areas by urban residents
6. Urban activities	Activities to preserve and restore seaweed beds and other shallow ocean regions near urban areas
7. Compound type	Composite activities to preserve ocean environments in areas in which river basins, urban areas, fishing villages, etc. overlap

Source: Ministry of the environment, Japan

gathering seafood along its shores, but there is also evidence which indicates its sea route were being used for trade and commerce.

The Seto Inland Sea has provided the people on its coast with the important services of "field (food, salt)", "garden (bathing, tourism)" and "road (ocean traffic, cargo transport)" for a long time. In particular, the Seto Inland Sea's provision of fish in its function as "field", which is a major ecosystem services, is exceptionally high even by international standards. Salt was also an important product. In addition, the tranquil archipelago landscape of the sea where the first Japanese national park was established has been internationally acclaimed.

2 Brief History of the Seto Inland Sea as the Background of Satoumi

Since *Satoumi* born in Japan is said to be grown up in the Seto Inland Sea (Fig. 1), or the Seto Inland Sea is sometimes called the birthplace of *Satoumi*. Brief history of the Seto Inland Sea as the background of *Satoumi* is described in this section. The Seto Inland Sea is the largest enclosed sea area in Japan. Approximately 30 million people inhabit in the coastal area of the sea, including the river watershed areas of rivers that flow into the Seto Inland Sea. For this reason, the area supports variety of industrial activities, and therefore, is susceptible to human behavior (Okaichi and Yanagi 1997; Okaichi 2003; Matsuda 2008).

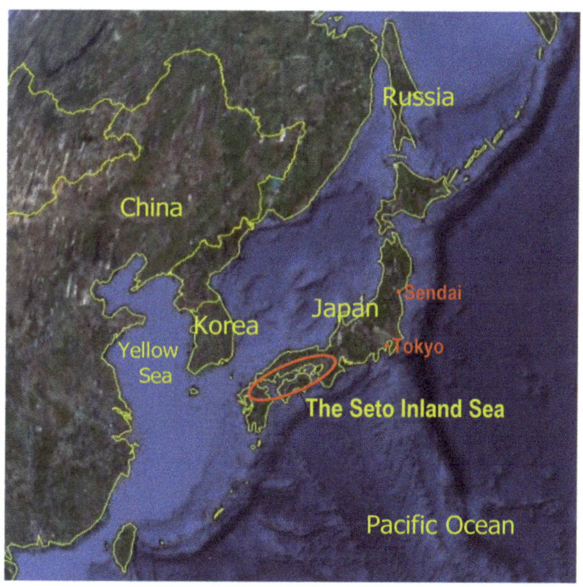

Fig. 1 Location of the Seto Inland Sea

The ecosystem services and landscapes provided by the Seto Inland Sea have been shaped by the long-standing interaction between human activities and the coastal waters reflecting regional characteristics and the historical background. However, the bountiful nature of the Seto Inland Sea as *Satoumi* maintained for a long time during a time when the influence of human activity on the marine environment and resources was relatively small, changed dramatically in the high economic growth period after World War II. The result was environmental degradation due to pollution and the disappearance of shallow neritic areas due to land reclamation.

These were accompanied by the deterioration of coastal ecosystems and decrease in marine resource levels. For this reason, the meaning of *Satoumi* has gradually changed from what it was initially, "the state that was", to a word indicating "a target for a lost environment to be restored" or "an ideal relationship between people and sea to be newly forged". This is the reason the term *Satoumi* is used in phrases such as "the creation of new *Satoumi*".

From the viewpoint of long-term history, the formation of the Seto Inland Sea as *Satoumi* is also closely linked to *Satoyama*. The beaches of white sand and green pines, which were once extolled as the typical landscape of the Seto Inland Sea, were formed as a result of salt industry benefiting from the region the tranquil archipelago landscape of the Seto Inland Sea where the first consumption of fuel to condense salt water. As a result, bare granite mountains appeared in the landscape, and eroded sand from those mountains accumulated to form the white beaches. The pine trees, which do well on poor soil, flourished on the granite sand beaches. The nationally famous oyster farming in Hiroshima Bay and the *Satoumi* landscape of rows of oyster rafts both benefit from *Satoyama* in the Ohta River basin. This is because the management of *Satoyama* and the forest directly reflects on the water quality and regime of the rivers, affecting the marine environment and the production of plankton, which is food for oysters. Since the bounties of *Satoumi* including seafood and salt have also been used inland, the relationship between *Satoumi* and *Satoyama* is basically bidirectional. From the relationship between *Satoumi* and *Satoyama*, combined management of *Satoumi* and *Satoyama* is more effective from the view point of material flow and integrated coastal management (ICM) (Fig. 2) (Matsuda 2012b; Matsuda 2014).

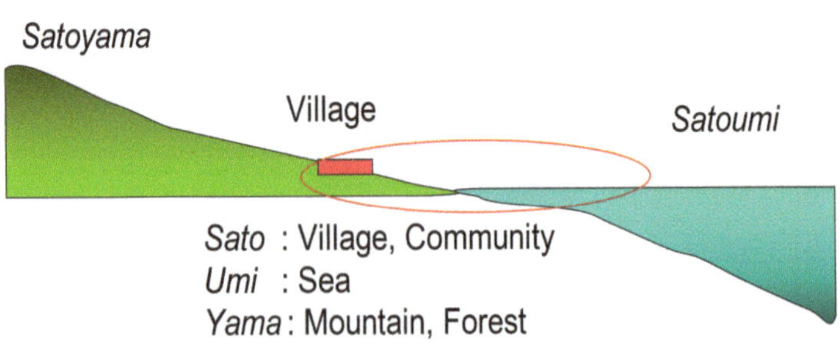

Fig. 2 Conceptual view of *Satoumi* and *Satoyama*

"The Seto Inland Sea Law" (The Law Concerning Special Measures for Conservation of the Environment of the Seto Inland Sea) enacted in 1973. Two major original functions of the law were as follow. One is area wide total pollution load control (TPLC) in terms of COD, Total Nitrogen (TN) and Total Phosphorus (TP). This has played an important role on the improvement of water quality. TPLC (COD, TN, TP) has been quite successful, as a result pollutant load in the Seto Inland Sea decreased (Fig. 3).

The other is suppression of land reclamation (landfill), effect of which was restricted because the law did not order total ban of land reclamation. Therefore, yearly area of land reclamation decreased but cumulative area has continuously increased (Fig. 4). As a result, natural shoreline entirely disappeared in the coastal area of Osaka Bay (Fig. 5). This suggests that tidal flat and sea grass bed as habitat, spawning ground and nursery were seriously damaged or disappeared.

From the long-term viewpoint, water quality has been improved, however, natural ecosystem of sea shore was seriously damaged. These measures as TPLC and suppression of landfill initiated in 1970s were categorized into preventive "passive conservation".

Recently conceptual shift of environmental policies in the Seto Inland Sea has occurred. That means change from passive conservation such as Total Pollution Load Control (TPLC) and suppression of land reclamation to active conservation such as restoration of biodiversity, biological productivity, habitat and well-balanced nutrient cycle between land and sea. So, the slogan of these movements also changed from "Stop excessive discharge" and "Decrease landfill" to "Restore tidal flat" and "Improve Sea grass bed".

Another conceptual shift might be the change from single issue approach such as water quality control to more holistic approach such as ecosystem-based management (EBM), integrated coastal management (ICM) and *Satoumi* related management including adaptive management.

3 Case Studies of Some Present *Satoumi* Creation Activities

Activities of three *Satoumi* creation sites in the Seto Inland Sea are introduced, which are Hinase area in the central part, Yamaguchi prefecture area in the western part and Kansai (Osaka) International Airport area in the eastern part of the Seto Inland Sea (Fig. 6).

3.1 Hinase Area (Bizen City, Okayama Prefecture)

Fisheries Cooperative of Hinase in Bizen city has impressive achievement of eel grass bed restoration for more than 35 years. And during the time, many outside people and groups of Hinase area such as urban citizen, students, companies, cooperatives have

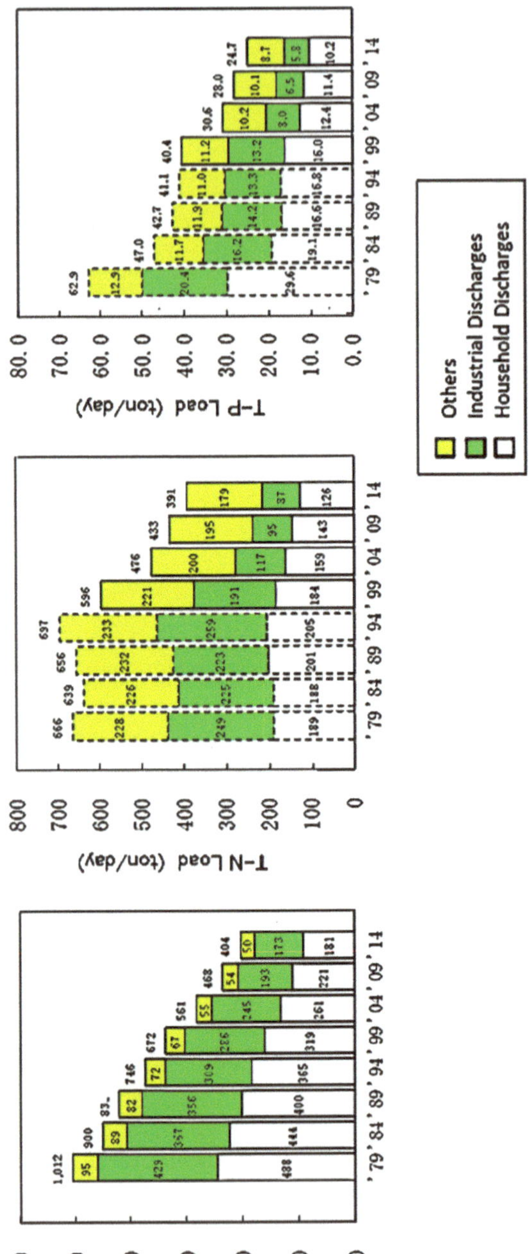

Fig. 3 General trend of pollutant load in the Seto Inland Sea

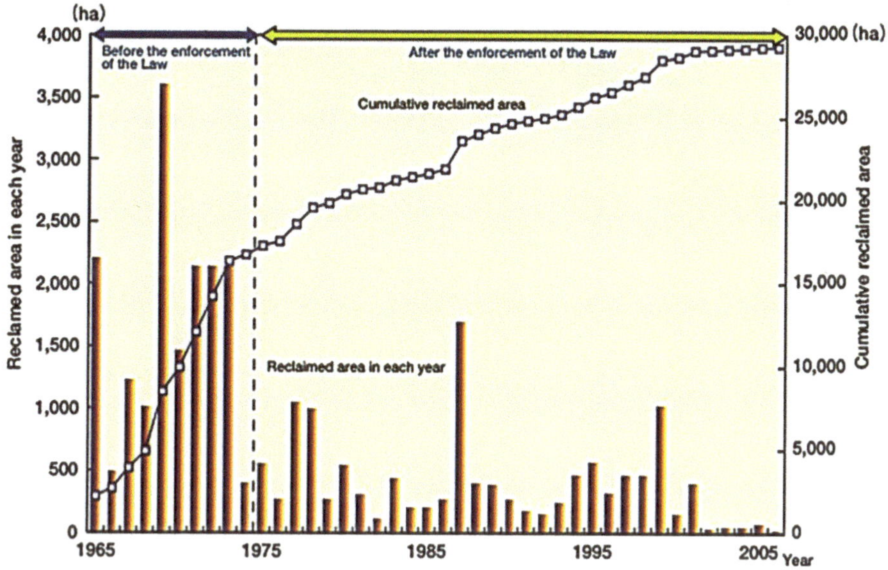

Fig. 4 Trends in reclaimed areas in the Seto Inland Sea-reclaimed areas in each year (bar graph) decreased after the enforcement of the Law but cumulative areas (line graph) increased

joined this activity gradually. Type of this *Satoumi* activity is originally classified in fishing village activities but is now more or less changing to compound type (Fig. 7).

Area of eel grass bed at Hinase area was once about 600 ha in1950 but almost disappeared in 1985 when only 12 ha of eel grass bed remained. Fisherman group developed seeding technology of eel grass and they recovered eel grass bed up to about 200ha in 2013 and more after that (Fig. 8). Restoration of eel grass bed gave positive effects on local fisheries and onsite local fisherman's sea food market "*Gomino-ichi* (5 tastes market)" is now quite popular with both local residence and tourist.

Since Hinase has been famous for eel grass bed restoration, "Eel grass summit" was held in June, 2016 in Hinase area of Bizen city for 3 days. Many kinds of activities were held during the days such as scientific symposium, fisher's talk, school student activities, field work, etc. Approximately 2000 people got together in this summit. Since many groups and people such as prefecture government, city office and varieties of citizen supported the meeting, communication among participants much improved. This might be the positive effects of *Satoumi* activities.

3.2 *Yamaguchi Prefecture*

As to Fushino river estuary and tidal flat restoration, the restoration council was established in 2007 based on "Promotion of Nature Restoration Act". The council contains

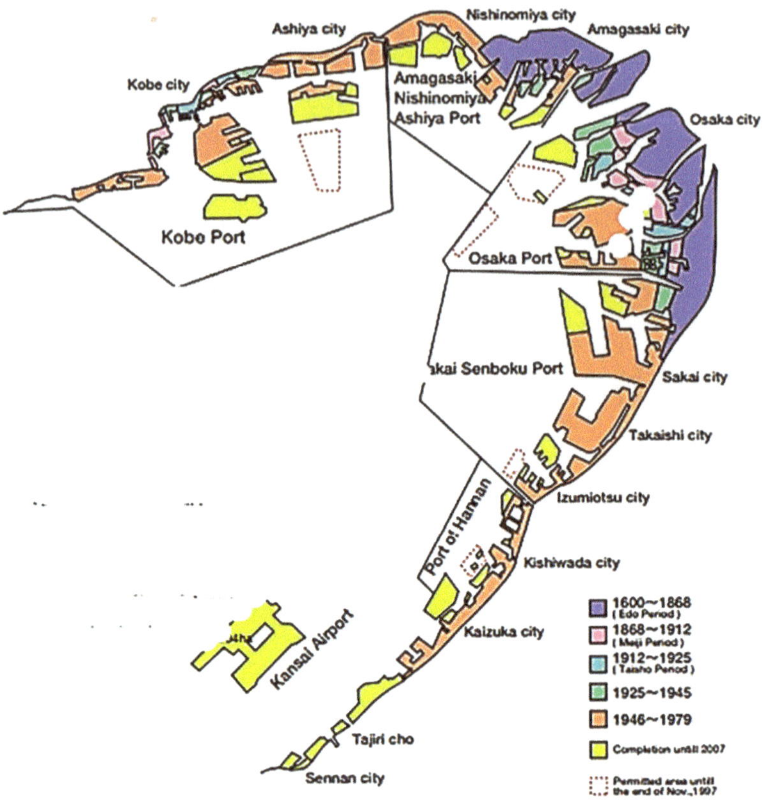

Fig. 5 Result of historical land reclamation in Osaka Bay

high variety of stakeholders such as academician (specialist), individual, group of organization, local government, administrative agency and central government. Type of this *Satoumi* activity is classified in typical river basin-integrated activities.

In this tidal flat restoration, promotion of clam shell production is one of main subjects of the restoration and many kinds of local people join this activity (Fig. 9). For example, bamboo stick provided by forest people upstream is used to promote settlement of clam shell larvae in tidal flat. The facility of "Kirara-hama Nature Park" located in the Fushino River estuary plays an important role on the restoration activity in many ways, for examples, scientific observation, environmental education and ecotourism.

In Ohshima area, Shunan city, Satoumi creation activity is being made at an artificial tidal flat which area is larger than 40ha. This tidal flat was constructed by the central government (MILT) with use of dredged sediment. But maintenance is being made by local group (Public–Private Cooperation). On this artificial tidal flat, naturally grown eel grass can be observed. These restored tidal flat and sea grass bed can play many roles on reduction of ocean risk. This artificial tidal flat is being used

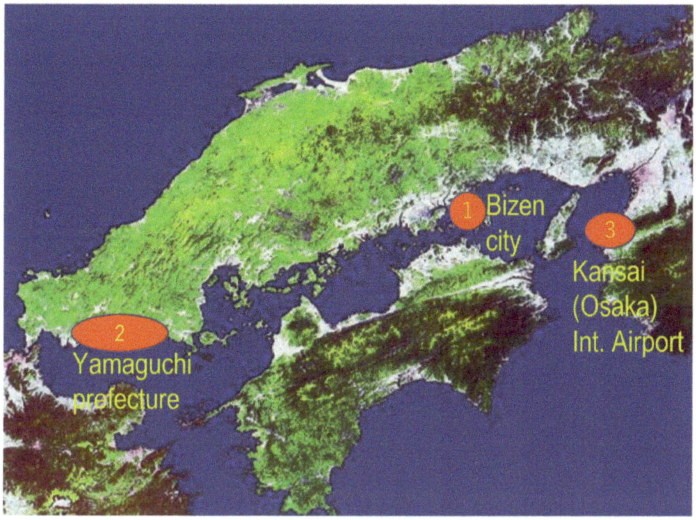

Fig. 6 Site of *Satoumi* activities as "active conservation" in the Seto Inland Sea

Fig. 7 Member of fisheries cooperative of Hinase in Bizen city who made impressive achievement of eel grass bed restoration for more than 35 years

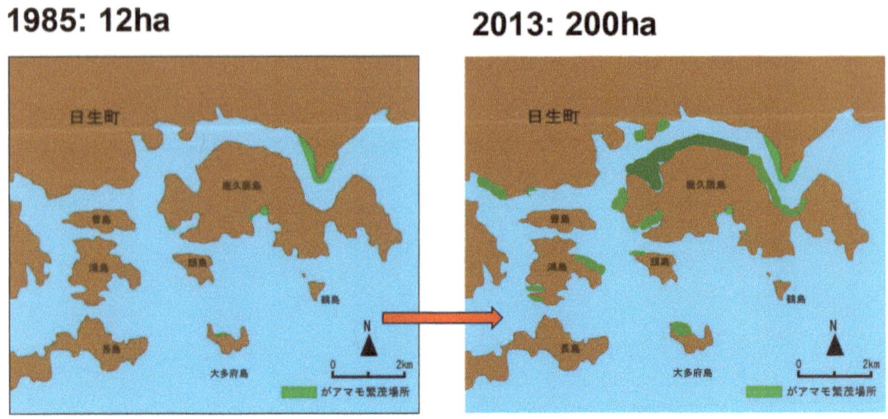

Fig. 8 Local people continued restoration of sea grass bed more than 35 years and almost disappeared eel grass bed was recovered up to about 200ha in 2013 (surface of sea grass beds in green)

Fig. 9 Promotion of clam shell production is one of main subjects of the restoration

Fig. 10 Naturally grown eel grass (*Zostera japonica*) on the artificial tidal flat

for community activities such as recreational clam shell hunting due to increased clam shell stock and educational activities (Figs. 10, 11, and 12).

3.3 Kansai International Airport Area

This unique activity means new creation of seaweed (algal) bed and habitat. Kansai (Osaka) International Airport is located on the artificial island. Gentle slope of natural rock and bio-philic protection block around the island promoted expansion of seaweed bed (ca. 55 h: more than 100 species). This indicates that muddy bottom without vegetation changed to dense seaweed bed (Figs. 13 and 14). This can be said to be new creation of seaweed (algal) bed by human activity.

This protection wall provided new habitat for many kinds of living organisms. Recent survey made it clear that newly created seaweed bed provided excellent habitat for many kinds of fishes, crustacean, mollusk etc. As a result of increased living resources, finless porpoise inhabits recently there and propagates. This is not restoration but creation of new habitat and ecosystem. Therefore, it is suggested that newly created sea weed bed around Kansai International Airport supplies living resources to Osaka Bay as a nursery ground.

Fig. 11 Recreational clam shell hunting due to increased clam shell stock

Fig. 12 A scene of environmental education at "Seashore Nature School"

Fig. 13 An artificial island on which Kansai international airport is located

Fig. 14 New creation of seaweed (algal) bed around the artificial island

4 Future Perspective

In the past, estimated landscape of the Seto Inland Sea in the nineteenth century (approximately 150 years ago) was amazing because Ferdinand von Richthofen, a well-known German geographer who named "Silk Road" (Seidenstrassen) visited the sea in 1860 and highly evaluated the scenic beauty and praised peaceful landscape as "almost heaven". He prayed the eternity of the beautiful Seto Inland Sea. But at the same time, he worried about extreme development of civilization and endless desire of human being in the coming future. It is regrettable that Richthofen's worry came true just after 100 years. The beautiful and rich Seto Inland Sea corrupted during late 1960s to 70s and was called "Dying Sea". The special law "The Seto Inland Sea Law" was enacted in 1973, about 50 years ago (Stickney and McVey 2002, Yanagi 2015).

Lessons learned from the history of the Seto inland Sea suggest that some problems can be solved by passive conservation, however, not all problems can be solved by only passive conservation. Not only passive conservation but also active conservation such as *Satoumi* is now very important. Since cause-effect relationship in coastal seas is so complicated, not only single approach is enough but integrated holistic approach is necessary.

In future 50–100 years, concentration of energy on restoration of landscape, healthy ecosystem and relationship between human and the sea is necessary. As a milestone of *Satoumi* related activity, the secretariat of the Convention om Biological Diversity (CBD) published a CBD Technical Series No.61 on "Biological and Cultural Diversity in Coastal Communities: Exploring the Potential of Satoumi for Implementing the Ecosystem Approach in the Japanese Archipelago" (Secretariat of CBD, 2011). This volume contains 10 case studies of *Satoumi* creation activities in Japan from Hokkaido, northern most area of Japan to Okinawa, southern most area of Japan, and related articles on the conservation and ecosystem approach to *Satoumi*. Among articles in this volume, "Relevance of *Satoumi* to the CBD mandate" written by Marijo Vierros is valuable because biodiversity issue is critical world concern at present and should be the same in future *Satoumi*.

In 2019, Elsevier publishing company published a book "Coast and Estuaries, the Future" in which the author and the late Tetsuo Yanagi wrote "Restoration of Estuaries and Bays in Japan—What's Been Done So Far, and Future Perspectives" (Matsuda and Yanagi 2019). In this article, we proposed "Clean, productive and prosperous coastal sea" as an aim of future *Satoumi* to be achieved. Among "Clean, productive and prosperous coastal sea", "prosperous coastal sea" means active and popular coastal sea in which many kinds of people and industries can participate.

At the first stage of *Satoumi* creation, natural science related approaches to biological productivity and biological diversity were essentially important (Kokubu et al. 2008, Matsuda 2010, Matsuda and Kokubu 2011), but social and anthropological science will also be important in future *Satoumi*. Our proposal included not only economic and cultural evaluation of ecosystem services but also implementation of PDCA cycle and adaptive management of *Satoumi* creation. Since Tetsuo Yanagi is the first proposer of the concept of *Satoumi*, economic and cultural aspects, and adaptive management is new and valuable to the future *Satoumi*. *Satoumi* as active conservation of coastal environment and resources can improve deteriorated environment and ecosystem toward SDGs-based human well-being and also towards beyond SDGs. It is now very essential to realize resilient coastal seas and pass it to next generation along the *Satoumi* context.

In many deteriorated enclosed coastal seas, only "passive (preventive) conservation" is not enough but "active conservation" such as *Satoumi* with human interaction is necessary. Among many types of "active conservation", *Satoumi*, community-based "active conservation" with people's participation, is vital to realize resilient coastal seas and to reconstruct better relationship between human and the coastal sea towards and beyond SDGs-based human well-being.

References

Berque J, Matsuda O (2013) Coastal biodiversity management in Japanese *Satoumi*. Mar Policy 39:191–200

Kokubu H, Okumira H, Matsuda O (2008) Historical changes in the tidal flat and its effects on benthos and sediment quality in ago Bay. J Jpn Soc Water Environ 31(6):305–311

Matsuda O (2010) Recent attempts towards environmental restoration of enclosed coastal seas: Ago Bay restoration project based on the new concept of *Satoumi*. Bull Fish Res Agency 29:9–18

Matsuda O, Kokubu H (2011) Towards *Satoumi* in Ago Bay, CBD technical series no. 61 (Biological and cultural diversity in coastal communities, exploring the potential of *Satoumi* for implementing the ecosystem approach in the Japanese archipelago), pp 62–69

Matsuda O, Yanagi T (2019) Restoration of estuaries and bays in Japan—what's been done so far and future perspectives. In: Coast and Estuaries-The Future", Elsevier.

Matsuda O (2008) Eutrophication and its cause/consequence: the case of the Seto Inland Sea. In: Mimura N (ed) Asia-Pacific coasts and their management. Springer, pp 78–92

Matsuda O (2012a) Western Japan cluster: Seto Inland Sea as *Satoumi*. In: Duraiappah AK et al (ed) *Satoyama-Satoumi* ecosystems and human well-beings. United Nations University Press, pp 381–402

Matsuda O (2012b) Combining activities of *Sato-Umi* and *Sato-Yama* in Japan: towards a new type of integrated coastal and watershed management. In: Taniguchi M, Shiraiwa T (ed) The Dilemma of boundaries-toward a new concept of catchment. Springer

Matsuda O (2014) Predicting and promoting the future state of coastal seas with special emphasis on *Satoumi*. In: Shimizu N et al (ed) Connectivity of hills, humans and oceans, challenge to improvement of watershed and coastal environments. Kyoto University Press

Okaichi, T. and Yanagi, T. ed. (1997), Sustainable Development in the Seto Inland Sea Japan-From the Viewpoint of Fisheries, 329 pp., Tera Scientific Publishing Company.

Okaichi, T. ed. (2003), Red Tides, 439 pp., Tera Scientific Publishing Company.

Secretariat of the Convention on Biological Diversity (2011), Biological and Cultural Diversity in Coastal Communities – Exploring the Potential of *Satoumi* for Implementing the Ecosystem Approach in the Japanese Archipelago, CBD Technical Series No. 61, Montreal.

Stickney, R. R. and McVey, J.P. ed. (2002), Responsible Marine Aquaculture, 391 pp., CABI Publishing.

Yanagi T (2013) Japanese Commons in the Coastal Seas: How the Satoumi Concept Harmonizes Human Activity in Coastal Seas with High Productivity and Diversity. Springer, Tokyo, p 113p

Yanagi T (ed) (2015) Eutrophication and Oligotrophication in Japanese Estuaries. Springer, Dordrecht, p 97p

Yanagi, T. (2007) Sato-Umi: A new concept for coastal sea management. Terrapubu Tokyo, 86p.

Coral Reef Satoumi in Okinawa, Japan

S. Kakuma

1 Introduction

This chapter describes the coral reef Satoumi in Okinawa, Onna Village and Okinawa City on Okinawa Island, and Shiraho district on Ishigaki Island (Fig. 1). The most commonly used definition of Satoumi in Japan is "high productivity and bio-diversity in the coastal sea with human interaction (intervention)" by Yanagi (2007). The author believes that the most important aspect (essence) of Satoumi, rather than a definition, is that the ecosystem functions of coastal waters are enhanced through environmental conservation and resource management in which local people are closely involved (Kakuma 2022b). This is in contrast to past Western concept that tended to eliminate people for the environmental conservation and resource management.

2 Three Important Points for Satoumi Creation

2.1 Residential Researchers and Bilateral Knowledge Translators

In Satoumi creation, there are two important actors: residential researcher and bilateral knowledge translator. The term residential researchers refers to researchers who reside and base themselves in a certain local community and become a member of the local community at the same time being experts in producing scientific knowledge.

S. Kakuma (✉)
Kumejima Okinawa, Japan
e-mail: kaku2789@cc.saga-u.ac.jp

Saga University, Saga, Japan

A.-C. Bennis et al. (eds.), *Constraints and Adaptations to Global Change at the Land-Sea Interface: For a Shared Ecological and Energy Transition*,
https://doi.org/10.1007/978-3-031-90050-1_20

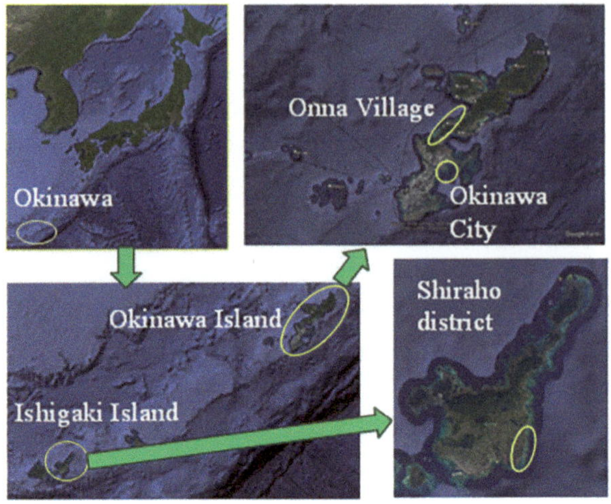

Fig. 1 Three Satoumi sites in Okinawa (Okinawa archipelago in the East China sea)

They are defined as "scientists and other types of knowledge producers living in a local community with the mission of facilitating and mediating the co-production of knowledge for collaborative decision making and actions to solve environmental problems" (Sato et al. 2018). "Bilateral knowledge translators are defined as individuals and/or groups able to bridge gaps between heterogeneous sets of knowledge (including specialized know-how) that emerge from different framings of knowledge, by creating new meanings of the knowledge from different perspectives or different epistemic communities and which are rooted in societal challenges" (Sato et al. 2018).

Figure 2 shows a schematic diagram of the residential researchers and bilateral knowledge translators. R represents the residential researcher and T represents the bilateral knowledge translator. Most of the residential researchers are the bilateral knowledge translators simultaneously, but most of the bilateral knowledge translators are not the residential researchers. There are cross-scale translations among e.g. local, national, and global scale, and horizontal translations between local communities and scientific societies or other local communities.

Residential researchers contribute to solving local problems by conducting comprehensive research directly related to the issues of the local community, such as the local environment and ecosystem services. Kikuchi (2016) compared the characteristics of visiting researchers and residential researchers and summarized them in Table 1.

Residential researchers are different from general researchers who produce knowledge in accordance with scientific and academic methods. In addition to knowledge production, residential researchers play complex roles in the practice of solving local problems by using the knowledge and in creating networks that promote collaborative problem-solving among diverse actors (Kanda and Shimizu 2022).

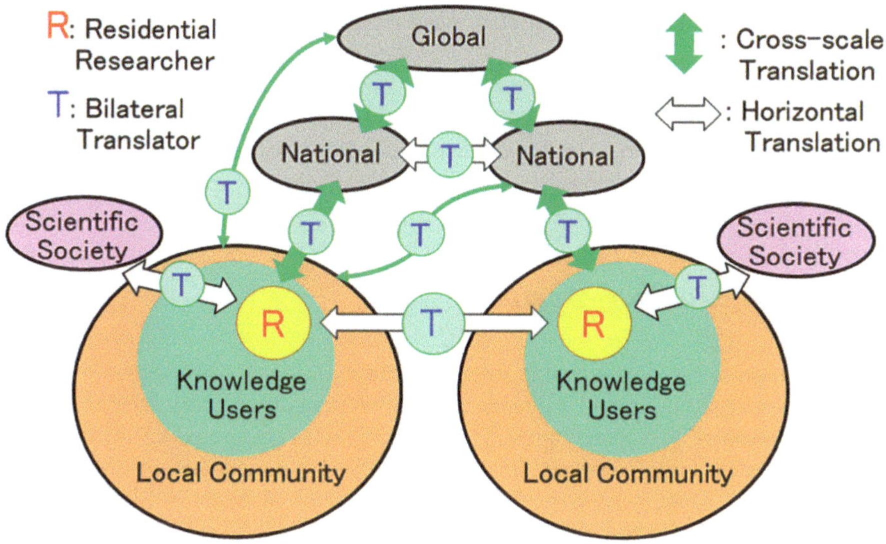

Fig. 2 Schematic diagram of the residential researchers and bilateral knowledge translators

Table 1 Comparison of visiting and residential researchers

	Visiting researchers	Residential researchers
Position in the local community	Someone who comes occasionally	Someone who is always present
Purpose of research	Progression of his or her own research	Solving local problems
Research method	Method commonly used in one's area of expertise	Transdisciplinary method tailored to the issues
Presentation of research	Academic society, Local community	Local community, academic society
Evaluation of research results	Academic society	Local community, academic society
Research and practice	Do not circulate	Circulate

Source Kikuchi (2016)

The bilateral knowledge translators facilitate the circulation and sharing of knowledge to promote collaborative decision making and actions among actors. Their roles are classified into two categories: (a) cross-scale and (b) horizontal translators.

Cross-scale translators bridge gaps between sets of knowledge/technologies produced at different spatial scales and/or governance levels, by creating their new meanings from the contexts of challenges at other scales and levels. Horizontal translators bridge gaps between heterogeneous sets of knowledge, including those brought by cross-scale translators, at a particular spatial scale and/or governance level (Sato et al. 2018).

In the three sites described in this paper, a staff member of a fisheries cooperative, a professional fisher (and the head of a non-profit organization) and a staff member of a non-governmental organization are the residential researchers, and they are also bilateral knowledge translators.

2.2 *Transdisciplinary Approach*

Satoumi creation requires not only interdisciplinary science that integrates natural sciences, social sciences, and humanities, but also transdisciplinary sciences incorporating co-design of research agenda, co-production of integrated knowledge, and co-delivery of research outcomes in collaboration with diverse actors/stakeholders outside academia (Sato et al. 2018). Social–ecological systems in Satoumi areas are extremely complicated and highly uncertain. This is why solving problems in Satoumi is not easy when scientists and experts from outside the community define the problems and implement research without the involvement of the local people. Scientists, experts, and other actors from outside the community should collaboratively think about the problems in Satoumi with local stakeholders and learn from each other to define the problems, produce knowledge, and put the research results into action together (Kakuma and Sato 2022).

2.3 *Active and Passive Measures*

The most important characteristic of Satoumi is the active measures that increase productivity and biodiversity through direct human intervention. Coral outplanting and coral-eating crown-of-thorns starfish (COTS: *Acanthaster planci*) control are examples in coral reef areas. However, these are not the only ways to create Satoumi. Indirect human interventions, passive measures, such as red soil pollution countermeasures and fishery resource management, are also deeply related to the essence of Satoumi. In Satoumi creation, it is important to adopt an approach that integrates active and passive measures (Kakuma 2022a).

3 Onna Village's Satoumi

Onna Village is on the northern west coast of Okinawa Island, the most southern part of Japan. The population is about 10,000, but 2.8 million tourists visit the village attracted by the beautiful sea, including coral reefs. The Onna Village Fisheries Cooperative (OVFC) had 114 full members and 142 associate members in 2017, mainly engaged in seaweed aquaculture and coastal fishing. Mozuku aquaculture accounts for approximately 80% of production and 40% of production value.

3.1 Fisheries Resource Management Plan

The OVFC developed a regional fisheries promotion plan in 1987. This plan was a five-year comprehensive plan to revitalize Onna's fisheries, and the main pillar of the plan was a management plan for coastal resources, which is a passive measure for the Satoumi creation. As a fisheries extension officer in Okinawa, the author supported the preparation of this plan using a transdisciplinary approach. The main tools for the resource management were marine protected areas for sedentary resources such as giant clams, troches, and sea urchins. At the time, the term "marine protected area" did not even exist in Japan. The plan has since been revised six times adaptively by 2018 (Kakuma 2022a).

3.2 Coral Culture and Outplanting

In Onna Village, coral culture and outplanting have been conducted since 1998 as part of the Satoumi creation activities. They help corals regenerate by spawning cultured and outplanted corals. These activities combine two methods: "coral stick culture," in which corals are grown on steel bars driven into the bottom, and "coral outplanting," in which fragments are collected from cultured corals and outplanted to the surrounding area (Higa et al. 2022).

Fishers in Onna Village had developed aquaculture techniques for all of Okinawa's major cultured seaweeds: Mozuku, Hiotoegusa (*Monostroma nitidum*), and Sea-grape (*Caulerpa lentillifera*). The coral stick culture technique was developed by applying the Mozuku stick culture method. The inspiration for this technique came from the fact that corals naturally grow on the tops of the steel sticks that connect the Mozuku nets. Corals are grown at a height of approximately 50 cm above the seafloor by driving steel sticks into the sand and gravel bottom, where corals normally do not grow (Fig. 3). This method allows corals to grow faster because the current is faster than the seafloor and the reflected light from the seafloor hits the underside of the corals (Higa et al. 2022).

Corals outplanted to the southern site in Onna Village bleached due to high water temperatures in 2016, resulting in the death of 85% of the corals (Nature Conservation Division 2017). Bleaching is a phenomenon in which symbiotic algae, zooxanthellae, in coral die or are expelled from the coral due to high water temperatures, causing the coral tissue to become transparent and the coral to turn white (the skeleton color). Corals obtain most of their nutrition from photosynthesis by zooxanthellae; therefore, if bleaching continues for a long time, the coral will die. However, in the area close to the outplanting area, 90% of the cultured corals survived (Higa et al. 2017). The

Fig. 3 Coral stick culture at Seragaki in Onna (©S. Kakuma)

reason for this high survival rate is not clearly understood, but it is assumed that the high current velocity increases coral metabolism and mitigates the effects of sediment at 50 cm above the seafloor.

OVFC had cultured 54 species of corals in 2016. Corals usually spawn both eggs and sperm in capsules called "bundle." A coral colony is the result of the iterated division of a single small polyp; therefore, all polyps are clones. So, eggs and sperm from the same coral colony cannot be fertilized. However, a study by the Okinawa Institute of Science and Technology (OIST) found that the cultured corals of OVFC were genetically diverse, even within the same species (Higa et al. 2022). In other words, the corals cultured by OVFC, as well as corals outplanted using the fragments of the cultured corals, supply coral larvae to the surrounding area.

In one study, eight colonies of 30-cm *Acropora tenuis* (a species of coral) reared in a tank on land yielded 926,000 larvae, or 115,000 per colony. OVFC had cultured 24,000 corals in 2016, so the total supply of larvae was calculated to be approximately 2.8 billion (Higa et al. 2017). The OIST's genetic survey also revealed that a high percentage of corals around the coral farm originated from the farm (Zayasu personal communication).

Furthermore, when 30 colonies of cultured corals were brought to land and examined for fish inhabiting them, 841 individuals of 33 species, mainly damselfish (*Chromis* spp.), were identified. The total number of fish inhabiting the entire farm (24,000 colonies) was calculated to be approximately 670,000 (Higa et al. 2017). Coral culture by OVFC is a good example of "increasing biodiversity through human intervention," as defined by Tetsuo Yanagi in his definition of Satoumi in 2007.

Coral outplanting in Onna Village began in 1989 when the youth group of OVFC conducted the outplanting on a trial basis. A coral seedling supply facility was established by OVFC without subsidy, and they began accepting coral outplanting tours (Higa et al. 2022).

Before the coral outplanting, fragments of cultured corals are cut out, attached to "Magwhite" substrates on land or in the sea, and grown for one to two years on shelfs set up on the seafloor. The reason for intermediate growing is that the survival rate is higher when larger corals are outplanted. Magwhite is an environmentally friendly material consisting of solidifying agents extracted from seawater and white sand. Initially, a plate type was used, but this was modified to a cylinder type with a diameter of approximately 3 cm for use with the tabular coral *Acropora hyacinthus*. When the coral covers the Magwhite surface, the "leg" of the table is formed and adheres to the bedrock. The energy required to form the skeleton of this part can be used for growth, resulting in faster growth. Next, the type was modified to a quadrangular prism for use with *A. tenuis*. This is to suppress the rotation of the Magwhite due to waves. A bolt is attached to the tip of the Magwhite, a hole is drilled into the bedrock of the seafloor, and the bolt of the Magwhite with the coral attached is inserted into the hole (Higa et al. 2022).

There are two methods of coral outplanting: the asexual reproduction method, in which coral fragments are cut and outplanted, and the sexual reproduction method, in which coral larvae are used. The OVFC method for securing coral seedlings using coral stick culture and outplanting them attached to Magwhite was the most efficient asexual reproduction method in 2016. Coral outplanting had been conducted using this method since 2013 as an Okinawa Government project, and by 2016, 120,000 corals had been outplanted in a 3-hectare area (Higa et al. 2022). This was the first coral outplanting of this scale in the world (Kakuma 2022a). OVFC is currently working on the sexual reproduction method, which is becoming mainstream in Okinawa.

3.3 Prevention of Red Soil Runoff

Red soil pollution is one of the most serious environmental problems in Okinawa. It has a negative impact on coral reefs that sustain coral reef fisheries and a direct impact on fisheries and aquaculture. When the red soil adheres to cultured Mozuku, it becomes unsaleable. When set nets and gillnets become dirty with red soil, fishers must wash their nets frequently because fish cannot be caught with dirty nets. Additionally, it is difficult for diving fishers to spear fish in the turbid waters of red soil (Kakuma 2022a).

Red soil runoff became a problem in Onna Village after Okinawa's reversion from the United States to Japan in 1972. Many development activities such as resort construction, the 1975 Okinawa International Ocean Expo, and expansion of the national highway, were conducted, and a large amount of red soil ran off into the ocean. In accordance with the "Okinawa Promotion and Development Special

Measures Law", agricultural land improvement projects had begun, which accelerated the runoff of red soil into the sea. The red soil runoff at that time was tremendous, and after heavy rains, mud accumulated on the tidal flats, creating rice paddies (Higa et al. 2022).

OVFC succeeded in Mozuku aquaculture in 1977, but the following year, 1978, cultured Mozuku was damaged by the red soil runoff. Although the fishers at the time strongly protested, the social trend was all for development. Thus, fishers protesting about red soil runoff felt a sense of alienation as if they were an obstacle to developing Okinawa. Around 1989, red soil runoff became a social problem throughout Okinawa, and a council led by fishers was formed at the initiative of fishery-related organizations. This council conducted surveys of red soil runoff in various parts of Okinawa, and with these results, requested the prefectural government to enact an ordinance to prevent red soil runoff, which was enacted in 1994 (Higa et al. 2022).

The concentration of red soil in seawater is unstable depending on the timing of rainfall and other factors, but the concentration of red soil in the bottom sediment of the seafloor is relatively stable. Therefore, in Okinawa, a simple method of measuring red soil concentration in bottom sediment (SPSS: Suspended Particles in Sea Sediment) became a standard method for monitoring red soil pollution (Omija 2004). In Onna Village, the youth group of the OVFC measures SPSS.

In addition, the OVFC established a system of "prior consultation" with private developers in Onna Village, whereby pollution prevention measures are discussed in advance, pollution prevention facilities are developed first, and the main construction work is started only after confirmation of these facilities. The prior consultation system was applied to the Village's agricultural land improvement projects in parallel with private sector constructions. In 1994, the effectiveness of red soil runoff prevention in Onna Village was estimated to be approximately 97% less than that of no countermeasures. The last serious red soil damage was the seaweed aquaculture damage caused by a tidal flat revetment construction in 1996. The production of Hitoegusa (*Monostroma nitidum*) on one of the tidal flats, which had been decreasing due to red soil runoff, increased from 11 t in 1991 to around 60 t in recent years as a result of red soil countermeasures, clearly indicating the effectiveness of such measures (Higa et al. 2022). Although red soil runoff from development sites has been reduced, it continues to occur on agricultural lands.

3.4 COTS (Crown-of-Thorns Star Fish) Control

Feeding damage by COTS is one of the most serious disturbing factors for coral reefs in Okinawa. The main goal of COTS control in Onna Village is to prevent periodic major outbreaks. Onna Village has had chronic small COTS outbreaks, with three major outbreaks occurring in a 12–13-year cycle in 1971, 1984, and 1996. Fishers in Onna Village have conducted starfish control activities for many years, and since 2002, they have set a goal to prevent the fourth major outbreak, and have decided to exterminate spawning-size COTS before the spawning season in the entire Onna Village (Higa et al. 2022).

In the 1970s and 1980s, corals in Okinawa were devastated by COTS. The corals recovered; however, around 2000, outbreaks began again on Okinawa Island and the Kerama Islands. The COTS control efforts of the 1980s eliminated as many starfish as possible, but as a result, the corals could not be saved. In 2002, the Okinawa Government established a COTS control committee (the author was included in the members) to discuss a new control policy. The goal of control was not to kill large numbers of COTS, but to protect valuable corals, and the policy was changed to establish relatively small most important conservation areas in each region and to concentrate the efforts there (Nature Conservation Division 2007).

There are two approaches to COTS control: the Kerama approach and the Onna approach. In the Kerama approach, the most important conservation areas are determined based on three criteria: "should be protected," "want to protect," and "can be protected." The "should be protected" criterion is determined on the basis of the presence of valuable corals in the area and whether the area is a source of coral larvae for other areas. The "want to protect" criterion is determined based on economic reasons, such as important dive sites or fish spawning areas. The "can be protected" criterion is determined based on the distance from the ports and the sea conditions during the winter season, because the COTS control is meaningless unless it is carried out continuously over a long time. The Onna approach reduces the density of spawning-size COTS before the spawning season over an area of approximately 30-km the Onna Village coastline. The approach is applied to a much wider area than the most important conservation areas in Kerama. This approach is possible in Onna Village because the control method devised by fishers there is extremely efficient. It is difficult to adopt the same approach in other areas.

In Onna Village, the entire area is divided into five to seven areas for the control of COTS, and the area to be worked on the particular day of the control is determined. Three to six fishers board a fishing boat, then at the control site, each fisher swims pulling a small boat, which is approximately 2 m long, and collects COTS by skin-diving with a harpoon or hook and puts them in the boat. This method allows for wide areas shallower than 5 m in depth. The date, time, location, and length composition in 5-cm increments of each starfish were recorded for each fishing boat. The results of the measurements were presented to the fishers the next day. The size composition graphs visualized the priorities of the areas to be exterminated, i.e., the areas where some starfish were still left to be removed, and this helped motivate the fishers. In Onna Village, the number of COTS exterminated has increased since 2011, but the fourth outbreak was prevented by deploying more than 500 control participants every year (Higa et al. 2022).

3.5 Mozuku Fund

In 2009, OVFC, a seaweed processing company in Tottori Prefecture, and consumer cooperatives nationwide collaborated to establish the "Mozuku Fund," a mechanism to secure funds for coral culture. Specifically, one to two yen per pack of processed Mozuku was set aside in the Mozuku Fund and used for coral culture. Twenty-eight co-ops (6.7 million members) across Japan cooperated, and by 2015, funds for 10 million packs had been accumulated, and 24,000 corals were cultured (Higa et al. 2022). The number of cultured corals now exceeds 40,000.

4 Okinawa City's Satoumi

Okinawa City is located on the middle east coast of Okinawa Island. It is the second largest city in Okinawa islands after Naha City. Reclamation of the shallow sea areas has progressed. Okinawa City's Satoumi creation has become an example of Satoumi creation in urban areas. The Okinawa City Fisheries Cooperative consists of 49 full members and 42 associate members in 2017, and is engaged mainly in Payao (fish aggregating device) fishery, Sodeika (*Thysanoteuthis rhombus*: large pelagic squid) fishery, and diving fishery. In the past, most of the fishers fished with diving fishery, hook and line fishery, bottom longline fishery, and small nets to catch coastal fish. However, due to various reasons, such as sedimentation caused by developments and overfishing, coastal fishery resources have decreased, and there has been a shift to offshore Payao and Sodeika fisheries (Yanagida and Kakuma 2022a, b).

4.1 Length Limits of the Most Important Fish

Akajin (Okinawa's local name of *Plectropomus leopardus*) and Makubu (Okinawa's local name of *Choerodon schoenleinii*) are the most important fishery resources in Okinawa (Fig. 4) because of their high prices, but their stocks are declining in many areas of Okinawa. As the stocks are declining in the fishing grounds of the Okinawa City Fisheries Cooperative, since 2014, the fishers there have voluntarily set length limits of 35 cm and 30 cm for Akajin and Makubu, respectively, to protect small fish and raise them to spawning sizes. Akajin and Makubu under the length limits are not allowed to be sold in the fishery markets (Yanagida and Kakuma 2022a, b).

Fig. 4 Akajin (left side: *Plectropomus leopardus*) and Makubu (right side: *Choerodon schoenleinii*) (©I. Ohota)

4.2 Marine Protected Area

Ippei Yanagida, a fisher and a leader of Okinawa City's Satoumi creation, succeeded in persuading fishers to establish a no-take MPA in 2015. Although the MPA is small (100-m square), it is rich in coral and fish. Buoys were placed at the four corners of the MPA to clearly indicate the extent of the MPA. Because the MPA rules are voluntary, no penalties have been set. In addition to coral planting in the MPA, giant clam restocking was conducted (Yanagida and Kakuma 2022a, b).

4.3 Coral Culture and Outplanting

Since 2010, Yanagida has cultured 13 species of *Acropora* corals on steel mesh tables set up in the sea. Cultured corals have been sold to companies that use coral outplanting for environmental education as corporate social responsibility (CSR) and have also been outplanted on the coastal waters of Okinawa City through various subsidy programs. Coral outplanting activities are currently being conducted in various parts of Okinawa with the aim of restoring coral reefs. As a result, the demand for coral for outplanting is increasing, and Yanagida is trying to supply the corals through aquaculture by the fishers' group. Although the scale of coral outplanting is still small, Yanagida's activities are frequently covered by the mass media, and are effective in educating children about the environment and spreading awareness on coral reef conservation throughout Okinawa (Yanagida and Kakuma 2022a, b).

4.4 Surveys by Fishers

In Okinawa City, fishers themselves conducted surveys of important resources on the recruitment of Akajin and Makubu. The survey method was guided by researchers from a local government research institute. At each survey point, three 50-m lines were set at depths of 5 m, 5–9 m, and 10 m or more. The number and length (nine classes of 5 cm intervals) of Akajin and Makubu were recorded by divers swimming along the lines (Yanagida and Kakuma 2022a, b).

Monitoring of fish and benthic organisms in the MPA was conducted periodically by fishers. Two methods of monitoring were selected, which are scientifically valid and feasible for the fishers to implement. One method is the line transect method, in which five 100-m-long lines are set at 25-m intervals in the MPA, and divers swim along the lines to measure the number and type of fish and other major organisms. The other method is the quadrat method, in which 17 square frames of 2 m × 2 m are set up on the lines. In Okinawa City, SPSS has also been monitored by fishers since 2013 (Yanagida and Kakuma 2022a, b).

5 Shiraho's Satoumi

In Shiraho district on Ishigaki Island, the World-Wide Fund for Nature (WWF) Japan and the local community have been collaborating on the common theme of Satoumi creation. Originally, there was a big difference in the way the two parties perceived local issues. WWF, a nature conservation organization, placed the conservation of globally precious coral reefs as the most important issue, while the Shiraho community, which had experienced community division due to the airport issue, emphasized cultural heritage and community revitalization. Collaboration between the two groups has been realized by filling in the gaps between these differing goals (Kamimura and Claus 2022).

5.1 Coral Village and Coral Reef Culture

In Shiraho, WWF opened the Coral Reef Conservation and Research Center (nicknamed Coral Village) in 2000 and has been conducting conservation activities in collaboration with local residents. At the time of its establishment, the impact of the construction of the new airport on the coral reef ecosystem was an important issue, and nature conservation activities were developed based on the concept of "protecting corals." Later, red soil runoff from farmland became an issue, and it became essential for land farmers, who make up most of the local population, to take measures to prevent red soil runoff. Coral Village focused on the lifestyle of farmers when they were deeply involved with coral reefs and shifted its approach for nature conservation to community development activities to perform "coral reef culture." Coral Village developed various cultural and economic activities, including community development that "resulted" in the conservation of coral reefs (Kamimura and Claus 2022).

5.2 Restoration of a Stone Tidal Weir

The stone tidal weir, a traditional stationary fishing equipment, is a semicircular or square stone wall built with boulders on shallow coast to catch fish as the tide ebbs and flows. They catch fish that come to shore at high tide but are blocked by the stone walls at low tide, preventing them from returning to offshore. In Shiraho, there were 13 stone tidal weirs in the past, but all were lost. Coral Village and the community decided to restore the stone tidal weir to use it as a place for cultural learning and nature experiences related to the sea (Fig. 5). This fishing equipment had been used by farmers in Shiraho, and it was hoped that it would provide an opportunity for the farmers to re-confirm their connection to the sea (Kamimura and Claus 2022).

The restoration work was a large-scale project involving the PTA (parent–teacher association) and students of Shiraho Elementary School and Junior High School. The stone tidal weir was approximately 200 m parallel to the coastline, 70 m offshore, and the height of the stone wall was approximately 1 m. Since its completion in October 2006, the site has been used as a place for local children to experience nature and for environmental education (Kamimura and Claus 2022).

Environmental surveys conducted by the Coral Village before and after the restoration of the stone tidal weir confirmed an increase in the number of fish species and individuals. This was due to the increase in fish that use the crevices of the stone structures as hiding places and those that feed on seaweed attached to the stones. The stone tidal weir drew attention as an example of the definition of Satoumi, "high productivity and biodiversity in the coastal sea with human intervention" (Kamimura and Claus 2022).

Fig. 5 The stone tidal weir used for environmental education (©S. Kakuma)

5.3 *Measures to Prevent Red Soil Runoff*

A 2007 fishing experience by junior high school students in the stone tidal weir resulted in a catch of only four fishes. The students believed that the cause of this small catch was red soil runoff from the farmland, and they decided to work on the prevention of red soil runoff and proposed an activity to plant green belts (shell ginger) around the farmland. At the time, the installation of green belts around farmland was considered as one of the most important measures for preventing red soil runoff in Okinawa. The Ishigaki City Government provided shell ginger seedlings, which were planted by junior high school students as volunteers (Kamimura and Claus 2022).

The green belt planting activities started with student volunteers, but later expanded to include WWF member tours, training for donor companies, coral reef conservation experience programs for junior high and high school excursions from the mainland Japan, and study tours. Participants were asked to contribute to an environmental cooperative fund to cover the cost of planting activities. This created a mechanism for conservation efforts while raising funds locally (Kamimura and Claus 2022).

5.4 *Restocking of Giant Clam*

The Shiraho community tried to increase the stock of giant clams, a local fishery resource that has been popular in the region. Giant clams used to be abundant in the waters of Shiraho and were familiar food items served at banquets after communal activities. However, the results of surveys conducted by Coral Village showed that the stock had declined significantly since 2003. Therefore, small giant clams were purchased from the Okinawa Government and restocked under the guidance of fisheries extension officers, including the author.

Restocking was conducted at snorkeling tourism points from the periphery of the stone tidal weir to the offshore area. A total of 9100 small giant clams (Fig. 6) were restocked in 2009 and 2010. By leaving them as spawning stocks instead of harvesting them, they will spawn and help increase the giant clam resources in the surrounding area. After discussions among Shiraho residents, it was agreed that the area should be closed to fishing throughout the year. Beautiful giant clams are also used as tourism resources (Kakuma 2022a).

Fig. 6 Restocked giant clam (*Tridacna gigas*) (©S. Kakuma)

6 Discussion

6.1 Residential Researchers and Bilateral Knowledge Translators

In Onna Village, Yoshimi Higa, a staff member of OVFC, played roles as a residential researcher to create Satoumi there. He conducted studies on coral culture and outplanting, efficient and effective COTS eradication, and red soil pollution monitoring. He received an excellent research paper award (for Higa et al. 2017) from the Japanese Caral Reef Society. He was also a bilateral knowledge translator, and bridged OVFC and the outside world. The author also acted as a bilateral knowledge translator and prefectural fisheries extension officer. The main role of fisheries extension officers is to go into fishing villages and provide fishers with technical guidance on fisheries, aquaculture, resource management, and fish distribution. Fisheries extension officers are potential bilateral knowledge translators. One direction of translation that they do is to organize information from research institutes in an easy-to-understand manner and convey it to fishers. In contrast, they translate the knowledge and experience of fishers into the language of science and introduce them to the world of science (Kakuma and Kitolelei 2018). There were 10 fisheries extension officers in Okinawa, and 433 in Japan in 2018.

Higa and the author acted multi-layered in the translation system in Onna village. The author brought knowledge/technologies to OVFC as a cross-scale translator. Higa stood between the author and the fishers as a horizontal translator, and translated the knowledge/technologies for the fishers. In an opposite way, the author translated and transmitted the fishers' knowledge and experience, through Higa, to the world as a cross-scale bilateral knowledge translator. The author also introduced the resource management efforts of OVFC at an international conference held in New Caledonia in 1995 as an excellent example of co-management (Kakuma and Higa 1995).

In Okinawa City, Yanagida played the roles of a residential researcher. He headed the NPO (Non-Profit Organization) INO. Ino, an Okinawa's local language, is a shallow, calm lagoon between the outer coral reef where waves break and the shore. INO has performed various activities such as preservation of marine culture, environmental education/research/conservation, and coral culture and outplanting. He was also a bilateral knowledge translator, and played the roles with the author as Higa and the author did in Onna village. In Shiraho, Masahito Kamimura, the head of the Coral Village, was the residential researcher and the bilateral knowledge translator. He led the activities of the Coral Village.

6.2 Transdisciplinary Approach

In Onna Village, the author supported the OVFC to develop the resource management plan with a transdisciplinary approach. He visited OVFC at least once a month for one year, bringing the information of research institutes needed for the management, and talked a lot with the fishers. The resource management plan was developed by integrating the information from research institutes and the knowledge and opinions of the fishers.

In Shiraho, giant clams (*Tridacna* spp.) were restocked. Using a transdisciplinary approach and as a fisheries extension officer, the author worked with fishers in various parts of Okinawa to develop techniques for restocking the smallest giant clam species. The young shellfish enter small holes in rocks and grow by scraping the rocks. Taking advantage of this ecology, a fisher in the northern part of Okinawa Island devised a method to make holes for the young shells by using an air-powered drill in the sea, which is usually used in car repair shops. However, the young shellfish would sometimes escape from the holes or be eaten by the fish. A fisher in a remote island came up with the idea of using a construction tacker (like a stapler) to block the holes where the young shells were embedded (the author improved this method by using the tacker to fasten small pieces of nets). Such ideas could not have come from researchers, and the richness of the ideas of the fishers is astonishing (Yanagida and Kakuma 2022a, b).

The author then disseminated the techniques to fishing villages throughout Okinawa, instructed fishers in the fields, and described techniques into a technical guide (Kakuma 1989). During 1988 and 1989, techniques were implemented in 12 districts of Okinawa. In Onna Village and Okinawa City, *T. crocea* were also restocked.

6.3 Active and Passive Measures

The author has studied various examples of Satoumi relating to active measures: (a) coral reef restoration, (b) mangrove planting, (c) seagrass restoration, (d) COTS

Table 2 Active and passive measures in the three sites

Sites	Active measures	Passive measures
Onna village	Coral culture and planting COTS control Restocking of giant clam	Resource management plan Prevention of red soil runoff
Okinawa city	Coral culture and planting Restocking of giant clam	Length limits for important fish MPA Surveys by fishers
Shiraho	Stone tidal weir Restocking of giant clam	Inheritance of coral reef culture Prevention of red soil runoff

control, (e) giant clam restocking, (f) squid spawning bed, (g) stone tidal weir, (h) artificial reefs, and (i) green snail restocking; and passive measures: (a) fisheries resource management, (b) pollution control measures, (c) fishers' surveys, (d) value chain promotion, (e) human & cultural exchanges, (f) marine protected areas, and (g) resource management plans. The active and passive measures in the three sites described in this paper are listed in Table 2 and classified according to active or passive measures.

6.4 Integration of Fisheries Resource Management and Ecosystem Conservation

6.4.1 Coral Reef Restoration

Coral reef restoration activities were undertaken in Onna Village and Okinawa City. These activities are considered as active measures of Satoumi creation that improve biodiversity through direct human intervention. Currently, coral restoration activities centering on outplanting are being undertaken worldwide.[1] However, there are many criticisms about coral outplanting. The first is that the area of coral reefs that can be directly restored through human outplanting is far smaller than the area of coral reefs that have been lost. Therefore, most coral outplanting activities in Okinawa do not directly increase the area of coral reefs by outplanting corals, but rather regenerate corals in the surrounding areas by allowing the outplanted corals to spawn (Kakuma 2022a).

There has been criticism of such coral reef restoration through spawning. The argument was that "there are far more corals left in the wild that are capable of spawning than corals that had been outplanted and survived to grow to the size to spawn." However, as the number of corals decreases and the distance between parent corals increases, the likelihood of fertilization failure increases. In one study, *Acropora yongei* (a species of coral) colonies that were artificially placed close together

[1] https://www.icriforum.org/restoration/coral-restoration-database/ Accessed 20 August 2024.

had a higher fertilization rate and were 1400 times more capable of supplying larvae than naturally surviving *A. yongei* (Zayasu and Suzuki 2018). Thus, by culturing and outplanting corals at appropriate intervals, it is possible to significantly enhance the regenerative function of corals (Kakuma 2022a).

6.4.2 Land-Based Pollution Countermeasures

In all three areas, measures were taken to prevent red soil pollution. These are passive measures of Satoumi creation. In Okinawa City, fishers themselves monitored red soil pollution by measuring SPSS. In Onna Village, the youth group of the fisheries cooperative measured SPSS, and a prior consultation system was adopted in which the fisheries cooperative and the developers discussed red soil runoff prevention measures before starting the construction. In Shiraho, green belts around farmland, promising measures to prevent red soil runoff, were initiated by volunteer junior high school students, which developed into self-sustainable activities such as coral reef conservation tours and study tour programs (Kakuma 2022a).

Red soil pollution is the biggest environmental problem in Okinawa, but excess nutrient influx is also a serious factor that disturbs coral reef ecosystems (Omija 2004). High concentrations of nutrients negatively affect corals for three indirect reasons: (a) increase seaweeds that compete with corals for sunlight (Bellwood et al. 2004); (b) increase phytoplankton and make seawater turbid, inhibiting photosynthesis of symbiotic algae in corals (Omija 2004); and (c) increase phytoplankton as a food source for COTS larvae, leading to outbreaks of them (Fabricius et al. 2010). As a direct physiological effect of excess nutrients on corals; phosphorus was recently found to inhibit the skeleton formation of juvenile corals (Iijima et al. 2019). The results of rearing juvenile *Acropora tenuis* in seawater with varying nutrient concentrations showed that coral survival rates decreased with increasing nutrient concentrations (Kakuma et al. 2019). Satoumi creation in Japan was initially most active in the Seto Inland Sea, an enclosed coastal sea. There, the management of material circulation, including nutrient control, was an important issue in Satoumi creation (Yanagi 2007). The problems of soil runoff and excess nutrients are not limited to Okinawa or Japan; examples of such problems and countermeasures in 33 districts throughout the world are provided (Wilkinson and Brodie 2011).

6.5 Surveys and Monitoring by Fishers

In Okinawa City, fishers conducted MPA monitoring, recruitment surveys of important resources, and monitoring of red soil pollution. Surveys by the fishers themselves have several advantages for resource management. First, the survey ability of the diving fishers is high. Diving fishers who always dive in the research area are familiar with the sea and can detect juveniles more efficiently than researchers if they are given the characteristics of the juveniles (juveniles often have a different morphology and color from adults). Another advantage is that information useful for resource management can be obtained inexpensively. If the survey is continued for a long time, fishers will be able to confirm the effectiveness of resource management with their own eyes. It also fosters a sense of ownership that the resource belongs to them. This sense of ownership is an important condition for the sustainability of resource management. Thus, scientific surveys and monitoring, traditionally conducted by professional scientists and experts, are increasingly being conducted by fishers and local residents in Satoumi through the transdisciplinary approaches, in collaboration with scientists and experts (Kakuma 2022a).

7 Conclusion

As I said previously, the most important aspect of Satoumi was "involvement of the local community" in the environmental conservation and resource management in the introduction of this paper. The involvement includes participation in not only active measures such as coral outplanting and COTS control, but also passive measures such as the prevention of red soil runoff and fisheries management. A variety of these measures were implemented in three cites, Onna, Okinawa-city and Shiraho, with the help of residential researchers and bilateral knowledge translators through transdisciplinary approaches.

The author believes that this important notion of Satoumi, involvement of the local community, and concepts of the active and passive measures, residential researchers and bilateral knowledge translators, and transdisciplinary approaches are effective not only in Satoumi areas but also in other coastal areas throughout the world. For example, the notion of "maritime territory" is similar to the notion of Satoumi. "Adopting a maritime territory approach is tantamount to integrating local cultures into collective decisions, which is essential both for the wellbeing of the actors and the ecosystems" (Prouzet et al. 2020). More than 600 fishing communities in Japan were implementing Satoumi processes through a program of the Fisheries Agency of Japan in 2018. Although most of the communities did not refer to their coastal areas as Satoumi, they conserved seagrass and argal beds, tidal flats, coral reefs, etc. (Kakuma 2022a). "Satoumi is also spreading overseas, and examples of Satoumi have been reported all over the world. Since the first international Satoumi workshop was held in China in 2008, international Satoumi conference had been held annually somewhere in the world for 12 consecutive years, in the Philippines, Japan (Kanazawa), the United States (Baltimore), the United States (Hawaii), Turkey, Japan (Tokyo), Vietnam, Russia, France, Thailand, and St. Lucia/Fiji" (Kakuma 2022a) before COVID-19. The author hopes the notion of Satoumi will continue to expand worldwide to conserve our precious coastal sea and contribute to the well-being of coastal communities.

References

Bellwood DR, Hughes TP, Folke C, Nyström M (2004) Confronting the coral reef crisis. Nature 429:827–832

Fabricius KE, Okaji K, De'ath G (2010) Three lines of evidence to link outbreaks of the crown-of-thorns seastar *Acanthaster planci* to the release of larval food limitation. Coral Reefs 29:593–605

Higa Y, Shinzato C, Zayasu Y et al (2017) Coral reef restoration efforts by a fisheries cooperative: the case of Onna Village in Okinawa prefecture. JPN Coral Reef SOC J 19(1):119–128 (in Japanese)

Higa Y, Takeuchi A, Yanaka S (2022) Connecting local regions and cities through Mozuku seaweed farming and coral reef restoration: Onna village, Okinawa. In: Kakuma S, Yanagi T, Sato T (eds) Satoumi science. Springer Nature, Singapore, pp 193–215

Iijima M, Yasumoto K, Yasumoto J et al (2019) Phosphate enrichment hampers development of Juvenile *Acropora digitifera* Coral by inhibiting skeleton formation. Mar Biotechnol 2019:291–300

Kakuma S, Teruya S, Ishikawa T (2019) Research on coral seedling production technology by sexual reproduction method using deep sea water: effects of high nutrient concentrations in deep sea water on juvenile corals. Okinawa Deep Sea Water Res. Cent Annu Rep 2019:24–26 (In Japanese)

Kakuma S, Higa Y (1995) Sedentary resource management in Onna village, Okinawa, Japan. In: Workshop on the management of South Pacific Inshore resource fisheries, South Pacific Commission, Noumea, vol 1, pp 427–438

Kakuma S, Kitolelei J (2018) Coastal resource management connected by Marine protected area networks. In: Sato T, Chabay I, Helgeson J (eds) Transformations of social-ecological systems: studies in co-creating integrated knowledge toward sustainable futures. Springer Nature, Singapore, pp 267–286

Kakuma S, Sato T (2022) Prologue: what is Satoumi? In: Kakuma S, Yanagi T, Sato T (eds) Satoumi science. Springer Nature, Singapore, pp 1–17

Kakuma S (1989) A guide to the release of giant clam *Tridacna crocea*. Okinawa prefectural fisheries extension center, Okinawa (in Japanese)

Kakuma S (2022a) Satoumi systems promoting integrated coastal resources management: an empirical review. Sustainability 14. https://doi.org/10.3390/su141811702

Kakuma S (2022b) Preface. In: Kakuma S, Yanagi T, Sato T (eds) Satoumi science. Springer Nature, Singapore, pp v–vi

Kamimura M, Claus A (2022) Enlivening ecosystems with human hands: building Satoumi through coral reef culture. In: Kakuma S, Yanagi T, Sato T (eds) Satoumi science. Springer Nature, Singapore, pp 69–87

Kanda M, Shimizu M (2022) Divers and fishermen working together to create Satoumi. In: Kakuma S, Yanagi T, Sato T (eds) Satoumi science. Springer Nature, Singapore, pp 217–238

Kikuchi N (2016) The multiple roles of the residential researcher: toward the creation of sustainable rural communities. Env Res Quarterly 180:80–88 (in Japanese)

Nature Conservation Division, Department of Environment, Okinawa Prefecture (2007) Guideline for the control of crown-of-thorns starfish (in Japanese)

Nature Conservation Division, Department of Environment, Okinawa Prefecture (2017) Okinawa prefecture coral reef conservation and restoration project summary report (in Japanese)

Omija T (2004) Terrestrial inflow of soils and nutrients. In: Coral reefs of Japan, Japan coral reef society and the ministry of environment, Tokyo, pp 64–68

Prouzet P, Monaco A, Sébastien L (2020) Management of the Marine environment: transdisciplinary and systemic approach. In: Maurice PA (ed) Encyclopedia of water: science, technology, and society. Wiley. https://doi.org/10.1002/9781119300762.wsts0067

Sato T, Chabay I, Helgeson J (2018) Introduction. In: Sato T, Chabay I, Helgeson J (eds) Transformations of social-ecological systems: studies in co-creating integrated knowledge toward sustainable futures. Springer Nature, Singapore, pp 1–7

Wilkinson C, Brodie J (2011) Catchment management and coral reef conservation. Global coral reef monitoring network and reef and rainforest research centre. Townsville
Yanagi T (2007) Sato-Umi: a new concept for coastal sea management. TERRAPUB, Tokyo
Yanagida I, Kakuma S (2022) Conserving multiple coral reef resources. In: Kakuma S, Yanagi T, Sato T (eds) Satoumi science. Springer Nature, Singapore, pp 123–143
Zayasu Y, Suzuki G (2018) Comparisons of population density and genetic diversity in artificial and wild populations of an arborescent coral, *Acropora yongei*: implications for the efficacy of artificial spawning hotspots. Restor Ecol 2018:1–7

Restoring the European Eel (*Anguilla Anguilla*) and Its Habitats in France: A Social, Economic and Environmental Dynamic Similar to the Sato-umi Concept

P. Prouzet, Ph. Boisneau, D. Macé, N. Porcher, N. Stolzenberg, A. Viera, V. Lonni, and F. Volage

1 Introduction

In 2019, there were 19 species of the genus *Anguilla* (Tsukamoto et al. 2020) in all the ecosystems investigated, 11 of which are widespread in the Atlantic Ocean (2 species) and the Indo-Pacific region. During their life cycles, they colonise a wide range of environments, from small streams to rivers, lakes and lagoons, estuaries and coastal areas, up to the ocean.

During their life cycles, these species, as well as many diadromous fish (using fresh, brackish and marine waters to achieve their biological cycles), make the link between environments presenting very different characteristics, demonstrating the need for continuity between the various ecological systems located in fresh, estuarine and lagoon, coastal and oceanic waters to ensure the future of their populations.

Some species of the *Anguilla* genus, such as *Anguilla anguilla*, *A. rostrata* and *A. japonica*, form the basis of a significant fishing and aquaculture industry. Two countries dominate this sector: China, the main exporter (US$209 million in 2022) and Japan, the main importer (US$216 million in 2022). The total value of exports is

Ph. Boisneau · D. Macé · N. Porcher · N. Stolzenberg
CONAPPED Comité National de la Pêche Professionnelle en Eau Douce, Saint-Viaud, France

P. Prouzet (✉)
SFJO—Société Franco-japonaise d'Océanographie France, Saint-Jean de Luz, France
e-mail: patrick.prouzet@orange.fr

V. Lonni · F. Volage
CNPMEM Comité National des Pêches Maritimes et des Elevages Marins, Paris, France

A. Viera
ARA France—Association pour le Repeuplement de l'Anguille en France, Paris, France

A.-C. Bennis et al. (eds.), *Constraints and Adaptations to Global Change at the Land-Sea Interface: For a Shared Ecological and Energy Transition*,
https://doi.org/10.1007/978-3-031-90050-1_21

estimated at US$498 million in 2022. On a global scale, it is a socially and economically important species, providing a livelihood for many fishing and aquaculture communities (Yuan et al. 2022).

It is also a fish appreciated for its gustatory value, particularly in Japan and Northern Europe, and for its symbolic value in the Indo-Pacific area. Kanak legends such as "Téâ Kanaké" bear witness to this. It is not highly valued in part of the Atlantic area, although it is the source of important economic activity in the continental and estuarine waters of this zone. This is the case in France, where the *Anguilla anguilla*, a snake-like fish, was considered an invasive and harmful species in salmonid waters. Until 1984, when the law 84–512 on freshwater fishing and the management of fish resources was introduced, this species was subject to regulatory electrofishing for eradication by the management authorities responsible for freshwater fishing.

As we shall see, this fish of the *Anguilla* genus is a perfect illustration of the necessary connectivity that must exist between the freshest waters and the most distant marine waters. Today, this continuity is jeopardised by climate change (modifying ocean circulation) and by human actions that impact aquatic environments: damming of rivers, artificialisation of coastal areas, obstacles to species migration and to ecological continuity, organic and chemical pollution, invasive species, depletion of water resources, fishing activities. The harmony between man and his terrestrial or aquatic environment has gradually been destroyed, even though it was vital to the survival and future of traditional fishing activities. It is the case of small-scale continental, estuarine and coastal fishing, which for many decades had managed to strike a balance between a way of life and sustainability of fish populations exploited in the interface zones between aquatic and terrestrial environments (Prouzet et al. 2010, 2019a, b; Boisneau et al. 2016). How can we recover this harmony between man and nature, the basis of the concepts defined in Japan under the names of "Sato-yama" and "Sato-umi"?

2 Sato-yama and Sato-umi

2.1 Sato-yama

As Nicolas Baumert (2010) notes, "at the dawn of the twenty-first century, few Japanese knew the meaning of this word" (and even fewer non-Japanese). The term was highlighted by the international press after the Nagoya Conference on Biodiversity in 2010. The term combines two words, "Sato" meaning village in Japanese and "yama" meaning mountain. "Sato-yama" is thus a humanised area at the interface of a natural and an urbanised zone, for example a rice paddy where a wide variety of flora and fauna thrive (Takeushi et al. 2003). Cultivated by man, this area becomes an anthropized element with high biodiversity, showing the possibility of harmonious coexistence between man and nature. As Nagahata Yosiyuki describes it, the "Sato-yama" is a socio-ecological area that has built up its biodiversity through the use of land for a variety of purposes, but respecting the rhythm of the seasons: harvesting wood, burning grasses, irrigating rice fields through natural channels, etc. But over the last 50 years, the "Sato-yama" has undergone major changes. The road network has intensified, the irrigation channels have been concreted, and the native grasslands have been abandoned and replaced by higher-yielding non-native grasses for fodder.

It is probably impossible to return to the original Sato-yama everywhere, but it is not too late to partially re-naturalize these interface areas between land and wetlands by using ancient knowledge and know-how before they disappear, in order to reconnect nature and culture, man and the environment.

2.2 *Sato-umi*

This term literally combines two words, "Sato", village or community, found in "Sato-yama" and "Umi", the sea or, more specifically, the coastal zone (Matsuda 2019).

Its main objective, as defined by Yanagi in (2007 and 2008) and again in (2012), is a desired state of the coastal zone characterised by high biodiversity and high productivity of coastal ecosystems maintained by human intervention in nature. Here, we advocate active restoration rather than passive restoration, which is linked to the enclosure of a natural environment that is almost impossible to achieve given the high level of human pressure in many natural areas, which have been conquered by our societies for economic reasons that often ignore the environmental dimension.

The "Sato-umi" concept is also linked to the Sato-yama, because the productivity of a coastal zone cannot be dissociated from the quality and protection of the catchment areas that connect to it and drain into it.

It is therefore a global vision that brings together a variety of natural and man-made ecosystems strongly influenced by human activity, whose ecological footprint must be controlled by including action of local communities to protect and restore environments and restock plant and animal populations. Action by local communities to restore ecosystems that were severely damaged, as was the case throughout the world, after the World War II is one of the fundamental bases of Sato-yama and Sato-umi, and aims to reconnect human communities with Nature for the sustainable use and conservation of coastal areas.

From this point of view, both "Sato-umi" and "Sato-yama" can be considered as socio-ecological systems (Uehara et al. 2019). The "Sato-umi" concept has been incorporated in Japan as one of the bases of environmental policy in Japan (Yamamoto et al. 2008) and in national regulatory systems such as the Basic Ocean Plan of 2008 (Matsuda 2019).

"Sato-umi" projects have been widely developed in Japan in recent decades (Komatsu and Yanagi 2015; Komatsu and Aoki 2020; Matsuda and Yanagi 2019; Tanaka 2014; Tanaka and Furukawa 2020; Uehara et al. 2019).

It's all about going back to the simple, common-sense principles set out by ecologists such as Tanaka Shozo: "Taking care of rivers means making sure that the garden is properly watered; getting rid of pollution is a simple measure of hygiene" (Strong 2015) or by men of culture like Albert Lataillade (former President of the Lous dou Gabe e de l'Adou association) "We must therefore do a better job of instructing and educating the young and not-so-young, especially in these times of accelerated globalisation, and show them the richness of our ancient identity. Certainly, without nostalgia for oil lamps, but by preserving the values of some of our traditions: respect for nature and water was one of them" (Prouzet, Lespine et Sébastien, 2009).

3 Implementing an Active Rather Than Passive Restoration Project Based on a Responsible Fishing Industry

3.1 The Eel Tree: A Fragile and Complex Structure

The eel is a robust species that can adapt and thrive in a variety of ecosystems: river, lake, lagoon, estuary and coastal areas. It can withstand high temperatures and migrate rapidly through ecosystems of very different salinities, but to do so it needs ecological continuity between the different habitats it can colonise (Tesch 1977; Adam et al. 2008).

The eel tree (Fig. 1) provides a good illustration of this ecological continuity between the different habitats at various scales: local, regional and distribution areas. It shows that the species' productivity and its maintenance are linked to the ecological connectivity existing at local level and between the production and reproduction areas, but also to the quality and number of productive areas (size and number of watersheds or lagoon and coastal areas).

Since the beginning of the twentieth century, the increasing number of obstacles to migration, some of which are totally impassable, particularly on major rivers, the reclamation of wetlands and the diking of estuarine areas to protect against flooding have fragmented the species' habitats and greatly reduced the size of its production area, whether for the European (*Anguilla anguilla*), American (*Anguilla rostrata*) or Japanese (*Anguilla japonica*) eels (Machut et al. 2007; Drouineau et al. 2018; Feunteun and Prouzet 2020; Prouzet et al. 2019b; Prouzet et al. 2023). Along with the degradation of continental ecosystems, this is one of the main reasons for the sharp fall in these populations from the late 1970s in the central part of the distribution area. The regulation of fishing for the species at European level through the introduction of Regulation 1100/2007 has helped to slow the decline in the abundance of the species and reverse the downward trend in glass eel recruitment (Fig. 2).

However, the process will be slow and can only be speeded up by the restoration of water quality and ecological continuity under the Water Framework Directive (WFD) and the Marine Strategy Directive (MSFD), the objectives of which are a long way from being achieved (Feunteun and Prouzet 2020; Prouzet et al. 2023).

This is borne out by the results of observation networks on the ecological status of surface waters in relation to WFD objectives in mainland France and the French

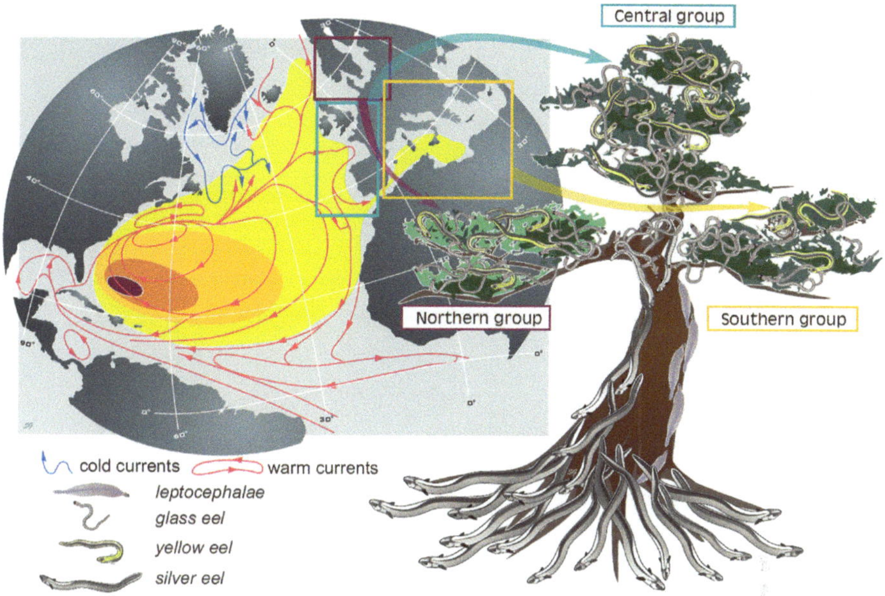

Fig. 1 The eel tree—image showing how the European eel population is renewed—from Adam et al, 2008. Comments: *The link between the spawning and production areas is materialised by the sap circulating in the trunk of the tree, which corresponds to the circulation of North Atlantic waters under the influence of climate change. Each regional eel group (northern, central, southern) corresponds to a part of the canopy, the productivity of which will depend on the number of branches and leaves that correspond to a watershed. By reducing the number of leaves and branches or weakening them, it is easy to understand why the eel population (canopy scale) is declining, whatever the efforts made by professional fishermen to conserve the population on the scale of the leaves that still exist*

overseas departments (DOM), where the ratio: number of bodies of water in good or very good status to bodies of water in poor or very poor status barely changed between 2013 and 2019: 0.78 in 2013; 0.80 in 2016 and 0.76 in 2019 (Fig. 3). It is hardly surprising that the objectives of these framework directives are constantly being pushed back. The European Environment Agency (EAA 2018) highlights the degradation of aquatic environments across Europe.

If the authorities at the various levels of management only implement a passive restoration policy, mainly by restricting fishing as is currently the case, it is not certain that significant restoration of the species will take place even on a very long-term scale.

It is for this reason that an active restoration policy using certain characteristics of the biology of this population needs to be put in place.

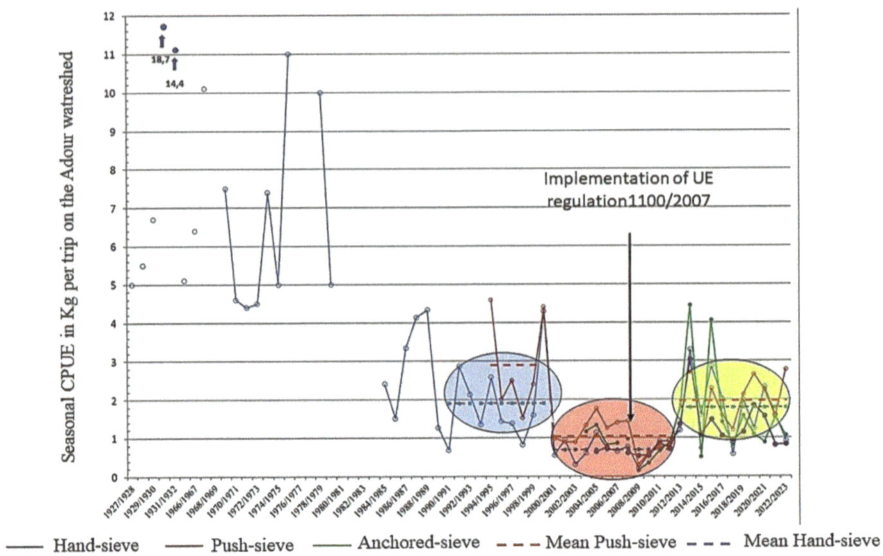

Fig. 2 Trend in the catch per unit of effort for the glass-eel fishery on the Adour river from Ifremer and CIDPMEM64-40 data. The effect of constraints on the sub-adult eel fishery from 2009 can be felt 3 years later, i.e. from the 2012–2013 season

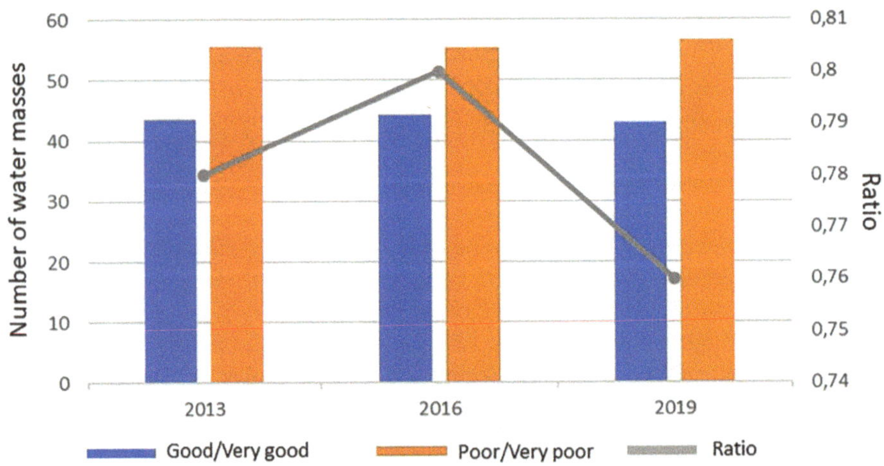

Fig. 3 Ecological status of surface waters with regard to WFD objectives in mainland France and the French overseas departments, according to data from the ministry for ecological transition

3.2 Implementation of a Range-Wide Active Restoration Programme

The historical data on the importance and geographical locations of glass eel fisheries, indicate that they are mainly developed in the central part of the distribution area centred on the Bay of Biscay. Of course, historically there were glass eel fisheries in the Mediterranean on large rivers such as the Ebro in Spain, the Tiber in Italy or in lagoons such as Albufera in Spain (Ciccotti and Morello 2023), but they were not as important as the fisheries developed on the major French rivers such as the Loire, the Gironde or the Adour (Prouzet 'Coord' 2003; Léauté and Caill-Milly 2003) located in the Bay of Biscay. In northern Europe, whether in the North Sea or the Baltic, no glass eel fisheries have developed. This location of the main fisheries in the centre of the range corresponds to a biological fact of this population, which is the arrival of a major flow of glass eels in the central part. As shown in Fig. 2, this is where the significant increase in estuarine recruitment linked to the constraints on fishing imposed by EU regulation 1100/2007 is the most apparent.

For all fish species, and even more so for diadromous fish species, the population can only develop if the surface area of available habitats is not too small and of sufficient quality to ensure high productivity. As mentioned in the national reports submitted to the ICES Eel Working Group (Eel working group national report 2023), the building of large dams for hydroelectric power generation from the early twentieth century in the north and centre of the range, then from the mid-twentieth century in the south (Mediterranean Sea) for water storage, has led to a fragmentation of the functional habitats colonized by the species. For this reason, many countries have used glass eel and small eel transfers to support inland eel fisheries since the beginning of the twentieth century (Le Clerc 1930). At the same time, water quality has declined, especially in the lower reaches of watersheds where young eels are produced (Prouzet et al. 2019b).

Hence the need to use a small part of the flow of glass eel received from the most irrigated central zone to supply glass eels to these peripheral areas, which are no longer colonized or only slightly colonized, but still of good quality, in order to boost the production of yellow and silver eels, whose production could be limited as a result of density-dependency factors in the central zone, whose habitats are severely degraded and limited.

3.3 A Responsible and Involved Profession at the Core of the Active Restoration Programme

Commercial eel fishing is carried out by two fishing communities with different social status. A sea fishers' community whose social status is that of a professional fisherman and a inland fishers' community whose social status is similar to that of a farmer. These two communities generally fish in different areas:

Table 1 Number of sea and inland professional fishers fishing for eel in 2023 (from GFCM report 2023 and CNPMEM and CONAPPED data). A: Atlantic area; M: Mediterranean sea

		Glass eel	Yellow and silver eel
Sea fishers	A	402	197
	M	0	221
Inland fishers	A	96	120
	M	0	3
Total		498	541

further downstream in catchment basins or on the coast for the marine fishermen, further upstream in freshwater for the inland fishermen, but they may share fishing areas (particularly for glass eel fishing). They belong to two separate national bodies: CNPMEM (National Committee for marine fisheries and sea farming) and CONAPPED (National Committee for inland professional fisheries) (Table 1).

They combine their efforts and thoughts within dedicated structures such as the CMEA (Commission pour le Milieu Estuarien et les Poissons Amphihalins) and ARA France (Association pour le Repeuplement Anguille en France) and participate within the COGEPOMIs (Comité de Gestion des Poissons Migrateurs) in the management of migratory fish (including eels) at river basin level with other users (notably leisure fishing).

The number of professional fishermen exploiting eels has been falling regularly since the middle of the twentieth century, given the numerous constraints linked to the degradation of aquatic environments and the increasing scarcity of fish resources. A sharp reduction has been observed since the implementation of the European Eel Regulation Plan 1100/2007, which has led to a marked reduction in fishing effort in a large number of fishing areas in continental and estuarine waters. This is particularly the case for glass eel fishing, one of the flagship economic activities of small-scale estuarine and inland fisheries (Fig. 4).

Faced with the degradation of aquatic environments and the use of fishing effort as a variable for adjusting to the scarcity of resources, professional fishing communities have strong doubts about the sustainability of their activities, which have not been taken into account in the negotiations for the implementation of a sustainable development policy (WCED 1987). That policy, in theory, is at the convergence of three spheres of equal importance: social, economic and environmental. In practice, the negotiations among actors favour activities that do not depend directly on the biodiversity and quality of aquatic environments: hydroelectric production, irrigation, drinking water supply, purification, protection against flooding, ….

Hence, the definition in recent years of so-called responsible fishing, based on the principle of responsibility (Jonas 1979; Prouzet et al. 2019b, 2023) the objective of which is to minimize the ecological footprint of the fishing activity (Anonymous 2020).

Responsible fishing aims to:

- Provide as many people as possible with high-quality local fish, through low-carbon footprint operations at an affordable price;

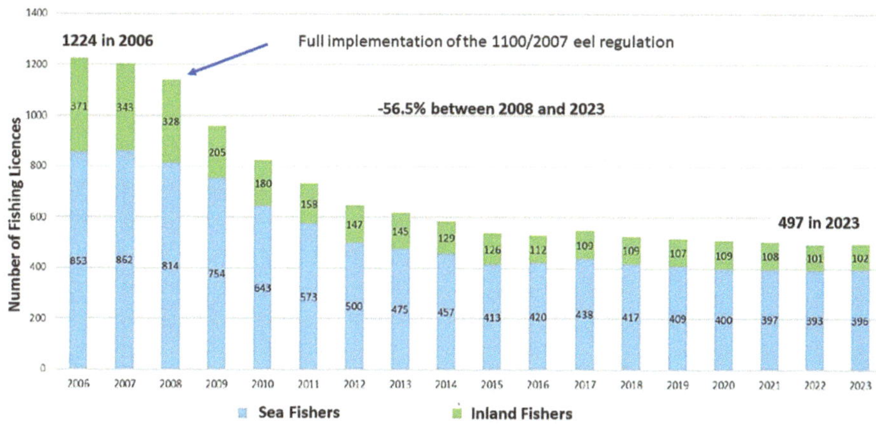

Fig. 4 Trend in the number of licences delivered for professional glass eel fishing. There was a sharp fall from 2008 to 2012, followed by a slower decline thereafter, according to CONAPPED and CNPMEM data

- Monitor the environment and act as whistleblowers to highlight problems caused by an overly sector-based approach to environmental management;
- Contribute to the management and restoration of fisheries resources and facilitate connectivity between key habitats for exploited migratory species.

The involvement of these fishing communities in active eel population restoration projects and their actions to alert managers to the degradation of habitats for diadromous species can be compared to Sato-umi and Sato-yama operations.

However, this approach is still poorly facilitated by local, regional and national administrations, which prefer a more centralized approach that is often based on criteria such as the species' listing in Appendix II of CITES, its inclusion on the IUCN red list and the often-unfounded application of the Precautionary principle (Gornet et al. 2020).

3.4 A Few Specific Examples Relating to Eels

3.4.1 Supplying the Eel Farming and Restocking Industry at European Level: Improving of Fishing Practices

France is the main producer of glass eels in Europe (90% of the European production since the Brexit). Fishing for this fry is authorised on France's Atlantic coast, but prohibited on the French mediterranean coast, where eels are mainly caught in the yellow and silver eel stages in mediterranean lagoons.

Photo 1 Push-sieve used in the Loire estuary and showing the reservoir at the bottom of the sieve which prevents the glass eels caught from being pressed against the net when the boat is moving. (copyright CONAPPED)

The great part of the landings is exported for direct consumption (in Spain mainly) or for rearing and restocking, which is particularly well developed in northern European countries (Netherlands, Germany, Denmark or Poland) and France (mainly for restocking).

In order to minimise mortalities after fishing and provide high-quality live glass eels for rearing and restocking, a guide to good practices for glass eel fishing and restocking has been published by CONAPPED, WWF France, CNPMEM and ARA France.[1] When they obtain their licence, each fisherman signs a charter of good practices and is required to comply with certain fishing conditions (boat speed, duration of trawling, mesh of the sieve) for the push-sieve and animal storage conditions. This has helped to improve the quality of glass eels caught by the push sieve fishery, and has resulted in a significant decrease of the mortality after fishing (Photo 1).

3.4.2 Environmental Watch and Whistleblowing

Professional fishing in continental and marine waters is a highly controlled activity, with licences, quotas and fishing seasons of increasingly restricted duration, as the management authorities see fishing as a variable for adjusting to the decline in fish resources. This is the case for eel fishing, where the number of fishers has fallen

[1] GBPCivelle-VF.pdf (repeuplementanguille.fr).

sharply in certain areas such as the Rhone. This is a general trend (see Fig. 4) and the environmental monitoring that these communities were able to carry out no longer has the coverage that it might have had a few decades ago.

Over the last few decades, professional fishermen have provided precise observations on the phenology of this species: rhythms and periods of downstream migration of silver eels; rhythms of upstream migration of glass eels and behaviour in relation to hydro-climatic characteristics (European INDICANG programme Adam et al. 2008; Boisneau et al. 2016; Prouzet et al. 2019a); participation in eel tagging in the Atlantic and Mediterranean to study eel migration routes in the North Atlantic (European EELIAD programme, Righton 2008, 2012; Prouzet et al. 2019a). They are also taking part in work on the impact of wels catfish populations on the mortality of migratory fish (including eels) (Trancart et al. 2023).

On a more general level, they intervene with non-governmental organisations at international events such as Fish Migration Day to denounce the implementation of overly sectoral governance leading to the endangerment of migratory fish populations, including eel (Gornet el al. 2020), or at the IUCN World Conservation Congress in 2021 with the publication of a leaflet: "Decline of highly migratory fish: Stop using commercial fishing as a scapegoat!—Reconciling sustainable development, environmental governance and responsible fishing". (Anonymous 2020).[2]

3.4.3 Contribute to the Eel Stock Management

Professional sea and inland fishing is subject to declaration of catches. This is particularly the case for glass eels, the eel biological stage for which fishing is strictly regulated and monitored.

Every sea or inland professional fisher must, as soon as he lands, fill in a fishing form giving information on the fishing area, the fishing gear, the weight of the catch and the use of the product caught (consumption or restocking). Photo 2 shows one type of form used by inland fishers. The fishing form used by sea fishers is different but provides the same pieces of information.

These fishing forms collect valuable information not only on catches, but also on fishing practices: gear used, area. They provide indicators such as CPUE for studying trends in estuarine recruitment (see Fig. 2), which are much more precise than those used by ICES expert groups. For reasons that have not been explained, the French government does not communicate this information and prefers to rely on data series whose representativeness is now more than questionable.

This is particularly true concerning the trend in the glass eel recruitment indicator at European level, which has been used in recent years by the ICES and which is based on data series that currently take very little account of the information gathered in the central part of the European eel's distribution area, the zone most irrigated by the flows of leptocephalus and then glass eels from the Sargasso Sea. This is more than damaging when we want to detect the first signal of an increase in estuarine

[2] LIvret-UICN-CONAPPED-2021-VF.pdf (lepecheurprofessionnel.fr).

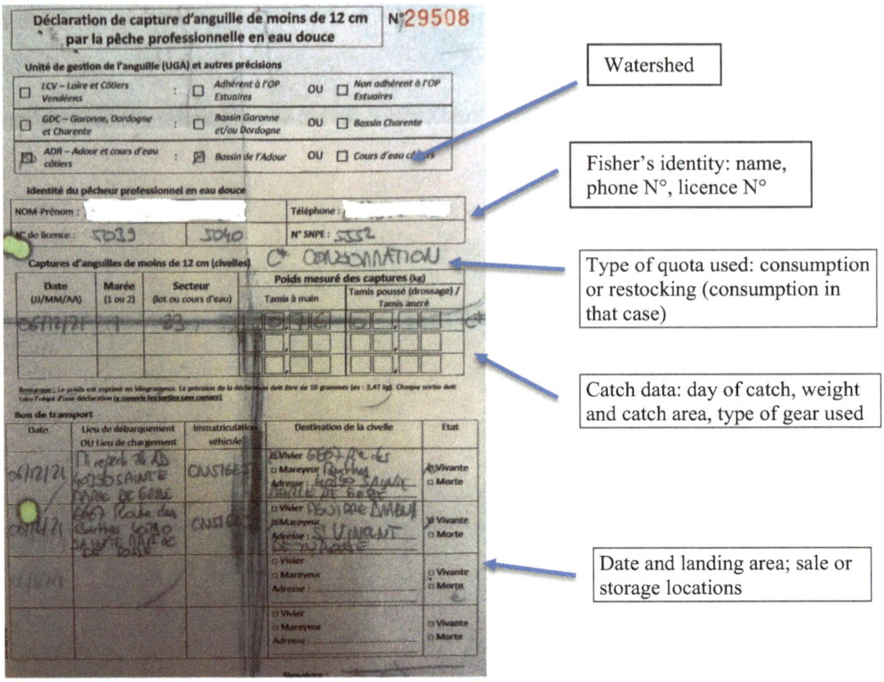

Photo 2 Example of fishing form fill in by an inland fisher on the Adour River

recruitment, which we know will occur first in this central part of the European eel's range.

Figure 5 adapted from the WGEEL 2021 (ICES 2021) shows the locations where the data used to plot the trend in glass eel recruitment on European Atlantic coasts, are gathered.

This figure shows that most of the data series collected in the Bay of Biscay on the French coast, the area with the highest glass-eel production in Europe, were stopped in 2008 on the three major French rivers along the Atlantic coast: the Adour River, the furthest south, the Gironde estuary (where only one data collection obtained during scientific monitoring at the Blayais nuclear power station remains) and the Loire River.

Without these series, how can be used the trend curve to correctly assess the effect of the restrictions on eel fishing in Europe, which are expected to be first seen in this central area of the European eel range from 2013 onwards? What is even more damaging is the fact that the data in these series is accurately gathered by professional fishermen, but not used by ICES experts. The consequences of this failure to take account of the data carefully collected by professional fishermen are significant in economic and social terms. DG MARE, taking the view from the analysis of the ICES and GFCM groups of experts that the efforts made by the fishing industry have not been sufficient, has further hardened its position by increasing the constraints on

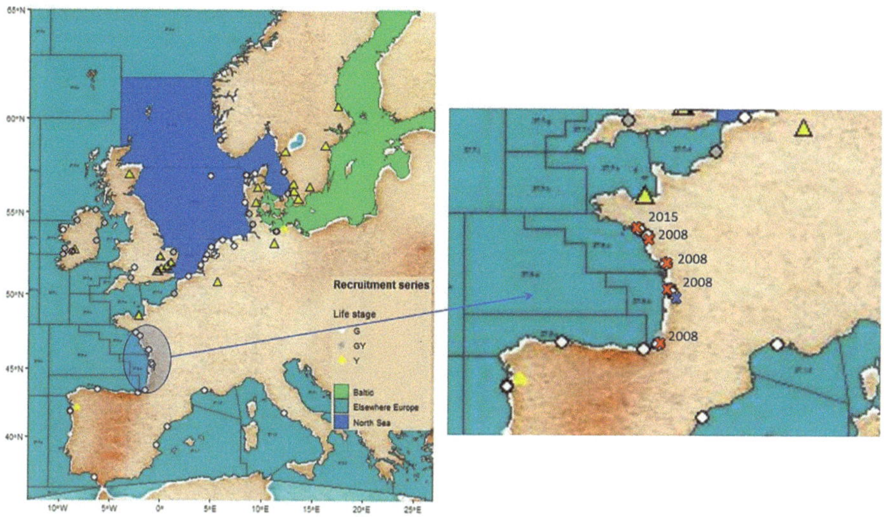

Fig. 5 Map showing the locations of observations made to assess the trend in estuarine glass eel recruitment. An enlarged view of the Bay of Biscay, the main site of glass eel arrivals on the Atlantic coast, is shown on the right. The red crosses indicate ceased monitoring, indexed by the year in which monitoring ceased, as communicated in the WGEEL 2021 database. (WGEEL report 2021—Annex 8—recruitment series)

this activity, despite the fact that the data from the fishing statistics as recorded by the professional fishermen show a more than significant increase in glass eels in the central zone of the colonisation area (see Fig. 2).

Another very valuable piece of information has been developed on the Loire with the use of anchored stow nets call "guideau" to estimate the abundance of silver eels migrating down one of the major French rivers along the Atlantic coast in using the mark-recapture method (Boisneau and Boisneau 2014, 2015).

That estimation method abandoned in the Loire due to a lack of funding, is now being proposed for the Rhône (Stohlzenberg 2023). The use of these anchored stow net fisheries also makes it possible to transfer populations upstream of a dam and release them downstream to limit mortality during downstream migration due to the passage of fish through turbines (see Sect. 3.4.4.).

3.4.4 Stocking and Transfer of Eel in Order to Restore Actively the Eel Stock

The transfer of glass eels from areas where they arrive in large numbers to areas on the northern and southern peripheral areas is not a new measure. It has been commonly used since the early twentieth century, for reasons that made managers fear (even

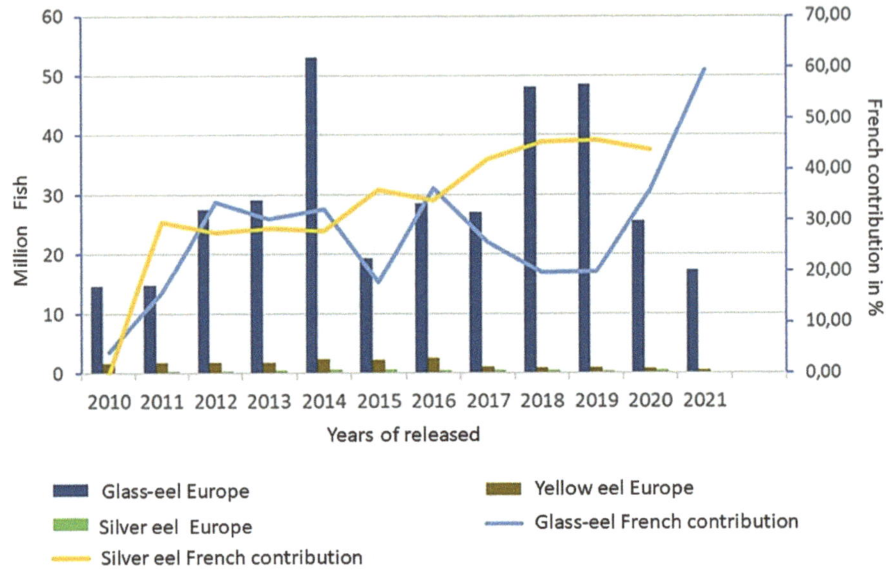

Fig. 6 Quantities of glass eels and yellow and silver eels released into European waters since the introduction of the eel regulation plan 1100/2007 (after WGEEL report 2022)

back then) that eel fry would be exploited too intensively.[3] On the Loire, one of France's great rivers, where eel fry arrived in abundance, the fishery first supplied the Spanish market for the end-of-year festivities, then the restocking operations at a time when glass eels were selling for almost nothing. By this time, Le Clerc (1930) had reported that France, the Netherlands, Germany, Poland and Italy were stocking several tens of millions of glass-eels a year.

Since then, many countries have continued to stock glass eels or young eels in various ecosystems, sometimes to support fisheries that would otherwise have been unable to sustain themselves (ICES 2023).

As part of the eel management plan for France as provided to the EU, a large-scale eel restocking and transfer programme both on the Atlantic coast and in the Mediterranean have been undertaken by professional sea and inland fishermen. France is the country devoting the most effort and investment to restocking in Europe (Fig. 6).

By 2021, 60% of the glass eels released in European waters will be in French waters under the control of ARA France.[4] In France, restocking operations have an annual budget of €2 million.

[3] At the same time, these same "Eaux et Forêts" officers were surprised that France had not developed silver eel fishing like Holland with "ankerkuils", as a result of "regulations that were not properly adapted to the economic interests of our country". Le Clerc (1930), L'anguille dans le basin de La Loire (suite et fin) Bull. Fr. Piscic., 19, 177–182.

[4] Association pour le repeuplement en anguille en France. (Repeuplement Anguille France I ARA France—Association Repeuplement Anguille France).

Photo 3 Glass-eels ready to be transferred to their release site. (copyright CONAPPED)

Since 2010, 14 tonnes of glass eels, around 45 million alevins, have been released into around ten rivers, from the Artois Picardie region to the Adour basin. This equates to 45 projects in 6 eel management units (UGAs) practising restocking (Photo 3).

To optimise the results of such restocking, an 8-stage specification has been defined[4]:

Stage 1: Choice, by the migratory fish management committee (COGEPOMI), of areas favourable to the species.
Stage 2: Selection of project promoters following the eel restocking call launched by the Directorate of Maritime Fisheries and Aquaculture in collaboration with the Directorate of Water and Biodiversity. The selection committee also includes the Office Français de la Biodiversité and the Muséum National d'Histoire Naturelle. Selection of fish mongers by the project promoter according to strict specifications (set out in the call for projects).
Stage 3: Collection of batches (over 5 to 10 days) by sea and inland fishermen.
Stage 4: Verification of the health quality of glass eels (lesions, parasites, viruses, etc.).
Stage 5: Marking the otoliths (bone parts), which will enable us to monitor the survival and growth of glass eels transferred.
Stage 6: Glass eels are packaged and transported by fishmongers according to methodology improved.

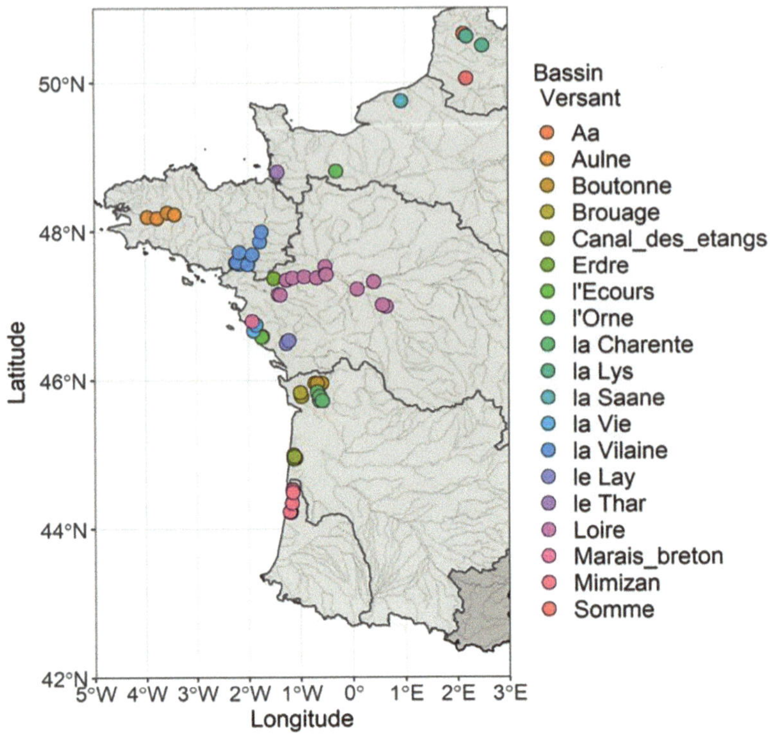

Fig. 7 Location of glass eel release sites from 2011 to 2021 (after Feunteun, Serranito et Le Peru 2023 and ARA France)

Stage 7: Once they have acclimatised to the receiving environment, glass eels are released on foot or by boat.
Stage 8: A strict scientific survey is carried out 6 months, 1 year and 3 years after stocking, using fyke nets, electro-fishing or flottang,[5] in order to assess the efficiency of stocking.

The first synthetic analysis (Feunteun, Serranito et Le Peru. 2023) carried out on the 70 restocking operations undertaken between 2011 and 2021 (Fig. 7) shows:

- The growth rate of transferred glass-eels increases with the distance to the sea and the number of obstacles downstream of the restocking site.
- The contribution of transferred glass-eels to the resident eel population is highly variable (between 10 and 90% of the population). Overall, it increased with distance from the sea and the number of obstacles downstream of the release point, demonstrating the effectiveness of restocking in certain parts of rivers where eels are absent or only marginally present.

---

[5] Superposition of 10 layers of 50 × 50 cm mesh, with buoys. Le FLOTTANG, quesako ? (observatoire-poissons-migrateurs-bretagne.fr).

Although this preliminary analysis does not provide any information on the survival of eels, it confirms that the density-dependency mechanisms widely described in the literature are at work within the population and shows that the release of glass eels in low-density areas upstream of rivers could ultimately lead to an increase in silver eel production.

The further aim of this work, ordered by the professional fishing organisations, is to estimate the quantity of spawners produced by the transfer of 28 t of glass eels (84 million individuals) between 2011 and 2021, following consensus in the scientific community on gains in survival, growth rate, sexual determination, size and age at silvering. According to the authors of that synthesis, the first silver eels resulting from restocking began to migrate down the Loire in 2018, and ongoing studies indicate that they could contribute between 5 and 10% of the migrating stock of silver eels. These first results confirm the observations made in other research carried out in Europe on eel transfers (Pedersen and Rasmussen 2015; Limburg et al. 2003; Lin et al. 2007; Rabic and Gornet, 2020).

In the Mediterranean Sea, to facilitate the downstream migration of silver eels, professional fishermen catch and then release fish caught in lagoons directly into the sea. 50% of silver eels transferred to sea in Europe are caught by French fishermen. Since the fishing season 2011–2012 about 300 tons of silver eels (296.41 t exactly[6]) have been released directly at the exit of the coastal lagoons in order to diminish the mortality during the downstream migration. This figure should be compared with catches of yellow and silver eels in French Mediterranean lagoons, which average 309 t of yellow eels and 81 t of silver eels per year over the period 2012–2020[5] (Photo 4).

4 Conclusion: Improve the Environment Whatever the Cost and Take Better Account of the Local Expertise of All Stakeholders

In November 2023, the rapporteur on the implementation of eel management plans under Regulation 1100/2007, Bert-Jan Ruyssen, MEP,[7] states that the objectives have been achieved in terms of reducing the ecological footprint associated with eel exploitation, but for the other uses is far from being reduced on the species' functional habitats.

It is clear that the conditions for setting up a Sato-yama and/or Sato-umi type initiatives for eel or for all the large diadromous species have not yet been met in the French politic and economic context, or even in the European context, as a result of the implementation of a sustainable development policy that does not respect the balance between the three spheres that make it up: environmental, social and economic. The environmental sphere has been neglected in France, and more widely

---

[6] Rapport de mise en œuvre du PGA—Bassin Rhône Méditerranée 2024 page 14.

[7] Member of European Parliament.

Photo 4 Release of silver eels in the Bouzigues Harbour—Hérault county in January 2023 (Copyright AFPMAR/Aguirrebarrena)

in Europe, in favour of uses that do not depend directly on the quality and productivity of aquatic environments (Prouzet et al. 2023).

The failure to take into account the local expertise of professional fishermen and a national and international expertise that does not use the best available data also result in problems, as the principle of prevention is not properly applied.

The existence of these fishing communities is important not only because of the quality products they generate, but also because of the knowledge and know-how they possess and pass on (Boisneau et al. 2016; Prouzet et al. 2019a, b). In a context of global change, characterised in particular by the fractioning of riverine, estuarine and coastal ecosystems, the transfer of glass-eel populations to optimise the use of habitats that are still functional for the species, either in the form of restocking or catch-and-release using non-traumatic and selective gears: stow nets or fyke nets for example, is a short-term solution that will make it possible to restore these populations on a local scale. This obviously involves that local traditional knowledge are taken into account alongside academic knowledge, which is the very basis of the "Sato-yama" and "Sato-umi" concepts for bottom-up rather than top-down management (Uehara et al. 2019; Toyoshima et al. 2022; Hsu and Chen 2023).

If a bottom-up management is implemented an "Eel Sato-umi" concept can be defined as shown by Fig. 8.

For diadromous species that use marine, transitional and continental ecosystems to complete their life cycle, the diversity of environments to be considered is greater than that described in the "Sato-umi" framework.

Therefore, it would be better to consider a joint implementation of a "Sato-umi" and "Sato-yama" experiment. For eel, this corresponds to a local approach taking into account not only the catchment area (Sato-yama) but also the coastal zone under

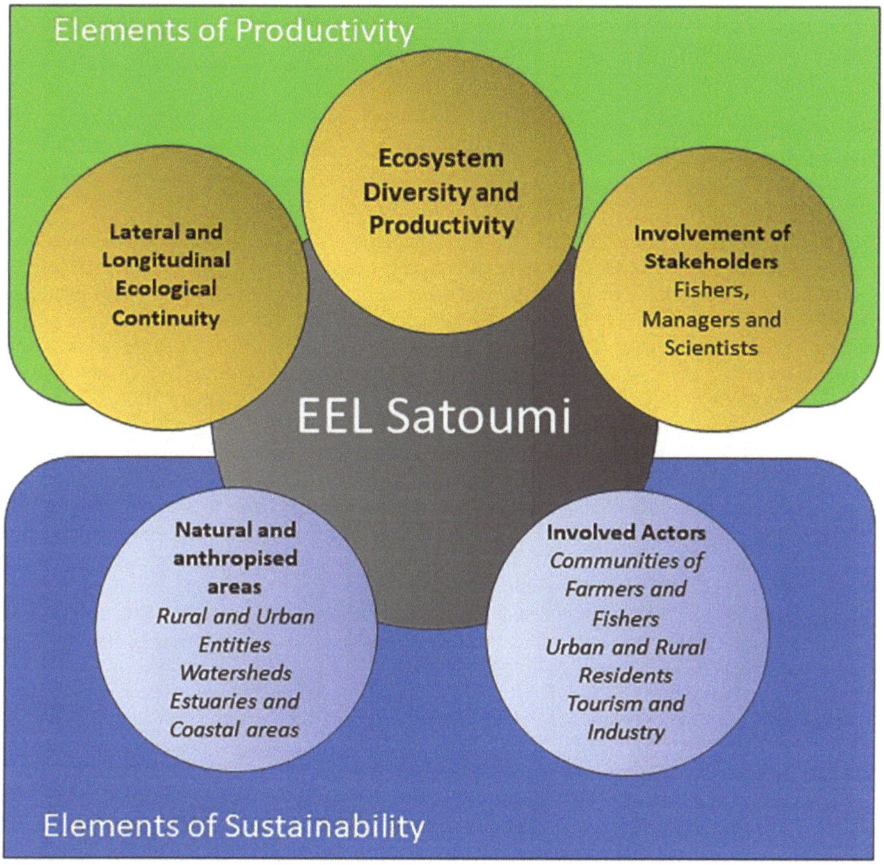

Fig. 8 The Eel Sato-umi concept adapted from What is Sato-umi? | satoumi.net [Ministry of the Environment] and Matsuda (2011)

the influence of the catchment area. This is particularly the case for Mediterranean lagoons, which are one of the main habitats occupied by this species.

Figure 8 shows two blocks called productivity elements (green) and sustainability elements (blue).

The sustainability elements depend on both the existing natural and man-made environments and the stakeholders involved in the use of these environments.

The productivity elements are made up of 3 sub-elements: ecological continuity, ecosystem diversity and productivity, and stakeholder involvement.

Given the actions taken by professional maritime and continental fishermen, such as minimising their ecological footprint, cooperating with administrative and scientific bodies to manage and restore this species, their environmental watch and whistleblowing, the implementation of Sato-umi/Sato-yama type initiatives for eels in Europe appears justified and certainly more effective than simple management centred on restricting or even eliminating fishing activity.

References

Adam G, Feunteun E, Prouzet P, Rigaud C (2008) L'anguille européenne—indicateurs d'abondance et de colonisation. Editions QUAE, collection Savoir-faire, p 393, ISBN 978-2-7592-0085-6 (in French) - see the English version at the following web address : https://www.researchgate.net/publication/266858339_The_European_Eel_abundance_and_colonization_indicators

Anonyme (2020) Decline of Diadromous species: stop using artisanal fishing as a scapegoat!—reconciling sustainable development, environmental governance and responsible fishing. Document CONAPPED, ARA France, SFJO, p 4. LIvret-UICN-CONAPPED-2021-VF.pdf (lepecheurprofessionnel.fr)

Baumert N (2010) Echo du Japon—Le Sato-yama à la conquête du monde. Géographie 2010/4 (No 1539). Editions Société de Géographie, pp 62–63. https://doi.org/10.3917/geo.1539.00Gr

Boisneau P, Boisneau C (2014) Long-term evolution of silver European eel indicator of abundance in the Loire watershed (1987–2013), a unique data set for Europe. In: Oral communication of the 144th American fisheries society (AFS). Annual meeting international eel symposium: are eels climbing back up the slippery slope? Aug 17–21, Quebec city, Canada

Boisneau C, Boisneau P (2015) Indice annuel d'abondance des anguilles d'avalaison du bassin de la Loire à partir des captures des pêcheurs professionnels au guideau. Rapport University. Tour, Plan Loire Grandeur Nature III, p 15 (in French)

Boisneau P, Stohlzenberg N, Prouzet P, Moreau D (2016) How to transmit information and maintain knowledge in the context of global change for French inland commercial fishers, pp 289–300. In: Taylor WW, Bartley DM, Goddard CI, Leonard NJ, Welcomme R (eds) Freshwater, fish and the future: proceedings of the global cross-sectoral conference, FAO of the United Nations, Rome; Michigan State University, East Lansing and American Fisheries Society, Bethesda, Maryland

Ciccotti E, Morello EB (eds) (2023) European eel in the Mediterranean Sea—outcomes of the GFCM research programme. Studies and reviews No. 103 (General fisheries commission for the Mediterranean). Rome, FAO. https://doi.org/10.4060/cc7252en

Drouineau H, Durif C, Castonguay M, Mateo M, Rochard E, Verreault G, Yokouchi K, Lambert P (2018) Freshwater eels: a symbol of the effects of global change. Fish and fisheries, Ed. Wiley, pp 1–28

EAA REPORT (2018) European waters. Assessment of status and pressures, 2018. EAA report 7/2018, p 90. https://www.eea.europa.eu/publications/state-of-water/at_download/file

Feunteun E, Prouzet P (2020) Forty years of decline and 10 years of management plan: are European eels (Anguilla Anguilla) recovering? In: Ceccaldi H-J et al (eds) Evolution of Marine coastal ecosystems under the pressure of global changes, pp 269–295. https://doi.org/10.1007/978-3-030-43484-7_31.

Feunteun E, Serranito B, Le Peru Y (2023) Etude ADRAF—analyse des données de 10 années de repeuplement anguille en France—2011/2021, p 48 (in French)

Gornet D, Lecomte M, Moreau D, Porcher N, Prouzet P, Stolzenberg N (2020) Poissons Migrateurs en Danger—Le cas de l'anguille européenne—La faillite d'une gouvernance environnementale trop sectorielle. Communication écrite world fish migration day. CNPMEM, CONAPPED et ARA France, p 27 (in French)

Hsu K, Chen J-L (2023) The coastal future we want: implications of coastal people's perceptions on Sato-umi actions and sustainable development in Northeastern Taiwan. Sutainability 15:7919. https://doi.org/10.3390/su15107919

ICES (2021) Joint EIFAAC/ICES/GFCM working group on eels (WGEEL). ICES scientific reports, vol 3, p 85, 205. https://doi.org/10.17895/ices.pub.8143

ICES (2022) Joint EIFAAC/ICES/GFCM working group on eels (WGEEL). ICES scientific reports. Report. https://doi.org/10.17895/ices.pub.20418840.v1
ICES (2023) Eel working group national report, p 358
Jonas H (1979) Das Prinzip Veranwortung. Versucheiner Ethik für die technologische Zivilisation, Francfort/Main, 1979 (trad. fr., J. Greisch), Le Principe de responsabilité. Essai d'une éthique pour la civilisation technologique, Paris, Le Cerf, 1990 (in French)
Komatsu T, Aoki N (2020) Japanese-type MPAs and self-organized MPAs by local communities in Japan. In: Ceccaldi H-J et al (eds) Evolution of Marine coastal ecosystems under the pressure of global changes, pp 457–477. https://doi.org/10.1007/978-3-030-43484-7_31
Komatsu T, Yanagi T (2015) Sato-umi: an integrated approach for sustainable use of coastal waters, lessons from human-nature interactions during the Edo-period of eighteenth-century Japan. In: Ceccaldi J-H, Hénocque Y, Koike Y, Komatsu T, Stora G, Tusseau-Vuillemin M-H (eds) Marine productivity: perturbations and resilience of socio-ecosystems. Springer, pp 283–290
Le Clerc J (1930) L'anguille dans le bassin de La Loire. Bull Fr Piscic 19:145–152
Léauté J-P, Caill-Milly N (2003) Caractéristiques des petites pêches côtières et estuariennes de la côte atlantique du sud de l'Europe. Synthèse du Contrat PECOSUDE. Ed Ifremer, Coll. Bilans and Prospectives, p 66 (in French)
Limburg KE, Wickström H, Svedäng H, Elfman M, Kristiansson P (2003) Do stocked freshwater eels migrate? Evidence from the Baltic suggests Yes. Am Fish Soc Symp 33:275–284
Lin Y-J, Lozys L, Shiao JC, Iizuka Y, Tzeng W-N (2007) Growth differences between naturally recruited and stocked European eel, *Anguilla Anguilla,* from different habitats in Lithuania. J Fish Biol 71:1773–1787
Machut LS, Limburg KE, Schmidt RE, Dittman D (2007) Anthropogenic impacts on American eel. Demographics in Hudson River tributaries. Trans Am Fish Soc 136:1699–1713
Matsuda O, Yanagi T (2019) Restoration of estuaries and bays in Japan—what's been done so far, and future perspectives. In: Wolanski E, Day JW, Elliott M, Ramachandran R (eds) Coasts and estuaries. Elsevier, Amsterdam, The Netherlands, pp 401–411
Matsuda O (2011) Evolution of Sato-umi frame during the recent international meetings in the context of sustainable coastal management. Oral presentation, EMECSS9 Sato-umi session—29 August 2011, Baltimore USA.
Matsuda O (2019) New direction of management policies in the Seto Inland Sea, Japan, in a changing environment. Bull Jap Fish Res Edu Agen 49:9–15
Pedersen MI, Rasmussen GH (2015) Yield per recruit from stocking two different sizes of eel (*Anguilla anguilla*) in the brackish Roskilde Fjord. ICES J Mar Sci. https://doi.org/10.1093/icesjms/fsv167
Prouzet P (Coord.) (2003) Historique des captures de civelles, intensité actuelle de leur exploitation, variation de leur capturabilité par la pêche maritime et indices de colonisation sur le bassin versant de l'Adour. Contrat EC/DG FISH (DG XIV) No 99/023, p 147 + annexes
Prouzet P, Sicot S, Véronneau F (2010) International meeting of small-scale professional sea and Inshore fishers: the official records. Edition CONAPPED, p 128
Prouzet P, Lespine J et Sébastien L, (2009) L'Adour - Un art de vivre et la transmisson des savoirs. Editeur A. Dewez, ISBN-13 : 9782952121514
Prouzet P, Amilhat E, Boisneau C, Boisneau P, Feunteun E, Michelet N (2019a) The European eel (*Anguilla anguilla*) in France: an example of close cooperation among researchers and fishers to study and manage and endangered species. Chapter 7. In: Komatsu T, Ceccaldi H-J, Yoshida J, Prouzet P (eds) Oceanography challenges to future earth: human and natural impacts on our seas. Springer, pp 69–93
Prouzet P, Monaco A, Sebastien L (2019b) Management of the Marine environment: transdisciplinary and systemic approach. In: Maurice PA (ed) Encyclopedia of water: science, technology and society. Wiley and sons, p 27
Prouzet P, Lecomte M, Herfaut J, Mas L, Porcher N (2023) Sustainable development and responsible exploitation. As an example: the management and exploitation of diadromous species in the

context of small-scale fisheries. La mer, p 61. https://doi.org/10.32211/lamer.61.1-2. Société franco-japonaise d'Océanographie, Tokyo
Rabic J, Gornet D (2020) Eel restocking in France: an emergency measure to save the European eel. In: Ceccaldi H-J et al (eds) Evolution of Marine coastal ecosystems under the pressure of global changes, pp 381–391. https://doi.org/10.1007/978-3-030-43484-7_31
Righton D (2008) Life in the big blue box: studying the marine life of European eels. Poster ICES CM 2008/P:18
Righton D (coord) (2012) EELIAD: European eels in the Atlantic: assessment of their decline. Project final report—grant agreement number 212133—UE FP7 program, p 40
Stolzenberg N (2023) Faisabilité de suivi de la dévalaison d'anguilles au guideau dans le Rhône. Présentation powerpoint au CTPOMI du 24/05/2023 (in French)
Strong K (2015) Un bœuf dans la tempête. Biographie de Tanaka Shozo, écologiste japonais. Traduction de Lucie Blanchard. Edition wild project. Fondation Franco-japonaise Sasakawa. ISBN 978-2-918-490-432
Takeushi K, Brown RD, Washitani I, Tsunekawa A, Yokohari M (eds) (2003) Sato-yama: the traditional rural landscape of Japan. Springer. ISBN 4-431-00007-0
Tanaka T (2014) Thinking sustainable circular regional society: through Sato-umi of eelgrass and oyster in Hinase, Okayama. EPCR 34(1):21–26
Tanaka T, Furukawa K (2020) Prospects for practical Sato-umi implementation for sustainable development goals: lessons learnt from the Seto Inland Sea, Japan. In: Ceccaldi H-J et al (eds) Evolution of Marine coastal ecosystems under the pressure of global changes, pp 353–364. https://doi.org/10.1007/978-3-030-43484-7_31
Tesch F-W (1977) The eel, biology and management of Anguillid eels. Chapman and Hall, London, p 434
Toyoshima J, Fujii I, Maekawa M, Tsunoda T, Kamada N, Hidaka H, Tojo Y, Ikeda K (2022) Assessing effectiveness of Sato-umi activities in Japanese coastal areas from ecological and socioeconomic perspectives. Ocean Coastal Manag 230:106354, 10
Trancart T, Robin E, Acou A et al (2023) GLANISPOMI: Etude globale de la prédation des migrateurs amphihalins par les silures (*Silurus glanis*) sur le bassin de la Loire. Trancart T, Robin E, Feunteun E (eds), p 287
Tsukamoto K, Kuroki M, Watanabe S (2020) Common names for all species and subspecies of the genus *Anguilla*. Environ Biol Fishes 103:985–991. https://doi.org/10.1007/s10641-020-009 88-3
Uehara T, Hidaka T, Matsuda O, Sakurai R, Yanagi T, Yoshioka T (2019) Sato-umi: re-connecting people to nature for sustainable and conservation of coastal zones. People Nat 2019(1):435–441
World Commission on Environment and Development (1987) Our common future. Report of the world commission on environment and development. Oxford University Press, Oxford, p 300
Yamamoto Y, Shinoda M, Miwa N (2008) Creation of Sato-umi as a policy in Japan. In: Worshop report Sato-umi, new concept that increases biological productivity and biodiversity. International EMECS Center Japan, pp 21–25
Yanagi T (2007) Sato-umi: a new concept for coastal management. Terra Scientific Publishing Company, Tokyo, p 94
Yanagi T (2008) Definition of Sato-umi. In: Workshop report: new concept that increases biological productivity and biodiversity. International EMECS center Japan
Yanagi T (2012) Japanese commons in the coastal seas: how the Sato-umi concept harmonizes human activity in coastal seas with high productivity and diversity. Tokyo, Japan, Editor Springer
Yuan Y, Yuan Y, Dai Y, Gong Y, Yuan Y (2022) Development status and trends in the eel farming industry in Asia. N Am J Aquac 84:3–17

Positive Effects of Seagrass Beds on Sanitary Conditions in Oyster Aquaculture

Masakazu Hori, Masami Hamaguchi, Marion Richard, Valérie Derolez, and Franck Lagarde

1 Introduction

1.1 Oligotrophication in Marine Coastal Areas

The effects of climate change have become apparent in many regions of the world, causing serious environmental changes in coastal marine ecosystems. There are concerns that sustainable seafood production in coastal areas will be endangered because of rapid increases in sea water temperature, known as marine heat waves, as well as a decrease in nutrient supply (Aoki et al. 2022). Proposals for immediate countermeasures are required in various regions.

In addition to the changes in coastal environments due to climate change, closed marine coastal areas have simultaneously been subjected to anthropogenic environmental change reflecting eutrophication over the years. Recent coastal management actions, which were mostly regulation of nutrient loading from land to marine coastal areas, have gradually and successfully decreased the nutrient contents of coastal waters in some regions (Collos et al. 2009; Yanagi 2015; Derolez et al. 2020). However, some stakeholders suggest that the improvement in water quality is now causing another issue in coastal ecosystem services called "oligotrophication". Oligotrophication could reduce pelagic productivity in coastal ecosystems, sometimes

M. Hori (✉)
Japan Fisheries Research and Education Agency, Yokohama, Japan
e-mail: hori_masakazu70@fra.go.jp

M. Hamaguchi
Fukui Prefectural University, Obama, Japan

M. Richard · V. Derolez · F. Lagarde
MARBEC, UM, CNRS, IFREMER, IRD, Sete, France

A.-C. Bennis et al. (eds.), *Constraints and Adaptations to Global Change at the Land-Sea Interface: For a Shared Ecological and Energy Transition*,
https://doi.org/10.1007/978-3-031-90050-1_22

resulting in the decrease of fishery catch because coastal fisheries have adapted to the previous eutrophic environment from the 1970s to the 2000s.

In particular, bivalve aquaculture that uses natural primary productivity is highly dependent on ecosystem condition. Therefore, although oyster farming is an important food production method being developed in coastal areas around the world, oligotrophication has recently raised concern about oyster production among some shellfish farmers (Bec et al. 2018; Derolez et al. 2019). Recent climate change has caused several negative effects on oyster aquaculture, such as poor spat collection due to oligotrophication, ocean acidification, and poor spat growth and survival due to frequent high seawater temperatures and anoxic events (Hori et al. 2018; Lagarde et al. 2018, 2020; Correia-Martins et al. 2022).

In contrast, oligotrophication with high water transparency has helped the recovery of benthic primary producers, including seagrass beds, which is important for climate change adaptation (i.e. limitation of anoxia and acidification) and mitigation such as sequestration and storage of blue carbon derived from atmospheric CO_2, and also as protection from sea-level rise and storm surges (Hori et al. 2018; Tsurita et al. 2017). Seagrass beds also function to improve water quality, including sanitary conditions (Larkum et al. 2006; Duarte et al. 2013; Groner et al. 2018). The recovery of these ecosystem services has been warmly welcomed by other stakeholders concerned with environmental issues.

1.2 Collaborative Research in the French Mediterranean Sea and the Seto Inland Sea, Japan

Because these environmental changes are now apparent in both the French Mediterranean Sea and the Seto Inland Sea, Japan, a project titled "Challenges toward sustainable oyster aquaculture and seagrass management under oligotrophication with climate change" has been promoted using financial support from a French–Japanese SAKURA-Hubert Curien project funded by the Japan Society for the Promotion of Science and Campus France (French Embassy in Japan), which was a collaboration between MARBEC (IFREMER with Montpellier University, CNRS, and IRD) and the Japan Fisheries Education and Research Agency (FRA).

The French–Japanese SAKURA project was developed to better understand the links between conservation and exploitation of shellfish-exploited-ecosystems (SEE) and their impacts on human well-being by developing an interdisciplinary approach with ecological and social dimensions (Fig. 1). The project aimed to (i) clarify and compare relationships between the dynamics of nutrient levels, phytoplankton and Pacific oyster (*Magallana gigas*) production in Thau lagoon (France) and Hiroshima and Aki Bays, Japan, using historical data analysis and carrying capacity models, (ii) highlight the contribution of eelgrass (*Zostera marina*) to the oyster life cycles by studying the variability of larval recruitment, survival, growth of juveniles and the trophic regime of oysters in the presence or absence of eelgrass meadows, and

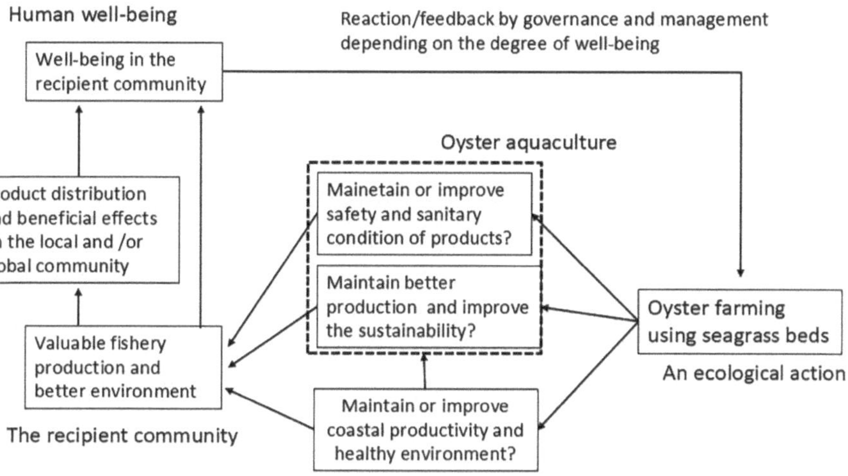

Fig. 1 Schematics of coastal integrated management that combines oyster farming and seagrass beds, modified from Hori et al. (2018)

(iii) describe and compare the dynamics of the socio-ecosystems of SEE under oligotrophication (Lagarde et al. 2020).

We have already performed research on coastal integrated management that combines oyster farming, an important fishery using pelagic primary production, and seagrass beds, which consist of effective ecosystem foundation species. The French Japanese project in the Seto Inland Sea/Hiroshima Bay and the Mediterranean Sea/ Thau lagoon aimed to (i) evaluate oyster farming methods using seagrass beds that mitigate the effects of oligotrophication and climate change on the ambient environment, oysters, and the local communities, (ii) clarify the feeding environment of oyster larvae and establish natural spat collection, especially in Thau lagoon (Lagarde et al. 2020), and (iii) understand the effects of changes in environmental microflora and phytoplankton under oligotrophication and marine heat waves (Richard et al. 2022).

In this study, we report the results of our eDNA analyses for the abundance of bacteria, including *Vibrio* spp., which was a factor in the mass mortality of oyster spat, in cultivated oysters and in the ambient seawater in Thau lagoon, France. Our analyses showed that the number of total bacteria and *Vibrio* spp. both inside shell and in the ambient waters were substantially lower at the sites with seagrass beds than at the sites without seagrass beds. We first compared the number of total bacteria and *Vibrio* spp. in the environmental water between the sites with and without seagrass beds in the summer before deploying oyster spat. We then deployed and cultivated spat until the winter. At the end of the cultivation period, oysters and the ambient water at both sites were sampled again and analyzed for the number of total bacteria and Vibrio spp.

2 Methods

2.1 Study Site

Thau lagoon (center coordinates: lat 43.41N, long 3.6241E) is the largest lagoon along the southern French coast in the Mediterranean Sea. The lagoon is famous for oyster farming using non-native Pacific oyster spat attached to ropes by a cement specifically for the spat. The ropes with spat are hung on oyster tables in the nearshore zone. About 10% of the French national production of oysters is cultivated there; it is the largest oyster farming area in the Mediterranean Sea. It has been suggested that the eelgrass beds in the lagoon appear to be recovering, and now the eelgrass cover up to 800 ha (Syndicat mixte du bassin de Thau, personal communication). The expansion of eelgrass beds was observed even within oyster farming areas in June 2016 (Fig. 2).

We established two study sites in the oyster farming area. Site A was in the Bouzigues production area, and Site B was in the production area between Meze and Marseillan. Seagrass coverage was different at Sites A and B, with dense coverage (more than 70%) and sparse coverage (less than 30%), respectively (Hori et al. 2021).

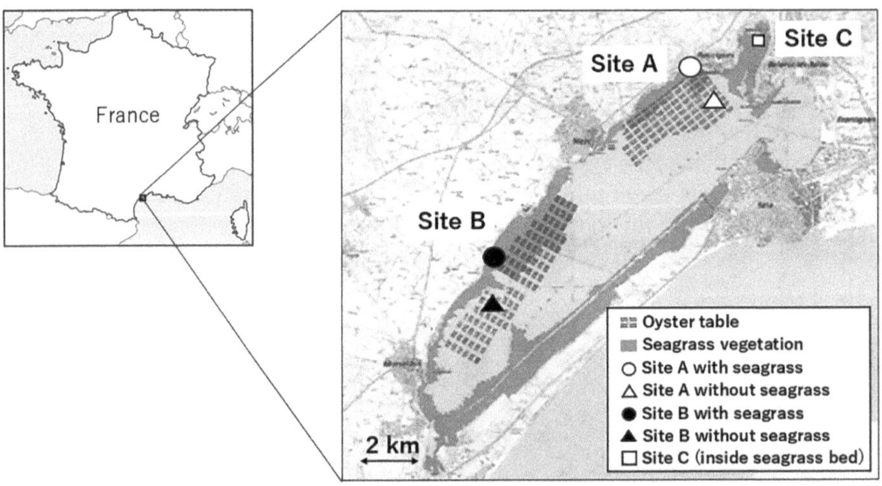

Fig. 2 Study sites in Thau lagoon, France. The coverage of seagrass was dense (more than 75%) around Site A and sparce (approximately 25%) around Site B. Site C was located in a small inlet, the bottom surface of which was completely covered with seagrass vegetation (more than 75%). This map of seagrass distribution was supplied by the Syndicat mixte du bassin de Thau in 2016

2.2 Field Deployment of Natural Spat

DNA analyses in this study were conducted using oysters collected at the end of the field experiment reported by Hori et al. (2021). For the field experiment, natural oyster spat collected from Thau lagoon in the summer of 2018 were deployed at Site A and B (see Fig. 2) (Lagarde et al. 2018). Each site consists of an area with seagrass beds and an area without. All oyster spat were mixed in a container and then separated into mesh bags (approximately 250 spat per bag). At each site, three mesh bags were hung on an oyster table at a depth above the seagrass canopy and three more inside the seagrass canopy in areas with seagrass beds (Fig. 3). Three mesh bags were also hung on an oyster table at the same two depths in the areas without seagrass beds. The oyster spat in mesh bags were cultivated for three months until December 2018. At the end of cultivation, the oysters in each mesh bag were measured for growth and survival.

Before the beginning of spat cultivation on 6 August 2018, Thau lagoon experienced hypoxic conditions. We used field probes to measure surface and bottom seawater temperature (°C), salinity (practical salinity scale), and dissolved oxygen concentration (DO, mg/L) in both sites. DO values allowed us to clarify any differences in DO conditions during anoxia between the area with and without seagrass beds.

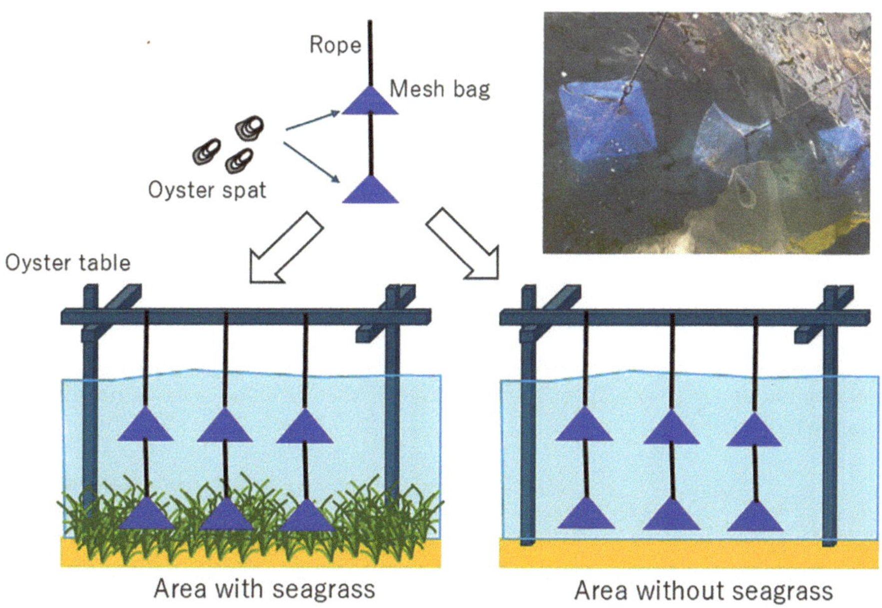

Fig. 3 Deployment of oyster spats at each site. At each site, mesh bags were set up at two depths: above and within the seagrass canopy

2.3 eDNA Sample Collection and DNA Extraction

To compare the bacterial flora in the water columns of the areas with and without seagrass beds, we collected 1000 mL of surface and bottom seawater at each study site using a Niskin water sampler at the beginning and the end of spat cultivation. Each 1000-mL seawater sample was filtered in the field using 100-mL syringes and Sterivex filter cartridges (pore size, 0.22 or 0.45 μm; Merck Millipore). The filter cartridges with samples were immediately preserved with RNA later solution and capped at both ends.

At the end of spat cultivation, surviving oysters were collected and brought back to the laboratory. First, the shell surfaces of all juvenile oysters from each area were rinsed with 1000 mL of Milli-Q water, and the rinse water was filtered using the same method as for ambient water. Second, water samples were collected from inside the shells (in-shell water samples, ISWS) of eight juvenile oysters from each mesh bag: those inside or above the seagrass canopy, or at the same depths, in areas with or without seagrass beds, at Sites A or B. We shucked each oyster using an oyster knife and used a sterile pipette tip immediately to collect the water sample. The knife was sterilized before each use. The ISWSs were also filtered using the same method as for the rinse water using Sterivex filter cartridges (pore size, 0.22 μm). All Sterivex filter cartridges were kept in a freezer until analysis. These methods follow those described in detail by Stoeckle et al. (2017).

The volume of ISWSs were small, ranging from 0.3 to 2.0 mL, but we still used Sterivex filter cartridges as for the seawater samples to provide the same conditions for quantitative counts of viruses and bacteria by quantitative PCR (qPCR). The volume of ISWS collected was recorded before being filtered through a Sterivex filter cartridges. Subsequent steps were the same as for the seawater samples. Disposable latex gloves were worn during field sampling and laboratory work. Disposable latex gloves were replaced with new ones at each sampling location and for each sample. The plastic tubes and pipette tips used in all experiments were of DNA low-bind type.

Sterivex cartridge filters holding samples were stored at −30 °C until DNA extraction. DNA was extracted from the Sterivex cartridge filters by the method of Miya et al. (2016). Extracted DNA and plasmids were quantified and qualified using a spectrophotometer (NanoDrop 2000, Thermo Fisher Scientific, or GeneQuant II, Amersham Pharmacia).

2.4 Determination of Total Bacteria and Total Oyster Pathogenic Vibrio spp.

Total bacteria, total Vibrio bacteria, and various oyster pathogenic *Vibrio* species were counted by qPCR using eDNA techniques and the primers listed in Table 1.

qPCR was performed with the CFX96 Touch Real-Time PCR detection system (Bio-Rad). qPCR was performed in a final volume of 10 μL, containing 5.0 μL iQ SYBR Green Supermix (Bio-Rad), 0.2 μL of each primer (200 nM final concentration), 1.0 μL template DNA, and 3.5 μL sterilized pure water. After the qPCR cycles a melting temperature curve of amplicons was generated to verify the specificity of the amplification. Herpes virus OsHV-1, total bacteria, total Vibrio and each oyster pathogenic Vibrio species were quantified by comparing the observed Cq values to a standard curve generated from the DNA of each virus and bacterium. The total bacteria count was based on *Escherichia coli* and the total *Vibrio* spp. count on *Vibrio splendidus* DNA.

3 Results and Discussion

3.1 Environmental Factors at the Beginning of Spat Cultivation

The environmental conditions in the areas with and without seagrass beds at each site are shown in Fig. 4. At both Sites A and B, the water temperature was lower at the bottom than at the surface, whereas at Site C, where seagrass beds were most widespread and dense, the water temperature and salinity were lower at the surface than at the bottom, suggesting that freshwater inflow had a greater effect at Site C.

The DO in the area with seagrass beds at Site A was similar to that at Site C at both the surface and the bottom, and the DO in the area without seagrass beds at Site A tended to be lower in the bottom layer. On the other hand, at Site B, the surface DO in the area with seagrass beds was lower than in the areas with seagrass beds at Sites C and A, and even the bottom DO in the area with seagrass beds at Site B was hypoxic, just like the areas without seagrass beds.

At both Sites A and B, the areas with seagrass beds were shallower nearshore than the areas without seagrass beds, so it would not be surprising if the DO tended to be higher at the surface in the areas without seagrass beds offshore than shallower part. In fact, the surface DO was higher in areas without seagrass at both Sites A and B. Nevertheless, at Site A, the bottom DO tended to be higher in the areas with seagrass beds than without, and there was no significant difference between areas with and without seagrass beds at Site B. We believe that the differences between Sites A and B reflect differences in seagrass density and biomass.

Although the climatic conditions of 2018 that led to heat waves and summer anoxia in Thau lagoon caused massive mortality of oysters and mussels (Le Ray et al. 2023), the survival rate of oyster spat with seagrass beds was interestingly higher than without seagrass in both sites A and B (Hori et al.. 2021). In particular, the oyster spat at Site A with more abundant seagrasses exhibited a survival rate around 90% (Hori et al. 2021). Anoxia reduces the growth of oysters even when its effects are not lethal, because oysters keep closing their shells during periods of

Table 1 Primers used for eDNA analyses in this study

Target species	Forward primer	GC	Tm	Length	Reverse primer	GC	Tm	Length	References
Herpes virus OsHV-1	ATTGATGATGTGGATAATCTGTG	35.0	60.1	23	GGTAAATACCATTGGTCTTGTTCC	42.0	63.1	24	Davison et al. (2005)
Total *Vibrio* species	GGCGTAAAGGCATGCAGGT	58.0	67.2	19	GAAATTCTACCCCCCTCTAACCAG	50.0	61.6	22	Mansergh and Zehr (2014)
Total Bacteria	CAGCMGCCGCGGTAA	66.7	56.7	15	GTGCTCCCCCGCCAATTCCT	65.0	64.2	20	Wang and Qian (2009)
Vibrio aestuarianus	GGGTGTCGTTAATAGCGGCATTTC	50.0	61.2	24	GACGTAAACTTTGACCGTTTG	42.9	54.5	21	In this paper
Vibrio ostreicida	TGGTAGTGTTAATAGCACCATCAT	37.5	56.8	24	ATCGTAAACTTTGACAGCCTG	42.9	55.3	21	In this paper
Vibrio europaneus	TGGTGTAGTTAATAACTGCATCAT	33.3	55.1	24	AACGTAAACTTTGACCGTCTG	42.9	55.4	21	In this paper
Vibrio crassostreae	GGGTAGTGTTAATAGCGCTATCTC	45.8	57.3	24	GACGTAAACTTTGACCACTTG	42.9	53.9	21	In this paper
Vibrio tubiashii	GGGTAACGTTAATAGCGTTATCTC	41.7	55.9	24	GACGTAAACTTTGACCACTTG	42.9	53.9	21	In this paper
Vibrio splendidus	TGGTGCAGTTAATAGCTGTATCAT	37.5	57.1	24	ATCGTAAACTTTGACCGCCTG	47.6	57.6	21	In this paper

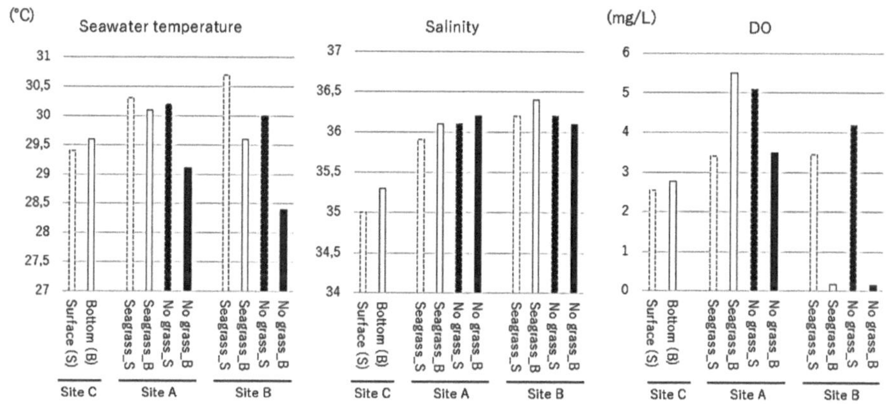

Fig. 4 Seawater temperature, salinity and dissolved oxygen concentration on 6 August 2018 during hypoxia. Site C was the control site with seagrass; the bottom there was completely covered with seagrass beds. Each environmental parameter was measured both at the surface (dotted lines) and at the bottom (solid lines) of the areas with seagrass (clear bars) and without seagrass (gray bars) at each site

anoxia (Coffin et al. 2021). The difference in the bottom DO at Site A between the areas with and without seagrasses suggests that the abundant seagrass beds mitigated the hypoxic event at the study site.

3.2 *Bacterial Flora in the Water Column*

There were clear differences in water-column microbial counts between areas with and without seagrass beds in summer (Fig. 5). *Vibrio* spp. were not detected in the dense seagrass beds of Site A, and even in the sparse seagrass beds of Site B, counts of *Vibrio* spp. were considerably lower than in the area without seagrass beds. At first, we should consider the effects of anoxia on bacterial community because these results corroborated with the observations of Le Ray et al. (2023) and Coffin et al. (2021), highlighting an increase of Total bacteria and *Vibrio* spp. abundances during anoxia. However, in the bottom water of Site B where DO was the lowest of all areas with and without seagrass beds, there was a clear difference in the amount of *Vibrio* spp. between these areas, suggesting that the presence of seagrass beds reduced the amount of *Vibrio* spp. and also the total amount of bacteria.

In the winter, after spat cultivation ended, our results for the total amount of bacterial DNA varied depending on the pore size of the Sterivex filter used (Fig. 6). Higher quantity of total bacteria was observed with 0.22 μm than 0.45 μm whatever the site and the presence or the absence of seagrass, that highlighted a large part of free-form-bacteria in seawater. In contrast, no difference was observed between *Vibrio* spp. abundances observed in shell-rinse seawater using 0.22 or 0.45 μm filter (Fig. 7), suggesting that Vibrio spp. was associated to suspended matter > 0.45 μm.

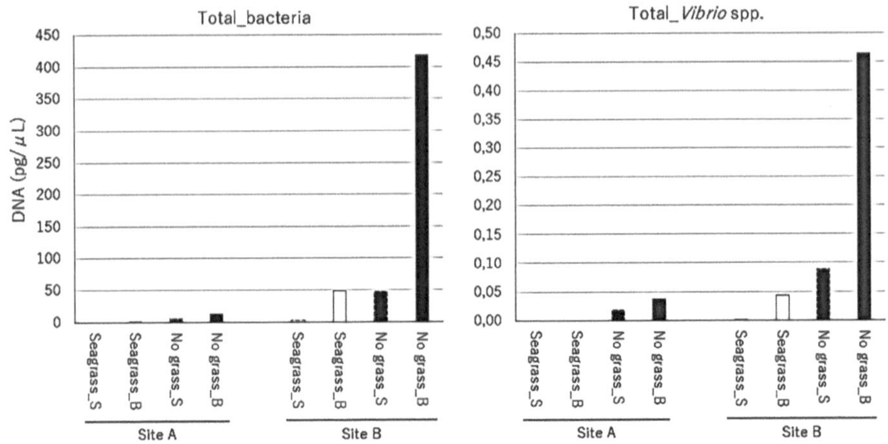

Fig. 5 Amount of total bacteria and of *Vibrio* spp. In the ambient seawater at the surface (S: dotted lines) and bottom (B: solid lines) of the areas with seagrass (clear bars) and without seagrass (gray bars) at each site during hypoxia. All seawater samples were filtered using 0.45-μm sterivex filter

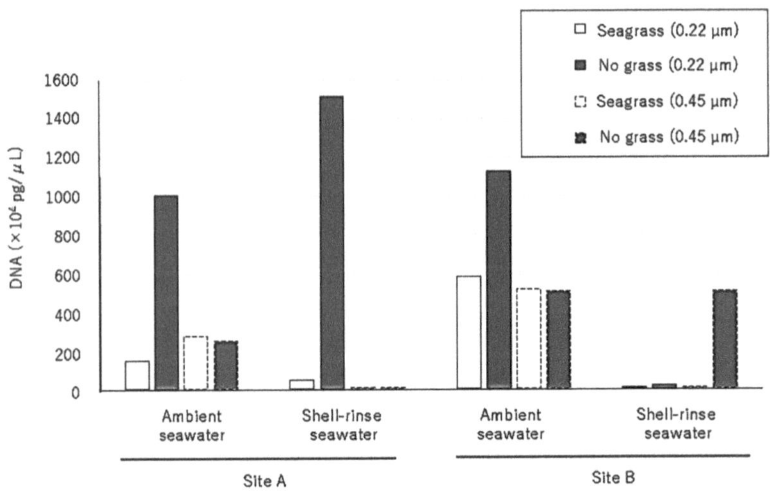

Fig. 6 Amount of total bacteria from ambient seawater and shell-rinse water in the areas with seagrass (clear bars) and without seagrass (gray bars) at each site. All samples were filtered using both 0.22-μm (solid lines) and 0.45-μm (dotted lines) Sterivex filters

We detected a clear difference in both ambient seawater and shell-rinse water between the areas with and without seagrass beds using a 0.22-μm filter. The total amount of DNA detected in the area with seagrass beds was considerably lower than in the area without seagrass beds (Fig. 6). In addition, the difference in the amount of DNA detected for Vibrio spp. varied between samples from ambient seawater and

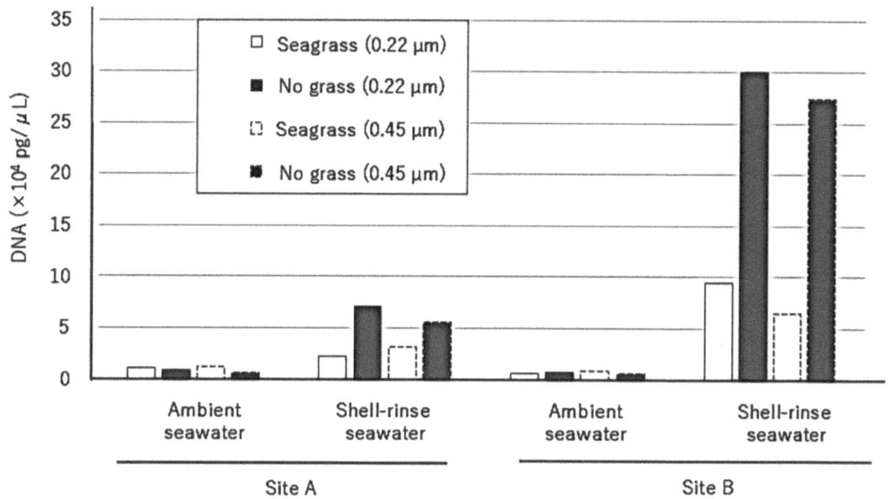

Fig. 7 Amount of *Vibrio* spp. from ambient seawater and shell-rinse water in the areas with seagrass (clear bars) and without seagrass (gray bars) at each site. All samples were filtered using both 0.22-μm (solid lines) and 0.45-μm (dotted lines) Sterivex filters

those from shell-rinse water with 41-fold higher abundances (Fig. 7), suggesting that Vibrio spp. is favored by oysters.

On the other hand, the amount of DNA detected from shell-rinse water showed clear differences between the area with and without seagrass bed at both sites. The amount of DNA in the shell-rinse water from areas with seagrass beds was less than half that from the areas without seagrass beds in samples from both the 0.22- and the 0.45-μm filters. These results suggest that seagrass beds have an ecological function to reduce both the total amount of bacteria and the number of harmful *Vibrio* spp. for juvenile oysters. In addition, the amount of *Vibrio* spp. detected at Site A, with more dense seagrass beds, was much lower than that detected at Site B, with sparse seagrass beds, suggesting that more dense seagrass beds might have a more effective ecological function. In fact, the survival rate of oyster spat at Site A was higher than that at Site B (Hori et al. 2021), even in the area without seagrass beds. This could indicate that the effects of more dense seagrass beds can extend to surrounding areas.

The comparison between ambient seawater and shell-rinse water (Fig. 7) suggests that *Vibrio* spp. were likely less abundant in the water column but instead attached directly to the surface of oyster shells. Some phytochemicals from tropical seagrass tissues have the potential to kill or inhibit bacterial pathogens including *Vibrio* sp. in vitro (Kannan et al. 2010), and the antibacterial function of tropical seagrasses can work even in natural coastal areas with the seagrass beds (Lamb et al. 2017). In addition, the antibacterial function was effective not only within the seagrass beds, but also in adjacent habitats. Our results indicate that temperate eelgrass species Z. marina also has an antibacterial function, and that the influence of dense seagrass beds at Site A might extend to the areas without seagrass beds.

Herpes virus OsHV-1 was not detected from all samples including ambient seawater, shell-rinse water, and inside-shell water.

3.3 Bacterial Flora Inside Juvenile Oysters

The same trends were observed for counts of both total bacteria and *Vibrio* spp. in the shell water at the end of the spat cultivation during winter (Figs. 8 and 9). At Site A, counts of both total bacteria and *Vibrio* spp. in the shell water were clearly lower in spat cultured both above and the within the seagrass canopy than in those cultured in the area without seagrass beds. In contrast, at Site B, total bacterial counts in spat from the area with seagrass beds were higher than from those in the area without seagrass beds.

These results indicate that the bactericidal effect of seagrass on bacterial communities was likely much weaker at Site B. However, there was no difference between two sites in the amount of total bacteria or *Vibrio* spp. in the shell water from the areas with seagrass beds, suggesting that the bactericidal effect of seagrass at Site B may not be weaker. On the other hand, those in the shell water at Site A were clearly higher than those at Site B, although those in the environmental water tend to be higher at Site B in both summer and winter.

There is at least a possible reason for the difference between the areas without seagrass at the two sites. It is likely that shell water samples were collected from survived oysters at the end of the experiment. At Site B, with higher amount of total bacteria in environmental water and of *Vibrio* on the shell surface, the mortality rate

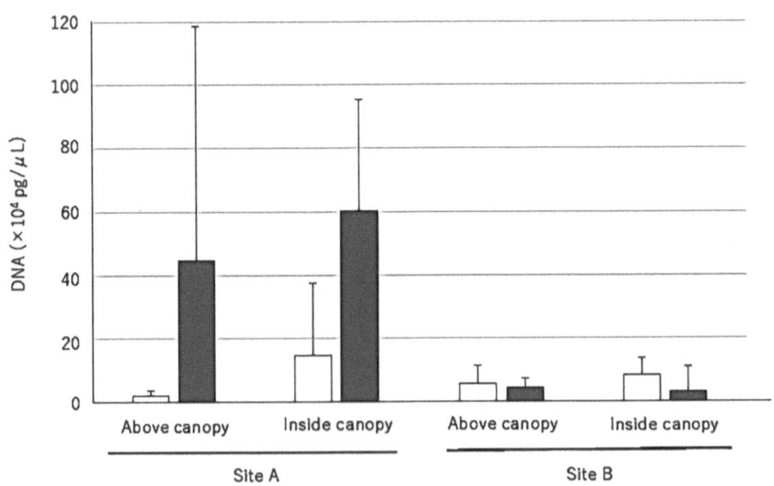

Fig. 8 Amount of total bacteria inside shells of juvenile oyster from mesh bags suspended in the areas with seagrass (clear bars) and without seagrass (gray bars) at each site at the end of the spat cultivation. Error bars indicate standard deviations

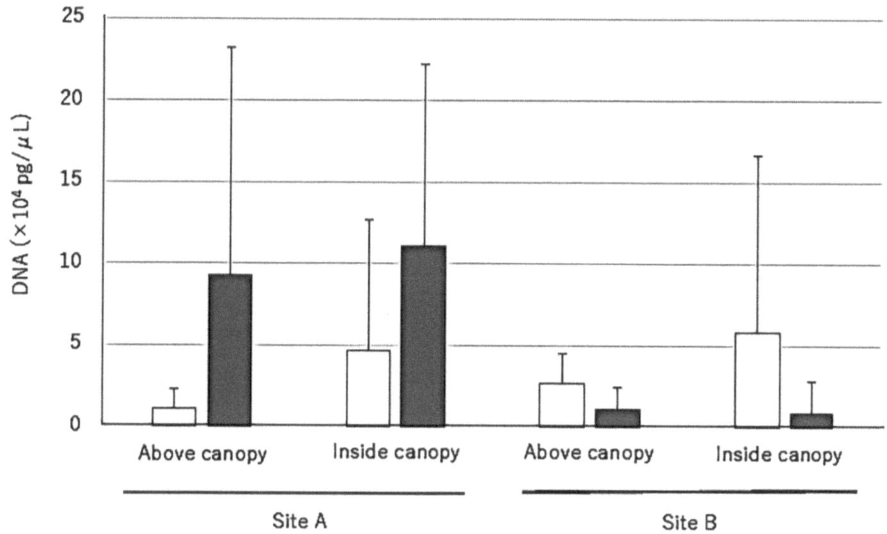

Fig. 9 Amount of *Vibrio* spp. inside shells of juvenile oysters from mesh bags suspended in the areas with seagrass (clear bars) and without seagrass (gray bars) in both sites at the end of the spat cultivation. Error bars indicate standard deviations

of cultivated oyster spat was higher than at Site A, and moreover, the survived oysters were significantly smaller and in worse condition than at Site A (Figs. 10 and 11, Hori et al. 2021). It is probable that the level of total bacteria and *Vibrio* spp. at Site A was non-lethal, not only in the area with seagrass beds but also in the areas without seagrass beds because of the dense seagrass coverage. On the other hand, at Site B, with sparce seagrass coverage, the mortality rate of oyster juveniles was higher especially in the area without seagrass beds, suggesting that highly infected individuals likely died and were not included in the shell water samples. At Site B, we detected a considerable amount of *Vibrio* spp. in the shell-rinse water from survived oysters (see Fig. 7), so the amount of total bacteria would be even higher in dead oysters.

4 Summary and Conclusions

The aims of the French Japanese project in the Seto Inland Sea/Hiroshima Bay and the Mediterranean Sea/Thau lagoon were to (a) evaluate oyster farming methods using seagrass beds that mitigate the effects of oligotrophication and climate change on the ambient environment, oysters, and the local communities, (b) clarify the feeding environment of oyster larvae and establish natural spat collection, especially in Thau lagoon (Lagarde et al. 2020), and (c) understand environmental microflora

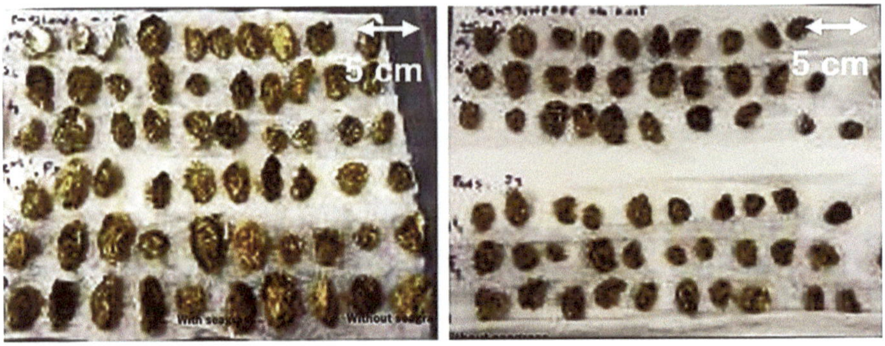

Fig. 10 Surviving oyster juveniles from each mesh bag in the area with seagrass at Site A (left) and at Site B (right)

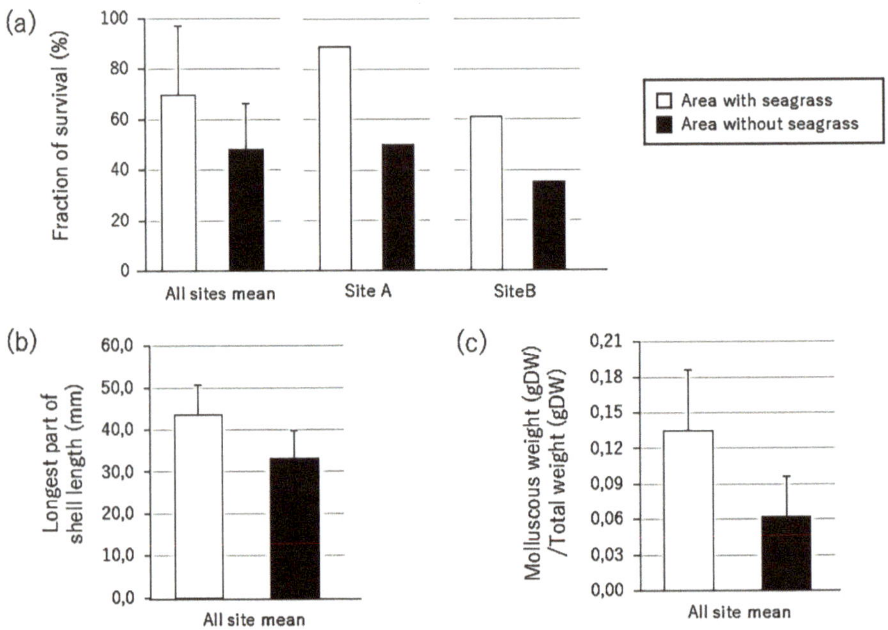

Fig. 11 Comparison of the **a** fraction of survival, **b** shell length, and **c** relative meat weight between oyster spat cultured in areas with (clear bars) and without seagrass (black bars). These results are from Hori et al. (2021). Error bars indicate standard deviations

and phytoplankton under oligotrophication and marine heat-wave conditions, using eDNA and metabarcoding (Richard et al. 2022).

For aim (a), based on returning to a traditional Japanese aquaculture using the natural biological interactions between oysters and seagrass beds (Hori et al. 2018), we have reported on field experiments to clarify the effect of eelgrass beds on the sustainability of oyster aquaculture in both the French Mediterranean Sea and the Seto

Inland Sea, Japan (Hori et al. 2021). The seagrass contribution to oyster production was divided into two processes in relation to stages of the oyster life cycle: recruitment processes from the larval stage to spat settlement, and post-recruitment processes associated with spat growth. A series of in situ experiments in Japan and France revealed that spat recruitment was significantly higher in areas without seagrass, presumably because of the weaker current and deeper conditions in seagrass beds than on intertidal oyster reefs, which were unsuitable for spat settlement.

On the other hand, spat growth and survival after settlement were higher in seagrass beds even when hypoxic events occurred at the study areas (Fig. 11). There are at least two possible hypotheses for why seagrass beds can provide a better environment for oyster spat: trophic support and the improvement of sanitary conditions. As a trophic support hypothesis, Hori et al. (2018) suggested that oyster spat of three endemic species (Pacific oyster, Kumamoto oyster *Magallane sikamea*, and Iwagaki oyster *M. nippona*) cultivated on the tidal flat of Hiroshima Bay in Japan can use not only pelagic production but also benthic production from seagrass beds, based on the results of carbon and nitrogen stable isotope analyses. Even Iwagaki oyster spat with poor survival exhibited better growth as relative meat weight, suggesting that a trophic function of seagrass beds would make some contribution to spat growth. These results are also supported in a study by Barbier et al. (2017) showing that bivalve recruitment is lower in *Zostera marina* beds than in other benthic habitats, such as subtidal coarse sand or maerl beds (Corallinophycidae, Rhodophyta). However, they did observe higher accumulation of fatty acids in the digestive glands of juveniles that settled in *Z. marina* beds, suggesting higher food availability.

The second hypothesis is that seagrass beds could maintain or improve environmental sanitary conditions for oysters (e.g., Lamb et al. 2017; Groner et al. 2018). There are some environmental factors that negatively affect oysters; some lethal viral or bacterial pathogens capable of causing oyster mortality are well known, and anoxia events have been serious recently, related to increasing summer seawater temperature. Lamb et al. (2017) reported that the presence of seagrass beds reduced the relative abundance of potential bacterial pathogens capable of causing disease in humans and marine organisms by 50%.

Although the trophic contribution of seagrass beds to bivalve mollusks is well known (Morimoto et al. 2017), there are also reports of some specific biochemicals produced by Zostera species and the antibacterial effects on the microbiome on leaves (Papazian et al. 2019; Jendresen and Nielsen 2019). However, there are few reports about the antibacterial effects of *Zostera* species on ambient waters or their contribution to fisheries organisms, such as the oysters in this study. Although we only compared the amount of bacteria with and without seagrass beds in this study, the bactericidal effect is presumed to be quite strong. There is no doubt that seagrass is functioning to improve the sanitary environment in Thau lagoon. Japanese fishers have cultivated oysters in areas with eelgrass beds for a long time because of better oyster growth and survival (e.g., Tanaka 2014; Tanaka and Furukawa 2020). Our results support the validity of this indigenous, local knowledge.

In conclusion, climate change countermeasures and adaptation have become urgent issues. Seagrass beds are important blue carbon ecosystems that act as greenhouse gas sinks, which are essential for mitigating climate change. In fact, Japan has officially included seagrass and macroalgal beds as CO_2 sinks in the national greenhouse gas inventory, submitted to the office of United Nations Framework Convention on Climate Change in 2024. At the same time, seagrass beds can perform a wide variety of functions, including food production, and improvement of water quality and sanitation. As a natural ecosystem, seagrass beds can adapt to environmental changes autonomously, with little artificial maintenance. Climate change mitigation and adaptation measures using such natural ecosystems would facilitate the sustainability of social communities.

Acknowledgements We thank all French and Japanese colleagues of the research project between FRA and IFREMER. We are grateful to the French–Japanese Society of Oceanography for their hospitality and for the fruitfulness of their scientific and friendly exchanges.

References

Aoki K, Shimizu Y, Yamamoto T, Yokouchi K, Kishi K, Akada H, Kurogi H (2022) Estimation of inward nutrient flux from offshore into semi-enclosed sea (Tokyo Bay, Japan) based on in-situ data. Estuar Coast Shelf Sci 274:107930

Barbier P, Meziane T, Forêt M, Tremblay R, Robert R, Olivier F (2017) Nursery function of coastal temperate benthic habitats: new insight from the bivalve recruitment perspective. J Sea Res 121:11–23

Bec B, Derolez V, Soudant D, Cesmat L, Pete R, Richard M (2018) Projet CAPATHAU : CAPacité Trophique de la lagune de THAU. Livrable I Capacité Trophique de la lagune de Thau. Evolution temporelle de l'état écologique de la lagune de Thau et des performances des coquillages en élevage au regard de la réduction des apports issus du bassin versant et des changements météorologiques, p 157. (in French)

Coffin MRS, Clements J, Comeau LA, Guyondet T, Mallet M, Steeves L, Winterburn K, Babarro JMF, Mallet MA, Hache R, Poirier LA, Deb S, Filgueira R (2021) The killer within: endogenous bacteria accelerate oyster mortality during sustained anoxia. Limnol Oceano 66:2885–2900

Collos Y, Bec B, Jauzein C, Abadie E, Laugier T, Lautier J, Pastoureaud A, Souchu P, Vaquer A (2009) Oligotrophication and emergence of picocyanobacterial and a toxic dinoflagellate in Thau lagoon, Southern France. J Sea Res 61:68–75

Correia-Martins A, Tremblay R, Bec B, Roques C, Atteia A, Gobet A, Richard M, Hamaguchi M, Miyajima T, Hori M, Miron G, Pouvreau S, Lagarde F (2022) Failure of bivalve foundation species recruitment related to trophic changes during an extreme heat wave event. Mar Ecol Prog Ser 691:69–82

Davison AJ, Trus BL, Cheng N, Steven AC, Watson MS, Cunningham C, Le Deuff RM, Renault T (2005) A novel class of herpesvirus with bivalve hosts. J Gen Virol 86(Pt 1):41–53

Derolez V, Bec B, Munaron D, Fiandrino A, Pete R, Simier M, Souchu P, Laugier T, Aliaume C, Malet N (2019) Recovery trajectories following the reduction of urban nutrient inputs along the eutrophication gradient in French Mediterranean lagoons. Ocean Coast Manag 171:1–10

Derolez V, Malet N, Fiandrino A, Lagarde F, Richard M, Ouisse V, Bec B, Aliaume C (2020) Fifty years of ecological changes: regime shifts and drivers in a coastal Mediterranean lagoon during oligotrophication. Sci Total Environ 732

Duarte CM, Losada IJ, Hendriks IE, Mazarrasa I, Marba N (2013) The role of coastal plant communities for climate change mitigation and adaptation. Nat Clim Chang 3:961–968

Groner ML, Burge CA, Cox R, Rivlin ND, Turner M, Van Alstyne KL, Wyllie-Echeverria S, Bucci J, Staudigel P, Friedman CS (2018) Oysters and eelgrass: potential partners in a high pCO_2 ocean. Ecology 99:1802–1814

Hori M, Hamaoka H, Hirota M, Lagarde F, Vaz S, Hori J, Makino M (2018) Application of the coastal ecosystem complex concept toward integrated management for sustainable coastal fisheries under oligotrophication. Fish Sci 84:283–292

Hori M, Hamaguchi M, Sato M, Tremblay R, Correia-Martins A, Delorez V, Richard M, Lagarde F (2021) Oyster aquaculture using seagrass beds as a climate change countermeasure. Bull Jpn Fish Res Educ Agency 50:123–133

Jendresen CB, Nielsen AT (2019) Production of zosteric acid and other sulfated phenolic biochemicals in microbial cell factories. Nat Commun 10:4071. https://doi.org/10.1038/s41467-019-12022-x

Kannan RRR, Arumugam R, Anantharaman P (2010) Antibacterial potential of three seagrasses against human pathogens. Asian Pac J Trop Med 3:890–893

Lagarde F, Richard M, Bec B, Roques C, Mortreux S, Bernard I, Chiantella C, Messiaen G, Nadalini JB, Hori M, Hamaguchi M, Pouvreau S, Roque d'Orbcastel E, Tremblay R (2018) Trophic environments influence size at metamorphosis and recruitment performance of the Pacific oyster. Mar Ecol Prog Ser 602:135–153

Lagarde F, Richard M, Delolez V, Bec B, Pete R, Hori J, Bayne C, Mortreux S, Correia-Martins A, Tremblay R, Hamaguchi M, Shoji J, Makino M, Sato M, Nakaoka M, Miyajima T, Poureau S, Hori M (2020) Integrated ecosystem management for exploited coastal ecosystem dynamics under oligotrophication and climate changes. In: Ceccaldi HJ, Henocque Y, Komatsu T, Prouzet P, Sautour B, Yoshida J (eds) Evolution of Marine coastal ecosystems under the pressure of global changes, pp 240–252

Lamb JB, Water JA, Bourne DG, Altier C, Hein MY, Fiorenza EA, Abu N, Jompa J, Harvell CD (2017) Seagrass ecosystems reduce exposure to bacterial pathogens of humans, fishes, and invertebrates. Science 355:731–733

Larkum AWD, Orth RJ, Duarte CM (2006) Seagrasses: biology, ecology and conservation. Springer, Dordrecht, p 676

Le Ray J, Bec B, Fiandrino A, Lagarde F, Cimiterra N, Raimbault P, Roques C, Rigaud S, Regis J, Mostajir B, Mas S, Richard M (2023) Impact of anoxia and oyster mortality on nutrient and microbial planktonic components: a mesocosm study. Aquaculuture 566:739171

Mansergh S, Zehr JP (2014) Vibrio diversity and dynamics in the Monterey Bay upwelling region. Front Microbiol 5:48. https://doi.org/10.3389/fmicb.2014.00048

Miya M, Minamoto T, Yamanaka H, Oka S-I, Sato K, Yamamoto S, Sado T, Doi H (2016) Use of a filter cartridge for filtration of water samples and extraction of environmental DNA. J vis Exp 117:e54741. https://doi.org/10.3791/54741

Morimoto N, Umezawa Y, San Diego-McGlone ML, Watanabe A, Siringan FP, Tanaka Y, Regino GL, Miyajima T (2017) Spatial dietary shift in bivalves from embayment with river discharge and mariculture activities to outer seagrass beds in northwestern Philippines. Mar Biol 164:1–16

Papazian S, Parrot D, Buryskova B, Weinberger F, Tasdemir D (2019) Surface chemical defence of the eelgrass *Zostera marina* against microbial foulers. Sci Rep 9:3323. https://doi.org/10.1038/s41598-019-39212-3

Richard M, Bec B, Bergeon L, Hebert M, Mablouke C, Lagarde F (2022) Are mussels and oysters capable of reducing the abundances of Picochlorum sp., responsible for a massive green algae bloom in Thau Lagoon, France?. J Exp Mar Biol Ecol 556:151797

Stoeckle MY, Soboleva L, Charlop-Powers Z (2017) Aquatic environmental DNA detects seasonal fish abundance and habitat preference in an urban estuary. PLoS ONE 12(4):e0175186. https://doi.org/10.1371/journal.pone.0175186

Tanaka T (2014) Hinase fishermen town as a Satoumi using eelgrass and oysters. Nippon Suisan Gakkaishi 80:72–75 (in Japanese)

Tanaka T, Furukawa K (2020) Prospects for practical Satoumi implementation for sustainable development goals: lessons learnt from the Seto Inland Sea, Japan. In: Ceccaldi HJ, Henocque Y, Komatsu T, Prouzet P, Sautour B, Yoshida J (eds) Evolution of Marine coastal ecosystems under the pressure of global changes, pp 320–326

Tsurita I, Hori M, Makino M (2017) Fishermen and conservation: sharing the case study of Hinase Bay, Japan. In: Marine protected areas: interactions with fishery livelihoods and food security, Westlund L, pp 43–50

Wang Y, Qian P-Y (2009) Consevative fragments in bacterial 16S rRNA genes and primer design for 16S ribosomal DNA amplicons in metagenomic studies. PLosOne 4(10):e7401. https://doi.org/10.1371/journal.pone.0007401

Yanagi T (2015) Estuaries of the world, eutrophication and Oligotrophication in Japanese estuaries. The present and future tasks. Springer, Kobe/Japan

Junior and Senior High School Students' Challenge to Solve the Problem of Marine Garbage in the Seto Inland Sea—Through the Practice of Citizen Collaboration Through "Civic Tech"-

T. Inoue

1 Introduction

Sanyo Gakuen Junior Senior High School is located in Okayama City, Okayama Prefecture, Japan. Okayama Prefecture is bordered by the Seto Inland Sea in the south, and the ocean is a familiar presence (Fig. 1).

The geography club is one of the school's club activities, with 28 junior high school students and 24 high school students (Fig. 2).

Approximately 80% of marine litter is of terrestrial origin and is domestic waste that has entered through rivers (Fujieda et al. 2010). Furthermore, the annual inflow and generation of waste from the four Setouchi prefectures, including Okayama Prefecture, is 388 tons, 86 tons of which are not collected (Shioiri et al. 2023). As a result of degradation caused by the litter drifting to the shore and depositing on the seafloor, it will become microplastics, and there are concerns about their future impact on the ecosystem and human health (Isobe 2020; Nishioka 2020).

Is the understanding and behavioral change of the citizens, the waste disposers, keeping pace with this current situation? Students have been working to solve the problem of marine garbage in the Seto Inland Sea by collecting it from fishing boats and on islands, and by spreading awareness of the need to reduce its generation (Fig. 3).

However, the majority of participating citizens were repeat visitors, and a survey of participants clearly showed that citizens were practicing solution behaviors, including the use of eco-bags (Fig. 4). In general, efforts to solve problems tend

T. Inoue (✉)
Sanyo Gakuen Junior Senior High School, Okayama, Japan
e-mail: takashi_size_5538@yahoo.co.jp

A.-C. Bennis et al. (eds.), *Constraints and Adaptations to Global Change at the Land-Sea Interface: For a Shared Ecological and Energy Transition*,
https://doi.org/10.1007/978-3-031-90050-1_23

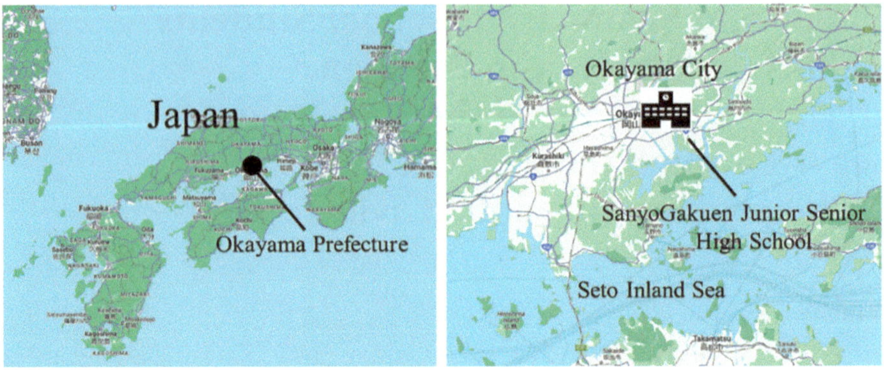

Fig. 1 On the left: Okayama Prefecture location; on the right: Sanyo Gakuen Junior Senior High School Location

Fig. 2 Geography club members

to be made by a few enthusiastic citizens, while many others have a sense of being someone else's problem and have little sense of being a party to the problem.

In order to curb the generation of waste, it is important to raise awareness of citizens by educating them about the importance of their involvement. Sugimoto (2021) promoted positive changes in society and citizens toward coral reef conservation through practices that broadened the base and lowered the hurdles to encourage

Fig. 3 Top left: collection of beach garbage on Teshima; top right: poster Session at the Science Council of Japan; bottom left: collecting garbage on the sea floor using a fishing boat: bottom right: study sessions at elementary schools

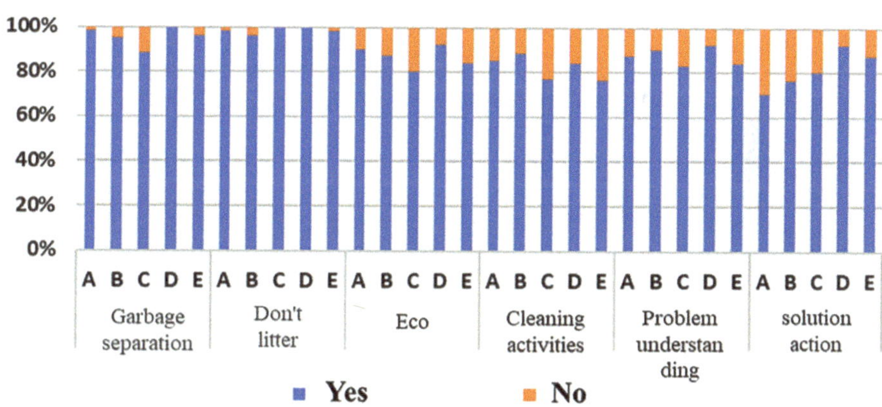

Fig. 4 Results of an awareness survey of citizens who participated in the awareness event. **a** Tamashima Community Center in Kurashiki City. **b** Fukuda Community Center in Okayama City. **c** Kibi Community Center in Okayama City. **d** Funao Community Center in Kurashiki City. **e** Mabi Community Center in Kurashiki City

citizen participation. The students engaged in a practice that can make progress toward solving the marine garbage problem through citizen collaboration by using methods that can be applied by all citizens, not just a few enthusiastic citizens.

2 Purpose and Significance of the Practice

The students' previous efforts to collect marine debris and raise awareness were limited in terms of date, time, location, number of participants, and awareness of participants, resulting in a narrow scope of practice and lack of breadth. In addition, there is repeated participation of conscious citizens with problem-solving behavior. If citizens, who are the disposers of the domestic waste that becomes marine garbage, are accompanied by a change in their awareness and behavior as parties concerned, i.e., solution providers, it is certain to be a major driving force in solving the problem.

It is not easy to involve all kinds of citizens with different lifestyles, residential areas, age groups, and values. However, we believe that if many citizens can participate through citizen collaboration and mobilize their power to solve problems, we can make great strides toward solving these issues. There are approaches to solving the marine garbage problem from every field, including the development of laws, the development of materials, and the enhancement of recycling systems. However, garbage becomes more serious when it leaves human hands, changing location and shape, becoming smaller, less visible, and less able to be picked up (Fig. 5).

Therefore, it is very important for people to be aware of and take action in the disposal and disposal of household waste. In order to raise awareness in this situation, we challenged the practice of promoting behavioral change by creating a mechanism that "anyone" can "easily" participate in and "do something worthwhile" for citizens, and by fostering a sense of ownership in solving the problem.

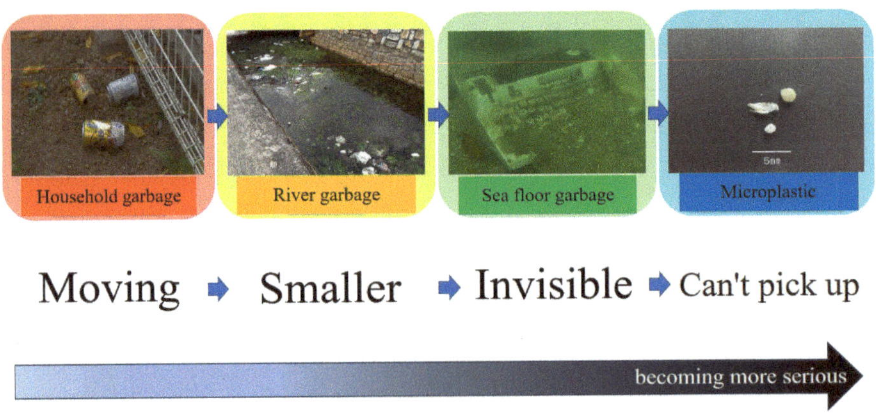

Fig. 5 Waste movement and change in size

3 Research Methods

The challenge for the students is the practice of having citizens contribute to solving problems. In order to encourage participation of all citizens, it is important that "anyone" can "easily" participate and have "something worthwhile" to do. As a fulfillment of these requirements, students utilized "civic tech". Civic Tech means that citizens use technology to solve problems. What we focused on as a tech is a smartphone that meets three criteria.

Smartphones have permeated every generation of citizens' lives, are easy to operate, and are now indispensable. We created garbage collection survey application that can be used on at least one smart phone per citizen, and made the public widely aware of the existence of the application so that they could engage in the project (Fig. 6). The advantages of our application are: (i) eliminates the need to enter personal information such as account and sign-in; (ii) allows easy entry of garbage information such as type and number of pieces of garbage; (iii) utilizes smartphone functions for sending date and location information; (iv) allows visualization (output) of data sent by citizens to maps, etc.; and (v) can be used offline as well as online, making it applicable in school education.

While it is often difficult for citizens to participate in garbage collection and educational events, this system makes it easy for citizens to contribute to solving the problem by using their spare time while commuting to work, walking, shopping, etc., to point their smartphones at garbage scattered in the city and survey and provide garbage information through a single simple operation. From the data sent by our citizens, we are able to determine where garbage is scattered and the type and number of pieces. The data collected by citizens became valuable data for understanding the actual state of garbage in the city, one of the origins of marine garbage. This is a practice that led to the solution of the marine garbage problem through data collected by citizens using smart phones.

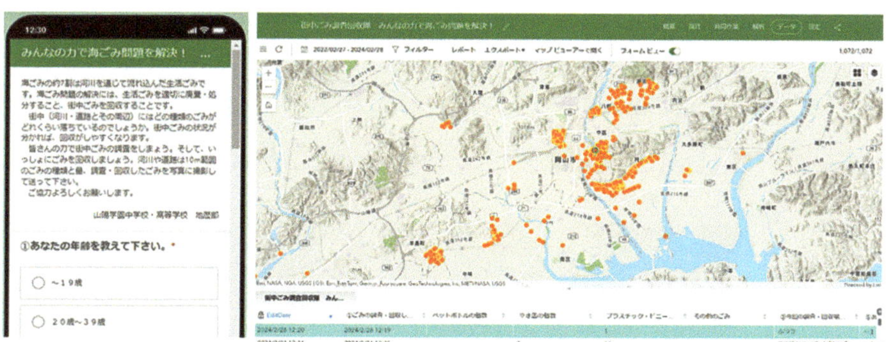

Fig. 6 On the left: app screen; on the right: aggregate screen of sent data

4 Results and Discussion

Our application has 1060 participants and 36,000 trash surveys received as of the end of February 2024 (see example in Fig. 7). Consideration of the content of the submissions clarified the following. (i) the majority of transmissions are from the younger generation, and the response from the smartphone generation is remarkable (Fig. 8); (ii) many of them are not using their days off to prepare tools for collection, but rather are using the time in between lunch shopping and other gaps in their workday (Fig. 9); (iii) citizens are becoming aware of garbage and changing their awareness and behavior, such as noticing scattered garbage that they normally do not notice; (iv) the areas around commercial facilities and irrigation canals were identified as garbage hot spots (accumulation areas), and the results could be "visualized" by mapping them (Fig. 10); (v) the project has fostered a sense of fulfillment and contribution among citizens, such as "I was able to contribute to the resolution of issues.

Rather than leaving the problem to only a few conscious citizens to solve, citizens themselves were able to step forward to take action that would lead to a solution to the marine garbage problem by using their familiar smartphones. We created a litter map from citizen collection data and distributed it to citizens in our coverage area. Citizens used the garbage map in their cleanup activities and designated collection points as priority items, which was not a one-time effort but an effective and sustainable practice.

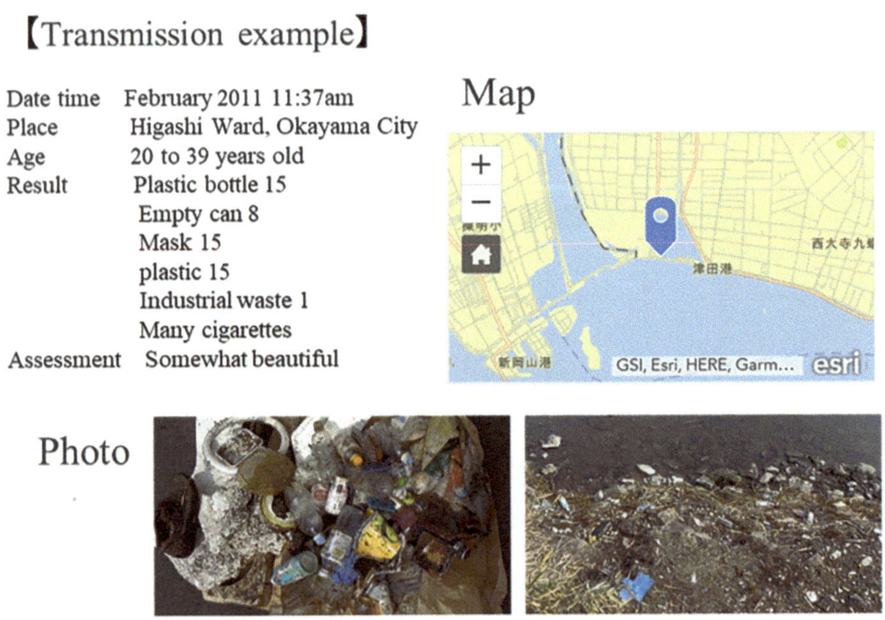

Fig. 7 Examples of submissions from citizens

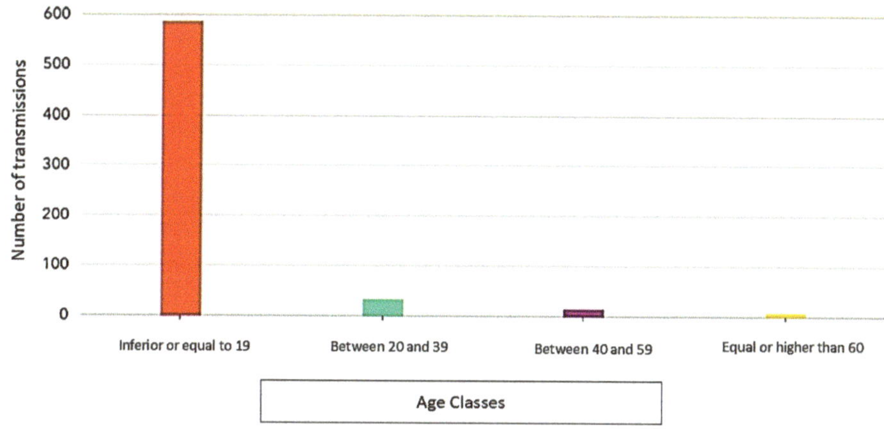

Fig. 8 Sender age range

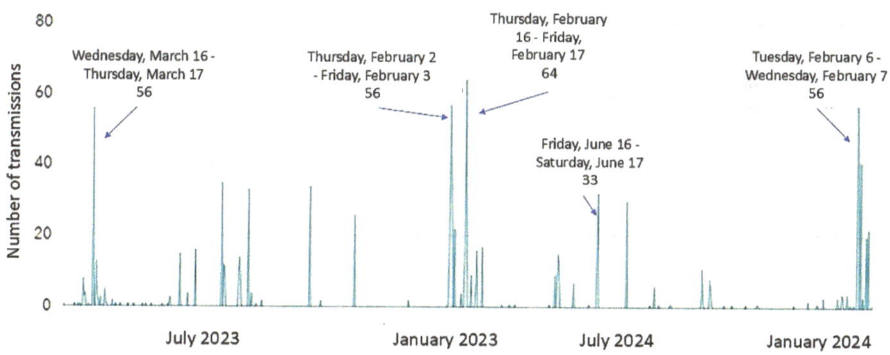

Fig. 9 Sender's primary date and time of transmission

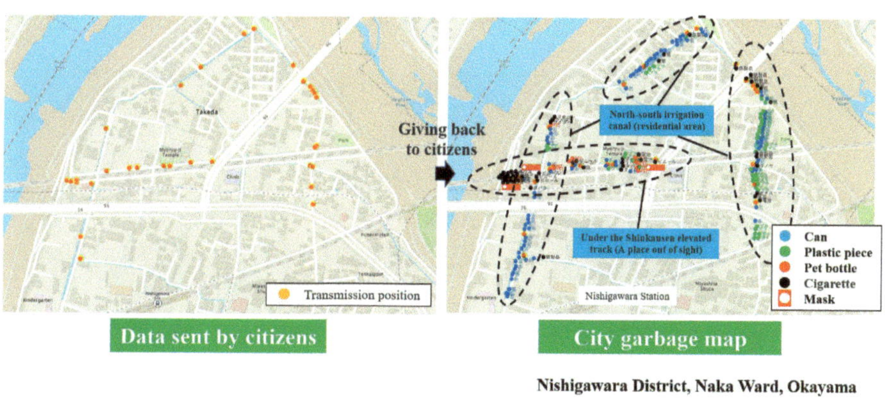

Fig. 10 Mapping data sent by citizens and returned to citizens

5 Conclusion and Future Prospects

The fact that marine debris is caused by human hands, and that the affected areas can be different places and sea areas, and may even cross-national borders, means that people's perceptions of the problem vary widely. Furthermore, indifferent people who consider the scale of the marine garbage problem to be someone else's problem, and irresponsible people who garbage and dump illegally are hindering the activities of conscious citizens with some solution actions. Citizens' use of the application using civic tech as a mechanism to involve all kinds of citizens expanded the base of citizen participation and lowered the hurdles, enabling citizens to change their awareness and behavior, contributing to solving problems and fostering a sense of contribution.

Data on waste sent by citizens is obligated to be returned to the citizens. Therefore, we consolidated the transmission data and created a "garbage map" so that the location, type, and number of pieces of garbage can be seen at a glance. This map clearly shows the characteristics of the distribution of garbage and its accumulation locations. And the residents of the area are using this map to engage in effective cleanup efforts.

Furthermore, the waste information collected by citizens proved to be a solution. The map was created from garbage information on irrigation canals around Naka-ku, Okayama City. In this case, by overlaying the location and information of the administrative water gates on this garbage map, it is clear that there is always a water gate in the irrigation canal where the garbage is accumulated (Figs. 11 and 12).

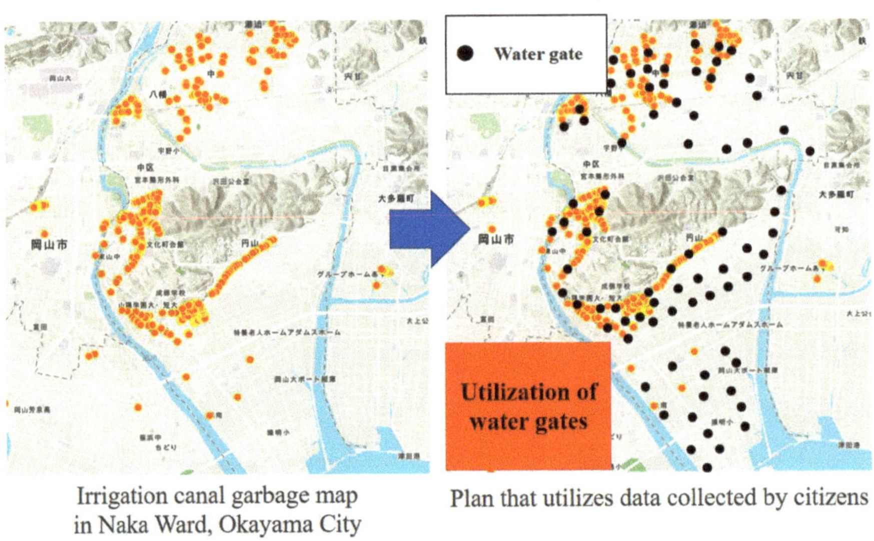

Fig. 11 Combine citizen-collected trash information with water gate data

Fig. 12 Water gate over irrigation channel

The original purpose of the water gate was to control flooding by using the water for agricultural purposes and to regulate the water level during heavy rainfall. However, if the opening and closing of the water gate can be successfully utilized for uses other than these, it is quite possible to collect garbage at the water gate. However, there are many issues that need to be resolved, such as collaboration with water gate installers and management and responsibility. However, the citizens' survey makes it clear that garbage is definitely accumulating in the irrigation canals. Collecting trash at the water gates of the irrigation channel will prevent the discharge of trash into the rivers and oceans beyond. The short time lapse from waste disposal also prevents waste deterioration and facilitates waste collection operations.

While the data collected by citizens is significant in its own right, we believe that further layering with other data will have a greater effect. The power of one citizen may be small, but when many citizens come together, they become a great force. Many citizens find it difficult to participate in garbage collection operations and educational events on their days off because of the hurdles involved. However, by widening the entry points for participation, lowering the hurdles, and above all, by making the activities interesting and worthwhile, many participants have provided us with information on their garbage. This is truly the realization of the practice of "not one person's 100 steps", but one person's 100 steps. And we believe that this practice has paved the way to transform citizens from "waste disposers" to "solvers" of the problem.

References

Fujieda et al (2010) Marine litter budget in the Seto Inland Sea. J Coast Estuar Res 22(4):17–29
Isobe A (2020) The truth of Marine plastic waste problem: the reality and future prediction of microplastics. Kagaku-dojin
Nishioka M (2020) Familiar plastics understand. Gijutsu Hyoronsha
Shioiri et al (2023) Inflows of plastic litter from land areas and amounts generated and collected in Marine Areas in four prefectures in the Seto Inland Sea. J Coast Area Stud 36(2):73–77
Sugimoto A (2021) Collaboration of community, science and art for social change: frontiers of coral reef conservation research. The Sasakawa Peace Foundation Institute for Ocean Policy Research. Ocean Newsletter No. 512

"*Satoumi*" Created by Eelgrass Beds and Oyster Farming, Hinase Town, the Seto Inland Sea, Japan

Takehiro Tanaka

1 What is "*Satoumi*"?

The term and concept of "*Satoumi*" was first proposed by Yanagi in (1998c), and defined as "a coastal marine area with high biodiversity and productivity due to human intervention (Yanagi 1998a, b). Since then, "*Satoumi*" has attracted international attention and is becoming established worldwide. This concept is rooted in the fisherman-led coastal management approach originating from Japan's ancient fishing system, and is an excellent ecosystem approach strategy to sustainably maintain Socio-Ecological Balance.

In contrast, what is the current state of the coastal environment in Japan? When fishing was the main source of livelihood in coastal areas, the coastal seas were healthy due to the proactive management of fishermen. However, the number of fishermen has decreased from 790,000 in 1953 to 130,000 in 2021 (White Paper on Fisheries 2021), and with the rapid aging of the fishing industry, fishermen alone are no longer able to manage the sea as a fishing ground. In Japan, fishermen have carefully managed the sea in front of them since the Edo period (1603–1868) as the sea inherited from their ancestors. This concept has been inherited in the current Fisheries Law, and has long been nurtured as a unique Japanese fishing rights system unparalleled anywhere in the world. However, the coastal management system based on the fishing rights system, in which fishermen are the "guardians of the sea," is now collapsing.

The ocean is an indispensable and precious asset shared by all people. How should people relate to the sea? The "*Satoumi*" is an effective approach and method to solve this difficult question. However, it is not easy to make it a reality because the actual situation of each coastal area varies depending on the actual condition of fisheries and

T. Tanaka (✉)
NPO Satoumi Research Institute, Okayama, Okayama, Japan
e-mail: satoumiken@gmail.com

A.-C. Bennis et al. (eds.), *Constraints and Adaptations to Global Change at the Land-Sea Interface: For a Shared Ecological and Energy Transition*,
https://doi.org/10.1007/978-3-031-90050-1_24

sea surface utilization, coastal topography, land use patterns of hinterland, industrial structure, culture and historical background, and so on, and the ideal state to aim for is also different. The most important common issue in realizing harmony between human activities and the coastal environment is how to sustainably secure marine products, which are necessary food for humans. In order to rediscover the characteristics and richness of the sea in each region and realize "satoumi," it is necessary to take this issue as the starting point and put the "people," mainly local fishermen, in the leading role, and to bring together the wisdom of local governments and people from various fields in the "industry," "academia," and "government. The realization of "satoumi" requires the collective wisdom of local governments, industry, academia, and government, with the "private sector" led by local fishermen.

2 Hinase Town as a "*Satoumi*"

Hinase Town in Bizen City consists of the mainland and 13 islands (Fig. 1). It has long been a fishing town with a thriving fishing industry, including small trawl net, small set nets, drift-net fishing, and Oyster farming. By the middle of the Muromachi Period (1336–1573), the fishing village had developed to the extent that it even sold fish to Kyoto and Osaka, and by the time of the Meiji Restoration (1868), 90% of the 830 households were engaged in the fishing industry. Fishermen in Hinase were so enthusiastic about fishing that they sought fishing grounds from Mie and Wakayama to Kyushu, and from the 1880s, they even went to the Korean Peninsula, Taiwan, Manila, and Singapore (data from the Hinase Fishermen's Cooperative Association). The small set nets called "*Tsuboami*" were invented by fishermen in Hinase and spread throughout western Japan, and some were even invited to teach the technique. In addition, in order to break the deadlock in the fishing industry in the Seto Inland Sea and develop it as a tourist fishery, fishermen from Hinase were among the first to start young yellowtail aquaculture, and by 1958 they were producing 400,000 fishes. The oyster aquaculture industry in Okayama Prefecture ranks third in Japan in terms of both production volume and production value (as of 2021), and Hinase has played a major role as a major production center. In 1996, the company joined with neighboring fishing cooperatives to achieve unified branding as "Okayama Oysters," contributing to improved quality and the development of sale channels. In 1971, they won the Emperor's Cup at the 10th Agricultural Festival for their efforts in Propagation of Japanese tiger prawn, and since then they have actively engaged in cultivation and fisheries resource management.

In the Seto Inland Sea Spanish Mackerel Stock Recovery Plan by 11 prefectures, they played a pioneering role by initiating the mesh regulation, releasing fertilized eggs, and Release of juvenile fish. In 1967, "*Gomino-Ichi*" was started to provide local residents with the leftover fish from market shipments at the Fisheries Cooperative Association's cargo handling facility after the auction was over. The "*Umi-no-Eki-Shioji*," a direct sales center with an oyster grill built in 1999, has become the centerpiece of the tourist facilities in the area. In this way, Hinase has always

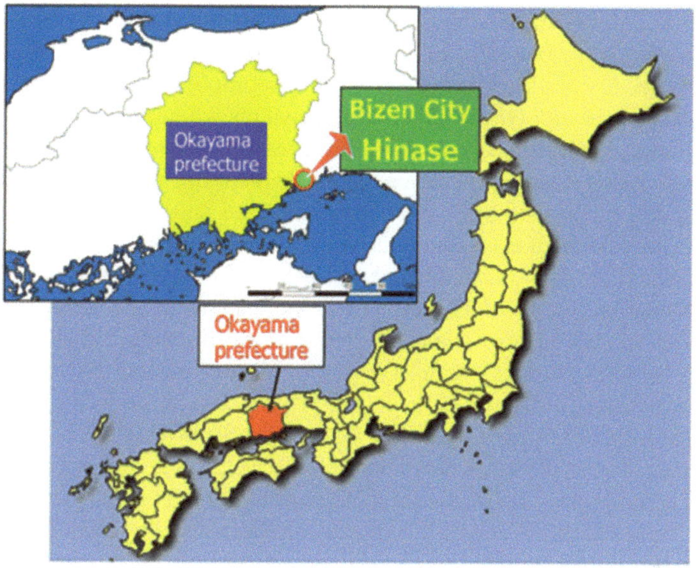

Fig. 1 Location of Hinase Town, Bizen City, Okayama Prefecture

played a leading and pioneering role in the conservation and restoration of the fishing environment and fishery resources, and has developed through cooperation between the fishing industry and the tourism industry to promote a sixth industrialization and the active addition of human resources. The coastal area is suitable to be called a "*Satoumi*" in that it has always played a pioneering role in the recovery of the area.

3 Roles and Functions of Eelgrass Beds

The roles and functions of eelgrass beds can be summarized as follows (Tanaka et al. 2017): (i) spawning grounds for fish and shellfish, (ii) nursery grounds for juvenile fish and shellfish (calm waters, shelter and refuge from predators), (iii) a variety of prey species (provision of hierarchical habitats for a variety of species, epiphytes, attached diatoms, interleaf animals and benthic organisms generated by the enormous surface area and space), (iv) promotion of material circulation by trapping and decomposing suspended organic matter, (v) a stocking ground for fishery resources, (vi) suppression of water temperature rise in summer, (vii) source of dissolved oxygen, (viii) absorption of nutrients, (ix) adsorption and storage of carbon dioxide, (x) supply of food through the food chain based on detritus of dead eelgrass leaves, and so on. The degree of contribution of eelgrass to primary production and its contribution to the upper trophic levels through the humus food chain is difficult to assess and is a matter of debate. For example, in Alaskan lagoons with large eelgrass beds, eelgrass-derived organic carbon contributes to a high proportion of all benthos and fish, but when the water body is more open and the area and cover of eelgrass beds are small, the contribution of eelgrass beds is hardly detected. However, it is generally accepted that the contribution of eelgrass to benthic production in the adjacent areas of eelgrass fields is significant because pentosan, a component of cellulose that makes up eelgrass tissue, is widely distributed even in waters where eelgrass does not grow, and detritus derived from seagrass has been found in bottom and surface mud far from eelgrass fields (Central Fisheries Research Institute, Fisheries Agency 1997). In recent years, further studies have been conducted and new findings have been obtained on the effects of eelgrass on the suppression of red tide (Imai 2017) and its role as a blue carbon that fixes the greenhouse gas CO_2 (Hori and Kuwae 2017).

Eelgrass bed restoration has been attempted mainly in the United States and Japan. In Japan, various efforts and studies have been conducted since the 1960s in Yamaguchi, Oita, Niigata, Aichi, Okayama, Hiroshima and Tottori prefectures, universities, national research institutes and private companies using methods such as, transplantation of nutritive stock, sowing, seedling transplantation, Input of flowering branches. In the 1980s, research on the causes of the decline and disappearance of eelgrass beds and the conditions for their establishment began to be conducted. In particular, the research on eelgrass bed creation methods conducted by the Central Research Institute of Electric Power Industry from 1982 to 1986 was epoch-making in that it not only clarified the ecological characteristics of eelgrass but also added coastal engineering knowledge to the conditions for the establishment of eelgrass beds, contributing greatly to the promotion of subsequent research (Central Research Institute of Electric Power Industry 1988). The results of this research have since been compiled and many technical manuals and guidelines have been published by the Ministry of the Environment, the Ministry of Land, Infrastructure, Transport and Tourism, the Fisheries Agency, and other organizations. In recent years, public works projects, fishermen, and citizen groups have been engaged in eelgrass bed restoration activities, but the reality is that some areas are still struggling to restore eelgrass beds.

4 The Beginning of Eelgrass Bed Restoration Activities

I first met Mr. Kazuo Honda (later to become the head of the Hinase Fishery Cooperative Association) in mid-summer 1981, when the national cultivated fisheries industry was in full swing and I was working as a fisheries extension agent for Okayama Prefecture as a fishery engineer. I was 27 years old, and I had to dive and remove the weights (iron chains) to pull up the 300-m-long net. When I came ashore after the diving work, I saw a man among the many people who was quickly giving instructions to others and performing the work in an orderly manner. This was Mr. Honda, who was the leader of "*Tsuboami*" team at the time. After all the work was completed, I said hello to him. After standing and talking for a while, Mr. Honda sat down, sighed heavily, and began to talk enthusiastically while looking out at the sea. He said, "Just releasing fry into the sea is not enough to bring the fish back. He said, "The sea in Nissei is essentially an eelgrass sea, and we must first restore it". "*Tsuboami*" is a method of waiting and catching fish by setting nets in the path of the fish. Through many years of experience fishing with "Tsuboami" widely scattered along the coast, "*Tsuboami*" fishermen are familiar with the life histories, growth, and seasonal migration paths of various fish and shellfish. Mr. Honda says that no matter how many fries are released, it is no good if the environment is not conducive to the fry's growth and the habitat they inhabit as they grow and migrate. Until the 1950s, there were about 590 ha of eelgrass beds in Hinase area, but by 1985 the number had plummeted to about 12 ha, and then further declined to only 5 ha.

Eelgrass bed research in Japan is said to have begun with a survey of fish growth in eelgrass beds conducted by the Okayama Prefectural Fisheries Experiment Station from 1922 to 1925. This study was triggered by a fishing dispute between a bivalve farmer and a trawler fisherman who tried to harvest the eelgrass beds, but the report's preface clearly stated the protection and nurturing of young fish and the relationship with the coastal environment. This is the first systematic and scientific clarification of the function of eelgrass beds as "cradles of the sea. The Okayama Prefectural Fisheries Experiment Station, which was well aware of the importance of eelgrass beds, also became increasingly concerned about the remarkable decline of eelgrass beds in the prefecture in the 1960s, and in 1979 began developing an eelgrass seed collection technique, which was put into practical use in 1985 (Fukuda et al. 1984). In 1985, a total of 26 people, including 19 fishermen of Hinase, four from the youth fishermen, and ourselves, including the Fisheries Experiment Station, began to work on the restoration of the eelgrass beds. The project has continued for about 40 years, and can be divided into three phases: the "initial phase," in which only volunteers including "*Tsuboami*" fishermen started and continued steady efforts; the "foundation building phase," in which the entire fishery cooperative started working on the project; and the "expansion phase," in which not only fishermen but also a wide range of related people joined and the activity body was organized.

5 Starting in the Dark—The Initial Phase

In the early summer of 1985, flowering branches were carefully picked from the few remaining eelgrass beds, placed in bags, and hung on oyster rafts for storage. In the fall, they were removed from the bags and washed repeatedly in seawater to obtain approximately 150,000 seeds. The seeds were sown in three locations that were considered suitable based on a preliminary survey. After that, seeds continued to be sown in 16 locations, but they did not thrive simply because they were sown.

Yonago Bay (Fig. 2) is a closed bay with water temperatures exceeding 30 °C in summer and silt and clay bottom sediments, making it a harsh environment for eelgrass.

However, the amount of eelgrass seeds formed here is overwhelmingly large, and the bay is important as a seedbed, so special emphasis was placed on it. Even in Yonago Bay, eelgrass did not thrive by seeding alone. In 1988, a test area was set up within a circle 20 m in diameter, sprinkled with zeolite. When the entire surrounding area was seeded, a dense eelgrass community of about 300 m^2 was formed only in this

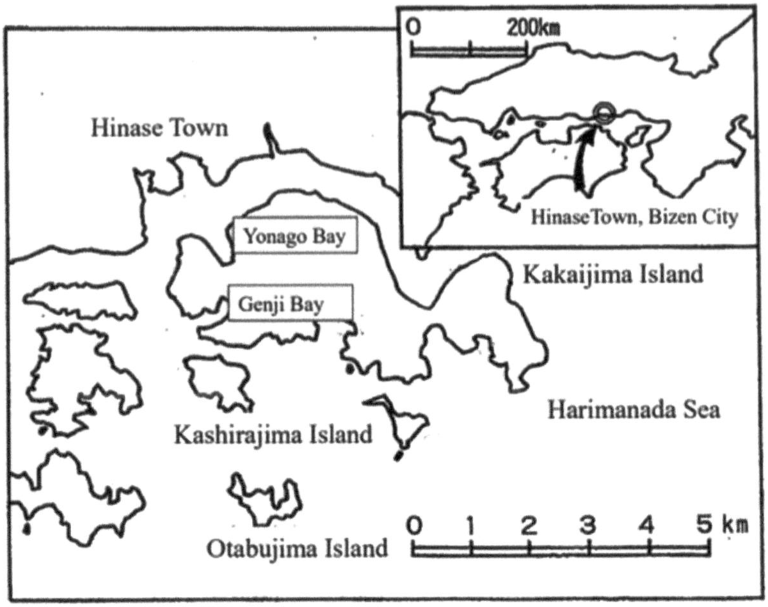

Fig. 2 Sea area, Hinase-town, Bizen City, Okayama Prefecture, Japan

test area (Fig. 3). This proved that bottom sediment improvement was effective. In 1990, with the help of university researchers, methods were repeatedly investigated in improving bottom sediments using various materials such as oyster shells, sea sand, steel slag, and zeolite, and it xas found that oyster shells were the most effective. In 1990, a "sandbag seeding method" was deviced to prevent seed loss and burial, and to ensure the successful implantation and establishment of eelgrass seedlings. In Yonago Bay, laying oyster shells and seeding was continuously carried out, and in May 1997, 6.8 ha of eelgrass beds were restored.

While fishermen continued their seeding activities, the Fisheries Agency provided support to conduct a basic survey for the development of a comprehensive coastal fishing area from 1994 to 1996 (subsidized by the Fisheries Agency). The government, university and private researchers, engineers, and fishermen worked together to elucidate the conditions for eelgrass growth and to accumulate knowledge on the improvement of growth limiting factors, such as bottom sediment improvement by oyster shell bedding (Tanaka 1998). Based on these results, a proposal was prepared and submitted to Marino Forum 21, and in 1998, a group of leading researchers on eelgrass beds at that time gathered to form the Shallow Sea Area Revegetation Technology Development Group, which conducted a series of demonstration tests and surveys of environmental conditions, mainly in the Hinase area.

In 2001, the group compiled the basic concept of eelgrass bed restoration, survey methods to identify limiting factors, and monitoring and maintenance techniques as "Technical Guidelines for Eelgrass Bed Creation". Until this point, we had been "happy when the eelgrass beds thrived" and "disappointed when they disappeared."

Fig. 3 Eelgrass beds thriving only in the Yonago Bay zeolite-sprayed test area (within a 20 m diameter circle)

However, the 15-odd years it took to reach the point of systematic technological development, with its ups and downs, could truly be called the "initial phase. In 2002, a marine ranching project centered on a large-scale eelgrass bed restoration project was launched in the Hinase area.

6 Expanding Circle of Activities—The Foundation Formation Period

In 2000, Mr. Kazuo Honda became the head of the fishermen's cooperative, and this marked the beginning of the foundation building period. 2005 saw the establishment of the Ocean Farm Appropriate Use Council, led by the fishermen's cooperative. In this year, 20 years after seeding, the eelgrass beds in the Hinase area finally recovered to approximately 38 ha (Fig. 4).

The number of Green tiger prawn (*Penaeus semisulcatus*) and Japanese blue crab (*Portunus trituberculatus*) also began to gradually increased, although the number fluctuates from year to year. Oyster shells were proven to be the most effective method of improving the bottom sediment for eelgrass bed restoration, and in 2006, the "Guidelines for Effective Use of Oyster Shells (Okayama Prefecture)" were completed to accelerate the bottom sediment improvement project. In 2007, the eelgrass bed area expanded to approximately 80 ha (Fig. 2), and in the spring of the following year, green tail prawn (*Metapenaeus moyebi*), which had disappeared for nearly half a century, returned to the area, and in the fall, the mottled spinefoot

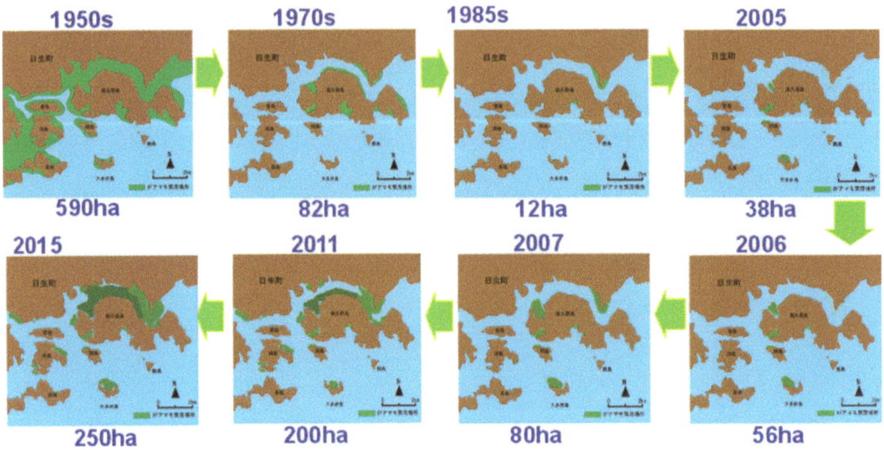

Fig. 4 40 years efforts for restoring eelgrass beds in the Hinase area

(*Siganus fuscescens*) returned. Eelgrass-fed, fatty mottled spinefoot are prized as a high-class fish in Hinase. Eelgrass bed restoration activities initiated by "*Tsuboami*" fishermen became a cooperative-wide project, and in 2009, the Hinase Seagrass Bed Creation Promotion Council was formed, consisting of 83 members from the Hinase Fisheries Cooperative (Fig. 5).

Fig. 5 The 83 fishermen who make up the Eelgrass bed Restoration Council of the Hinase Fishing Cooperative, which was established in April 2011

In 2010, they began receiving grants from the Fisheries Agency for environmental and ecosystem conservation, but this was the first time they were able to obtain funding for their activities, and the fishermen of Hinase had been working on a volunteer basis for 25 years.

7 Expansion from Fisheries Cooperative Activities to Community Activities and Marine Education: The Expansion Phase

In 2010, these activities entered a period of expansion, and began to receive support from a consumer cooperative, the Okayama Co-op, NPOs, and several universities, etc. In 2011, the Ocean Policy Research Foundation certified Hinase Town as an ICM (Integrated Coastal Management) model site, and it developed into the Bizen City Integrated Coastal Management Study Group, a cooperative effort of various stakeholders, including Bizen City, Okayama Prefecture, commerce and industry, tourism associations, and universities.

By the spring of 2011, the eelgrass beds had recovered to more than 200 ha (see Fig. 4). After reaching 80 ha in 2007, the oyster cultivation production, which had been repeatedly poor every year or two, remained stable at high levels for five consecutive years (Fig. 6), and the importance of eelgrass beds for oyster cultivation, which Mr. Honda had been advocating, became widespread.

As if relieved to find this out, President Honda passed away suddenly in November of the same year. Mr. Fuchimoto, who succeeded Mr. Honda as head of the cooperative, rebuilt a strong pyramid-shaped organization with a core of four fishermen who were then in the youth club and had supported eelgrass bed restoration activities for many years since the early days. In May 2012, a four-party agreement was concluded

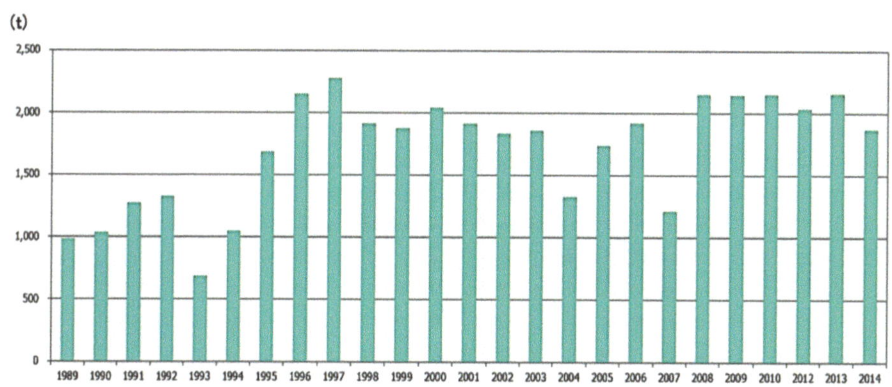

Fig. 6 Change of the oyster aquaculture production in Hinase Fisheries Cooperative Association

Fig. 7 Eelgrass bed restoration activities by Okayama citizens started in 2012

between the Hinase Fisheries Cooperative, the Okayama Co-op, Okayama Prefecture, and the NPO Satoumi Research Institute Council, strengthening the support system (Fig. 7).

However, while the eelgrass beds have recovered and the results of the fishery have bounced back, a new and troubling problem has emerged. Eelgrass drifted ashore in large quantities and became entangled in the propellers of vessels, impeding their navigation. In 2013, children from Hinase Junior High School began to play an active role in the project. The Hinase Junior High School, which had been incorporating oyster farming as part of its integrated learning program, was first introduced to the eelgrass beds by the Hinase Fisheries Cooperative Association. Now, about 200 children from three grades board the fishermen's boats and go out to sea to collect driftweed (Fig. 8). This has also become an activity to collect eelgrass seeds, so that all the seeds needed for sowing can be provided by collecting the driftweed without having to pick the flowering branches. The marine ranching that began in 2002 was completed in 2013, creating a "satoumi of eelgrass and oysters" (Fig. 9) in the sea of Hinase, and more than 100 million seeds have been sown over the past 30 years, with the eelgrass bed area reaching approximately 250 ha in spring 2015 (see Fig. 4).

In the course of a series of research studies, it became clear that oyster shells are not only useful for improving the bottom sediment for eelgrass growth, but also effective for improving the habitat of various benthic and attached organisms. These technologies are elemental technologies to promote a wide-area material cycle achieved by utilizing the characteristics of oyster shells and other shells to cultivate

Fig. 8 Marine learning with a focus on eelgrass bed restoration activities at Nissei Junior High School

Fig. 9 Satoumi created with eelgrass and oysters (completed in 2013)

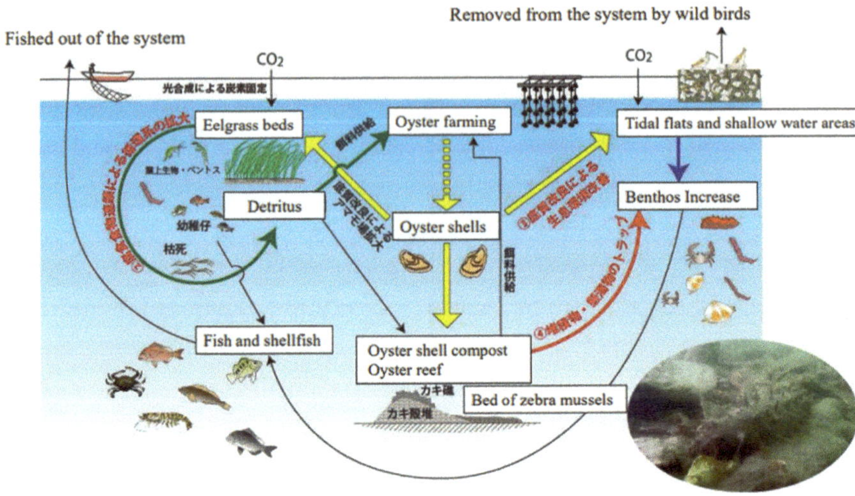

Fig. 10 Satoumi ecosystem (thick, long and smooth material circulation) using oyster shells

and propagate small animals, which feed on organic matter, leading to the expansion of the corrosion chain and predation by higher fish and shellfish, which in turn leads to the biotic food chain (Tanaka 2015).

The basis of satoumi development is to realize and maintain a thick, long, and smooth material cycle (Yanagi 2006). Hinase is a "satoumi of eelgrass and oysters". Oyster shells generated by oyster cultivation are used to restore eelgrass beds, tidal flats, oyster reefs, and zebra mussel (*Dreissena polymorpha*) beds, and to create oyster shell beds (a sandy seafloor topography created by heaping up oyster shells). The mosaic of these features, including oyster culture rafts as habitats, can create a diverse habitat for living organisms. From 2015 to 2022, Okayama Prefecture took the initiative in public works projects to expand thick, long, and smooth material circulation through the "creation of mosaic-like ecotones" (Fig. 10).

The circle of activities further expanded, and the participation of the general public grew as well. The passion of the fishermen in Hinase was conveyed to Kasaoka City and Yorishima Town in the western part of Okayama Prefecture, and eelgrass bed restoration activities by the fishermen began in earnest, spreading to the entire coastal area of Okayama Prefecture. Fishermen have continued to climb mountains to plant trees, but people engaged in forestry in the northern part of Okayama Prefecture have also begun to participate in the "marine forest creation" activities in Hinase, and exchanges with Maniwa City, a leader in "*Satoyama*"(a woodland area), have been strengthened.

8 Development of Eelgrass Bed Restoration Technology

The basic premise of eelgrass bed restoration is that "the sea area in question used to be a place where eelgrass grew in the past. It is extremely difficult to create a new eelgrass bed in a place where eelgrass has never thrived. Even if eelgrass grows temporarily by seeding or transplanting, it will eventually disappear unless the environment is improved and the limiting factors are removed. The "Technical Guidelines for Eelgrass bed Creation (Marino Forum 21 2001)" were developed based on this concept. The trigger for the large-scale decline of eelgrass beds in the Hinase area was a typhoon that hit the Japanese archipelago in 1961, and the rapid deterioration of water quality and sediment quality that accompanied the rapid economic growth of Japan. By maintaining stable vegetation, eelgrass beds mitigate the flow environment by acting as breakwaters and fluid resistance, ensure transparency by trapping suspended particles, and regulate their own growth environment through decomposition of organic matter by epiphytes, interleaf animals, and benthic animals, as well as through material circulation functions. If eelgrass beds decline rapidly in a short period of time, light conditions, hydrodynamic forces, and the bottom sediment environment will change drastically, and conditions for thriving will be damaged. Environmental factors related to eelgrass vegetation include water temperature and salinity, soil elevation, light flux transmission, waves and currents, bottom sediment, and seafloor movement, etc. These factors act in relation to each other, and the dominant factors vary depending on the environmental conditions at the site. For example, light is a limiting factor for eelgrass growth in deep water, but in shallow water, drying due to ebb and flow and movement of the seafloor due to waves are considered to be important. In order to restore eelgrass beds, it is necessary to identify and remove limiting factors that cause the eelgrass beds to become unstable.

The following are the guidelines for the establishment of eelgrass beds: Significant wave height less than 0.5 m, bottom current velocity 0.6 m/s, physics shields number less than 0.2, and minimum net photosynthesis greater than 0 mol quanta m^{-2} day^{-1}. The same guidelines have been used for the restoration of eelgrass beds in the Karakoto Eco-coast project in Kurashiki City, Okayama Prefecture, and for the transplantation and expansion of eelgrass beds in the Shiraishi Island Nishinoura coastal conservation project in Kasaoka City, Okayama Prefecture, and are used as a practical guide for eelgrass bed restoration.

In 1990, he devised the "soil seeding method1 (Ueki and Mitsunaga 1996)" to prevent seed loss and burial, and to ensure the implantation and establishment of eelgrass seedlings. The initial stages of eelgrass vegetation are germination, implantation, and establishment. Germination is promoted in an environment with low redox potential, and seeds and young shoots are easily lost or buried. This technique is also effective in determining suitable sites for eelgrass bed restoration projects.

In the case of the Hinase area, a major limiting factor was the bottom sediment environment. Oyster shells, which are readily available in Hinase, where oyster cultivation is very active, provided a solution to this problem. The functions of oyster shells can be summarized in three categories. The first function is as an anchor

Fig. 11 Eelgrass with roots entangled in oyster shells

material for eelgrass. Hinase is a town with extremely soft ground due to the silt–clay sediments of the rhyolite strata. If the original ground is left as it is, even if seeds germinate and implant, they will be swept away by the slightest current at the young seedling stage. When oyster shells are laid 10–20 cm thick, the oyster shells penetrate into the original substrate and interlock with it to form a layer, stabilizing the substrate. Eelgrass grows by extending its roots from the underground stem and intertwines with the oyster shells to stabilize the grass body (Fig. 11).

The second is the promotion of photosynthesis by improving transparency. Oyster shells are laid to cover the bottom mud, preventing a decrease in transparency. The third is the reduction of photosynthetic inhibition caused by floating mud, which is prevented by the same mechanism as the second one. The most effective oyster shells for bottom sediment improvement materials are unprocessed oyster shells, and oyster shells generated from oyster shell peeling are stored at a marine sedimentation site for at least six months to decompose and remove attached organisms. The "Guidelines for Effective Utilization of Oyster Shells (Okayama prefecture 2006)" summarizes these specific methods, which have cleared regulations under the Waste Disposal and Public Cleansing Law and made it possible to legally utilize oyster shells.

9 To a New Stage—Connecting Satoumi, Satoyama and Urban Areas

The 9th National Eelgrass Summit "From Bizen! Creation of Satoumi and Satoyama Brand—Connecting Communities and Generations" was held in Hinase Town, Bizen City, June 3–5, 2016. The number of participants from all over Japan, from Hokkaido to Okinawa, reached 2000 during the exhibition, which led to the establishment of the "Bizen City Satoumi Satoyama Brand Promotion Council with ICM" on the 6th of February 2017. In January 2019, the basic concept of the "Project for the Promotion of Exchange Centers Connecting with «Satoumi," "Satoyama," and "urban areas" was formulated. The project is designed to promote the exchange of people and goods through activities to restore eelgrass beds, marine education, marine experiences, and marine research that transcend regional, generational, and positional boundaries. The collaboration with Maniwa City, which is a leader in satoyama, has been strengthened, and with the help of the Nippon Foundation, the "Hinase-Umilabo", marine education and research center, opened in September 2021 on Kasirajima Island, Bizen City and the "Satoyama and Satoumi Exchange Center "Shinpio" in Hokubo, Maniwa City, in June 2022 (Fig. 12).

The "Satoumi" project in Hinase, Bizen City, involves not only fishermen, but also a wide range of people, including elementary, junior high and high school students, children, city residents, people from the satoyama, and people involved in agriculture. They see value in these connections and work together to enhance each other's activities. The "Satoumi" project in Hinase, which transcends regional, generational and social boundaries to protect the sea, could become a model for the ideal form of coastal zone management in the future. Originally, humans cannot

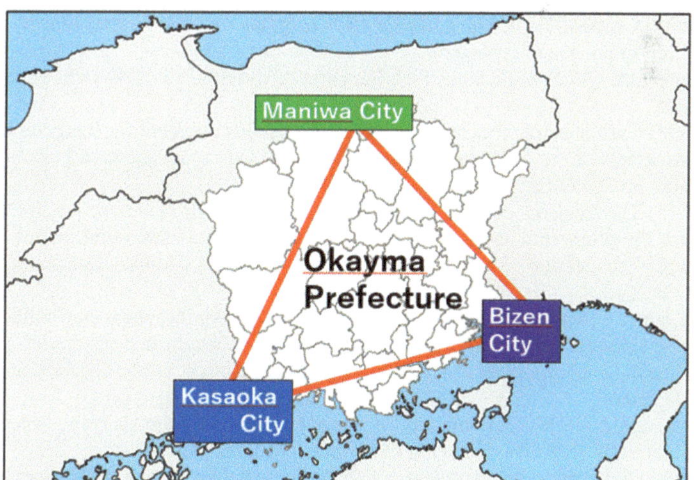

Fig. 12 Marine education triangle promotion initiative

survive without water and salt, so in the past, there were "salt roads" all over Japan, which people living in the mountains used to obtain salt from the sea. The basic principle is that forests, villages, rivers and the sea are connected by the river basin, but just as the sea and mountains were once connected by "salt roads", it is important that they are connected by the exchange of people and goods. From satoyama to satoumi, from satoumi to satoyama, and from "town" to satoumi and satoyama, as people and goods move, the forests, villages, rivers and sea become connected. We must continue to send out this "Message from Satoumi" in the future.

Acknowledgements The author would like to thank all stakeholders, including the fishermen of Hinase Town, for their efforts towards the realisation of Satoumi activities. The bulk of this manuscript is the result of the combined experience and knowledge of these stakeholders. The author would also like to thank Dr. Yumeng Pang of the University of Washington for his expert advice on the scientific English translation.

References

Central Fisheries Research Institute, Fisheries Agency (1997) Functions of seaweed beds (in Japanese)

Central Research Institute of Electric Power Industry (1988) Study on methods of creating eelgrass beds (in Japanese)

Fukuda T, Yasuie S, Tsuchiya Y, Terashima T (1984) Studies on the development of eelgrass beds-I. Methods of seed collection and preservation. Cultivat Technol Res Inst 13(2):77–84 (in Japanese)

Hori M, Kuwae T (2017) Blue carbon, CO_2 sequestration and storage in shallow water and its utilization. Chijin Shokan, Tokyo (in Japanese)

Imai I (2017) Studies on the mechanism of development and outbreak control of noxious toxic plankton. Nippon Suisan Gakkaishi 83(3):314–324 (in Japanese)

Marino Forum 21 (2001) Technical guidelines for eelgrass bed restoration. Marino Forum 21, Technical report no. 49 (in Japanese)

Okayama Prefecture (2006) Guidelines for utilization of oyster shells. Okayama Prefecture, 15p (in Japanese)

Tanaka T (1998) Current status and issues of technological development for eeelgrass bed restoration. In: Abstracts of the 11th open symposium of the Kansai aquatic environmental research organisation, pp 25–47 (in Japanese)

Tanaka T (2015) The forefront of shell utilization as a coastal environment restoration technology II—towards the promotion and improvement of material circulation—, promotion of material circulation by the creation of mosaic-like ecotone using shells. Monthly Ocean 47(3):115–125 (in Japanese)

Tanaka T, Furukawa K, Kuwae T, Imai I (2017) Report of the 32nd joint symposium of liaison council of academic societies on coastal environment "Road of Zostera bed restoration in Japanese Coastal Zones"—past processes and the future prospects. Nippon Suisan Gakkaishi 83(6):1042–1053 (in Japanese)

Ueki N, Mitsunaga T (1996) Development of sowing method of eelgrass using several kinds of substrate materials. Rep Okayama Fish Exper Stat 111:23–26 (in Japanese)

Yanagi T (1998a) Civil engineering projects and environmental protection in the inner bay. J JSCE 83–11:32–33 (in Japanese)

Yanagi T (1998b) The "Satoumi" of coastal waters. J Jpn Soc Water Environ 21:703 (in Japanese)

Yanagi T (1998c) Satoumi theory. Kouseisha Kouseikaku, 168p (in Japanese)

The Culture's Interface Between Human's Economy and Ecosphere's Integrity—The EcoSensible Sphere, a New Concept to Study an Art-Mediated Salt Marsh Restoration

N. Lamontagne, A. Collin, D. James, G. Hervouët, T. Durozier, E. Feunteun, and L. Wedding

1 Introduction

1.1 General Context: From Holocene to Anthropocene

Since the beginning of Holocene era, 12,000 years ago, *Homo sapiens*, with one million individuals, mostly shifted his hunter-gather lifestyle to a sedentary lifestyle based on farming and breeding, whose direct consequences altered the landscapes and seascapes, and the greenhouse gas concentrations (Bellwood 2004). Several centuries later, human population reached one billion in 1800 along with the start of the industrial revolution in the Western region (Ritchie et al. 2023). Some scholars consider that period as the commencement of the Anthropocene (e.g., Ruddiman 2013). However, a very recent study has shown through fossils that the Anthropocene began in 1945–1950 and its authors proposed the term of "Crawfordian", which is the name of a Canadian lake in Ontario (McCarthy et al. 2023). At that point, human population was estimated at 2.5 billion to reach, in 2023, 8 billion individuals (Elhacham et al. 2020).

N. Lamontagne (✉) · A. Collin · D. James · E. Feunteun
Coastal GeoEcological Lab, Ecole Pratique des Hautes Etudes–PSL, Dinard, France
e-mail: nancylamontagne.art@gmail.com

G. Hervouët
Conservatoire du Littoral–Délégation Bretagne, Plérin, France

T. Durozier
Conservatoire du Littoral–Délégation Normandie, Hérouville-Saint-Clair, France

L. Wedding
Oxford Seascape Ecology Lab, University of Oxford, Oxford, UK

A.-C. Bennis et al. (eds.), *Constraints and Adaptations to Global Change at the Land-Sea Interface: For a Shared Ecological and Energy Transition*,
https://doi.org/10.1007/978-3-031-90050-1_25

1.2 Human Needs: The "Great Acceleration"

This post-World War II population surge, coined as "the great acceleration", generated exponential growth in 11 other socio-economic indicators, such as urban population, international tourism, transportation, telecommunication, water use, large dam construction, fertilizer consumption, paper production, primary energy use, foreign direct investment, real gross domestic product, as well as 12 earth trend indicators, CO_2 and CH_4 emissions, NO_2 and O_3 concentrations, coastal nitrogen, surface temperature, ocean acidification, marine fish capture, shrimp aquaculture, tropical deforestation, domesticated land, biosphere degradation (Steffen et al. 2015a). In 2020, the human-made mass, estimated at 1.1 teratonnes, even surpassed the global biomass, quantified as 1 tera tonne (Elhacham et al. 2020). This hegemonic footprint reflects the expansion of the human sphere, called "the anthroposphere", which corresponds to the Earth's surface needed to nurture and fuel mankind.

1.3 Management for Mitigation: Panels, Concepts and Prospects

The "great acceleration" has also provoked some 5 dramatic improvements, such as life expectancy, child mortality, education, poverty, gender equality and democracy (Gapminder.org). The rise in global education, shifting from 10% literacy rate in 1900 to 86% in 2020, has fostered an increase in the global ecological awareness, about the unsustainability use of the ecosphere by the anthroposphere. In response, three international platforms have worked for the last 30 years to build a comprehensive framework to understand and manage the crucial trajectories at stake. The International Panel on Climate Change (IPCC) has released its first assessment report in 1990 and has recently published its 6th report in 2023 (Lee et al. 2023). Last conclusions indicate that global CO_2 emissions must be halved by 2030 and reach net-zero by 2050 to avert disastrous climate disruptions. In 2015, the United Nations carved 17 sustainable development goals.[1] In 2019, the Intergovernmental Science-Policy Platform on Biodiversity and Ecosystem Services (IPBES), in its first global assessment report, highlighted anthropogenic erosion of ecosystem and biodiversity at such a rate that the 6th biological crisis has emerged (IPBES 2019). From scholars, a "safe operating space for humanity" identified and quantified planetary boundaries that must not be transgressed (Rockström et al. 2023; Steffen et al. 2015b). Its last update, in 2023, demonstrated that six of nine planetary boundaries were crossed (Richardson et al. 2023).

In 2020, a 3-year research study proposed a "global safety net" (approximately 50% of the terrestrial surface) to reverse biodiversity loss and stabilize Earth's climate by compiling 5 main geospatial layers, such as species rarity sites, distinct species

[1] https://sdgs.un.org/fr/goals.

assemblages, large mammal assemblages, intact/remote land areas, and additional carbon storage areas. Last year, researchers enhanced the planetary boundaries with the notion of human justice (Rockström et al. 2023). They indeed used modelling and literature assessment to quantify safe and just planetary boundaries for the climate, biosphere, water and nutrient cycles, and aerosols at global and sub-global scales.

1.4 Specific Context: The Coastal Zone

The coastal realm fractally reflects planetary issues, insofar as the emergency increases as the product of coastal and urgency (Lenton et al. 2019). As the urgency corresponds to the ratio of the reaction and intervention time, coastal risks combine hazards (cyclone/storm, inundation), exposure (people living close to the shoreline and over lowlands), and vulnerability (sensitivity and adaptative capacity). That is why worldwide initiatives were specifically translated for the coast, such as the Future Earth Coasts, created in 2012, in the form of a global research network aiming at supporting society transformation to sustainability through action. Regionally, such a platform can be echoed with "coastal resilience", from the USA-based Nature Conservancy, and Japanese-driven "Sato-umi", 里海, (Yanagi 1998).

Coastal resilience, underpinned by a community engagement approach, has strived, for the last decade, to examine nature's role in reducing coastal flood risk, through science, tool and technology, a global network of practitioners, policy and field-based conservation and restoration actions.

Sato-umi, literally meaning the sea-village, consists of a coastal area where biological productivity and biodiversity harmonizes with human interaction. It is built with three components that support the preservation and regeneration of ocean regions elements (substance circulation, ecosystems, contact with people), and two other components that bolster practical activities (activity spheres, entities that perform activities).

1.5 The Issue

Following the planetary awareness and emergency, translated on the coastal zone, the stewardship is necessary for a sustainable, just and resilient future to anthroposphere and ecosphere. The ecosphere-anthroposphere, hereinafter referred to as Nature-Culture, continuum usually creates the stewardship's consensus to provide a conceptual framework for addressing today's planetary crisis. However, the role of the Culture, as the knowledge and its production (e.g., Lamontagne and Collin 2020, 2022a), remains too vague or site-specific, hindering its multi-scale transferability (as sketched in the symposium's poster, Lamontagne et al. 2023).

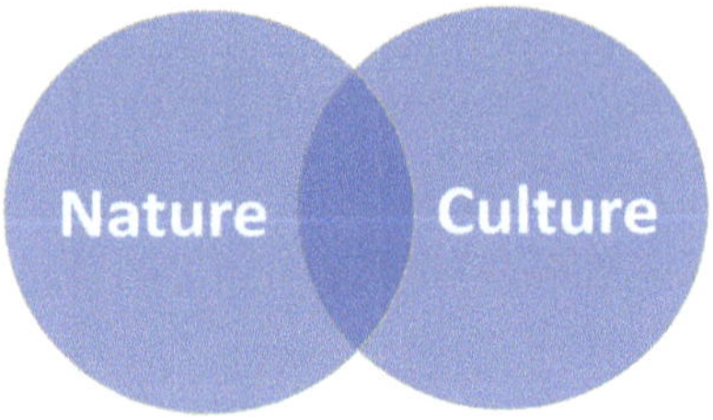

Fig. 1 The conventional vision of the Nature-Culture continuum

2 The Conceptual Solution: The EcoSensible Sphere

2.1 *The Nature-Culture Continuum: A Paradigm's Shift*

2.1.1 Conventional Vision: The Polar Sphere

The Nature-Culture duality was of key importance and first philosophical writings come from the Antiquity: Confucius, Lao Tseu, Buddha, Zoroaster, Socrate via Platon, etc. That period of global search of wisdom tended to "elevate" the spirituality of *Homo sapiens* above the animality. The superiority of the spirit/mind onto the body can be schematized by a pyramidal concept. Even if those philosophers invited people to think by themselves without religious dogma, the hierarchical concept underlined the primacy of the verticality, tainted with paternalism and colonialism.

From the end of the Middle-Age, the Nature-Culture continuum has been somewhat horizontalized and conventionally conceptualized by a binary vision, in the form of, either, a segment separating two poles (points), or two circles/disks/spheres, that can be distant, tangential or secant (Fig. 1). This representation underlies a kind of opposition, what is still usually conveyed when the solutions for the environmental issues seem to be detrimental to the society and its economy.

2.1.2 Innovative Vision: The EcoSensible Sphere

We propose, here, a novel conceptual scheme, that essentially transforms the polarized exclusive vision of the "Polar Sphere" into an inclusive outlook, coined as the "EcoSensible Sphere" (Fig. 2). Instead of an underlying conflict, the ecosphere integrates the anthroposphere, composed of its inner needs, reflected by "economy", and its outer interface, symbolized by "culture". This triptych can be represented, logically-speaking, as a three-layer sphere: socio-economy $\subseteq$ socio-culture $\subseteq$ Ecosphere (aka Nature).

The main features of this conceptualization of the linkage between human and environment/milieu reside in (i) the inclusion of the anthroposphere into the ecosphere, (ii) the finite aspect of the ecosphere, and (iii) the emphasis on the human culture as the exchange sphere.

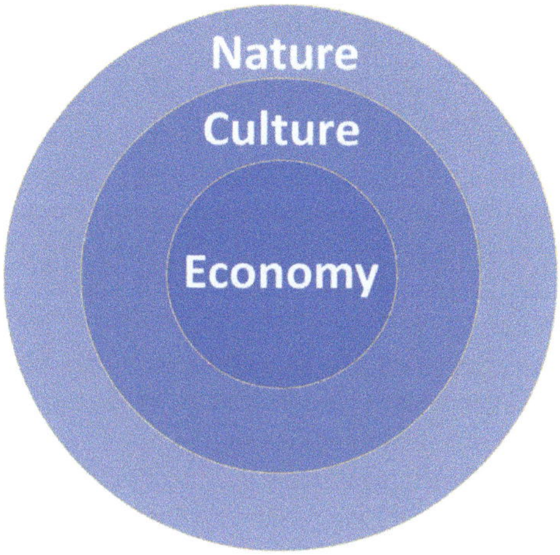

Fig. 2 The proposed vision integrating the anthroposphere (inner economy and outer culture) into the ecosphere (aka Nature): the EcoSensible sphere

The re-integration of the human (both economy and culture) is a keystone to re-connect *Homo sapiens* to its living system, unique in the known universe, namely Earth. Its disconnection from its animal status was promoted intrinsically by the patriarchal monotheisms and the mechanistic scientism (Metzner 1999). The direct consequence was the intentional negligence of Earth's finite processes, leading to over-consumption, exemplified by the "Earth overshoot day".

The finite feature of the EcoSensible Sphere is crucial to apprehend the planetary boundaries, which have the potential to expedite the sustainable decisions and lifestyles, contrary to the bi-polar spheres that are not delimited.

The distinction between the human economy and its culture reveals the role of the latter one as the interface between human fundamental needs and the ecosphere's integrity. Human primary needs, like other species, consist of optimizing their survival and reproduction in a given spatial–temporal window, featured by other species and abiotic factors. This sub-sphere of the anthroposphere can be deemed as centripetal (anthropocentric) as this socio-economic core tends to expand at the detriment of the non-economy remaining space. Contrarywise, the sub-sphere of the culture integrates the economy and can be readily perceived as centrifugal as it thrives within the ecosphere, that is to say non-human. It is therefore crucial to bear in mind that the proposed concept of the culture interface thrives with the non-human whereas the economy expands on (or instead of) both the culture and the non-human.

Finally, the human sub-spheres are, on the one hand, centered in the ecosphere, and on the other hand, geometrically perfect, only for didactic stakes. Their position can be de-centered and their contour can display irregularities, such as villosities.

3 The Case Study of Beaussais' Marsh

3.1 Study Site

The innovative EcoSensible Sphere concept will be applied to a coastal site presenting human economy, culture and nature sub-spheres in transformation. Beaussais' marsh consists of a land-sea interface, of approximately 60 ha, lying along the upper bay of Beaussais in Brittany (48.582016°, −2.160007°) at an average 5 m altitude (Fig. 3). It is composed of a seaward salt marsh, punctuated by hunter's ponds, a dyke, and a landward pastureland—fodder crop in transformation.

This representative site of a bay's marsh belongs to the *Conservatoire du littoral*, the French national administration tasked with coastal protection. It was selected by the European Life project "Adapto" as a pilot site to experiment a soft management of the shoreline based on nature-based solutions (https://www.lifeadapto.eu/home.html). Historically, this salt marsh site was indeed disconnected from the sea by a dyke construction in the XIXth century, then reconnected to the sea in 2020 by a dyke breach voluntarily unrepaired. The polderization—depolderization process was

Fig. 3 The study site of Beaussais' marsh: a depolderization led in the European Life Program "Adapto" proposed vision integrating the anthroposphere (inner economy and outer culture) into the ecosphere (aka Nature): the EcoSensible Sphere

Fig. 4 The artwork « *Relier: à l'abri dans la marée*» from Nancy Lamontagne, lying on Beaussais' marsh at high spring tide

partly monitored by ultra high-resolution drone monitoring from 2021 to 2023 at the season scale (James et al. 2022) and by high resolution satellite monitoring from 2017 to 2023 at the month scale using the very high temporal resolution Planetscope constellation (Collin et al. 2021, 2023, 2024).

In 2021–2023, the jury of the "*Mondes nouveaux*" program,[2] funded by the French Ministry of Culture, selected the project "*Relier: à l'abri dans la marée*" (Lamontagne and Collin 2022b, c, d), proposed by the artist Nancy Lamontagne: https://nancy-lamontagne.wixsite.com/artiste/a-labri-dans-la-maree (Fig. 4). The project led to the creation of a work of art, inspired by the environment and elements of local landscape: breakwaters, wooden trellis, agricultural posts, waves, tidal curves, etc.

3.2 *Trajectory of Beaussais' Marsh in the Form of EcoSensible Spheres*

Following the design and the explanation of the EcoSensible Sphere, its application to the depolderization process occurring at Beaussais has been intended (Fig. 5).

[2] https://mondes-nouveaux.culture.gouv.fr/fr.

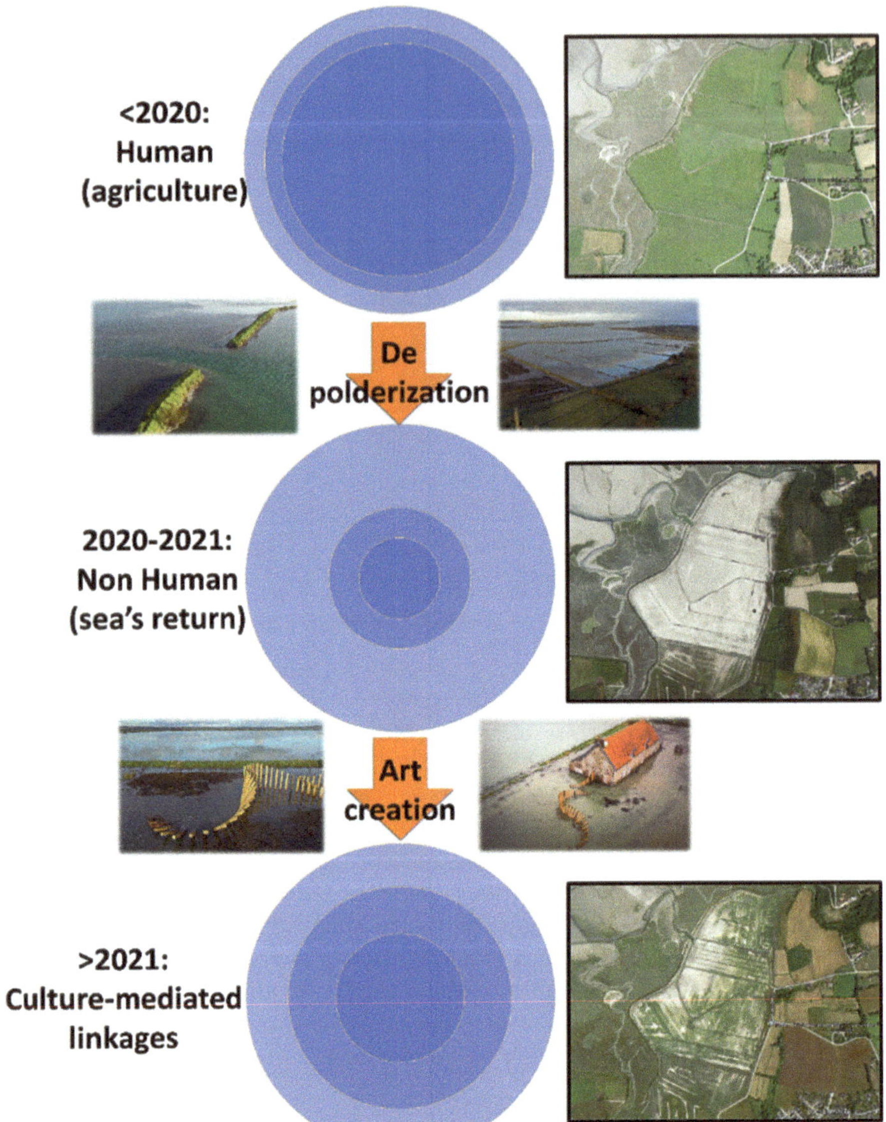

Fig. 5 The EcoSensible Sphere's trajectory of the culture-mediated de-polderisation in Beaussais' marsh. Prior to 2020, the agriculture-dominated polder landscape is featured by a EcoSensible Sphere marked with the economy. In 2020–2021, the breach in the dyke lets the sea and the flow-transported sediment re-enter and deposit in its previous space, what is reflected by a non-human (ecosphere)-dominated EcoSensible Sphere. After 2021, the concomitant lower marsh vegetation and artwork "*Relier: à l'abri dans la marée*" enables to balance the three components of the EcoSensible Sphere

Prior to the breach, that happened on March 2020, the 150-year-old polder was dominated by a pastureland—fodder crop dug with linear draining channels. This economy-based landscape has been satisfactorily represented by a dominant economy sub-sphere, surrounded by a thin culture sub-sphere, integrated itself into a thin ecosphere (Bazin and Olivry 2017). When the breach happened, after spring tides, the depolderization process started in its physical aspect, letting the seawater re-enter and fill its ancient domain until the ancient shoreline (2020–2021). The novel landscape was featured by the sedimentary deposits of mud, transported by the tidal currents, over the terrestrial vegetation associated with the pastureland—fodder crop. The (non-human) ecosphere drastically constrained the anthroposphere, be it economic or cultural. From 2021, the salt marsh vegetation has begun to recolonize the salt-roasted grassland while the depolderization commenced in its cultural aspect. For the last 150 years, the studied marsh has not received the marine sediments, and has undergone soil drainage and compaction, thus the colonization by the lower level of the salt marsh vegetation. In parallel, the artist has wandered and immersed the transforming marsh, by capturing the pioneer glassworts (*Salicornia spp.*) thriving on the mud, the abandoned wooden posts of ancient fences, the tidal waves shaping the land-sea continuum, etc.

3.3 *Artwork: Relier: À l'abri Dans La Marée*

3.3.1 Description

"*Relier: à l'abri dans la marée*'s" first intention was to create an artwork that could reflect the changes that the site was experiencing. It relied on history and the change of practices in order to highlight the evolution of the relationship between human and non-human. It also dealt with defending an inclusive and peaceful relationship with the environment (Fig. 6).

First, the artist collected information and images. She observed the site and she watched the agricultural poles, formerly straight, slowly move and fall to the ground under the action of seawater. Thinking about the general context of the site, the Brittany's *Côte d'Émeraude*, she made a linkage with different types of wooden fences, mussel's breeding posts "*bouchots*" and breakwaters, facing stormy waves. When thinking about fences—straight and blocking access to a site—and the same elements on the ground, the result was that a palisade and a promenade were similar. Thus, she created what could be seen as a palisade on the ground or an erected promenade. It is on a human scale, to establish links with the body, between hindrance and connection, protection and openness. It is also linking a house, symbol of humanity, and a natural zone, at the end of what was the house's garden. The shape of a wave was used to merge the vertical and horizontal ends. Seen from above, it takes the shape of a tide curve. This floodable sculpture is in two parts: the small parts is a reference to the architecture of the house and its "*penty*"—a lean-to—. The two parts create a

Fig. 6 The drone-based oblique view of the artwork as an artistic extension of a house, considered uninhabitable given its recurrent tidal inundation

breach, which evokes the breach of the dyke. With time, the artwork will change: it is always in progress.

In the artist's poetic definition: "*Relier*" ("Reconnecting"): A word, a verb, an action, an invitation.

A wall that comes down or a bridge/walk that goes up: both at the same time. This work explores the return of the sea in the human space, the place we concede to the other.

3.3.2 Governance

The realization of the artwork was achieved thanks to the political, administrative financial, scientific, conception and earthwork assistance of a mosaic of actors from both public and private sectors. The genesis of the artwork started in 2021 with the call from the "*Mondes nouveaux*" program, issued by the French Ministry of Culture, and funded by the "*Plan de Relance*", following the Covid-19 crisis. Once written by the artist, the project was officially supported by the marsh's proprietary (the *Conservatoire du littoral*), the marsh's manager (the local inter-municipality, the *Communauté de Communes Côte d'Émeraude),* and the local research center that studies the salt marsh re-generation (the Coastal GeoEcological Center). Following the project's acceptance by the jury of "*Mondes nouveaux*", the local municipality (Beaussais-sur-Mer), became another public stakeholder along with the two private companies: the designer (Métalobil), and the earthworker (Rance Travaux Publics).

3.3.3 Artwork Impact

The artwork has nurtured the cultural component of the anthroposphere within the EcoSensible Sphere to better link the economy sub-sphere with the ecosphere.

Inspired by (1) the decadal history of the human conquest over the sea, in the form of reclaimed (or "gained") land, (2) the recent reconnection of the sea, collegially decided, and (3) the aesthetic power of this mutating marsh, the artist, as the culture's vector, transforms the dyke's breach into a bridge between the anthroposhere and the ecosphere. The dyke, that symbolizes a wall, separating the land from the sea and disconnecting the human from environment, has artistically been twisted in a walkway, thus connecting the (re-)nascent salt marsh, hosting multiple ecological niches, with a sustainable economy.

That economy is no longer characterized by short-term pastureland or fodder crop, subjecting soil and water to narrow yields, that do not optimize biodiversity and ecosystem services, but short-term art sightseeing, mid-term eco-tourism and long-term ocean-climate regulation. The artwork, through its size, wave shape, wooden structure, symbolism, constitutes a site of interest for people curious to look at and stroll around contemporary in situ installation at the edge of the transformation. The travel guide, denominated as *Le Petit Futé*, promotes the artwork as a sculpture in its 2024 release.[3] The internet geographic applications, such as Google Maps[4] and Apple Plans, identify the artwork as a sculpture and public art, respectively, to their users. The renewal of the salt marsh besides provides habitats to its inherent fishes, insects or birds, such as the spoonbill, what creates a hotspot for eco-tourists, such as professional or amateur ornithologists. In addition to those cultural services, the salt marsh also ensures some regulation functions such as the nutrient recycling, the wave dampening, the sediment trapping, the carbon sequestration, or the shoreline protection.

4 Discussion and Perspectives

4.1 The Culture: Adaptation and Ecotone

The cultural linkage between the human-centered economy and the non-human plays a key role in the sustainable future. By exacerbating the socio-economy sub-sphere, at the detriment of the socio-culture's one, the interface with the ecosphere is eroded, resulting in the misunderstanding of the non-human. This deep and alarming lack of knowledge can be jocularly reflected by TV programs where western selected people, used to focus on the socio-economy, are challenged to survive in an unknown

[3] https://www.petitfute.com/v8716-beaussais-sur-mer-22650/c1173-visites-points-D-interet/c999-galerie-D-art-lieu-D-exposition-fondation-centre-culturel/2087501-sculpture-relier-A-L-abri-dans-la-maree.html.

[4] https://maps.app.goo.gl/zipeYtDavEJdAEbVA.

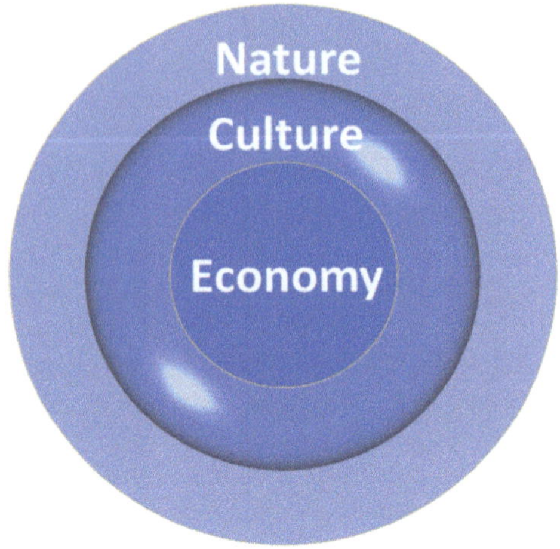

Fig. 7 The representation of the thickened culture sub-sphere in the EcoSensible sphere

type of ecosphere with no food or clothing. Beyond the voyeurism-guided media consumption, this kind of experiment may exemplify the need for the culture to understand then adapt to the human's environment. This cultural interface ought to be pivotal in the current context of the climate change (rise in sea level, sea and land surface temperature, wind and wave intensification, etc.) and biodiversity erosion, by offering a reading grid of the changes and especially a novel vision of the issues to build nature-based solutions.

By analyzing the EcoSensible Sphere as a conceptual landscape, the culture can be appreciated as a kind of ecotone, that is to say a transitional edge between the socio-economy and non-human. The science of the landscape ecology indicates that an ecotone may host more (species) richness than the sum of the richness associated with both adjacent ecosystems. The so-called "edge effect" represents that synergistic phenomenon, as an emergent property of the system. By translating that effect to the culture's ecotone, it is getting noticeable to thicken the importance of the culture, which could be illustrated on the EcoSensible Sphere by a third dimension, especially observable for that interface (Fig. 7).

4.2 *EcoSensible Sphere's Variations*

The malleable concept of the EcoSensible Sphere invites the policy-maker, proprietary, manager, visitor, artist or scientist to propose three schematic variations based on (i) a diffuse instead of a marked boundary between sub-spheres, (ii) a villosity-like pattern, and (iii) a diffuse villosity-like pattern (Fig. 8).

The diffuse boundary between the EcoSensible Sphere's components emphasizes on the porosity/permeability of this interface, thus augmenting the amount of information per time, that is to say the flux, between the socio-ecological components.

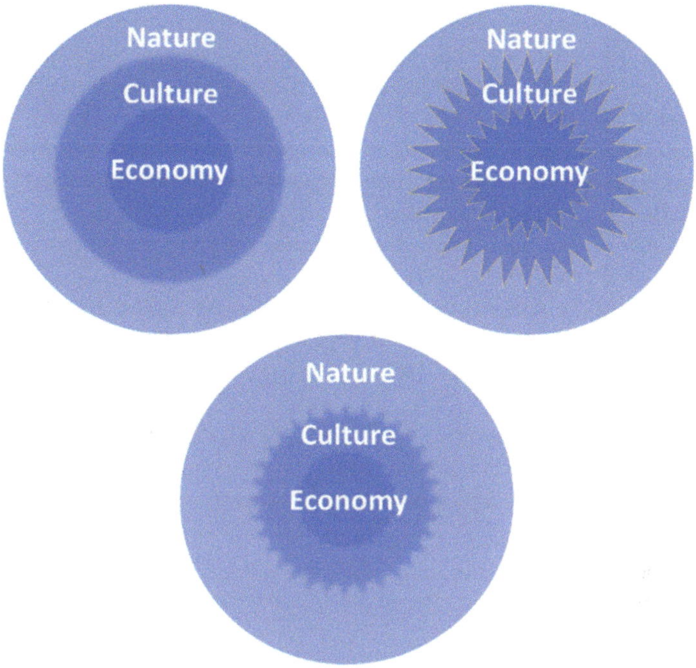

Fig. 8 The representation of the diffuse, villosity-like, and diffuse villosity-like boundaries between the sub-spheres of the EcoSensible sphere

The villosity-like spatial pattern, inspired from the shape of the intestinal wall, increases the exchange surface area, thus increasing the amount of information shared.

The combination of both previous variations gives birth to the diffuse villosity-like pattern that consequently boosts both the amount and the circulation of information: the more porous and convoluted the interface's pattern is, the richer and more complex the socio-ecological processes are.

Further investigation will attempt to answer this question: what would be the optimal proportions of the spheres to obtain maximum robustness/resilience of the EcoSensible Sphere?

5 Conclusions

Today's planetary emergency requires informed and proactive multi-scale stewardship. From Holocene, the secular environment's exploitation by *Homo sapiens* has led to a clear disconnection between the ecosphere and anthroposphere, that can be schematized by a polarized Nature-Culture continuum, more or less distant/ secant. Since the "great acceleration", dated back to the end of the World War

II, the continuum however mutated with a domination of the anthroposphere onto the ecosphere. We advocate that the systematic erosion of the culture component by the economy is responsible for that planetary over-exploitation given the misunderstanding.

The introduction of the conceptual scheme, coined as the EcoSensible Sphere, has the potential to reflect the re-integration of the anthroposphere into the ecosphere, and the re-balancing of the culture versus the economy.

That framework has been applied to a sustainably-managed coastal marsh, in which an artwork ("*Relier: à l'abri dans la marée*") as the cultural vector, has successfully linked the socio-economy with the ecosphere. The artwork has indeed encouraged the social acceptability of the sea's and salt marsh's return, as a nature-based adaptation to the sea level rise, but also of a long-term economy, such as eco-tourism.

The proposed concept and its coastal application hold great promise to be transferable to other coastal sites at risk/emergency, whose nature-based management can be considerably underpinned by the culture, in general, and the art, in particular.

References

Bazin P, Olivry D (2017) Focus-Le programme Adapto et les premières réflexions menées sur le site de la baie de Lancieux. Sci Eaux Territoires 24:54–59

Bellwood P (2004) First farmers: the origins of agricultural societies. Wiley-Blackwell, p 384

Collin A, Durozier T, Hervouet G, Costa S (2021) Détecter le processus de dépoldérisation par l'observation satellite à très haute résolution temporelle. In: Proceedings of the 16th SAGEO, 5–7 May, Virtual, pp 250–252

Collin A, James D, Lamontagne N, Feunteun E, Durozier T, Hervouet G (2023) Before/after the submersion process using planetscope monthly time-series at high spatial resolution: the case study of Beaussais' Marsh. In: Proceedings of coast, 23–27th October, Caen, p 99

Collin A, James D, Lamontagne N, Feunteun E, Durozier T, Hervouët G (2024) Monitoring the submersion using monthly time-series at high spatial resolution. In: Proceedings of the 19th Franco–Japanese oceanography symposium: submitted. Springer International Publishing, Cham

Elhacham E, Ben-Uri L, Grozovski J, Bar-On YM, Milo R (2020) Global human-made mass exceeds all living biomass. Nature 588:442–444

IPBES (2019) Global assessment report on biodiversity and ecosystem services of the intergovernmental science-policy platform on biodiversity and ecosystem services. In: Brondizio ES, Settele J, Díaz S, Ngo HT (eds) IPBES secretariat. Bonn, Germany, p 1148

James D, Collin A, Mury A, Letard M, Legal O, Le Quilleuc A (2022) Very high-resolution monitoring and evaluation of tidal and ecological restoration in Beaussais' Bay. In: European spatial data for coastal and Marine remote sensing: proceedings of international conference EUCOMARE 2022-Saint Malo, France. Springer International Publishing, Cham, pp 207–211

Lamontagne N, Collin A (2020) Paysage sensible art-science d'un confinement sur le littoral breton: Mise en relief d'un calligramme géopoétique. L'information Géographique 84(3):48–55

Lamontagne N, Collin A (2022a) Pulsation spatiale: le sédiment au cœur de la métamorphose du paysage. L'information Géographique 86(2):100–112

Lamontagne N, Collin A (2022b) Inoculation. L'information Géographique 86(4):107–109

Lamontagne N, Collin A (2022c) Inoculation. C. ART Cartographies artistiques, Exarmas#19, Bordeaux, 9th November
Lamontagne N, Collin A (2022d) Evolution of the sensitive connectivity in a salt marsh under metamorphosis. International geographic union, 18–22 July, Paris
Lamontagne N, Collin A, James D, Feunteun E (2023) The cultural interface between human needs and nature's integrity: the case study of Beaussais' Marsh. Coast, 23–27th October, Caen
Lee H, Calvin K, Dasgupta D, Krinner G, Mukherji A, Thorne P et al (2023) IPCC, 2023: climate change 2023: synthesis report, summary for policymakers. In: Core Writing Team, Lee H, Romero J (eds) Contribution of working groups I, II and III to the sixth assessment report of the intergovernmental panel on climate change. IPCC, Geneva, Switzerland
Lenton TM, Rockström J, Gaffney O, Rahmstorf S, Richardson K, Steffen W, Schellnhuber HJ (2019) Climate tipping points—too risky to bet against. Nature 575(7784):592–595
McCarthy FM, Patterson RT, Head MJ, Riddick NL, Cumming BF, Hamilton PB et al (2023) The varved succession of Crawford Lake, Milton, Ontario, Canada as a candidate global boundary Stratotype section and point for the Anthropocene series. Anthropocene Rev 10(1):146–176
Metzner R (1999) Green psychology: transforming our relationship to the earth. Simon and Schuster, p 240
Richardson K, Steffen W, Lucht W, Bendtsen J, Cornell SE, Donges JF et al (2023) Earth beyond six of nine planetary boundaries. Sci Adv 9(37):eadh2458
Ritchie H, Rodés-Guirao L, Mathieu E, Gerber M, Ortiz-Ospina E, Hasell J, Roser M (2023) Population growth. Published online at OurWorldInData.org. Retrieved from: https://ourworldindata.org/population-growth [Online Resource]
Rockström J, Gupta J, Qin D et al (2023) Safe and just earth system boundaries. Nature 619:1–10
Ruddiman WF (2013) The Anthropocene. Annu Rev Earth Planet Sci 41:45–68
Steffen W, Broadgate W, Deutsch L, Gaffney O, Ludwig C (2015a) The trajectory of the Anthropocene: the great acceleration. Anthropocene Rev 2(1):81–98
Steffen W, Richardson K, Rockström J et al (2015b) Planetary boundaries: guiding human development on a changing planet. Science 347(6223):1259855
Yanagi T (1998) Satoumi creation in the coastal sea. J Jpn Water Environ Soc 21:703

History and Prospects of French-Japanese Collaboration and Friendship in Oyster Farming

Y. Koike, T. Seki, and P. Prouzet

1 Overview of Oyster Farming in Japan and France

Oysters are a useful fishery resource in both France and Japan and have been farmed for a long time. Oyster farming was started in Japan in the eighteenth century and in France in the mid-nineteenth century (Ogasawara 1980; Buestel et al. 2009). In recent years, production, in both countries, has been around 100,000 tonnes in France and 180,000 tonnes in Japan (Koike and Seki 2020).

The natural habitat of the main cultured species of oyster is the intertidal mid-oceanic zone, where it lives by adhering to rocks and other substrates. At low tide the oyster is exposed to the air. Often extensive oyster reefs form in favourable conditions (Fig. 1).

Oyster farming takes place in coastal areas in both countries, but there are significant differences in their marine environments. The tidal range in Japan is roughly within 2 m (exceptionally 7 m in the west coast of Kyushu), while in France it reaches between 4 and 10 m or more (14 m in the Mont Saint-Michel Bay). Because of these differences in marine environments, farming methods are also adapted to each marine environmental condition.

Y. Koike (✉)
Tokyo University of Marine Science and Technology (TUMSAT). SFJO-Japon, Koenji, Suginami-ku, Tokyo, Japan
e-mail: oreillemer@ybb.ne.jp

T. Seki
Japan Fisheries Science and Technology Association Aoyama, Rifu-Cho, Miyagi-gun, Miyagi, Japan

P. Prouzet
SFJO France, Saint-Jean de Luz France, France

A.-C. Bennis et al. (eds.), *Constraints and Adaptations to Global Change at the Land-Sea Interface: For a Shared Ecological and Energy Transition*,
https://doi.org/10.1007/978-3-031-90050-1_26

Fig. 1 Left side: natural oyster reef at île d'Aix, during low tide on the Atlantic side of France. Right side: natural oyster reef at Kenzaki, at the entrance of Tokyo Bay. Both images highlight the natural habitat and marine life of intertidal zones. Copyright Y. Koike

In Japan, where the tidal range is smaller, the subtidal zone is used vertically. In other words, rafts or longlines are floated on the surface of the water, and collectors with seed oysters (spats) are hung down with ropes (Fig. 2).

On the north and west coasts of France, where the difference in tidal range is greater, tidal flats of intertidal zone are used horizontally. Tables are set up in the vast intertidal zone exposed at low tide and seed oysters are enclosed in baskets on the tables (Fig. 3). Only, in mediterranean lagoons with a small tidal range, the hanging type is used as in Japan. Oysters in France are grown exposed at low tide, while in Japan they are kept submerged at all times.

Fig. 2 Left side: oyster farming using the sub tidal zone vertically in Matsushima, Japan. Right side: oysters are fixed on the collectors and suspended by ropes. Copyright Y. Koike

Fig. 3 Left side: oyster farming, using the large tidal flat in Northern Brittany, on the Atlantic coast of France. Right side: oysters are enclosed in plastic bags fixed on tables. Copyright Y. Koike

2 Consumption Habits in France and Japan

In both countries, records of oysters as a food source are ancient, having been consumed since the Jomon period in Japan and Roman times in France (Nomura et al. 1995).

The painting, Oyster Lunch (*Ostrea edulis*), once hung in the dining room of the Palace of Versailles, was painted by the artist Jean-François de Troyes at the order of King Louis XV of France in 1735. Dozens of raw oysters are served on a silver plate, the porcelain dishes is made in Japan, and the accompanying liquor is identified as champagne from the cork flying in the air, depicting a luxurious lunch for the aristocracy (Fig. 4 on the left). The picture of dish shows a serving of oysters for guests in a restaurant in today's Arcachon (Fig. 4 on the right).

Fig. 4 On the left: example of oyster dishes in France are shown in the painting "Le déjeuner d'huîtres" (oyster lunch), by Jean-François de Troy in 1735 (musée Condé Chantilly). Oysters are served raw in half shell. On the right: a dish of raw oysters served in its half shell in a contemporary restaurant. Copyright Y. Koike

These painting and picture show that the same raw eating custom has been followed in France for over 300 years. For this reason, in France, oysters are distributed raw with their shells and are chucked just before eating at restaurants or at homes (Fig. 5).

In Japan, oysters have traditionally been eaten cooked rather than raw (Fig. 6), so they were mostly shucked for distributing (Fig. 7). The majority is distributed as shucked, although nowadays it is more often in with shell for raw consumption.

There is a big difference in taste between the two conditionings. The differences between the French and Japanese oyster preparation methods are closely related to the environment of natural oyster habitat and the oyster farming methods adapted to their environments. In French culture style, oysters are exposed to the air at low tide, just like their natural habitat. During this time, the oysters keep their shells tightly

Fig. 5 On the left: cleaning of oysters before packing. On the right: packing of a basket of living oysters for transport at oyster farm of Marennes. Copyright Y. Koike

Fig. 6 On the left: traditional cooking, oyster pot. On the right: fried oysters, a very popular cooking. Copyright T. Seki

Fig. 7 On the left: shucking of oysters before packing at Matsushima (Sendai, Japan). On the right: packing of shucked oysters for transport at Matsushima. Copyright T. Seki

closed to prevent evaporation of water, which makes them muscular and gives them the crisp, clean taste favoured by the French people. In other words, this is the method of growing oysters suitable for eating raw. And they take about 3 years to shipping size.

In Japanese culture style, oysters are sunk in seawater at all times grow to shipping size only in 1.5–2 years, because they can spend 24 h to feed and grow fast. In the water, they do not need to close their shells as much, so their muscles become softer. They are also able to feed constantly, which allows them to grow larger and produce the creamy taste that the Japanese people prefer. This is the best way to raise oysters for cooking.

Although both sides prefer oysters, they actually appreciate the opposite tastes. For the Japanese, oysters are an ordinal side dish of winter seasons and they seem to have a mild taste, whereas for the French, oysters eaten mainly during Christmas and New Year's feasts, and they prefer a distinctive taste, characteristic of the oyster's origine.

Mutual understanding of these different food cultures is difficult and requires knowledge of their long history and culture background. These differences in food culture are also reflected in differences in growing methods and distribution.

3 History of French-Japanese Mutual Assistance

The endemic French oyster declined sharply in the early 1900s due to overfishing and disease, and was replaced by the Portuguese oyster (*Magallana angulata*), but this species was also devastated by disease in the mid-1900s. The director of the Institut Scientifique et Technique des Pêches Maritimes (ISTPM), Dr. Trochamp, then asked Professor Takeo Imai of Tohoku University in Japan, with whom he had already been in contact, to try to transplant Japanese oyster seed. Figures 8 and 9

Fig. 8 Top left: professor Takeo Imai, Tohoku University at the Oyster Research Institute. Top right: Dr. Akimitu Koganezawa, Miyagi Fisheries experimental Station. Below: Mr. Kunio Goto, Miyagi Prefecture fisheries experimental station

show the production of Japanese oyster single seeds by Prof. Takeo Imai and his research team in Ishinomaki, Miyagi Prefecture.

Fig. 9 On the left: Seed oyster culture system at Mangoku-Ura, Miyagi Prefecture. On the right: seed oysters prepared to Arcachon, France. Copyright Kunio Goto

Figure 10 shows the evolution of French oyster production, which has seen a very significant rise in production since the 1970s after the transplantation of Japanese oysters (Mariojouls and Prou 2015).

About forty years later (on March, 2011), oyster farming on the Sanriku coast in Tohoku area was devastated by the huge Tsunami (Fig. 11).

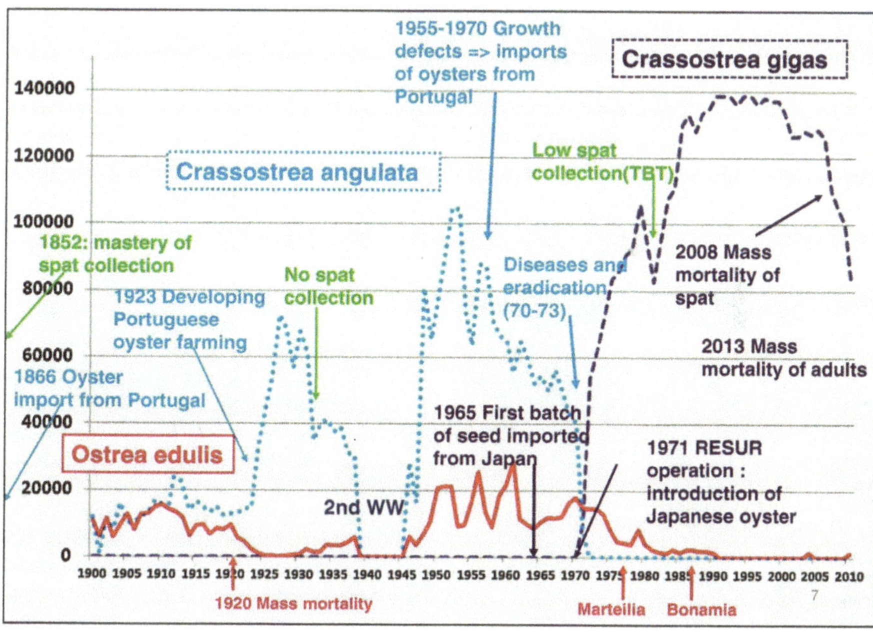

Fig. 10 Production in tons of oyster culture in France and main changes (Mariojouls and Prou 2015). Red line shows *Ostrea edmortality,ulis*, blue dots show *Crassostrea angulata* and black dots show *Crassostrea gigas* inported from Japan. Notable events include the 1920 mass mortality, 1955-1970 growth defects, 1965 seed import from Japan. The chart also notes historical developments like the 1866 oyster import from Portugal and the 1971 introduction of Japanese oyster spats

Fig. 11 Fishing gears and debris of aquaculture facilities affected by huge Tsunami in 2011 at Sanriku coast, Japan. Copyright Sasaki

To help in this crisis, research institutes and the Committees of oyster farmers from all over France provided research equipment such as microscopes and plankton nets for collecting oyster spats and aquaculture materials such as ropes, buoys and life jackets (Fig. 12).

Fig. 12 Donation of microscopes—on the left: Professor H.-J. Ceccaldi (former President of SFJO France), Professor C. Mariojouls (President ADA) with the director of Miyagi Prefecture Fisheries Technology Institute. On the right: professor Y. Koike and Professor T. Komatsu (SFJO Japan) with the Director of Iwate Prefectural Fisheries Research Center

Fig. 13 On the left: seminar at FRA Tohoku on 4th October 2012. Presentation by Dr. Tetsuo Seki, "Improving the value of oysters in Japan: traditions and future opportunities". On the right: a view of participants, with Mr. Kunio Goto (foreground right) and members of French delegation

The tsunami aid triggered mutual assistance and technical exchanges between the two Japanese-French Oceanographic Societies (Japan and France), which were more active than before. At first, on December 2011, Prof. Catherine Mariojouls, President of French Association on Development of Aquaculture (ADA) and former vice-president of SFJO-France, visited to encourage oyster farmers of disaster area and then on February 2012, Professor Hubert J. Ceccaldi, former President of SFJO-France, and Professor Georges Stora, from Aix-Marseille University have also made a visit and provided encouragement to regional fishermen's cooperations.

The cooperation of our two French-Japanese Societies also carried out joint seminars for oyster farming recovery. Starting with a seminar at *Tohoku* Fisheries Research Agency on 4th October 2012 (Fig. 13).

These events were followed in November 2012 by the 1st World Oyster Congress in Arcachon, France, attended by Yasuyuki Koike, Tetsuo Seki, Kunio Goto and two oyster farmers from Ishinomaki city. At the Congress, Mr. Kunio Goto was presented as a member who prepared the first oyster spats from Japan to Arcachon in the seventies.

During the visits, Japanese delegation was widely reported as the members who revived oyster farming in France (Fig. 14).

Subsequently, mutual seminars between two countries were held at Boulogne-sur-Mer and Marseille, France in 2013, and went on at Shiogama, Japan in October 2015 (Fig. 15) followed with open public seminar, titled "Oyster culture of France and Japan towards the recovery of Sanriku" at Nichifutsu Kaikan (Maison Franco Japonaise: MFJ) in Tokyo. At the Open public seminar, the hall was full to capacity with a general audience (Fig. 15). This exchange was further continued and took place in Bordeaux, France, in 2017.

Although the symposium was subsequently interrupted by the global corona pandemic, the 60th Anniversary Symposium could be held finally online at the Maison Franco-Japonaise, Tokyo, in October 2021. And the Oyster Fisheries Associations of two countries have also started exchanges in aquaculture technology.

Fig. 14 On the left: key note presentation by Pr. Y. Koike on the French-Japanese cooperation on oyster culture. On the right: group of Japanese participants, Mr. Kunio Goto (center), was presented by a regional TV station as a person who saved oyster farming in France

Fig. 15 On the left: French delegation visiting oyster farm of Mr. Masanori Hatakeyama at Kesennuma, Miyagi Prefecture on November 2015 before the 15th Symposium. On the right: open Seminar at the "Maison Franco-Japonaise" in Tokyo, on November 2015, the conference room was full by the public audience

4 Transplantation of French Oyster, *Ostrea edulis*, to Tohoku Region

The Oyster Research Institute (non-profitable private foundation) introduced the French oyster (*Ostrea edulis*) in 1959 with the aim of enabling stable aquaculture production even in low-temperature waters, based on the oyster larvae tank nursery technology developed by Professor Takeo Imai of the Faculty of Agriculture, Tohoku University.

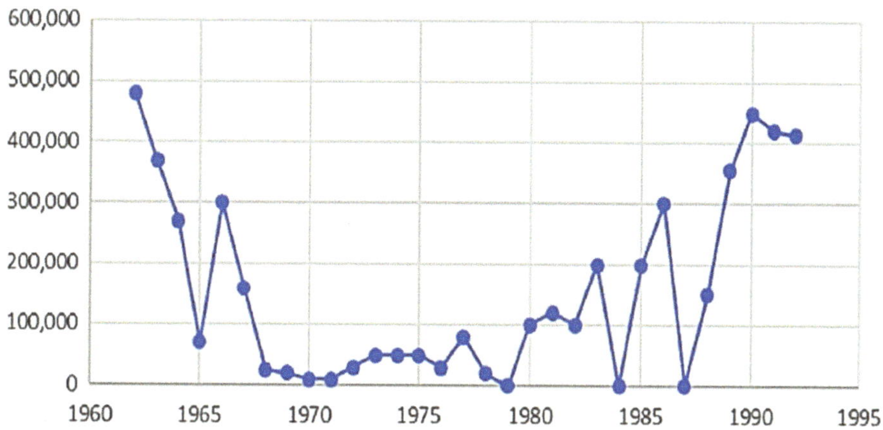

Fig. 16 Production trend of French oyster seed at the Oyster Research Institute. Line chart showing data trends from 1960 to 1992

In 1960, with the support and contributions of volunteers from the business community in Sendai, the 'French Oyster Association' was founded and the 'Oyster Research Foundation' was established.

At the Oyster Research Institute, the production of French oysters continued from 1962 until 1992, when Dr. T. Seki retired (Fig. 16).

The French oysters that grew to adulthood during this period were highly appreciated by visitors to the institute and by those who offered them for tasting. However, it was not customary to eat oysters on the half shell. In Japan 40 years ago, most of the oysters were distributed in shucked style, and there were no chefs who could serve oysters in the shell. There were no oyster bars that accepted and served oysters in the shell. Besides such demand less condition at the market, due to the very low survival rate of the seedlings that were distributed, requests from oyster farmers gradually ceased. Subsequently, all raft facilities of the Oyster Research Institute were lost due to the Tsunami damage caused by the Great East Japan Earthquake in March 2011, and the French oysters in Japan were lost.

5 New Technique of Oyster Farming

It is known that the shape of oyster growth and the characteristics of the sense of taste differ depending on the oyster cultivation technic. This is the case of the hanging technic of oyster farming in Matsushima Bay, Miyagi Prefecture, where the growth rates, shell shape and taste characteristics differ depending on the depth position of the oysters hanging in the intertidal zone (Fig. 17).

With the increase of the number of oyster bars, the oyster farmers are looking for ways to differentiate themselves and increase their income from good-quality

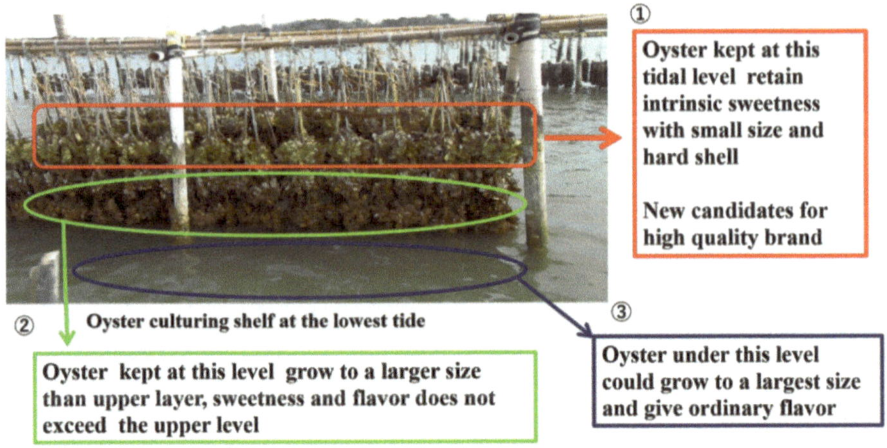

Fig. 17 Difference of the oyster characteristics resulting from the level in suspending rope of the intertidal zone aquaculture. The top level outlined in red, notesthat oysters retain intrinsic sweetness with small size and hard shell, suitable for high-quality branding. The middle level, outlined in green, indicate oysters grow larger the upper layer, with moderate sweetness and flavor. The lowest level, outlined in blue, suggests oysters grow to the largest size wirh ordinary flavor

oysters, and are developing methods of cultivating oysters with good-looking, fat and tasty oysters. The first Japanese oyster bar opened in Tokyo in 1999, and raw oyster eating by the half shell style started (Fig. 18).

Since then, the number of outlets has rapidly increased, growing to a turnover of USD 84 million by 2015. General Oyster, the company with the largest number of outlets, has opened 28 outlets across the country and continues to operate (Fig. 19).

This aquaculture technique makes use of the intertidal zone, mainly in Kyushu, where there are large tidal flats. Oysters are housed in a single-seeded, well-meshed plastic basket with good drainage, and the rocking of the waves causes the oysters to rub against each other, preventing them from sticking to each other. The oysters are also characterized by the shaved ends of the shell, resulting in a uniformly shaped oyster (Fig. 20). Not only oyster bars, but also in conventional shelled oyster farming, efforts have begun to produce high quality oysters with large soft body parts and thick interstitial connective tissue rich in glycogen and rich in flavor.

Mr. Masanori Hatakeyama (oyster farmer at Karakuwa, Kesennuma, Miyagi) and his colleagues are committed to sustaining an attractive aquaculture industry by providing consumers with quality oysters they can be proud of. They use hot water treatment to remove stuck organisms from the oyster hanged brocks and produce well-grown oysters. In October, prior to shipment, hot water treatment (70 °C during sixty seconds) thermally kills all other organisms except oysters, stimulating the release of the few remaining unleased sperm and eggs in the oyster gonads, resulting in a faster filling (Fig. 21).

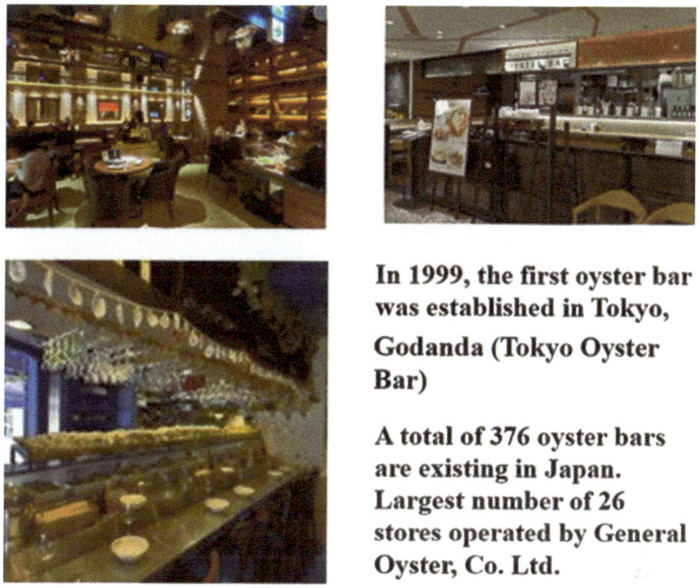

Fig. 18 Start of raw oyster eating culture in Japan by the prosperity of oyster bars

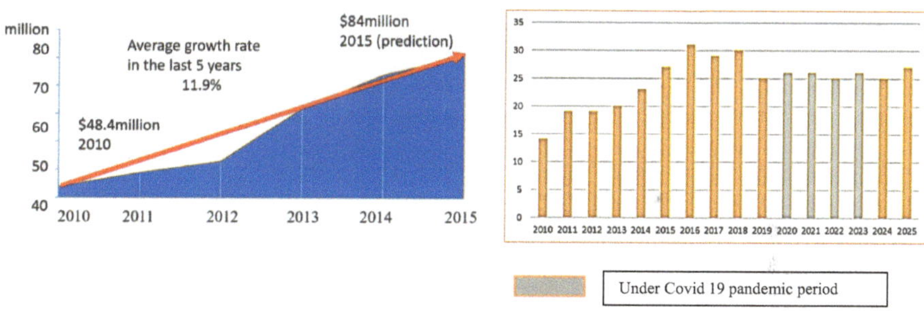

Fig. 19 On the left: oyster bar turnover trend in Japan. The left panel shows a line graph from 2010 to 2015, indicating a growth from $48.4 million in 2015, with an average growth ratee of 11.9%. On the right: number of oyster bars of General Oyster Co. Ltd. Operating from 2017 [from food service industry marketing data book 2014 Fuji Keizai marketing research and consulting group and authors data collected from URL 1. https://irbank.net/3224/140120191105418896140120191105418896.pdf (irbank.net)]

6 Blending of Different Food Cultures

The half-shell oyster eating culture that spread in Japan started with a drink-type taste and brought the pleasure of tasting and comparing the characteristics of different regions, but there is a sense that customers at oyster bars are gradually returning to the

Fig. 20 Oysters served in half shell style at the oyster bars

traditional oyster eating culture, where they expect a large soft part, rich in glycogen and thick taste, which are considered characteristics of good quality oysters.

Recently, T. Seki (one of the co-authors) had the opportunity to taste oysters from a land-based aquaculture project which he provides technical guidance. The oysters fed only three types of cultured microalgae as feed were extremely tasty. In future land-based aquaculture, it may be possible to control the taste depending on the feedstuff (Fig. 22).

When T. Seki attended the Oyster World Congress in Arcachon in 2012, he was surprised to see young French people enjoying grilled oysters with "foie gras" on top (Fig. 23). When he asked if farmed oysters are ever cooked in France other than in half-shells, the reply was that it was unthinkable to cook such fresh oysters with the flavor of the sea.

However, when a French delegation visited Mr. Masanori Hatakeyama's oyster hut, they were impressed and praised the cooked oysters, including fried oysters and grilled oysters. This made us realize that when we take our respective long-cultivated food cultures as our daily routine, we do not realize the advantages to other food

Fig. 21 At the top: uniformly shaped oysters grown in the newly developed intertidal aquaculture; at the bottom: hot seawater bathing treatment to eliminate fouling organisms. Copyright T. Seki

cultures. The significance of exchange is to come into contact with different cultures and realize the value of each other, like the Japanese who discovered the pleasure of half-shells with French oysters and the French visitors who experienced the cooked taste of large, fat oysters (Fig. 24).

Fig. 22 *Magallana gigas* grown in the land-based tank utilizing deep seawater fed with three species of cultured microalgae (at the presentation party of the GO Farm Co. Ltd., August 4th 2013)

Fig. 23 Grilled oysters with foie gras on top. Copyright T. Seki

Fig. 24 Possible oyster cooking trend in the future France through the interchange between Japanese oyster eating culture (after girl eating oyster from Jan Steen)

7 Conclusion

The French-Japanese Societies of Oceanography have strengthened mutual understanding and technical exchange in their fields of research, such as oceanography and fisheries science. Currently, there is a need for mutual understanding of the challenges facing the aquaculture sector in the face of global climate change, and for cooperative measures to address these challenges.

In the field of oyster farming, lectured here, it is desirable to understand the food culture of background to improve mutual technology. And in order to improve the technology between oyster farmers and researchers in both countries who have obtained agreements by this time, there is a strong need to open, observe and experience technical contents of each other.

During the French-Japanese symposium in Bordeaux in 2017, Mr. Masanori Hatakeyama, former chair of *Miyagi,* met with Mr. Gérald Viaud, the chair of Marennes, to confirm future technical exchanges (Hatakeyama and Seki 2020).

In February 2024, Mr. Olivier Laban, Chair of Arcachon, attended the French-Japanese seminar on *Satoumi* held in Hinase, Okayama, Japan. Some important points immerged from these discussions. The experience of the Seto Inland Sea clearly indicates that there is no opposition between the development of fishing and oyster farming and the restoration of seagrass beds, quite the contrary. The existence of these traditional activities that strengthen the identity of the territories is necessary to carry out this restoration by involving the local communities. The use of oyster shells to allow a better implantation of eelgrass sprouts is a particularly remarkable example of the use of an oyster by-product which allows an improvement of the substrate for the development of the seagrass beds in the Seto Inland Sea. Similarly, the actions of oyster farmers in the Arcachon Basin to limit the development of wild oyster beds, which modifies the environment and limits the growth of farmed oysters, are interventions that allow better use of the productivity of a semi-enclosed environment subject to numerous constraints. At the end of this seminar Mr. Hatakeyama and Mr. Laban expressed the hope that exchanges of young oyster farmers would take place and that stronger technical cooperation would be implemented in the future. Unfortunately, the recent Corona disaster and problems linked in France link to the degradation of water quality have delayed theses exchanges, but it is strongly expected to be achieved at the earliest opportunity.

References

Buestel D, Ropert M, Prou J, Gouletquer P (2009) History, status and future of oyster culture in France. J Shellfish Res 28(4):813–820

Fuji Keizai marketing research and consulting groupe 2014: Food service industry maeketteng data book. URL 1. https://irbank.net/3224/140120191105418896140120191105418896.pdf(irbank.net)

Hatakeyama M, Seki T (2020) Innovation role of restoration support for Tsunami-affected oyster farming in the Karakuwa district, Miyagi Japan, contribution to the rationalization of fishing ground management. In: Proceedings of Coast Bordeaux symposium and of the 17th French-Japanese oceanography symposium. Springer, pp 419–429

Koike Y, Seki T (2020) Innovation and adaptation of recent oyster culture techniques in Japan. In: Proceedings of Coast Bordeaux symposium and of the 17th French-Japanese oceanography symposium. Springer, pp 409–418

Mariojouls C, Prou J (2015) Changes, adaptations, and resilience: the case of Franch oyster farming. In: Proceedings of the 15th French-Japanese oceanography symposium. Springer, pp 299–307

Nomura T, Sugawara Y, Mori K, Takeuchi M, Numachi K, Matsutani T (1995) Oyster・scallop・abalone. Koseisha-Koseikaku, pp 249–260 (In Japanese)

Ogasawara Y (1980) Oyster and their culture. In: Fisheries in Japan (Abalone and oyster). Japan Marine Products Photo Materials Association. Tokyo, pp 201–221

Author Index

A.-C. Bennis et al. (eds.), *Constraints and Adaptations to Global Change at the Land-Sea Interface: For a Shared Ecological and Energy Transition*,
https://doi.org/10.1007/978-3-031-90050-1

Kolibri 360 GmbH
Ferdinand-Jühlke-Straße 7
99095 Erfurt, Deutschland
produktsicherheit@kolibri360.de